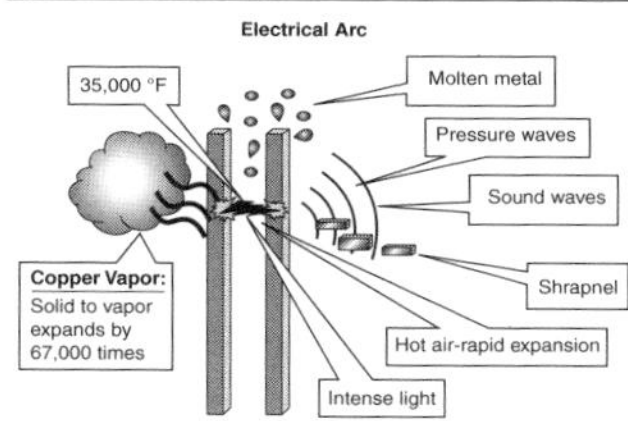


*EXHIBIT 100.4 Results of an arc flash incident. (Courtesy of Bussman Division, Cooper Industries)*

The commentary text in this handbook is intended to assist users in understanding and applying *NFPA 70E.* To distinguish the Standard text from the commentary text, the Standard and Annexes are printed in black ink.

**Arc Flash Hazard Analysis.** A study investigating a worker's potential exposure to arc-flash energy, conducted for the purpose of injury prevention and the determination of safe work practices, arc flash protection boundary, and the appropriate levels of PPE.

An *arc flash hazard analysis* determines the flash protection boundary and the amount of incident energy that could be impressed on an employee as a work task is being performed. The analysis can take one of several different forms. An arc flash hazard analysis is necessary regardless of any label or marking on the surface of electrical equipment. Referring to a warning label could constitute one step in the analysis; however, the analysis also must consider risk. When the analysis is complete, the worker has sufficient information for selecting the necessary arc flash protective personal protective equipment (PPE) and the work practices necessary to minimize any exposure to a thermal hazard. Part of the analysis includes determining the arc flash protection boundary and the incident energy. See Annex D for information on incident energy arc flash calculation methods.

**Arc Flash Suit.** A complete FR clothing and equipment system that covers the entire body, except for the hands and feet. This includes pants, jacket, and beekeeper-type hood fitted with a face shield.

The PPE industry has experienced a dramatic evolution of protective schemes and equipment for workers. During this time, the term *arc flash suit* has not been used consistently throughout the industry. This definition clarifies the specific components that comprise a flash suit and points out that a face shield alone cannot be used when a flash suit is required.

Although the term *arc flash suit* does not provide protection for an employee's hands and feet, all other body parts are covered by protective equipment. The hands and feet also must be protected.

**Arc Rating.** The value attributed to materials that describes their performance to exposure to an electrical arc discharge. The arc rating is expressed in cal/cm$^2$ and is derived from the determined value of the arc thermal performance value (ATPV) or energy of breakopen threshold ($E_{BT}$) (should a material system exhibit a breakopen response below the ATPV value)

holes in
the ma-

Workplace



*CASE STUDY*

On an October Thursday, Ryan went to work anticipating the sixth World Series game.* Houston was down three games to two. Ryan had four tickets to the game and would be leaving Beaumont at 2:00 P.M. to drive to Houston with his family. He had planned to work in the morning before leaving for the game. He was as excited as his sons were.

Ryan was a senior electrician at a large chemical plant near Beaumont. He had "paid his dues" at the plant and felt that he had earned an easy morning. He was assigned to the electronics crew in the maintenance organization.

The solid-state drive room was air-conditioned in an effort to remove heat generated within the power electronics in the drives. Each section of the drive line-up had an exhaust duct at the top of the unit and a louvered door with an air filter on the inside of the door. This solid-state drive was old equipment. The equipment was physically very large. Electrically, the equipment was very large as well; the transformer supplying the rectifier was rated at 2000 kVA. The rectifier was close-coupled to the transformer, so there was no secondary overcurrent protection on the 480-volt bus.

The supervisor knew that Ryan was going to the game and wanted him to enjoy the outing. Ryan was spared all the heavy work on this Thursday. Instead, his work task was to change all of the air filters in the drive room. Ryan knew that the filters were all the same size: 12 by 18 inches. The filter material was man-made fiber, constrained in shape by aluminum screen, and the filters were light in weight. Ryan knew that the drive equipment would be running, but he also knew that would cause no problem. The filters were always changed while the equipment was running. All he had to do was open the door, change the filter, then close the door. There were no interlocks to defeat.

New to the 2009 edition of the handbook are Case Studies. The Case Studies illustrate the importance of implementing safe and appropriate processes in the workplace. As is the case throughout the commentary text, the Case Studies are outlined and printed in blue ink.

Ryan picked up the filters from the storeroom and put the box on his tool cart. Yessir, he would have an easy morning. He proceeded to the drive room and opened the first door. He changed the filter and discarded the dirty filter in the trash container nearby.

Ryan moved on to the second unit. He sure hoped that Houston would win the game. He hoped someone in his family would catch a foul ball. He knew the bat boy and would be able to get the ball autographed. As he reached for the door handle to open the door, he heard a noise. It seemed to be coming from the large rectifier at the other end of the room.

Ryan moved closer to the rectifier. Yes, the loud humming was coming from inside the rectifier. Instinctively, Ryan opened the door of the unit to look inside. He had done that a thousand times before. This time, however, there was a problem. He never learned the source of the humming. The explosion was immediate, as soon as the door was moved. A short circuit occurred somewhere on the bus from the transformer. Ryan's cotton clothing caught fire. No one else was in the room, but other workers came quickly when the noise from the explosion reverberated around the facility.

The pressure forces from the explosion blew Ryan against the adjacent wall. His lifeless body was transported to the hospital.

After an in-depth analysis of the incident, the chemical plant relocated the filters to the outside of the door. In the future, changing filters would become a job for two people. Arc-flash-resistant clothing would be provided for all employees who were required to enter the room. No rectifier or transformer door would be opened with the equipment running.

That action was too late to help Ryan. He never knew that Houston won the ball game.

# Handbook for Electrical Safety in the Workplace

# Handbook for Electrical Safety in the Workplace

## SECOND EDITION

Edited By

E. William Buss     Mark W. Earley     Ray A. Jones

With the complete text of the 2009 edition of *NFPA 70E*®, *Standard for Electrical Safety in the Workplace*®

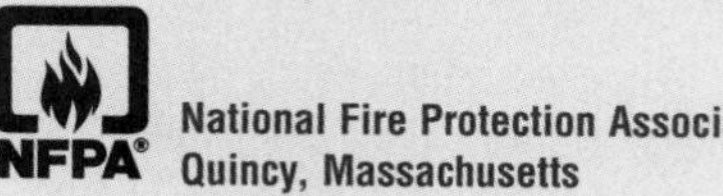

National Fire Protection Association®
Quincy, Massachusetts

Product Manager: Debra Rose
Developmental Editor: Khela Thorne
Project Editor: Irene Herlihy
Permissions Editor: Josiane Domenici
Composition: Modern Graphics, Inc.

Art Coordinator: Cheryl Langway
Cover Design: Cameron Inc.
Manufacturing Manager: Ellen Glisker
Printer: Courier/Westford

**Notice Concerning Liability:** Publication of this handbook is for the purpose of circulating information and opinion among those concerned for fire and electrical safety and related subjects. While every effort has been made to achieve a work of high quality, neither the NFPA® nor the contributors to this handbook guarantee the accuracy or completeness of or assume any liability in connection with the information and opinions contained in this handbook. The NFPA and the contributors shall in no event be liable for any personal injury, property, or other damages of any nature whatsoever, whether special, indirect, consequential, or compensatory, directly or indirectly resulting from the publication, use of, or reliance upon this handbook.

This handbook is published with the understanding that the NFPA and the contributors to this handbook are supplying information and opinion but are not attempting to render engineering or other professional services. If such services are required, the assistance of an appropriate professional should be sought.

NFPA codes and standards are made available for use subject to Important Notices and Legal Disclaimers, which appear at the end of this handbook and can also be viewed at *www.nfpa.org/disclaimers.*

**Notice Concerning Code Interpretations:** This 2009 edition of *Handbook for Electrical Safety in the Workplace* is based on the 2009 edition of *NFPA 70E®, Standard for Electrical Safety in the Workplace®*. All NFPA codes, standards, recommended practices, and guides are developed in accordance with the published procedures of the NFPA by technical committees comprised of volunteers drawn from a broad array of relevant interests. The handbook contains the complete text of NFPA70E and any applicable Formal Interpretations issued by the Association. These documents are accompanied by explanatory commentary and other supplementary materials.

The commentary and supplementary materials in this handbook are not a part of the Standard and do not constitute Formal Interpretations of the NFPA (which can be obtained only through requests processed by the responsible technical committees in accordance with the published procedures of the NFPA). The commentary and supplementary materials, therefore, solely reflect the personal opinions of the editor or other contributors and do not necessarily represent the official position of the NFPA or its technical committees.



NFPA No.: 70EHB09
ISBN-10: 0-87765-828-5
ISBN-13: 978-0-87765-828-3

Printed in the United States of America
09 10 11 12 13 5 4 3 2

# Dedication

This edition of the handbook is dedicated to Kenneth G. Mastrullo of the New England office of the Occupational Safety and Health Administration (OSHA). Ken was formerly a Senior Electrical Specialist who served as the NFPA staff liaison for the Committee on Electrical Safety in the Workplace. He is passionate about electrical safety and has always had a deep concern for electrical workers. During his tenure at NFPA, he became a spokesperson for electrical safety and spread the word about electrical safety through a variety of forums, including the IEEE Electrical Safety Workshop, the Voluntary Protection Program Association, and the NFPA seminars on *NFPA 70E*®. His dedication had measurable results, with a noticeable increase in the prominence of *NFPA 70E* and electrical safety programs around the United States and elsewhere.

Ken knows electrical safety first-hand. He has had many years of experience as an electrician and a plant engineer, and he understands the real cost of electrical accidents to the worker, the coworkers, and the employer. While Ken can quote the statistics on electrical accidents, he knows that behind those cold hard numbers are real people. He has dedicated his career to protecting those people. Ken left NFPA to dedicate his career to electrical safety on the staff of OSHA, but he left his mark on NFPA and electrical safety materials such as this handbook.

# Contents

# Preface

The 2009 edition of the *Handbook for Electrical Safety in the Workplace* contains the latest information on electrical safety. More than 112 years have passed since March 18, 1896, when a group of 23 persons representing a wide range of organizations met at the headquarters of the American Society of Mechanical Engineers in New York City. Their purpose was to develop a national code of rules for electrical construction and operation. This was the first national effort to develop electrical installation rules for the United States. This successful effort resulted in the *National Electrical Code*® (*NEC*®), the installation code used throughout the United States and in many countries around the world.

With the implementation of the Occupational Safety and Health Act, it became apparent that a separate standard would be necessary to provide requirements for safe work practices for people who might be exposed to electrical hazards. On January 7, 1976, the Standards Council of the National Fire Protection Association appointed the Committee on Electrical Safety Requirements for Employee Workplaces. The Standards Council recognized the importance of the creation of a document that could be used in conjunction with the *National Electrical Code*. To keep these documents well coordinated, the Standards Council decided that the new committee should report to the association through the National Electrical Code Technical Correlating Committee. Although the committee recognized the importance of compliance with all of the requirements of the *NEC*, the first edition of *NFPA 70E*®, *Standard for Electrical Safety in the Workplace*®, dealt primarily with those electrical installation requirements from the *NEC* that were most directly tied to worker safety. In subsequent editions, the document expanded to include safety-related work practices, safety-related maintenance requirements, and safety requirements for special equipment. For the 2009 edition, the installation requirements were removed because OSHA no longer believed that they were necessary because the *NEC* is now widely adopted and used.

For the first few editions, *NFPA 70E* was a four-part document that was essentially four books bound together. Beginning with the 2004 edition, *NFPA 70E* adopted the *NEC Style Manual*, which provided a simple means to integrate the parts of the document into a comprehensive and cohesive standard. Since the *NEC* requirements were deleted from the standard, there are now three chapters. However, the handbook continues to include some of the highlights of the *NEC* to assist the user in understanding how the installation requirements of the *NEC* can make a safer work environment.

Until the 2000 edition of *NFPA 70E*, most believed that the only electrical hazard was electric shock. The 2000 edition brought attention to the hazards of arc flash phenomena. The use of the standard has grown tremendously as workers and their employers try to provide protection from this dangerous hazard.


## Acknowledgments

Electricity can be very dangerous occupational hazard. Almost all members of the workforce are exposed to electrical energy as they perform their duties every day. Since the creation of *NFPA 70E*, electrical workers have become more aware of the hazards of electricity and of how to protect themselves. The 2000 and 2004 editions of this standard increased the awareness of arc flash phenomena. The increased use of this standard resulted in a significant increase in public input in the form of proposals and comments. This edition would not be possible without the tireless work of the dedicated professionals who serve on the Committee on Electrical Safety in the Workplace. Their work will save countless lives.

The editors have learned much from the deliberation and discussions of the technical committee. We hope our work in this book accurately reflects the wisdom we accumulated from the committee.

Handbooks are a team effort, and the editors of this book have been supported by an outstanding team of professionals. We wish to acknowledge with thanks the wonderful work of Kim Fontes, Division Manager, Product Development; Debra Rose, Product Manager; Khela Thorne, Senior Developmental Editor; and Irene Herlihy, Senior Project Editor, who supported us every step of the way. We acknowledge their patience and professionalism.

The editors wish to note special acknowledgement for the outstanding support of Sylvia Dovner. Sylvia retired from NFPA as we were putting the finishing touches on this handbook. She has been a dedicated publishing professional who has taught us all how to be better editors. Some of us had the privilege of working with her on other projects, including the *National Electrical Code Handbook.* She was always patient and understanding and never lost her sense of humor. We will miss working with her.

The editors also wish to acknowledge the contributions of former NFPA staff who served as staff liaisons to the committee: Dick Murray, Ken Mastrullo, and Joe Sheehan. They were responsible for educating their successors about the importance of electrical safety.

This is the second edition of the *Handbook for Electrical Safety in the Workplace*. The editors gratefully acknowledge the work of the team that assembled the first edition, including Ken Mastrullo and Jane Jones. All of the professionals that we have worked with on this project have made a difference in making the world a safer place.

# About the Editors

### E. William Buss

Bill is a Senior Electrical Engineer at NFPA. Prior to joining NFPA in 2006, he was in the electrical industry for more than 38 years as an electrical engineer and electrical engineering manager. He has worked in the electric utility and large chemical process industries. His utility experience includes coal, oil and nuclear power generation facilities, substations, protective relaying systems, and underground coal mining. His chemical industry experience includes large process and co-generation facilities built around the world. He is the staff liaison for *NFPA 70E*, writes the *70E Connection* column for the *NFPA Journal,* and responds to *NFPA 70E* advisory service e-mails and phone calls. He was a member of the working group for *IEEE 1584-2002-IEEE Guide for Performing Arc-Flash Hazard Calculations* and has completed arc flash studies for many large facilities. He is a Senior Member of the IEEE and has presented papers and participated on IEEE standards committees. He is a registered professional engineer.

### Mark W. Earley

Mark is Assistant Vice President and Chief Electrical Engineer at NFPA. He has served as Secretary of the *NEC* since 1989 and is the co-author of NFPA's reference book, *Electrical Installations in Hazardous Locations*. Prior to joining NFPA, he worked as an electrical engineer at Factory Mutual Research Corporation. Additionally, he has served on several of NFPA's electrical committees and *NEC* code-making panels. Mark is a registered professional engineer and a member of IAEI, IEEE, SFPE, the Standards Engineering Society, the Automatic Fire Alarm Association (AFAA), the UL Electrical Council, U.S. National Committee on the International Electrotechnical Commission, and the Canadian Electrical Code Committee. He also serves on the IEEE/NFPA Arc Flash Project Steering Committee.

### Ray A. Jones

Ray has spent most of his working life, including 35 years with the DuPont Company, working with industrial electrical systems and installations as a professional electrical engineer. He has served on several national consensus committees and panels, was a charter member of the Petroleum and Chemical Industry Electrical Safety Subcommittee, and is currently chairman of the *NFPA 70E* Technical Committee. The author of many published articles in industry journals, Ray has also presented numerous papers and tutorials on electrical safety issues at technical conferences. Ray coauthored *Electrical Safety in the Workplace*, published by NFPA. He is currently president of Electrical Safety Consulting Services, Inc., a consulting business that helps companies improve their electrical safety programs. Ray is a registered professional engineer.

# PART ONE

# *NFPA 70E®, Standard for Electrical Safety in the Workplace®,* with Commentary

Part One of this handbook includes the complete text and figures of the 2009 edition of *NFPA 70E®, Standard for Electrical Safety in the Workplace®*. The text, tables, and figures from the standard are printed in black and are the official requirements of *NFPA 70E*. Illustrations from the standard are labeled as "Figures."

In addition to standard text and annexes, Part One includes explanatory commentary that provides the history and other background information for specific paragraphs in the standard. This insightful commentary takes the reader behind the scenes, into the reasons underlying the requirements.

Commentary text, captions, and tables are printed in blue, to clarify identification of commentary material. So that the reader can easily distinguish between the illustrations of the standard and those of the commentary, line drawings, graphs, and photographs in the commentary are labeled as "Exhibits."

This edition of the handbook includes a table of contents at the beginning of each chapter to explain what is contained in each chapter. Each chapter also has a summary of changes for the 2009 edition of *NFPA 70E*.

ARTICLE 90

# Introduction

**Summary of Changes**

- **90.1:** Added language to explain that the purpose of the standard is to provide a practical safe working area for employees.
- **90.2:** Added language to the scope of the standard indicating that it applies to activities such as installation, operation, maintenance, and demolition.
- **90.3:** Revised to remove reference to Chapter 4 and to illustrate how each chapter of the standard is intended to apply.
- **90.4:** Revised to add references to the three new Annexes M, N, and O.
- **90.5:** Added a new section to explain the difference between permissive and mandatory rules.
- **90.6:** Added a new section to explain that general NFPA rules governing Formal Interpretations apply to the standard.

**Contents**



*NFPA 70E®*, *Standard for Electrical Safety in the Workplace®*, provides requirements for workers who are or might be exposed to electrical hazards as they provide the essential services of installing, operating, or maintaining electrical equipment. The requirements defined in the body of the standard are intended to be adopted by any employer or agency that desires to improve the protective nature of a safety program. Article 90 serves as an introduction to the main body of the standard.

*NFPA 70E* contains no installation requirements. Installation requirements are found in *NFPA 70*®, *National Electrical Code*® (*NEC*®). The requirements of *NFPA 70E* are intended to apply to work practices as workers accomplish an installation, operate equipment, or maintain a facility. Requirements contained in the standard consider exposure to electrical hazards regardless of the industrial, commercial, or institutional segment in which an exposure might occur. These requirements are predicated on an installation in accordance with the *NEC*. Previous editions of this standard extracted a few safety related requirements from the *NEC* to assist the user of this document. However, since compliance with other requirements from the *NEC* was necessary, it was decided to refer directly to the code as the source of those requirements. *NFPA 70E* has been used in countries that do not use the *NEC*. The safety principles of *NFPA 70E* require the use of installation rules that maintain fundamental installation safety principles.

*NFPA 70E* contains no installation requirements; therefore, it does not include design information. However, equipment installation and system design have a significant bearing on when and how workers are exposed to electrical hazards. The design of an installation could provide equipment to enable *NFPA 70E* requirements to be met easily. For instance, circuit design might include only equipment that accepts lockout devices or enables selective lockouts. Although not intended as a design specification, consideration of *NFPA 70E* requirements during the design phase might provide significant cost savings over the life of a facility. A new annex, Annex O, has been added to the 2009 edition of *NFPA 70E* to illustrate design concepts that can reduce exposure to electrical hazards.

The Technical Committee on Electrical Safety in the Workplace reports to the Association through the Technical Correlating Committee of the *National Electrical Code*. This reporting structure provides consistency between installation rules and safety related work practice requirements for workers.

The content of *NFPA 70E* was the basis for requirements contained in OSHA 29 CFR 1910, Subpart S. Requirements in *NFPA 70E* are consistent with OSHA 29 CFR 1926, Subpart K.

### 90.1 Purpose

**The purpose of this standard is to provide a practical safe working area for employees relative to the hazards arising from the use of electricity.**

The purpose of *NFPA 70E* is to provide requirements to reduce worker exposure to electrical hazards. By reducing exposure to electrical hazards, injuries and fatalities can be reduced or eliminated. When a facility is in a normal condition and meets the following requirements, workers are not usually exposed to shock, electrocution, arc flash, or fire:

- Complies with installation codes, such as the *National Electrical Code*
- Meets the manufacturer's installation, maintenance, and operating requirements
- Operates normally

When any of these requirements are not met, the facility is not in a normal condition, and workers might be exposed to hazards that are associated with electrical energy.

The work practices in this standard provide practical safeguarding from electrical hazards. The term *practical safeguarding* suggests that electrical hazards may not be completely eliminated. Unless an electrically safe work condition exists, electrical energy remains in the circuit, equipment, or system. Avoiding contact with any remaining energy requires analyzing the work steps and making a judgment about how to minimize risk of injury from any remaining energy. *NFPA 70E* defines processes that must be applied to reduce exposure.

*NFPA 70E is not intended to be used by untrained persons.* Appropriate application of this standard requires a qualified person to recognize hazardous conditions, analyze the potential for injury, and take steps to minimize the potential for injury. Practical judgment is an inherent requirement needed to avoid injury from electrical energy or any other energy source.

Article 110 discusses work practices that workers should use to avoid injury from a release of electrical energy as they perform their work task(s). Article 120 focuses on eliminating the potential for injury from an electrical energy source by creating an electrically safe work condition. However, some tasks require that equipment be energized. Therefore, Article 130 provides strategies to control exposure to electrical hazards.

The requirements in *NFPA 70E* apply to all types of employers or facilities, including in-house employees, contractor employees, general industrial workplaces, and construction workplaces.

The requirements in *NFPA 70E* are based on the following four protective strategies, which are listed in order by their decreasing protective nature:

1. Creating an electrically safe work condition
2. Training workers
3. Planning the task
4. Selecting and using personal protective equipment (PPE)

The most important and most protective strategy is eliminating all sources of electrical energy by creating an electrically safe work condition. If the source of energy is removed, then no risk of injury exists. The second most important strategy is training. If workers are trained to recognize and avoid the potential for injury, the chance of injury is reduced. The third most important strategy is planning. If workers take the time to consider each step in the process of executing a work task and consider all potential results from each action, they are less likely to be surprised by a result. The fourth, last, and least effective protective strategy is selecting and using personal protective equipment (PPE). Although PPE is the least effective strategy in *preventing* an electrical injury, it is an essential component of *avoiding* injury.

### 90.2 Scope

At the request of the United States Occupational Safety and Health Administration (OSHA), the NFPA Standards Council chartered the *NFPA 70E* Technical Committee and assigned it the responsibility of defining safe work practices that should be used in general industry. The scope of *NFPA 70E* is similar to the scope of the *National Electrical Code*. *NFPA 70E* is intended to be in harmony with the *NEC* and other NFPA standards and codes.

A subtle but important difference between the scope of the *NEC* and the scope of *NFPA 70E* is that the *NEC* applies to *installations*, whereas *NFPA 70E* applies to *workplaces*. To ensure consistency between the requirements contained in the *NEC* and *NFPA 70E*, all *NEC* panels and the *NFPA 70E* Technical Committee report to the Association through the same Technical Correlating Committee.

**(A) Covered.** This standard addresses electrical safety requirements for employee workplaces that are necessary for the practical safeguarding of employees during activities such as the installation, operation, maintenance, and demolition of electric conductors, electric equipment, signaling and communications conductors and equipment, and raceways for the following:

(1) Public and private premises, including buildings, structures, mobile homes, recreational vehicles, and floating buildings
(2) Yards, lots, parking lots, carnivals, and industrial substations
(3) Installations of conductors and equipment that connect to the supply of electricity

(4) Installations used by the electric utility, such as office buildings, warehouses, garages, machine shops, and recreational buildings, that are not an integral part of a generating plant, substation, or control center

Section 90.2(A) describes facilities that are intended to be within the scope of the standard. Although the scope described in this section is similar to the scope of the *NEC*, the purpose and intent of these documents are considerably different. Whereas the *NEC* applies to equipment and circuits, *NFPA 70E* applies to work practices associated with a workplace that contains electrical hazards both during and after the installation is complete.

Section 90.2(A) clarifies which portions of electric utility facilities are covered by *NFPA 70E*. [See 90.2(B) and the related commentary for information on facilities and specific lighting that are not covered by this standard.] The distinction between electric utility facilities to which this standard does and does not apply is illustrated in Exhibit 90.1.

***EXHIBIT 90.1*** *Typical electric utility complexes showing examples of facilities covered and not covered by the NEC.*

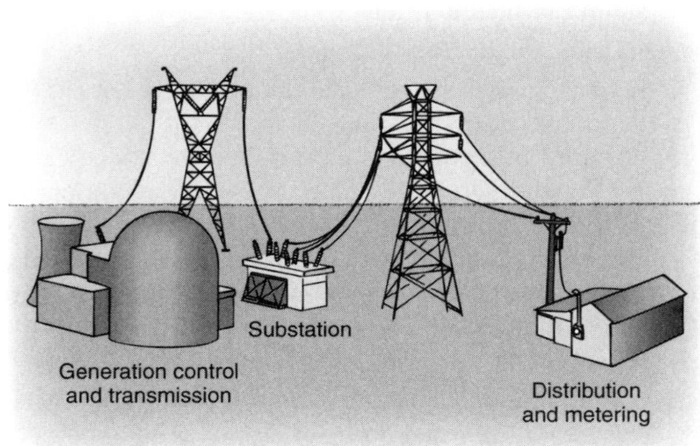


*NEC* does not apply.

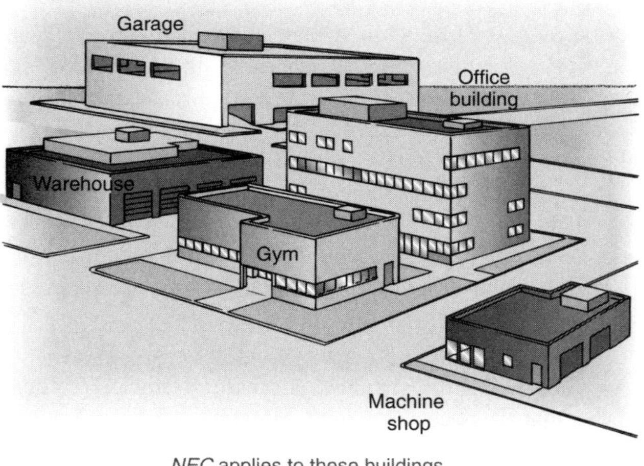


*NEC* applies to these buildings.

Industrial and multibuilding complexes and campus-style wiring often include substations and other installations that employ construction and wiring similar to those of electric utility installations. These installations are on the load side of the service point, and the installation is usually an owner-maintained substation, so their work practices are clearly within the purview of *NFPA 70E*.

**(B) Not Covered.** This standard does not cover the following:

(1) Installations in ships, watercraft other than floating buildings, railway rolling stock, aircraft, or automotive vehicles other than mobile homes and recreational vehicles
(2) Installations underground in mines and self-propelled mobile surface mining machinery and its attendant electrical trailing cable
(3) Installations of railways for generation, transformation, transmission, or distribution of power used exclusively for operation of rolling stock or installations used exclusively for signaling and communications purposes
(4) Installations of communications equipment under the exclusive control of communications utilities located outdoors or in building spaces used exclusively for such installations
(5) Installations under the exclusive control of an electric utility where such installations:
  a. Consist of service drops or service laterals, and associated metering, or
  b. Are located in legally established easements or rights-of-way designated by or recognized by public service commissions, utility commissions, or other regulatory agencies having jurisdiction for such installations, or
  c. Are on property owned or leased by the electric utility for the purpose of communications, metering, generation, control, transformation, transmission, or distribution of electric energy.

Section 90.2(B) describes installations that are not necessarily covered by *NFPA 70E*. However, the work practices discussed in this standard could be adopted and implemented by employers that might otherwise be excluded.

### 90.3 Standard Arrangement

This standard is divided into the introduction and three chapters, as shown in Figure 90.3. Chapter 1 applies generally for safety-related work practices; Chapter 3 supplements or modifies Chapter 1 with safety requirements for special equipment.

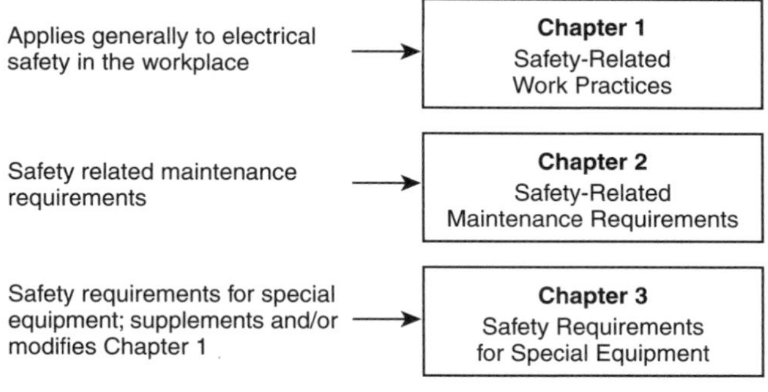


***FIGURE 90.3*** *Standard Arrangement.*

Chapter 2 applies to safety-related maintenance requirements for electrical equipment and installations in workplaces.

Annexes are not part of the requirements of this standard but are included for informational purposes only.

Figure 90.3 illustrates that the requirements defined in Chapter 1 apply generally to all facilities. The maintenance requirements contained in Chapter 2 are intended also to apply generally to all facilities. However, Chapter 3 contains requirements that apply to special systems, facilities, or equipment. In some instances, requirements defined in Chapter 1 are, or might be, unsafe in special circumstances. In other instances, work practices defined in Chapter 1 must be modified to preserve the intent of the requirements because of equipment construction or process operation. Chapter 3, Special Systems, is intended to identify such practices.

### 90.4 Organization

This standard is divided into the following three chapters and fifteen annexes:

(1) Chapter 1, Safety-Related Work Practices
(2) Chapter 2, Safety-Related Maintenance Requirements
(3) Chapter 3, Safety Requirements for Special Equipment
(4) Annex A, Referenced Publications
(5) Annex B, Informational References
(6) Annex C, Limits of Approach
(7) Annex D, Incident Energy and Flash Protection Boundary Calculation Methods
(8) Annex E, Electrical Safety Program
(9) Annex F, Hazard/Risk Evaluation Procedure
(10) Annex G, Sample Lockout/Tagout Procedure
(11) Annex H, Simplified, Two-Category, Flame-Resistant (FR) Clothing System
(12) Annex I, Job Briefing and Planning Checklist
(13) Annex J, Energized Electrical Work Permit
(14) Annex K, General Categories of Electrical Hazards
(15) Annex L, Typical Application of Safeguards in the Cell Line Working Zone
(16) Annex M, Layering of Protective Clothing and Total System Arc Rating
(17) Annex N, Example Industrial Procedures and Policies for Working Near Overhead Electrical Lines and Equipment
(18) Annex O, Safety-Related Design Requirements

Section 90.4 provides an overview of how *NFPA 70E* is structured. The standard is organized to emphasize important requirements at the beginning of the standard and cover issues of decreasing importance as the standard develops. For instance, establishing an electrically safe work condition is very important; consequently, Article 120 discussed this practice. Article 250 covers personal protective equipment because PPE is the last line of protection that might prevent an injury.

Chapter 3 covers some unique industrial facilities and equipment and contains requirements that apply in those unique situations. Some workplaces contain unique processes and electrical systems, so work practices that are appropriate for normal facilities might not be appropriate for circuits in unique circumstances. Requirements contained in Chapter 3 modify general requirements defined in Chapter 1, exempt one or more general requirements, or provide new requirements that are not included in the general requirements.

Several annexes are also included in the standard. Each annex illustrates an important aspect of the general requirements or a specific industrial, commercial, or institutional workplace. All annexes provide supplementary material and examples, are placed at the end of the standard, and are not mandatory.

### 90.5 Mandatory Rules, Permissive Rules, and Explanatory Material

**(A) Mandatory Rules.** Mandatory rules of this standard are those that identify actions that are specifically required or prohibited and are characterized by the use of the terms *shall* or *shall not*.

Mandatory rules are characterized by the use of the terms *shall* or *shall not* as indicated in 90.5(A).

**(B) Permissive Rules.** Permissive rules of this standard are those that identify actions that are allowed but not required, are normally used to describe options or alternative methods, and are characterized by the use of the terms *shall be permitted* or *shall not be required.*

Permissive rules are intended to identify choices to accomplish an intended objective. Permissive rules are characterized by the permissive terms *shall be permitted* or *shall not be required* as indicated in 90.5(B). Although safe alternative practices might exist, employers must ensure that any alternative practice does not increase exposure to injury.

**(C) Explanatory Material.** Explanatory material, such as references to other standards, references to related sections of this standard, or information related to a Code rule, is included in this standard in the form of fine print notes (FPNs). Fine print notes are informational only and are not enforceable as requirements of this standard.

Brackets containing section references to another NFPA document are for informational purposes only and are provided as a guide to indicate the source of the extracted text. These bracketed references immediately follow the extracted text.

FPN: The format and language used in this standard follow guidelines established by NFPA and published in the *NEC Style Manual*. Copies of this manual can be obtained from NFPA.

Fine Print Notes are not statements of intent and do not contain requirements. Fine Print Notes are characterized by the term *FPN* preceding an indented paragraph or sentence. Fine Print Notes contain explanatory material intended to assist in understanding a requirement or to provide reference to additional important material.

Another part of *NFPA 70E* that should not be confused with the Fine Print Note is a table footnote. Some tables have footnotes that immediately follow the table. A footnote is a part of the standard and is intended to be enforceable content. Generally, footnotes that follow a table describe limitations or boundaries in which the table content applies. Tables apply only within the limits described by the footnotes. Although they are in small print, footnotes are not the same as Fine Print Notes. Footnotes are enforceable.

Another example of explanatory material in *NFPA 70E* is the annex material. The standard contains 15 annexes following Chapter 3. Annexes do not contain enforceable requirements. Instead, each annex contains one or more illustrations of how to implement or accomplish requirements defined in Chapters 1, 2, or 3.

### 90.6 Formal Interpretations

To promote uniformity of interpretation and application of the provisions of this standard, formal interpretation procedures have been established and are found in the NFPA Regulations Governing Committee Projects.

The procedures for implementing Formal Interpretations of provisions of this standard are outlined in the NFPA Regulations Governing Committee Projects. These regulations are included in the *NFPA Directory*, which is available from NFPA. The Formal Interpretations procedure is found in Section 6 of the regulations.

Formal Interpretations will be provided as described in the Regulations. Although informal interpretations may be requested, only Formal Interpretations establish the official position of the Association. Committee members and NFPA staff members may offer personal opinions, but only opinions that are processed in accordance with Section 6 of the Regulations Governing Committee Projects represent the official position of the Technical Committee.

Most interpretations of *NFPA 70E* are rendered as the personal opinions of NFPA electrical engineering staff members or of an involved member of the *National Electrical Code* Committee because the request for interpretation does not qualify for processing as a Formal Interpretation in accordance with NFPA Regulations Governing Committee Projects. Such opinions are rendered in writing only in response to written requests. The correspondence contains a disclaimer indicating that it is not a Formal Interpretation issued pursuant to NFPA regulations and that any opinion expressed is the personal opinion of the author and does not necessarily represent the official position of NFPA or the *NFPA 70E* Technical Committee.

CHAPTER 1

# Safety-Related Work Practices

### Summary of Changes

**Article 100**

- **Balaclava:** Added to distinguish the balaclava or sock hood from other types of protective hoods, such as switching hoods.
- **Current-Limiting Overcurrent Protective Device:** Added to reflect the 2008 *National Electrical Code* definition.
- **Live Part:** Deleted to reflect the replacement throughout the standard of variations of the term *energized electrical conductor and circuit parts* with the term *live parts.*
- **Selective Coordination:** Added to reflect the definition from the 2008 *NEC.*
- **Switchgear, Arc-Resistant:** Added to reflect the definition found in IEEE standards and other standards for arc-resistant equipment.
- **Switchgear, Metal-Clad:** Added to Article 100 to define the term in accordance with IEEE C37.20.7.
- **Switchgear, Metal-Enclosed:** Added to Article 100 to define the term in accordance with IEEE C37.20.7.
- **Working Near:** Deleted to eliminate the implication that only working on live parts is dangerous, and to reflect that non-work activities can be equally dangerous when carried out within the limited approach boundary.
- **Working On:** Revised to indicate that any task that requires direct contact with an exposed energized conductor, including "diagnostic" and "repair" tasks, is included within the definition.

**Article 110**

- **110.4:** Revised to indicate that Chapter 1 is divided into four articles.
- **110.5:** Revised to identify specific information that must be exchanged and to clarify responsibility of host and contractor employers.
- **110.6(C):** Revised to require annual re-certification of CPR trained employees.
- **110.6(D)(d):** Revised to require retraining if an employee performs the work practice less frequently than one year.
- **110.6(D)(e):** Added a new requirement for employees to be trained on how to select and use a voltage detector.

- **110.6(D)(e):** Added a new FPN that suggests employment records constitute an acceptable training record.
- **110.7:** Revised to clarify that the electrical safety program must be documented.
- **110.7(F):** Added a new FPN suggesting that the hazard/risk analysis should determine whether a second employee is necessary.
- **110.7(H):** Added a new requirement for electrical safety program procedures to be audited. The audit must include the content of the electrical safety program procedures and how employees implement them.
- **110.8(B):** Revised to replace the term *working near* with the term *within the limited approach boundary*.
- **110.9(A)(1):** Added a new FPN that refers to the American National Standard for voltmeters, ANSI/ISA/UL-61010-1.
- **110.9(A)(4):** Added a new requirement to verify proper operation of instruments used to test for absence of voltage both before and after each use.

**Article 120**

- **120.1(5):** Added a new FPN that refers to ANSI/ISA/UL-61010-1.
- **120.2(D)(3):** Revised to clarify that each complex lockout/tagout plan must be in writing.

**Article 130:** Revised the title of this article to remove the term *live parts.*

- **130.1(A):** Revised to remove the term *live parts* and replace it with the term *energized electrical conductors and circuit parts.*
- **130.1(B)(3):** Revised to add an exemption from the energized work permit requirement when crossing the limited approach boundary for visual inspection only.
- **130.2(C)(1):** Revised to clarify that the requirements of 130.2(C)(1), (2), and (3) are intended to protect uninsulated body parts.
- **130.3:** Revised to require that the arc flash hazard analysis be revised at intervals not exceeding 5 years.
- **130.3:** Added exception indicating certain conditions under which an arc flash hazard analysis might not be necessary.
- **130.3:** Added exception indicating that 130.7(C)(9), (10), and (11) can be used in lieu of a detailed arc flash hazard analysis.
- **130.3(A):** Added a new section indicating that when the voltage level is more than 600 volts, the arc flash protection boundary can be calculated.
- **130.3(B):** Added a new section that permits 130.7(C)(9), (10), and (11) to be used for selection of PPE.
- **130.3(C):** Added a new requirement that a field-installed label be present that provides either available incident energy or the necessary arc-rated FR protective equipment.
- **130.5:** Revised to replace the term *working near* with the term *within the limited approach boundary.*

- **130.7(A):** Added FPNs to clarify that although following the PPE requirements of 130.7 protects an employee against flash or shock hazard, that employee may still be exposed to trauma or injury, particularly at higher voltages.
- **130.7(B):** Revised to identify storage requirements for PPE.
- **130.7(C)(3):** Revised to include the requirement that hairnets or beard nets be nonmelting and flame resistant.
- **130.7(C)(6)(c):** Added a new section requiring that protective equipment be maintained in accordance with ASTM F 496 and inspected before each use.
- **130.7(C)(9):** Added a new FPN that explains how the tables were derived.
- **Table 130.7(C)(9):** Revised to align with experience associated with protective equipment. Some new work tasks such as performing infrared thermography inspections and use of arc-resistant switchgear were added.
- **Table 130.7(C)(10):** Revised the content and format of the table based on field experience. Added Note 10, which permits a face shield with a balaclava to be used for Category 2 head protection.
- **130.7(C)(15):** Revised to clarify that hardhat liners and hairnets must be arc-rated FR material, if worn.
- **130.7(C)(16):** Added a new subsection, (d), which describes cleaning and repair of protective clothing.
- **130.7(D)(1)(a)(3):** Added a new section requiring inspection before each use of insulated tools and equipment.
- **130.7(E)(4):** Added a new section covering look-alike equipment.

**Contents**

*NFPA 70E* provides requirements to eliminate or minimize exposure to electrical hazards in the workplace. *NFPA 70E* is intended to work in harmony with *NFPA 70®*, *National Electrical Code®* (*NEC®*), and NFPA 70B, *Recommended Practice for Electrical Equipment Maintenance*. The *NEC* provides the requirements for a safe electrical installation. It includes workplaces as well as residential applications. NFPA 70B provides guidelines for maintenance of electrical equipment and for creating a preventive maintenance program for electrical equipment. *NFPA 70E* provides requirements for selecting and implementing work practices when the electrical system is in an abnormal condition, such as during maintenance, testing, or repairs.

Workers can be killed and can suffer a lost time injury due to electric shock and burns. Often, these injuries are due to unsafe conditions, unsafe equipment, or poor work practices. The objective of *NFPA 70E* is to provide requirements for safe work practices, including the use of protective equipment that will eliminate or minimize exposure to electrical hazards and resulting injuries.

Chapter 1 has four articles that cover the fundamentals of safety related work practices. Article 100 provides definitions of terms that are used in multiple articles of the standard. Common terms that are defined in English language dictionaries are not normally defined in Article 100. Article 110 contains the general requirements for establishing safety related work practices. It covers training programs, work permit systems, relations among multiple crafts, and multiple employers. Article 120 provides the basis for establishing an electrically safe work condition. It includes planning and lockout/tagout procedures, and its focus is on working on deenergized equipment. Article 130 addresses work involving electrical hazards where personnel can be exposed to a hazard. This can also involve work where the equipment must remain energized.

# ARTICLE 100
## Definitions

In order to ensure electrical safety in the workplace, it is critical for workers to have a common understanding of key terms. This understanding will then limit communication disagreements and misunderstandings that can often result in workers' increased exposure to hazards. Article 100 lists the definitions that are essential for workers to understand.

Some of the definitions in Article 100 are extracted from the *NEC*. These definitions are indicated with a bracketed reference to *NFPA 70* at the end of the definition. The remaining definitions are determined by the *NFPA 70E* Technical Committee on Electrical Safety in the Workplace.

**Scope.** This article contains only those definitions essential to the proper application of this standard. It is not intended to include commonly defined general terms or commonly defined technical terms from related codes and standards. In general, only those terms that are used in two or more articles are defined in Article 100. Other definitions are included in the article in which they are used but may be referenced in Article 100. The definitions in this article shall apply wherever the terms are used throughout this standard.

Common general terms, including those defined in general English language dictionaries, are not defined in *NFPA 70E* unless they are used in a unique or restricted manner. Commonly defined technical terms, such as volt (abbreviated V) and ampere (abbreviated A), are found in the *Authoritative Dictionary of IEEE Standards Terms.*

Not all definitions are listed in Article 100, but can be found instead in their appropriate article. These articles follow the common format according to the *NEC Style Manual* and list the section number as XXX.2, Definition(s). For example, the definitions applicable to electrolytic cells can be found in Section 310.2, Definitions.

**Accessible (as applied to equipment).** Admitting close approach; not guarded by locked doors, elevation, or other effective means. [**70,** 2008]

Exhibit 100.1 illustrates a few examples of equipment considered to be *accessible (as applied to equipment).* The main rule for switches and circuit breakers used as switches is shown as Exhibit 100.1(a) and is according to 404.8(A) of the *NEC.* In Exhibit 100.1(b), the busway installation is according to *NEC* 368.10. The exceptions to the main rule are illustrated in Exhibit 100.1(c), and the installation of busway switches are according to *NEC* Section 404.8(A), Exception No. 1. Exhibit 100.1(d) shows a switch installed adjacent to a motor according to *NEC* 404.8(A), Exception No. 2. Exhibit 100.1(e) shows a hookstick-operated isolating switch installed according to *NEC* Section 404.8(A), Exception No. 3.

**Accessible (as applied to wiring methods).** Capable of being removed or exposed without damaging the building structure or finish or not permanently closed in by the structure or finish of the building. [**70,** 2008]

Wiring methods located behind removable panels designed to allow access are not considered permanently enclosed and are considered exposed as applied to wiring methods. See 300.4(C) of the *NEC* regarding cables located in spaces behind accessible panels.

Exhibit 100.2 illustrates examples of wiring methods and equipment that are considered to be *accessible (as applied to wiring methods).*

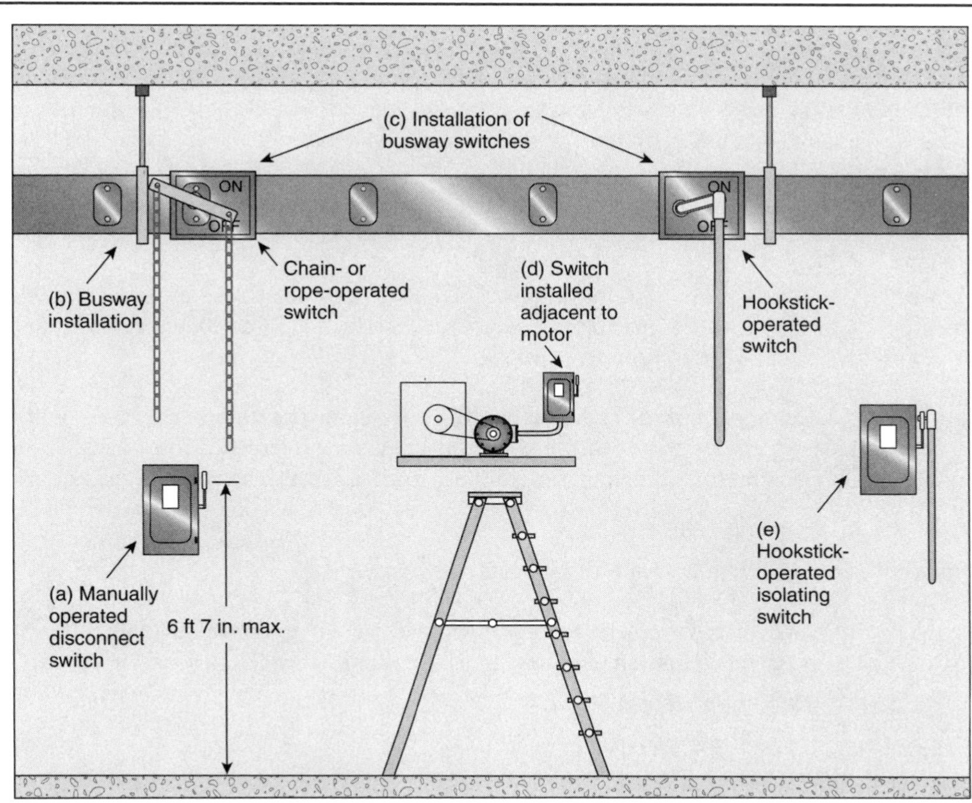


***EXHIBIT 100.1*** *Examples of busway and of switches considered accessible, even if located above 6 ft, 7 in.*

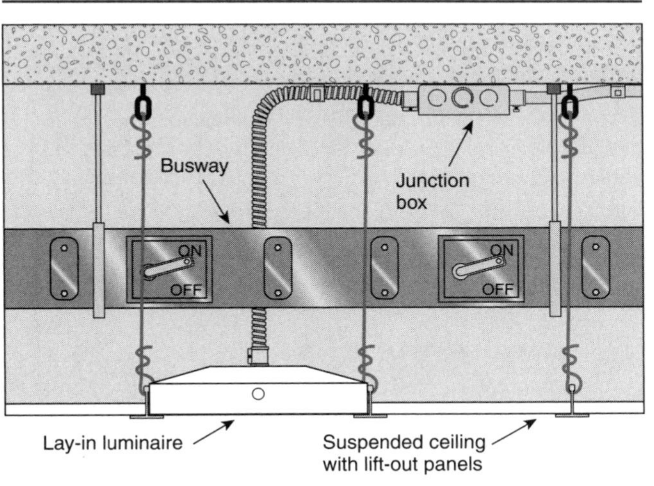


***EXHIBIT 100.2.*** *Examples of busways and junction boxes considered accessible, even if located behind hung ceilings having lift-out panels.*

**Accessible, Readily (Readily Accessible).** Capable of being reached quickly for operation, renewal, or inspections without requiring those to whom ready access is requisite to climb over or remove obstacles or to resort to portable ladders, and so forth. [**70,** 2008]

**Ampacity.** The current, in amperes, that a conductor can carry continuously under the conditions of use without exceeding its temperature rating. [**70,** 2008]

The definition of the term *ampacity* states that the maximum current a conductor carries varies continuously with the conditions of use, as well as with the temperature rating of the conductor insulation. For example, ambient temperature is a condition of use. A conductor with insulation rated at 60°C, installed near a furnace where the ambient temperature is continuously maintained at 60°C, has no current-carrying capacity. Any current flowing through the conductor will raise its temperature above the 60°C insulation rating. Therefore, the ampacity of this conductor, regardless of its size, is zero. See the Correction Factors section for temperature at the bottom of *NEC* Table 310.16 through Table 310.20, or see Annex B of the *NEC*.

Another condition of use is the number of conductors in a raceway or cable. [See 310.15(B)(2) of the *NEC*.] Increasing the number of conductors in a restricted space limits the ability of the conductor to dissipate heat.

**Appliance.** Utilization equipment, generally other than industrial, that is normally built in standardized sizes or types and is installed or connected as a unit to perform one or more functions such as clothes washing, air conditioning, food mixing, deep frying, and so forth. [**70,** 2008]

**Approved.** Acceptable to the authority having jurisdiction.

See the definition of *authority having jurisdiction (AHJ)* for a better understanding of the approval process and of the definition of *approved.* Understanding *NEC* terms such as *listed, labeled,* and *identified (as applied to equipment)* also can assist the user in understanding the approval process.

The important role of the AHJ cannot be overstated in the current North American safety system. The basic role of the AHJ is to verify that an installation complies with the code. The definition of AHJ and the accompanying explanation (the FPN) are very helpful to understand code enforcement, the inspection process, and the definition of the term *approved.*

**Arc Flash Hazard.** A dangerous condition associated with the possible release of energy caused by an electric arc.

FPN No. 1: An arc flash hazard may exist when energized electrical conductors or circuit parts are exposed or when they are within equipment in a guarded or enclosed condition, provided a person is interacting with the equipment in such a manner that could cause an electric arc. Under normal operating conditions, enclosed energized equipment that has been properly installed and maintained is not likely to pose an arc flash hazard.

FPN No. 2: See Table 130.7(C)(9) for examples of activities that could pose an arc flash hazard.

FPN No. 3: See 130.3 for arc flash hazard analysis information.

An *arc flash hazard* exists if a person is or might be exposed to a significant thermal hazard. If the thermal hazard is of a severity that might expose a person to 1.2 calories per square centimeter (cal/cm$^2$) (or more) of thermal energy, the thermal hazard is considered to be significant. Protective equipment with a rating that exceeds the thermal hazard must be worn.

In certain conditions, an arcing fault contained within equipment could generate a pressure wave and destroy the integrity of the enclosure. The technical committee suggests that the term *interacting with the equipment* could mean opening or closing a disconnecting means, pushing a reset button, or latching the enclosure door. However, if equipment is installed in accordance with the requirements of the *NEC*, maintained adequately, and operating normally, the chance of one of these actions initiating an arcing fault is remote.

If the door of functioning electrical equipment is open or not completely closed and latched, an arc flash hazard is more likely to exist. An arc flash hazard analysis can determine the flash protection boundary.

Exhibit 100.3 shows examples of an arc flash incident. An arc flash incident results in intense heat, light, and pressure waves as can be seen in Exhibit 100.4.

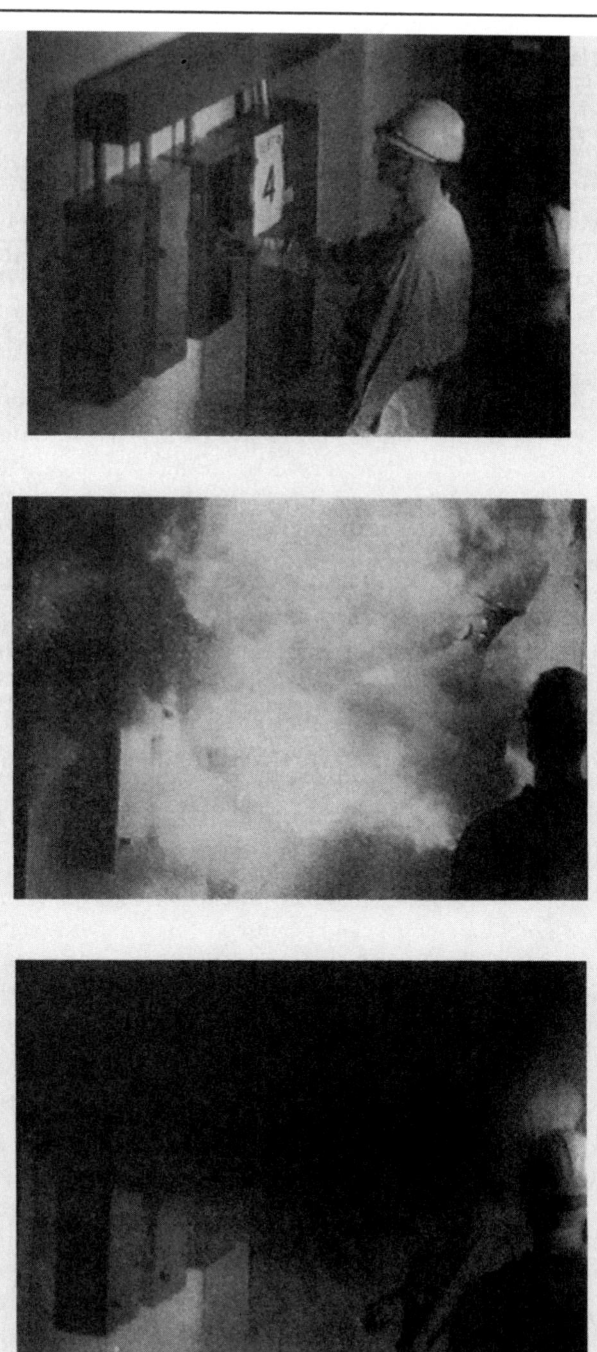


***EXHIBIT 100.3*** *Photos of an arc flash. (From Electrical Safety in the Workplace, Jones and Bartlett, 2000)*

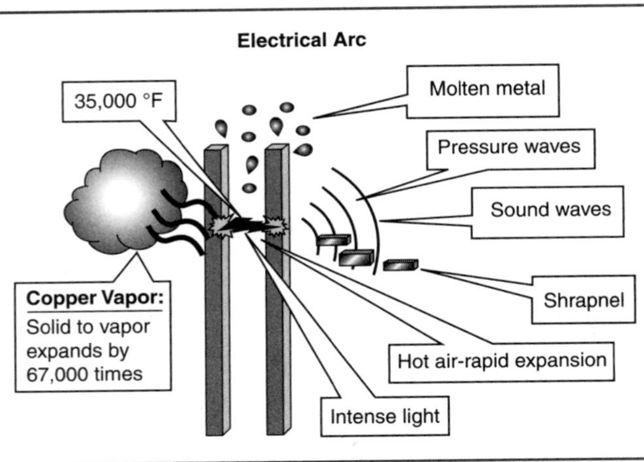


***EXHIBIT 100.4*** *Results of an arc flash incident. (Courtesy of Bussman Division, Cooper Industries)*

**Arc Flash Hazard Analysis.** A study investigating a worker's potential exposure to arc-flash energy, conducted for the purpose of injury prevention and the determination of safe work practices, arc flash protection boundary, and the appropriate levels of PPE.

An *arc flash hazard analysis* determines the flash protection boundary and the amount of incident energy that could be impressed on an employee as a work task is being performed. The analysis can take one of several different forms. An arc flash hazard analysis is necessary regardless of any label or marking on the surface of electrical equipment. Referring to a warning label could constitute one step in the analysis; however, the analysis also must consider risk. When the analysis is complete, the worker has sufficient information for selecting the necessary arc flash protective personal protective equipment (PPE) and the work practices necessary to minimize any exposure to a thermal hazard. Part of the analysis includes determining the arc flash protection boundary and the incident energy. See Annex D for information on incident energy arc flash calculation methods.

**Arc Flash Suit.** A complete FR clothing and equipment system that covers the entire body, except for the hands and feet. This includes pants, jacket, and beekeeper-type hood fitted with a face shield.

The PPE industry has experienced a dramatic evolution of protective schemes and equipment for workers. During this time, the term *arc flash suit* has not been used consistently throughout the industry. This definition clarifies the specific components that comprise a flash suit and points out that a face shield alone cannot be used when a flash suit is required.

Although the term *arc flash suit* does not provide protection for an employee's hands and feet, all other body parts are covered by protective equipment. The hands and feet also must be protected.

**Arc Rating.** The value attributed to materials that describes their performance to exposure to an electrical arc discharge. The arc rating is expressed in cal/cm$^2$ and is derived from the determined value of the arc thermal performance value (ATPV) or energy of breakopen threshold ($E_{BT}$) (should a material system exhibit a breakopen response below the ATPV value) derived from the determined value of ATPV or $E_{BT}$.

FPN: *Breakopen* is a material response evidenced by the formation of one or more holes in the innermost layer of flame-resistant material that would allow flame to pass through the material.

The definition of *arc rating* provides consistency in the selection of protective apparel, as it correlates with ASTM F 1506, *Standard Performance Specification for Flame Resistant Textile Materials for Wearing Apparel for Use by Electrical Workers Exposed to Momentary Electric Arc and Related Thermal Hazards,* for protective apparel and ASTM F 1891, *Standard Specification for Arc and Flame-Resistant Rainwear,* for protective raingear. The definition is consistent with the terminology in the applicable ASTM standards and in the selection of PPE.

Manufacturers determine the arc rating for protective equipment. Labels on arc-rated clothing should include the rating, as shown in Exhibit 100.5. The flash suit arc rating should be visibly marked to ensure the correct rating is being used.

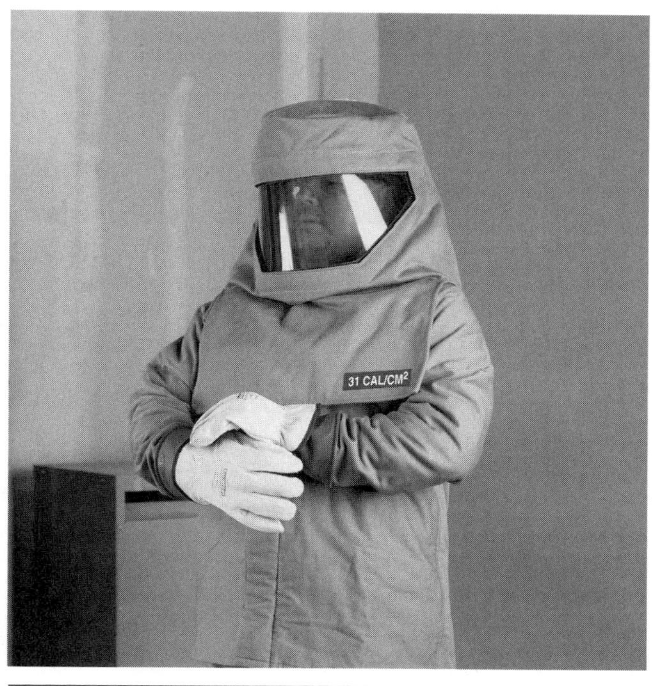


***EXHIBIT 100.5*** *Arc flash suits with the rating clearly marked on the suit. (Courtesy of Salisbury Electrical Safety, LLC)*

**Attachment Plug (Plug Cap) (Plug).** A device that, by insertion in a receptacle, establishes a connection between the conductors of the attached flexible cord and the conductors connected permanently to the receptacle. [**70,** 2008]

Standard *attachment plugs* are available with built-in options, such as switching, fuses, or even ground-fault circuit interrupter protection. Attachment plug contact blades have specific shapes, sizes, and configurations so that a receptacle or cord connector cannot accept an attachment plug of a different voltage or current rating than that for which the device is intended.

**Authority Having Jurisdiction (AHJ).** An organization, office, or individual responsible for enforcing the requirements of a code or standard, or approving equipment, materials, an installation, or a procedure.

FPN: The phrase "authority having jurisdiction," or its acronym AHJ, is used in NFPA documents in a broad manner, since jurisdictions and approval agencies vary, as do their responsibilities. Where public safety is primary, the authority having jurisdiction may be a federal, state, local, or other regional department or individual such as a fire chief; fire marshal; chief of a fire prevention bureau, labor department, or health department; building official; electrical inspector; or others having statutory authority. For insurance purposes, an insurance inspection department, rating bureau, or other insurance company representative may be the authority having jurisdiction. In many circumstances, the property owner or his or her designated agent assumes the role of the authority having jurisdiction; at government installations, the commanding officer or departmental official may be the authority having jurisdiction.

**Automatic.** Self-acting, operating by its own mechanism when actuated by some impersonal influence, as, for example, a change in current, pressure, temperature, or mechanical configuration. [**70,** 2008]

An *automatic* function or operation requires no manual action to cause the function or operation to occur. An automatic function might be initiated by a mechanical or non-mechanical operation. For instance, an under- or over-temperature measurement might cause a heater to increase or decrease output.

**Balaclava (Sock Hood).** An arc-rated FR hood that protects the neck and head except for facial area of the eyes and nose.

A new requirement in the 2009 edition of *NFPA 70E* is that a flame-resistant (FR) *balaclava* sock hood with a face shield (both with at least an arc rating of 8 cal/cm$^2$) can be used as an alternate to an arc flash suit hood for Hazard/Risk Category 2* PPE per Table 130.7(C)(10), Note 10.

A *balaclava*, as shown in Exhibit 100.6, is knitted from yarn and fits tightly against the wearer's head and neck. These protective devices have a similar appearance and feel to a knitted sock. Little, if any, air pocket exists between the balaclava and the wearer's skin. Some balaclavas are flame resistant and have an established arc rating in calories per square centimeter. They are intended for use as protection from an arcing fault when worn with other appropriately selected protective equipment. Note that some balaclavas are intended for warmth only and must not be worn as arc flash protection unless having a defined arc rating.

**Bare-Hand Work.** A technique of performing work on energized electrical conductors or circuit parts, after the employee has been raised to the potential of the conductor or circuit part.

The expertise necessary to perform *bare-hand work* can be acquired only through specialized training. The technique is not readily applicable to work tasks other than working on bare overhead transmission lines. The technique sometimes requires the worker to wear special conductive mesh over his or her clothing to avoid touch potential conditions across his or her body. The bare-hand work method is usually for electric utility installations, rather than facilities constructed under the requirements of the *NEC.*

**Barricade.** A physical obstruction such as tapes, cones, or A-frame-type wood or metal structures intended to provide a warning about and to limit access to a hazardous area.

A *barricade* might consist of "yellow warning tape" or similar materials. A barricade, as shown in Exhibit 100.7, provides warning to approaching individuals about a condition that exists. Barricades do not prevent approach to an unsafe condition. Their purpose is warning only.

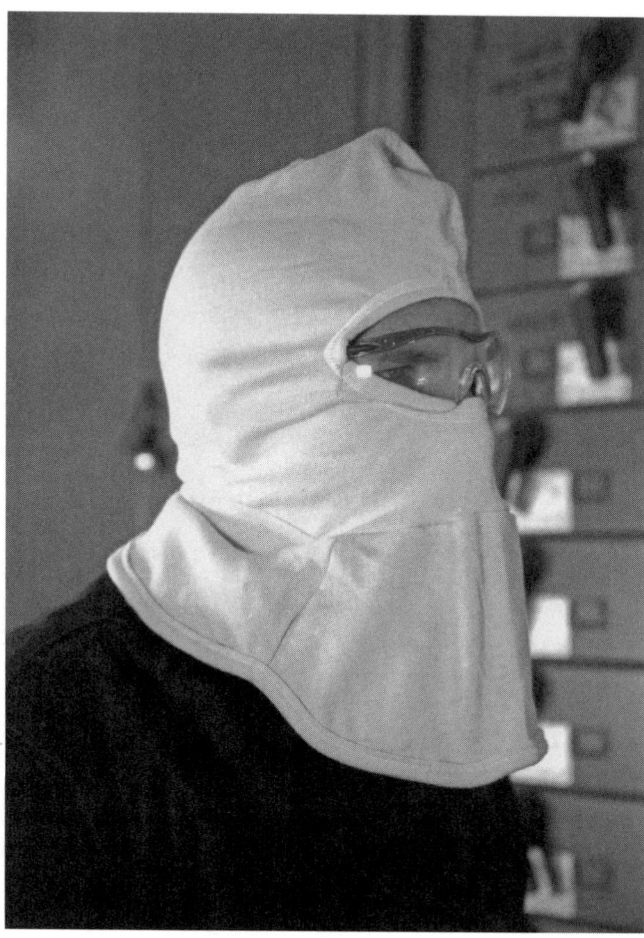

***EXHIBIT 100.6*** *A balaclava sock hood rated 8 cal/cm$^2$. A hard hat and face shield will complete the protection. See Exhibit 130.9(a). (Courtesy of Salisbury Electrical Safety, LLC)*

**Barrier.** A physical obstruction that is intended to prevent contact with equipment or energized electrical conductors and circuit parts or to prevent unauthorized access to a work area.

A *barrier* is intended to prevent contact. A barrier must be of sufficient integrity to eliminate the chance of unsafe contact. A barrier might be constructed from voltage-rated materials and be in physical contact with an energized conductor. A barrier might be constructed from wood or metal and installed with a safe distance between an energized conductor and a worker.

**Bonded (Bonding).** Connected to establish electrical continuity and conductivity. [**70,** 2008]

The purpose of *bonding* is to establish an effective path for fault current that, in turn, facilitates the operation of the overcurrent protective device.

**Bonding Jumper.** A reliable conductor to ensure the required electrical conductivity between metal parts required to be electrically connected. [**70,** 2008]

A *bonding jumper* is an electrical conductor that is installed to "jump" around discontinuous or potentially discontinuous portions of an intentionally conductive path. For instance, both concentric- and eccentric-type knockouts can impair the electrical conductivity between the

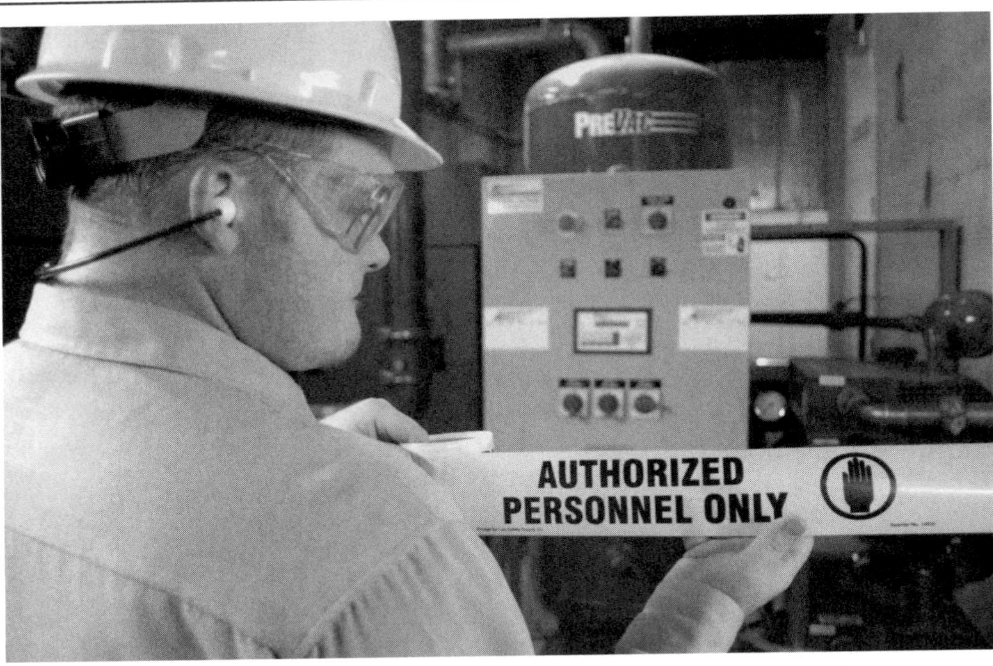


*EXHIBIT 100.7* *Yellow tape used to make sure unqualified people are warned about a work area. The tape is considered a barricade.*

metal parts and can actually introduce unnecessary impedance into the grounding path. Installing bonding jumper(s) is one method often used between metal raceways and metal parts to ensure continuous electrical conductivity. Exhibit 100.8 illustrates the difference between concentric- and eccentric-type knockouts as well as one method of applying bonding jumpers at these types of knockouts.

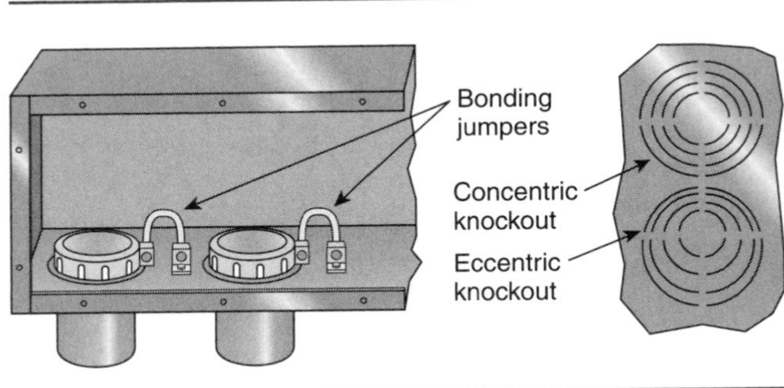


*EXHIBIT 100.8* *Bonding jumpers installed around concentric or eccentric knockouts.*

**Boundary, Arc Flash Protection.** When an arc flash hazard exists, an approach limit at a distance from a prospective arc source within which a person could receive a second degree burn if an electrical arc flash were to occur.

The *arc flash protection boundary* is an imaginary boundary that separates an area in which a person is exposed to a second-degree burn injury from an area in which the potential for injury does not include a second-degree burn. All body parts closer to an arc flash hazard than the arc flash protection boundary must be protected from the potential thermal effects of the hazard. The arc flash protection boundary does not depend upon the amount of incident energy except that if the possible incident energy is less than 1.2 cal/cm$^2$, no second-degree burn can occur.

The arc flash protection boundary is the first issue to be determined in an arc flash hazard analysis. The arc flash protection boundary defines the point at which FR protection is necessary to avoid a second-degree burn. All body parts of a worker are required to be protected. For example, if a worker's hand and arm are within the flash protection boundary, the hand and arm must be protected from the thermal hazard. If a worker's head is within the flash protection boundary, the worker's head (including the back of the head) must be protected from the thermal hazard. (See Section 130.7 for information on personnel protective equipment.)

**Boundary, Limited Approach.** An approach limit at a distance from an exposed energized electrical conductor or circuit part within which a shock hazard exists.

The *limited approach boundary* is not related to arc flash or incident energy. The limited approach boundary is a shock protection boundary and is intended to define the approach limit for unqualified workers and to eliminate the risk of contact with an exposed energized electrical conductor. The term is used to identify an imaginary distance beyond which special considerations are necessary to protect the worker. If an unqualified worker is required to work within the limited approach boundary, he or she must be directly and continuously supervised by a qualified worker.

**Boundary, Prohibited Approach.** An approach limit at a distance from an exposed energized electrical conductor or circuit part within which work is considered the same as making contact with the electrical conductor or circuit part.

The *prohibited approach boundary* is a shock protection boundary that is not related to arc flash or incident energy. The distance determined to be the prohibited approach boundary must not be crossed without PPE that protects the worker from the full circuit voltage. Any person crossing the prohibited approach boundary must possess an authorized energized work permit. He or she also must comply with all other restrictions and controls defined in the electrical safety program.

**Boundary, Restricted Approach.** An approach limit at a distance from an exposed energized electrical conductor or circuit part within which there is an increased risk of shock, due to electrical arc over combined with inadvertent movement, for personnel working in close proximity to the energized electrical conductor or circuit part.

The *restricted approach boundary* is a shock protection boundary that is not related to arc flash or incident energy. It is the approach limit for qualified workers. Qualified workers should have the knowledge and ability to avoid unexpected contact with an exposed energized conductor. If necessary for a qualified worker to cross the restricted approach boundary, he or she must be protected from unexpected contact with the conductors that are energized and exposed. The restricted approach boundary is related to shock.

**Branch Circuit.** The circuit conductors between the final overcurrent device protecting the circuit and the outlet(s). [**70,** 2008]

Exhibit 100.9 illustrates the difference between *branch circuits* and feeders. Conductors between the overcurrent devices in panelboards and duplex receptacles are branch-circuit

*EXHIBIT 100.9.* *Feeder (circuits) and branch circuits.*

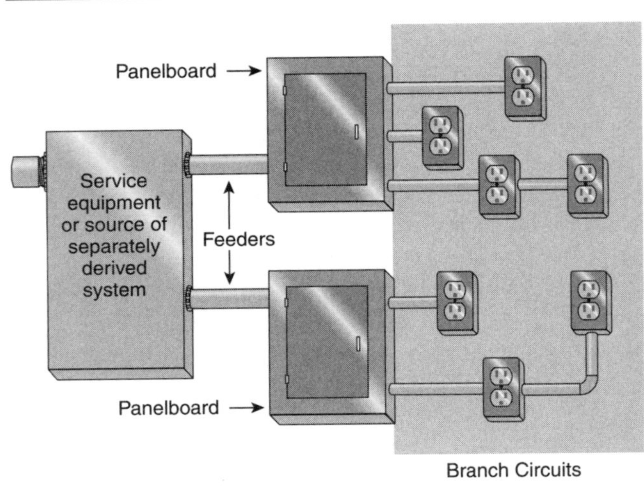


conductors. Conductors between service equipment or the source of separately derived systems and the panelboards are feeders.

**Branch-Circuit Overcurrent Device.** A device capable of providing protection for service, feeder, and branch circuits and equipment over the full range of overcurrents between its rated current and its interrupting rating. Branch-circuit overcurrent protective devices are provided with interrupting ratings appropriate for the intended use but no less than 5,000 amperes. [**70,** 2008]

*Branch-circuit overcurrent devices* must have an interrupting rating appropriate for the intended use; however, the interrupting rating may not be less than 5 kA.

**Building.** A structure that stands alone or that is cut off from adjoining structures by fire walls with all openings therein protected by approved fire doors. [**70,** 2008]

A *building* generally is considered to be a roofed or walled structure that can be used for supporting or sheltering any use or occupancy. However, a building also might be a separate structure such as a pole, billboard sign, or water tower.

Definitions of the terms *fire walls* and *fire doors* are the responsibility of building codes. Generically, a fire wall can be defined as a wall that separates buildings or subdivides a building to prevent the spread of fire and that has a fire-resistance rating and structural stability. Fire doors (and fire windows) are used to protect openings in walls, floors, and ceilings against the spread of fire and smoke within, into, or out of buildings.

**Cabinet.** An enclosure that is designed for either surface mounting or flush mounting and is provided with a frame, mat, or trim in which a swinging door or doors are or can be hung. [**70,** 2008]

*Cabinets* are designed for surface or flush mounting with a trim to which a swinging door(s) is hung. Cutout boxes are designed for surface mounting with a swinging door(s) secured directly to the box. Panelboards are electrical assemblies designed to be placed in a cabinet or cutout box.

**Circuit Breaker.** A device designed to open and close a circuit by nonautomatic means and to open the circuit automatically on a predetermined overcurrent without damage to itself when properly applied within its rating. [**70,** 2008]

A *circuit breaker* is not designed to be operated as a switch unless rated for the appropriate service.

FPN: The automatic opening means can be integral, direct acting with the circuit breaker, or remote from the circuit breaker. [**70,** 2008]

**Conductive.** Suitable for carrying electric current.

As used in this standard, the term *conductive* refers to any material that intentionally or unintentionally conducts electrical current. If a material does not have an established voltage rating, such as voltage-rated rubber products, the material should be considered to be conductive.

**Conductor, Bare.** A conductor having no covering or electrical insulation whatsoever. [**70,** 2008]

*Bare conductors* are visibly copper colored or aluminum colored.

**Conductor, Covered.** A conductor encased within material of composition or thickness that is not recognized by this standard as electrical insulation. [**70,** 2008]

Typical *covered conductors* are the green-colored equipment grounding conductors contained within a nonmetallic-sheathed cable or the uninsulated grounded system conductors within the overall exterior jacket of a Type-SE cable. Covered conductors should always be treated as bare conductors for working clearances because they are really uninsulated conductors.

**Conductor, Insulated.** A conductor encased within material of composition and thickness that is recognized by this standard as electrical insulation. [**70,** 2008]

For the covering on a conductor to be considered insulation, the conductor with the covering material generally is required to pass minimum testing required by a product standard. One such product standard is UL 83, *Standard for Safety Thermoplastic-Insulated Wires and Cables.* To meet the requirements of UL 83, specimens of finished single-conductor wires must pass specified tests that measure (1) resistance to flame propagation, (2) dielectric strength, even while immersed, and (3) resistance to abrasion, cracking, crushing, and impact. Only wires and cables that meet the minimum fire, electrical, and physical properties required by the applicable standards are permitted to be marked with the letter designations found in Table 310.13 and Table 310.61 of the *NEC.* See *NEC* 310.13 for the exact requirements of insulated conductor construction and applications.

**Controller.** A device or group of devices that serves to govern, in some predetermined manner, the electric power delivered to the apparatus to which it is connected. [**70,** 2008]

A *controller* can be a remote-controlled magnetic contactor, switch, circuit breaker, or device that normally is used to start and stop motors and other apparatus and, in the case of motors, is required to be capable of interrupting the stalled-rotor current of the motor. Stop-and-start stations and similar control circuit components that do not open the power conductors to the motor are not considered controllers.

**Coordination (Selective).** Localization of an overcurrent condition to restrict outages to the circuit or equipment affected, accomplished by the choice of overcurrent protective devices and their ratings or settings. [**70,** 2008]

The main goal of selective coordination is to isolate the faulted portion of the electrical circuit quickly while at the same time maintaining power to the remainder of the electrical system. The electrical system overcurrent protection must guard against short circuits and ground faults to ensure that the resulting damage is minimized while other parts of the system not directly involved with the fault are kept operational until other protective devices clear the fault.

Overcurrent protective devices, such as fuses and circuit breakers, have time/current characteristics that determine the time it takes to clear the fault for a given value of fault current. Selectivity occurs when the device closest to the fault opens before the next device upstream operates. For example, any fault on a branch circuit should open the branch-circuit breaker rather than the feeder overcurrent protection. All faults on a feeder should open the feeder overcurrent protection rather than the service overcurrent protection. When selectivity occurs, the electrical system is considered to be coordinated.

With coordinated overcurrent protection, the faulted or overloaded circuit is isolated by the selective operation of only the overcurrent protective device closest to the overcurrent condition. This isolation prevents power loss to unaffected loads. Examples of overcurrent protection without coordination and coordinated protection are illustrated in Exhibit 100.10.

*EXHIBIT 100.10 Overcurrent protection schemes without coordination and with coordination.*

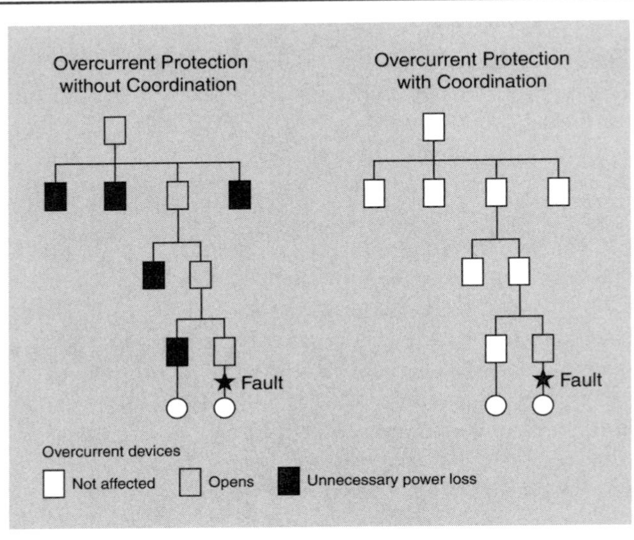


**Current-Limiting Overcurrent Protective Device.** A device that, when interrupting currents in its current-limiting range, reduces the current flowing in the faulted circuit to a magnitude substantially less than that obtainable in the same circuit if the device were replaced with a solid conductor having comparable impedance.

By limiting the amount of current (energy) that is permitted to flow through an overcurrent device during a faulted condition, the overcurrent device serves to reduce the amount of available incident energy during an arcing fault. One important circuit characteristic that impacts incident energy is the duration of the faulted condition. Installing a *current-limiting overcurrent protective device* is one method to reduce incident energy.

**Cutout.** An assembly of a fuse support with either a fuseholder, fuse carrier, or disconnecting blade. The fuseholder or fuse carrier may include a conducting element (fuse link), or may act as the disconnecting blade by the inclusion of a nonfusible member.

A *cutout* is commonly called a *fuse cutout*. A cutout is usually associated with protection of a distribution conductor.

**Cutout Box.** An enclosure designed for surface mounting that has swinging doors or covers secured directly to and telescoping with the walls of the box proper. [**70,** 2008]

**Deenergized.** Free from any electrical connection to a source of potential difference and from electrical charge; not having a potential different from that of the earth.

The term *deenergized* describes an operating condition of electrical equipment. The term should be used for no other purpose. *Deenergized* does not describe a safe condition.

**Device.** A unit of an electrical system that carries or controls electric energy as its principal function. [**70,** 2008]

Components (such as switches, circuit breakers, fuseholders, receptacles, attachment plugs, and lampholders) that distribute or control but do not consume electricity are considered to be *devices*.

Electrical equipment is usually an assembly of several devices. For instance, a unit starter might contain devices such as terminal blocks, fuse, fuse holder, disconnect, and contactor.

**Disconnecting Means.** A device, or group of devices, or other means by which the conductors of a circuit can be disconnected from their source of supply. [**70,** 2008]

*Disconnecting means* can be one or more switches, circuit breakers, and other rated devices that are or might be used to disconnect electrical conductors from their source of energy. Only disconnecting means that are load rated should be used to disconnect an operating load.

**Disconnecting (or Isolating) Switch (Disconnector, Isolator).** A mechanical switching device used for isolating a circuit or equipment from a source of power.

These devices are intended to be operated after interrupting and removing the load current. Locks and tags can be installed on these devices.

**Electrical Hazard.** A dangerous condition such that contact or equipment failure can result in electric shock, arc flash burn, thermal burn, or blast.

> FPN: Class 2 power supplies, listed low voltage lighting systems, and similar sources are examples of circuits or systems that are not considered an electrical hazard.

Fire, shock, and electrocution have been considered to be *electrical hazards* for many years. Since the 1995 edition of *NFPA 70E*, arc flash has been recognized as an electrical hazard. The arc flash hazard currently is defined to consider only the thermal aspects of an arcing fault. Other hazards include flying parts and pieces and the pressure wave (blast) that is generated in an arcing fault. Other electrical hazards also might be associated with an arcing fault.

Electrical equipment that is energized at 50 volts or less is not normally considered to be an arc flash hazard. However, workers should recognize that the effects of an arcing fault are related to available energy and are voltage independent. In some instances, an arcing fault hazard might be significant. If increased exposure to electric burns or to explosion hazards due to electric arc exist, an electrically safe work condition and PPE could be needed.

**Electrical Safety.** Recognizing hazards associated with the use of electrical energy and taking precautions so that hazards do not cause injury or death.

*Electrical safety* is a condition that can be achieved by doing the following:

- Identifying all of the electrical hazards
- Generating a comprehensive plan to mitigate exposure to the hazards
- Providing protective schemes including training for both qualified and unqualified persons

**Electrically Safe Work Condition.** A state in which an electrical conductor or circuit part has been disconnected from energized parts, locked/tagged in accordance with established standards, tested to ensure the absence of voltage, and grounded if determined necessary.

Establishing an *electrically safe work condition* is the only work practice that ensures that an electrical injury cannot occur. However, workers should recognize that operating disconnecting means and verifying absence of voltage might be hazardous work tasks. Until the electrically safe work condition exists, a risk of injury from electrical energy exists.

**Enclosed.** Surrounded by a case, housing, fence, or wall(s) that prevents persons from accidentally contacting energized electrical conductors or circuit parts. [**70,** 2008]

Equipment or devices that are *enclosed* cannot be directly contacted unintentionally. However, open conductors that are enclosed within a fence structure could be touched with long equipment.

**Enclosure.** The case or housing of apparatus, or the fence or walls surrounding an installation to prevent personnel from accidentally contacting energized electrical conductors or circuit parts or to protect the equipment from physical damage. [**70,** 2008]

Commentary Table 100.1 summarizes the intended uses of the various types of *enclosures* for nonhazardous locations. Enclosures that comply with the requirements for more than one type may be marked with multiple designations. Enclosures marked with a type can also be marked as follows:

- Type 1: "Indoor Use Only"
- Types 3, 3S, 4, 4X, 6, or 6P: "Raintight"
- Type 3R: "Rainproof"
- Type 4, 4X, 6, or 6P: "Watertight"
- Type 4X or 6P: "Corrosion Resistant"
- Type 2, 5, 12, 12K, or 13: "Driptight"
- Type 3, 3S, 5, 12K, or 13: "Dusttight"

For equipment that is designated as *raintight,* testing designed to simulate exposure to a beating rain does not result in entrance of water. For equipment designated as *rainproof,* testing designed to simulate exposure to a beating rain does not interfere with the operation of the apparatus or result in wetting of live parts and wiring within the enclosure. *Watertight* equipment is constructed so that water does not enter the enclosure when subjected to a stream of water. *Corrosion-resistant* equipment is constructed so that it provides a degree of protection against exposure to corrosive agents such as salt spray. *Driptight* equipment is constructed so that falling moisture or dirt does not enter the enclosure. *Dusttight* equipment is constructed so that circulating or airborne dust does not enter the enclosure. Commentary Table 100.1, which is Table 110.20 in the *NEC*, is a table of the standard enclosure types for electrical equipment.

**Energized.** Electrically connected to, or is, a source of voltage. [**70,** 2008]

***COMMENTARY TABLE 100.1*** *Enclosure Selection*

| *Provides a Degree of Protection Against the Following Environmental Conditions* | *For Outdoor Use* | | | | | | | | | |
|---|---|---|---|---|---|---|---|---|---|---|
| | *Enclosure-Type Number* | | | | | | | | | |
| | ***3*** | ***3R*** | ***3S*** | ***3X*** | ***3RX*** | ***3SX*** | ***4*** | ***4X*** | ***6*** | ***6P*** |
| Incidental contact with the enclosed equipment | X | X | X | X | X | X | X | X | X | X |
| Rain, snow, and sleet | X | X | X | X | X | X | X | X | X | X |
| Sleet* | — | — | X | — | — | X | — | — | — | — |
| Windblown dust | X | — | X | X | — | X | X | X | X | X |
| Hosedown | — | — | — | — | — | — | X | X | X | X |
| Corrosive agents | — | — | — | X | X | X | — | X | — | X |
| Temporary submersion | — | — | — | — | — | — | — | — | X | X |
| Prolonged submersion | — | — | — | — | — | — | — | — | — | X |
| *Provides a Degree of Protection Against the Following Environmental Conditions* | *For Indoor Use* | | | | | | | | | |
| | *Enclosure Type Number* | | | | | | | | | |
| | ***1*** | ***2*** | ***4*** | ***4X*** | ***5*** | ***6*** | ***6P*** | ***12*** | ***12K*** | ***13*** |
| Incidental contact with the enclosed equipment | X | X | X | X | X | X | X | X | X | X |
| Falling dirt | X | X | X | X | X | X | X | X | X | X |
| Falling liquids and light splashing | — | X | X | X | X | X | X | X | X | X |
| Circulating dust, lint, fibers, and flyings | — | — | X | X | — | X | X | X | X | X |
| Settling airborne dust, lint, fibers, and flyings | — | — | X | X | X | X | X | X | X | X |
| Hosedown and splashing water | — | — | X | X | — | X | X | — | — | — |
| Oil and coolant seepage | — | — | — | — | — | — | — | X | X | X |
| Oil or coolant spraying and splashing | — | — | — | — | — | — | — | — | — | X |
| Corrosive agents | — | — | — | X | — | — | X | — | — | — |
| Temporary submersion | — | — | — | — | — | X | X | — | — | — |
| Prolonged submersion | — | — | — | — | — | — | X | — | — | — |

*Mechanism shall be operable when ice covered.

FPN: The term *raintight* is typically used in conjunction with Enclosure Types 3, 3S, 3SX, 3X, 4, 4X, 6, and 6P. The term *rainproof* is typically used in conjunction with Enclosure Types 3R, and 3RX. The term *watertight* is typically used in conjunction with Enclosure Types 4, 4X, 6, 6P. The term *driptight* is typically used in conjunction with Enclosure Types 2, 5, 12, 12K, and 13. The term *dusttight* is typically used in conjunction with Enclosure Types 3, 3S, 3SX, 3X, 5, 12, 12K, and 13.

*Source:* Table 110.20, *NFPA 70®*, *National Electrical Code®*, 2008 edition, National Fire Protection Association, Quincy, MA.

A conductor that is connected to a source of electricity is considered *energized*. Equipment or devices that are connected to a source of electricity are considered energized. A device, such as a capacitor or battery, that is a source of electricity is considered energized. Electrolytic processes are considered energized.

**Equipment.** A general term, including material, fittings, devices, appliances, luminaires, apparatus, machinery, and the like used as a part of, or in connection with, an electrical installation. [**70,** 2008]

The term *equipment* refers to a broad range of items from simple assemblies of devices to cabinets containing complex circuits, test devices, and many other electrical accessories.

**Explosionproof Apparatus.** Apparatus enclosed in a case that is capable of withstanding an explosion of a specified gas or vapor that may occur within it and of preventing the ignition of a specified gas or vapor surrounding the enclosure by sparks, flashes, or explosion of the gas or vapor within, and that operates at such an external temperature that a surrounding flammable atmosphere will not be ignited thereby. [**70,** 2008]

FPN: For further information, see ANSI/UL 1203-2006, *Explosion-Proof and Dust-Ignition-Proof Electrical Equipment for Use in Hazardous (Classified) Locations.*

*Explosionproof apparatus* must be rated to be used within a specific environment. The temperature generated in the explosion and the pressure generated by an internal explosion depends on the chemical composition of the gases, vapors, or liquids in the environment. Explosionproof apparatus is capable of reducing the temperature of the internal explosion without igniting the environment in the area. The apparatus contains a flame path that permits the gases to cool below the ignition temperature of the external environment before exiting the enclosure. Exhibits 100.11 through 100.13 show typical explosionproof equipment.

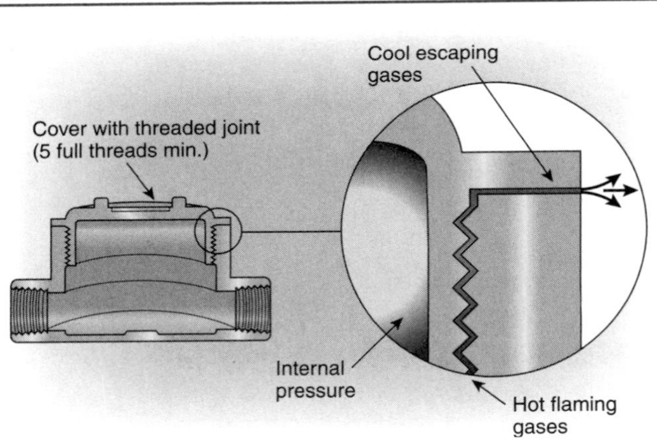


***EXHIBIT 100.11*** *Cooling of hot gases as they pass through the threads of a screw-type cover of an explosionproof junction box.*

***EXHIBIT 100.12*** *An explosionproof panelboard. (Courtesy of Appleton Electric LLC., Emerson Industrial Automation)*

**Exposed (as applied to energized electrical conductors or circuit parts).** Capable of being inadvertently touched or approached nearer than a safe distance by a person. It is applied to electrical conductors or circuit parts that are not suitably guarded, isolated, or insulated.

***EXHIBIT 100.13*** *An explosionproof enclosure for a motor control starter and circuit breaker. (Courtesy of Appleton Electric LLC, Emerson Industrial Automation)*

Conductors that are unguarded and/or uninsulated are considered to be *exposed*. Some electrical equipment contains conductors that are uncovered and guarded only by the enclosure. If the equipment has ventilation openings, wires and tools could be inserted through the ventilation holes and contact energized conductors. Wiring methods that are behind lift out ceiling panels are considered to be *exposed*. The *NEC* would consider this "exposed (as applied to wiring methods)."

**Exposed (as applied to wiring methods).** On or attached to the surface or behind panels designed to allow access. [**70,** 2008]

**Externally Operable.** Capable of being operated without exposing the operator to contact with energized electrical conductors or circuit parts.

Disconnecting means, push buttons, selector switches, and similar equipment are considered to be externally operable if no doors must be opened to achieve access to the equipment. When a door must be opened or a cover removed to gain access to a control device, the device is not externally operable. For instance, some secondary switchgear has trip push buttons located behind a door. These push buttons are not *externally operable*.

**Feeder.** All circuit conductors between the service equipment, the source of a separately derived system, or other power supply source and the final branch-circuit overcurrent device. [**70,** 2008]

**Fitting.** An accessory such as a locknut, bushing, or other part of a wiring system that is intended primarily to perform a mechanical rather than an electrical function. [**70,** 2008]

Examples of *fittings* include condulets, conduit couplings, EMT connectors and couplings, and threadless connectors.

**Flame-Resistant (FR).** The property of a material whereby combustion is prevented, terminated, or inhibited following the application of a flaming or non-flaming source of ignition, with or without subsequent removal of the ignition source.

> FPN: Flame resistance can be an inherent property of a material, or it can be imparted by a specific treatment applied to the material.

A material could be *flame resistant* and still incapable of protecting a person from the high temperature of an arcing fault. Some flame-resistant products are intended for protection from fuel-based fires. Although the temperature of fuel-based fires might reach a few thousand degrees Fahrenheit, an arcing fault produces plasma that might reach 30,000 degrees Fahrenheit. Arc plasma reaches maximum temperature in one or two seconds, whereas the maximum temperature of a fuel-based fire may take several minutes. For protection from an arcing fault, the PPE must be tested and rated for protection from an arcing fault.

Both *flame resistant* and *flame retardant* have been used to describe clothing characteristics. To establish consistency with other standards, the technical committee accepted the term *flame resistant.* This definition is modified from NFPA 2112, *Standard on Flame-Resistant Garments for Protection of Industrial Personnel Against Flash Fire,* and is very similar to ASTM F 1891, *Standard Specification for Arc and Flame Resistant Rainwear.*

**Fuse.** An overcurrent protective device with a circuit-opening fusible part that is heated and severed by the passage of overcurrent through it.

> FPN: A fuse comprises all the parts that form a unit capable of performing the prescribed functions. It may or may not be the complete device necessary to connect it into an electrical circuit.

*Fuses* are active components of an electrical circuit. Fuses act directly on the current flowing in the circuit. Fuse action does not depend on generating or receiving a signal from another circuit element. Exhibit 100.14 shows two examples of Class G fuses rated 300 volts. Note the plainly marked barrels.

**Ground.** The earth. [**70,** 2008]

The term *ground* is in common use in North America. It is intended to refer to a large body that is used as a reference point to establish circuit potential. In power circuits, ground means earth. However, in some electronic circuits, the term does not necessarily refer to earth.

**Grounded (Grounding).** Connected (connecting) to ground or to a conductive body that extends the ground connection. [**70,** 2008]

**Grounded, Solidly.** Connected to ground without inserting any resistor or impedance device. [**70,** 2008]

**Grounded Conductor.** A system or circuit conductor that is intentionally grounded. [**70,** 2008]

***EXHIBIT 100.14*** *Two Class G fuses rated 300 volts. (Courtesy of Bussmann Division, Cooper Industries)*

In most instances, one conductor of an electrical circuit is intentionally connected to earth. That conductor is the *grounded conductor*. An effective system ground enables the overcurrent protection to function.

**Ground Fault.** An unintentional, electrically conducting connection between an ungrounded conductor of an electrical circuit and the normally non–current-carrying conductors, metallic enclosures, metallic raceways, metallic equipment, or earth.

Any fault is unintentional and normally unexpected. A fault results in an electrical current flowing in an unintended circuit. The unintended circuit could include a person. The primary purpose for grounding an electrical system is to provide a path for fault current to flow that excludes a person.

Two types of faults can occur in an electrical circuit, and each type results in different hazardous conditions: a bolted fault and an arcing fault. A bolted fault can exist in equipment. For instance, if a circuit is deenergized and safety grounds installed for maintenance or repair purposes, then reenergized with the safety grounds still in place, a bolted fault exists. Most electrical equipment is tested under bolted-fault conditions.

The most common type of fault is an arcing fault. An arcing fault might result from a conductive object falling into the circuit or a component failure. Unless the equipment is rated as *arc resistant*, it is not tested under arcing fault conditions.

Bolted faults and arcing faults exhibit different characteristics and present different hazards. The primary hazard associated with a bolted fault is shock/electrocution. If a bolted fault exists too long the mechanical force or pressure wave produced by the current can cause conductors to move significantly and result in an arcing fault. On the other hand, an arcing fault normally is associated with a thermal and physical hazard. Arcing faults also can present a shock/electrocution hazard.

**Ground-Fault Circuit-Interrupter (GFCI).** A device intended for the protection of personnel that functions to de-energize a circuit or portion thereof within an established period of time when a current to ground exceeds the values established for a Class A device. [**70,** 2008]

FPN: Class A ground-fault circuit-interrupters trip when the current to ground is 6 mA or higher and do not trip when the current to ground is less than 4 mA. For further information, see UL 943, *Standard for Ground-Fault Circuit Interrupters.*

Using a *GFCI* on cords and equipment with cords is the best practice. GFCIs are reliable measures to limit the flow of fault current and prevent electrocutions. The range of current permitted in a GFCI-protected circuit is much below that necessary for an electrocution to occur. GFCIs are required for all temporary installations and in wet conditions. However, these devices should be used in every instance.

Exhibits 110.15 through 110.18 show some examples of ways to implement the ground-fault circuit-interrupter requirements specified in Section 590.6(A) of the *NEC* for temporary installations.

**Grounding Conductor.** A conductor used to connect equipment or the grounded circuit of a wiring system to a grounding electrode or electrodes. [**70,** 2008]

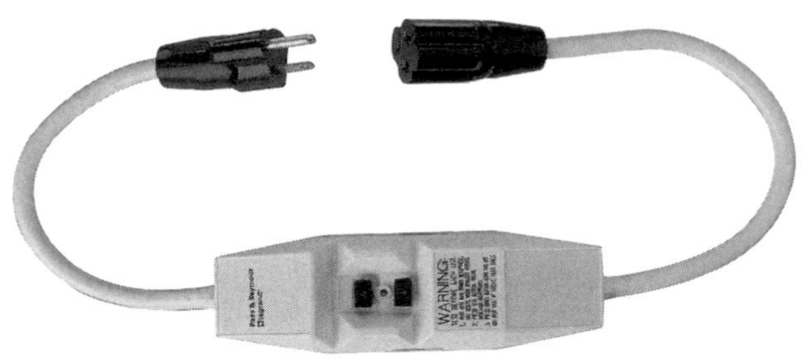

***EXHIBIT 100.15*** *A raintight GFCI with open neutral protection that is designed for use on the line end of a flexible cord. (Courtesy of Pass & Seymour/Legrand®)*

***EXHIBIT 100.17*** *A watertight plug and connector used to prevent tripping of GFCI protective devices in wet or damp weather. (Courtesy of Hubbell Wiring Device–Kellems)*

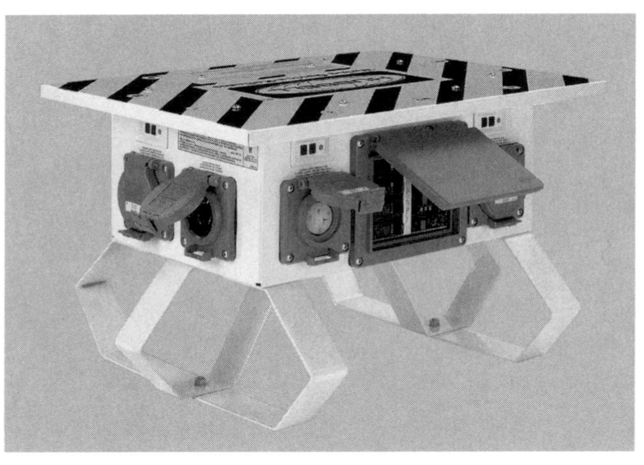

***EXHIBIT 100.16*** *A temporary power outlet unit commonly used on construction sites with a variety of configurations, including GFCI protection. (Courtesy of Hubbell Wiring Device–Kellems)*

***EXHIBIT 100.18*** *A 15-ampere duplex receptacle with integral GFCI that also protects downstream loads. (Courtesy of Pass & Seymour/Legrand®)*

**Grounding Conductor, Equipment (EGC).** The conductive path installed to connect normally non–current-carrying metal parts of equipment together and to the system grounded conductor or to the grounding electrode conductor, or both. [**70,** 2008]

FPN No. 1: It is recognized that the equipment grounding conductor also performs bonding.

FPN No. 2: See *NFPA 70*, Section 250.118 for a list of acceptable equipment grounding conductors.

The primary purpose for an *equipment-grounding conductor* (EGC) is to reduce the possible potential difference (voltage) between surfaces that a worker is likely to touch. By reducing the possible potential difference, the chance of electrocution is reduced. If the integrity of the EGC is satisfactory, the risk of shock injury or electrocution is eliminated. For an equipment-grounding conductor to be effective, the integrity of the conductor must be maintained.

A person must be exposed to a potential difference of 50 volts or more for a shock hazard to exist. The purpose of establishing an effective *ground-fault current path* is to eliminate the possibility that a potential difference of 50 volts or more can exist.

**Grounding Electrode.** A conducting object through which a direct connection to earth is established. [**70,** 2008]

A *grounding electrode* provides the electrical connection to earth. The grounding electrode might take several forms, such as a ground rod or butt ground on a pole. However, all forms serve to connect the grounding conductor to earth. See Exhibit 100.19.

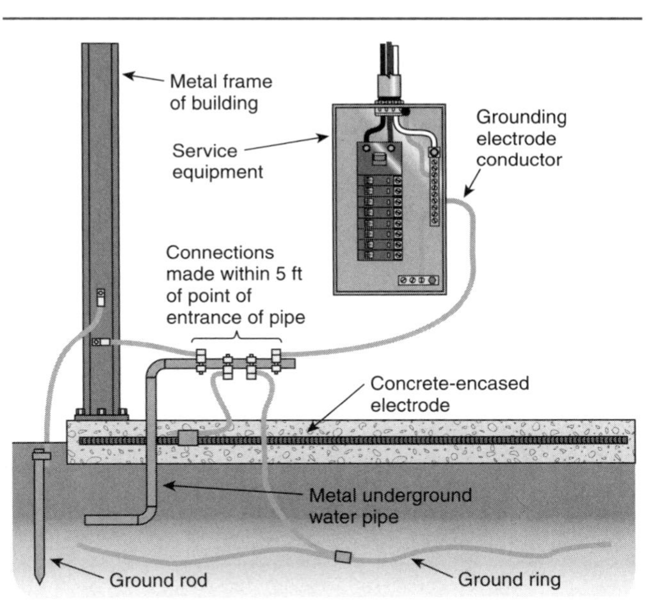


*EXHIBIT 100.19 A grounding electrode system that uses the metal frame of a building, a ground ring, a concrete-encased electrode, a metal underground water pipe, and a ground rod.*

**Grounding Electrode Conductor.** A conductor used to connect the system grounded conductor or the equipment to a grounding electrode or to a point on the grounding electrode system. [**70,** 2008]

The grounding electrode conductor is covered extensively in Article 250 of the *NEC*. The *grounding electrode conductor* is required to be copper, aluminum, or copper-clad aluminum. It is used to connect the equipment-grounding conductor or the grounded conductor (at the service or at the separately derived system) to the grounding electrode or electrodes for either grounded or ungrounded systems. See Exhibit 100.19.

**Guarded.** Covered, shielded, fenced, enclosed, or otherwise protected by means of suitable covers, casings, barriers, rails, screens, mats, or platforms to remove the likelihood of approach or contact by persons or objects to a point of danger. [**70,** 2008]

An exposed conductor is *guarded* when a person who is approaching the exposed conductor is unlikely to contact the conductor. A person must be exposed to a potential difference of 50 volts or more for a shock hazard to exist.

A person might be exposed to hazards associated with an arcing fault even when the conductor is guarded. A guarded conductor protects a person from exposure to shock or electrocution but not to arc flash hazards.

**Incident Energy.** The amount of energy impressed on a surface, a certain distance from the source, generated during an electrical arc event. One of the units used to measure incident energy is calories per centimeter squared (cal/cm$^2$).

Incident energy could be expressed in several different terms, such as calories per square centimeter, joules per square centimeter, or calories per square inch. However, incident energy must be expressed in the same terms in which the PPE is thermally rated. ASTM standards require PPE to be rated in calories per square centimeter, which enables a worker to select adequate PPE.

Physical characteristics of materials vary and react differently when exposed to elevated temperature. Some man-made materials melt before igniting when exposed to the thermal energy generated in an arcing fault. Some other materials ignite and burn when exposed to an arcing fault. The most severe injuries occur when clothing melts onto a worker's skin or when a worker's clothing ignites and burns. Melting and igniting of many materials occur when heated to a few hundred degrees Fahrenheit.

Incident energy raises the temperature of a worker's clothing or skin when exposure to an arcing fault exists. Predicting the amount of available incident energy is critical to prevent injury from melting or burning clothing or from direct skin exposure to incident energy.

**Insulated.** Separated from other conducting surfaces by a dielectric (including air space) offering a high resistance to the passage of current.

FPN: When an object is said to be insulated, it is understood to be insulated for the conditions to which it is normally subject. Otherwise, it is, within the purpose of these rules, uninsulated.

To be *insulated*, the protective covering on the electrical conductor must be rated according to the environment in which it is installed. Some conductors are covered by materials intended to protect the conductor from degradation due to the environment. In this instance, the conductor is considered uninsulated.

**Interrupter Switch.** A switch capable of making, carrying, and interrupting specified currents.

*Interrupter switches* that serve as disconnecting means might be rated to interrupt load current. Some of these switches do not carry such a rating. The label installed by the manufacturer contains information about the interrupting rating of the switch.

**Interrupting Rating.** The highest current at rated voltage that a device is intended to interrupt under standard test conditions. [**70,** 2008]

FPN: Equipment intended to interrupt current at other than fault levels may have its interrupting rating implied in other ratings, such as horsepower or locked rotor current.

Disconnecting means that have an *interrupting rating* can be opened at current that is equal or below its rating. The interrupting rating might be stated directly in amperes or some other descriptive term, such as horsepower rating.

**Isolated (as applied to location).** Not readily accessible to persons unless special means for access are used. [**70,** 2008]

A conductor, device, or equipment is considered *isolated* if it is protected from accidental contact by an unqualified person. If special means, such as a tool or key, is required to gain access to the conductor, device, or equipment, then the conductor, device, or equipment is considered isolated.

**Labeled.** Equipment or materials to which has been attached a label, symbol, or other identifying mark of an organization that is acceptable to the authority having jurisdiction and concerned with product evaluation, that maintains periodic inspection of production of labeled equipment or materials, and by whose labeling the manufacturer indicates compliance with appropriate standards or performance in a specified manner.

*Labeled* equipment has an identifying mark installed by the manufacturer indicating that the equipment, device, or conductor meets all requirements defined by the appropriate consensus standards including construction and performance. The label might be from Underwriters Laboratories Inc. (UL) or any other nationally recognized testing laboratory. The label must identify the organization that evaluated the product. Equipment or conductors required or permitted by the *NEC* to be labeled are acceptable only if they have been approved for the specific application or environment.

**Listed.** Equipment, materials, or services included in a list published by an organization that is acceptable to the authority having jurisdiction and concerned with evaluation of products or services, that maintains periodic inspection of production of listed equipment or materials or periodic evaluation of services, and whose listing states that either the equipment, material, or services meets appropriate designated standards or has been tested and found suitable for a specified purpose.

> FPN: The means for identifying listed equipment may vary for each organization concerned with product evaluation, some of which do not recognize equipment as listed unless it is also labeled. Use of the system employed by the listing organization allows the authority having jurisdiction to identify a listed product.

Listing is the most common method of third party evaluation of the safety of a product. Laboratories that list products evaluate the product in accordance with appropriate product standards. The label must include the mark of the organization that evaluated the product. In order for the listed product to be approved, the product listing must be acceptable to the authority having jurisdiction.

**Live Parts.** Energized conductive components. [**70,** 2008]

The *NEC* defines the term *live parts*. When the term is used in the context of the *NEC*, the definition is complete. However, when used in the context of work practices, additional descriptors are needed. Historically, the term *live parts* was associated with a hazard. Although being associated with a hazard seems to have been removed by the current definition, including the word *energized* carries a similar connotation, as in *energized live parts*. Although the enclosure of an energized electrical panel might be conductive, if the enclosure is complete, no *live part* exists.

**Luminaire.** A complete lighting unit consisting of a lamp or lamps, together with the parts designed to distribute the light, to position and protect the lamps and ballast (where applicable), and to connect the lamps to the power supply. It may also include parts to protect the light source or the ballast or to distribute the light. A lampholder is not a luminaire. [**70,** 2008]

**Motor Control Center.** An assembly of one or more enclosed sections having a common power bus and principally containing motor control units. [**70,** 2008]

A *motor control center* might contain starters, disconnect switches, power panels, solid-state drives, and similar components.

**Neutral Conductor.** The conductor connected to the neutral point of a system that is intended to carry current under normal conditions. [**70,** 2008]

A *neutral conductor* is connected to earth ground only at one point in the electrical system. The neutral current must flow through the neutral conductor. It is important to remember that the neutral conductor is a current-carrying conductor. Many believe that, because the neutral conductor is a grounded conductor, it is safe to work on it while it is energized. This is a very dangerous practice that has led to many serious electric shocks.

**Neutral Point.** The common point on a wye-connection in a polyphase system or midpoint on a single-phase, 3-wire system, or midpoint of a single-phase portion of a 3-phase delta system, or a midpoint of a 3-wire, direct-current system. [**70,** 2008]

FPN: At the neutral point of the system, the vectorial sum of the nominal voltages from all other phases within the system that utilize the neutral, with respect to the neutral point, is zero potential.

Exhibit 100.20 shows examples of a neutral point.

*EXHIBIT 100.20* *Four examples of a neutral point.*

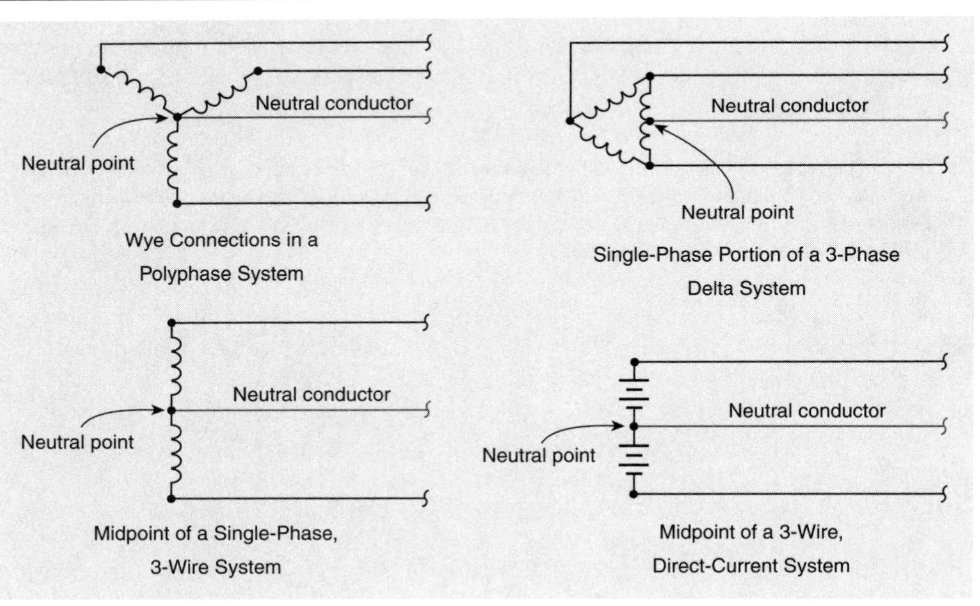


**Open Wiring on Insulators.** An exposed wiring method using cleats, knobs, tubes, and flexible tubing for the protection and support of single insulated conductors run in or on buildings.

**Outlet.** A point on the wiring system at which current is taken to supply utilization equipment. [**70,** 2008]

An *outlet* might be a receptacle or lighting outlet.

**Overcurrent.** Any current in excess of the rated current of equipment or the ampacity of a conductor. It may result from overload, short circuit, or ground fault. [**70,** 2008]

FPN: A current in excess of rating may be accommodated by certain equipment and conductors for a given set of conditions. Therefore, the rules for overcurrent protection are specific for particular situations.

Conductors and other conductive elements have an inherent capacity to dissipate heat generated when current flows through them. The ampacity of a conductor or other conductive element is determined by the ability of the element to dissipate heat. When current flow generates more heat than the circuit element can dissipate without overheating, excess current exists in the circuit. The excess current commonly is called *overcurrent*.

**Overload.** Operation of equipment in excess of normal, full-load rating, or of a conductor in excess of rated ampacity that, when it persists for a sufficient length of time, would cause damage or dangerous overheating. A fault, such as a short circuit or ground fault, is not an overload. [**70,** 2008]

When circuit conditions develop that result in an overcurrent condition, an *overload* exists. The environment in which the circuit element is installed has an impact on the ability of a conductor or other circuit element to conduct current without overheating. However, a short circuit or other faulted condition does not constitute an overload.

**Panelboard.** A single panel or group of panel units designed for assembly in the form of a single panel, including buses and automatic overcurrent devices, and equipped with or without switches for the control of light, heat, or power circuits; designed to be placed in a cabinet or cutout box placed in or against a wall, partition, or other support; and accessible only from the front. [**70,** 2008]

**Premises Wiring (System).** Interior and exterior wiring, including power, lighting, control, and signal circuit wiring together with all their associated hardware, fittings, and wiring devices, both permanently and temporarily installed. This includes: (a) wiring from the service point or power source to the outlets; or (b) wiring from and including the power source to the outlets where there is no service point.

Such wiring does not include wiring internal to appliances, luminaires, motors, controllers, motor control centers, and similar equipment. [**70,** 2008]

All system wiring electrically downstream of the service point is considered to be *premises wiring*. However, wiring installed within equipment is not considered to be premises wiring.

**Qualified Person.** One who has skills and knowledge related to the construction and operation of the electrical equipment and installations and has received safety training to recognize and avoid the hazards involved. [**70,** 2008]

For a person to be considered qualified, he or she must understand electrical hazards associated with the contemplated work task. He or she also must understand the limitations of PPE before selecting the necessary protective equipment. A qualified person must have the ability to recognize all electrical hazards that might be associated with the work task being considered.

A worker could be qualified to perform one work task and not qualified to perform a different task. A qualified worker must understand the construction and operation of the equipment or circuit associated with the contemplated work task.

The latest revision of the OSHA definition for qualified person (1910.399 8/07) includes the phrase "has demonstrated skills." To meet this, the person has to actually demonstrate that he/she can perform the task. A dress rehearsal using appropriate PPE for the task will ensure that the worker can perform the task with the lighting limitations of the flash suit hood and the dexterity limitations of voltage-rated gloves with leather protectors.

A *qualified person* must understand how to select appropriate test equipment and apply that equipment to the work task. He or she must be trained to understand and apply the details of the electrical safety program and procedures provided the employer.

A qualified person must be able to perform a hazard/risk analysis and to react appropriately to all hazards associated with the work task.

**Raceway.** An enclosed channel of metal or nonmetallic materials designed expressly for holding wires, cables, or busbars, with additional functions as permitted in this standard. Raceways include, but are not limited to, rigid metal conduit, rigid nonmetallic conduit, intermediate metal conduit, liquidtight flexible conduit, flexible metallic tubing, flexible metal conduit, electrical metallic tubing, electrical nonmetallic tubing, underfloor raceways, cellular concrete floor raceways, cellular metal floor raceways, surface raceways, wireways, and busways. [**70,** 2008]

Cable trays are not considered *raceways*. They are a support method.

**Receptacle.** A receptacle is a contact device installed at the outlet for the connection of an attachment plug. A single receptacle is a single contact device with no other contact device on the same yoke. A multiple receptacle is two or more contact devices on the same yoke. [**70,** 2008]

A receptacle is an outlet. However, not all outlets are receptacles. The *NEC* defines an outlet as "a point on the wiring system at which current is taken to supply utilization equipment." This term is frequently misused. Common examples of outlets include lighting outlets, receptacle outlets, and smoke alarm outlets. See Exhibit 100.21.

***EXHIBIT 100.21***
*Receptacles.*

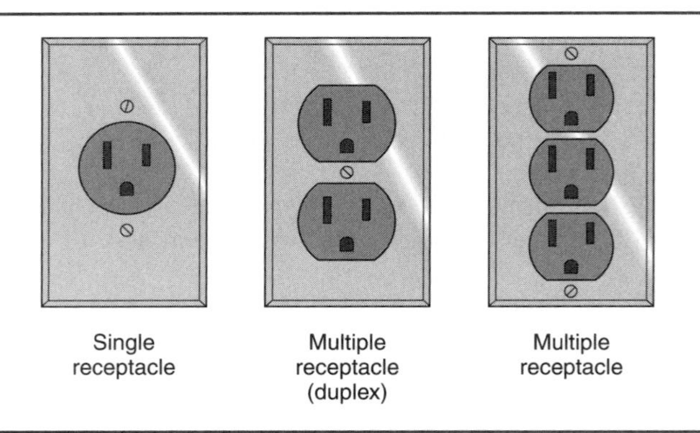


**Separately Derived System.** A premises wiring system whose power is derived from a source of electric energy or equipment other than a service. Such systems have no direct electrical connection, including a solidly connected grounded circuit conductor, to supply conductors originating in another system. [**70,** 2008]

**Service.** The conductors and equipment for delivering electric energy from the serving utility to the wiring system of the premises served. [**70,** 2008]

A *service* is the source of electrical energy for a facility from a utility. If the electrical energy is not supplied by a serving utility, the supply conductors and equipment are considered to be feeders, not a service.

**Service Conductors.** The conductors from the service point to the service disconnecting means. [**70,** 2008]

The term *service conductors* is a broad term that can include service drops, service laterals, and service-entrance conductors, but this term specifically excludes any wiring on the supply side (serving utility side) of the service point.

If the utility has specified that the service point be at the utility pole, the service conductors from an overhead distribution system originate at the utility pole and terminate at the service disconnecting means. If the utility has specified that the service point be at the utility manhole, the service conductors from an underground distribution system originate at the utility manhole and terminate at the service disconnecting means. Where utility-owned primary conductors are extended to outdoor pad-mounted transformers on private property, the service conductors originate at the secondary connections of the transformers only if the utility has specified that the service point be at the secondary connections.

**Service Drop.** The overhead service conductors from the last pole or other aerial support to and including the splices, if any, connecting to the service-entrance conductors at the building or other structure. [**70,** 2008]

Overhead *service-drop* conductors run from the utility pole and connect to the service-entrance conductors at the service point. See Exhibit 100.22. Conductors on the utility side of the service point are not covered by the *NEC*. Instead, the utility specifies the location of the service point, which varies from utility to utility, as well as from occupancy to occupancy.

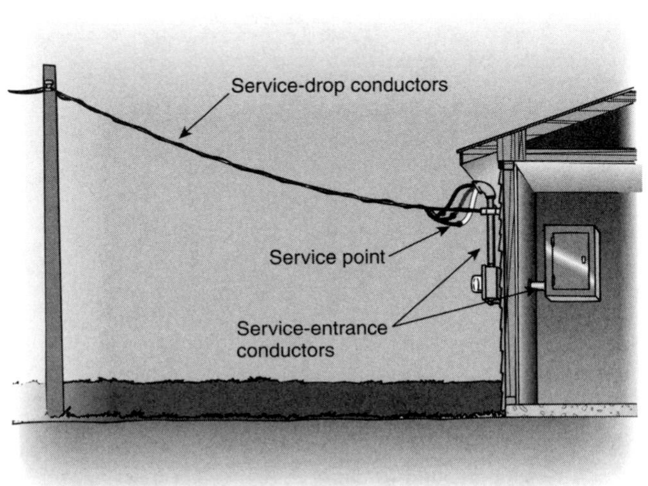


***EXHIBIT 100.22*** *Overhead system showing a service drop from a utility pole to attachment on a house and service-entrance conductors from point of attachment (spliced to service-drop conductors), down the sides of the house, through the meter socket, and terminating in the service department.*

**Service-Entrance Conductors, Overhead System.** The service conductors between the terminals of the service equipment and a point usually outside the building, clear of building walls, where joined by tap or splice to the service drop. [**70,** 2008]

**Service-Entrance Conductors, Underground System.** The service conductors between the terminals of the service equipment and the point of connection to the service lateral. [**70,** 2008]

FPN: Where service equipment is located outside the building walls, there may be no service-entrance conductors, or they may be entirely outside the building.

See Exhibit 100.23 for an illustration of *service-entrance conductors in an underground system.* As illustrated, the underground service laterals may be run from poles (top) or from transformers (bottom) and with or without terminal boxes, provided they begin at the service point. Conductors on the utility side of the service point are not covered by the *NEC.* The utility specifies the location of the service point, which vary from utility to utility, as well as from occupancy to occupancy.

***EXHIBIT 100.23***
*Underground systems showing service laterals run from a pole and from a transformer.*

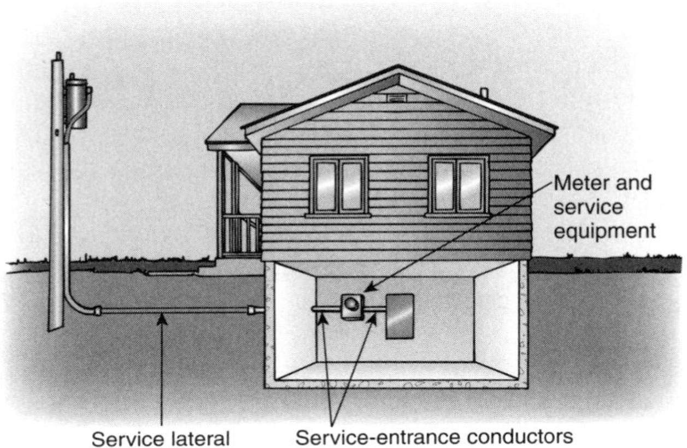


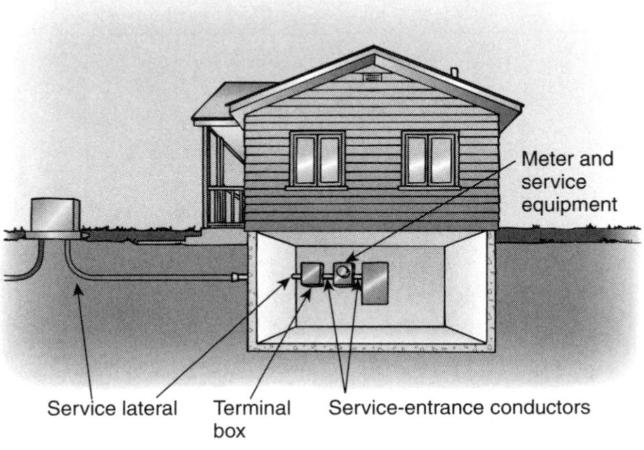


**Service Equipment.** The necessary equipment, usually consisting of a circuit breaker(s) or switch(es) and fuse(s), and their accessories, connected to the load end of service conductors to a building or other structure, or an otherwise designated area, and intended to constitute the main control and cutoff of the supply. [**70,** 2008]

**Service Lateral.** The underground service conductors between the street main, including any risers at a pole or other structure or from transformers, and the first point of connection to the service-entrance conductors in a terminal box or meter or other enclosure, inside or outside the building wall. Where there is no terminal box, meter, or other enclosure, the point of connection is considered to be the point of entrance of the service conductors into the building. [**70,** 2008]

**Service Point.** The point of connection between the facilities of the serving utility and the premises wiring. [**70,** 2008]

The *service point* is the point of demarcation between the serving utility and the premises wiring. The service point is the point on the wiring system where the serving utility ends and the premises wiring begins. The serving utility generally specifies the location of the service point.

Because the location of the service point generally is determined by the utility, the service-drop conductors and the service-lateral conductors might or might not be part of the service as covered by the *NEC.* Only conductors physically located on the premises wiring side of the service point are covered by the *NEC.* Conductors located on the utility side of the service point are not covered in the definition of service conductors; therefore, they are not covered by the *NEC.*

Generally, based on the definitions of the terms *service point* and *service conductors,* any conductor on the serving utility side of the service point is not covered by the *NEC.* For example, a typical suburban residence has an overhead service drop from the utility pole to the house. If the utility specifies that the service point be at the point of attachment of the service drop to the house, the service-drop conductors are not considered service conductors because the service drop is not on the premises wiring side of the service point. Alternatively, if the service point is specified as "at the pole" by the utility, the service-drop conductors are considered service conductors, and the *NEC* would apply to the service drop.

Exact locations for a service point may vary from utility to utility, as well as from occupancy to occupancy.

**Shock Hazard.** A dangerous condition associated with the possible release of energy caused by contact or approach to energized electrical conductors or circuit parts.

Tolerance of electrical current flow varies from one person to another. Although not technically substantiated, tolerance seems to be related to current density. However, existing documentation indicates that any person might receive a shock if the amount of current exceeds 0.020 amperes. Any exposure to contact with a source of electrical energy that might result in this current flow is a *shock hazard.* When the voltage is 50 volts or greater, in general, a shock hazard exists.

**Short-Circuit Current Rating.** The prospective symmetrical fault current at a nominal voltage to which an apparatus or system is able to be connected without sustaining damage exceeding defined acceptance criteria. [**70,** 2008]

The ability of an overcurrent device to safely interrupt fault current depends on several factors, including the ability of the overcurrent device to quench any arc that might be associated with operation of the overcurrent device. The maximum amount of current that an overcurrent device can safely interrupt is the *short-circuit current rating.*

**Single-Line Diagram.** A diagram that shows, by means of single lines and graphic symbols, the course of an electric circuit or system of circuits and the component devices or parts used in the circuit or system.

A *single-line diagram* illustrates a complete system or a portion of a system. Consisting of graphic symbols, a single-line diagram illustrates all disconnecting devices. Single-line diagrams must be kept current to adequately indicate disconnecting devices. Record copies of these drawings should be marked by "red-lining" or other similar mechanisms to illustrate all changes made to the system. See Figure D.3 in Annex D for an example of a single-line diagram.

**Special Permission.** The written consent of the authority having jurisdiction. [**70,** 2008]

**Step Potential.** A ground potential gradient difference that can cause current flow from foot to foot through the body.

Exhibit 100.24 illustrates how current might flow to a human body as a result of step and touch potential. Body impedance is illustrated in the figure. Contact resistance depends on the insulating quality of gloves or footwear. However, the gloves or footwear must be assigned a voltage rating by the manufacturer.

***EXHIBIT 100.24*** *Step and touch potential for current flow path. (From Safe Work Practices for Electricians, by Ray A. Jones and Jane G. Jones, published by Jones and Bartlett, 2008)*

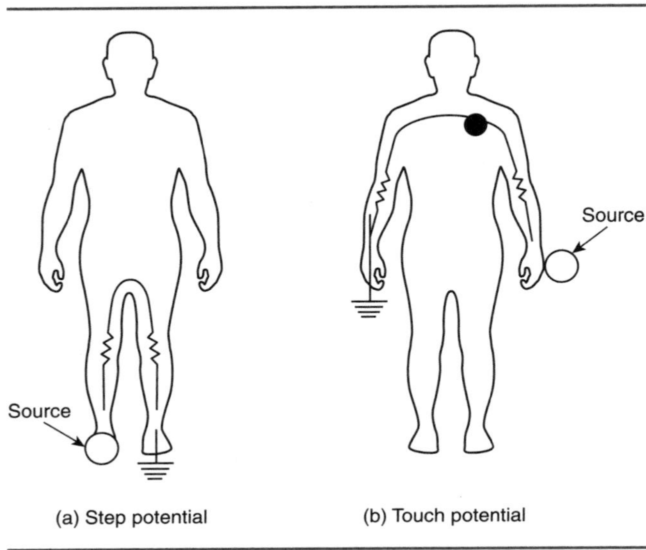


**Structure.** That which is built or constructed. [**70,** 2008]

**Switchgear, Arc-Resistant.** Equipment designed to withstand the effects of an internal arcing fault and that directs the internally released energy away from the employee.

*Arc-resistant switchgear* provides adequate protection from any internal arcing fault when the equipment is closed and operating normally. If doors and covers (including fasteners) are not completely closed, workers are exposed to the hazards associated with an arcing fault just as if no arc-resistant rating existed.

**Switchgear, Metal-Clad.** A switchgear assembly completely enclosed on all sides and top with sheet metal, having drawout switching and interrupting devices, and all live parts enclosed within grounded metal compartments.

**Switchgear, Metal-Enclosed.** A switchgear assembly completely enclosed on all sides and top with sheet metal (except for ventilating openings and inspection windows), containing primary power circuit switching, interrupting devices, or both, with buses and connections. This assembly may include control and auxiliary devices. Access to the interior of the enclosure is provided by doors, removable covers, or both. Metal-enclosed switchgear is available in non-arc-resistant or arc-resistant constructions.

**Switch, Isolating.** A switch intended for isolating an electric circuit from the source of power. It has no interrupting rating, and it is intended to be operated only after the circuit has been opened by some other means. [**70,** 2008]

*Isolating switches,* which normally are inadequate to break load current, provide a means to install lockout devices after the load current has been removed.

**Switchboard.** A large single panel, frame, or assembly of panels on which are mounted on the face, back, or both, switches, overcurrent and other protective devices, buses, and usually instruments. Switchboards are generally accessible from the rear as well as from the front and are not intended to be installed in cabinets. [**70,** 2008]

In the past, the term *switchboard* was used to refer to a structure comprising disconnecting and control components in a construction that might be open in the back and/or sides. Switchgear does not meet that definition, but many electrical workers incorrectly apply the term *switchboard* to switchgear and motor control centers.

**Switching Device.** A device designed to close, open, or both, one or more electric circuits.

A *switching device* can be any device that is rated for this use. Splicing devices located in mid-conductor sometimes are rated for use as a disconnecting device and, therefore, are acceptable as switching devices.

**Touch Potential.** A ground potential gradient difference that can cause current flow from hand to hand, hand to foot, or another path, other than foot to foot, through the body.

Any current path through a victim's body other than from one foot to the other foot is called *touch potential.* (See Exhibit 100.24.)

**Ungrounded.** Not connected to ground or to a conductive body that extends the ground connection. [**70,** 2008]

If voltage is measured between a conductor in an ungrounded circuit and ground, the indication will be zero; however, the conductors might still be energized with respect to each other. A voltage measurement of zero does not suggest that no electrical hazard exists. In fact, the risk of injury might be elevated because no potential difference exists between any circuit conductor and ground.

**Unqualified Person.** A person who is not a qualified person.

Workers who might be exposed to an electrical hazard as a work task is performed must be trained to recognize that a hazard exists and how to avoid that hazard. Any person who has not received specific training is an *unqualified person.* A worker who has been trained to perform a task might be qualified to perform that task and still be unqualified to perform any other task. The characteristics of being qualified and unqualified are task dependent.

**Utilization Equipment.** Equipment that utilizes electric energy for electronic, electromechanical, chemical, heating, lighting, or similar purposes. [**70,** 2008]

Any equipment designed to convert electrical energy to other forms of energy that is directly usable is *utilization equipment*. For instance, a motor converts electrical energy to mechanical energy.

**Ventilated.** Provided with a means to permit circulation of air sufficient to remove an excess of heat, fumes, or vapors. [**70,** 2008]

*Ventilated* equipment usually relies on convection to provide air at ambient temperature through openings at or near the bottom of the equipment and discharges heated air through openings at or near the top. Some ventilated equipment relies upon forced convection from fans and similar equipment. The proper PPE must be worn when working with equipment with ventilation openings. A worker cannot assume a "closed door" condition because the ventilation openings provide an escape for the hot gases and fumes.

**Voltage (of a Circuit).** The greatest root-mean-square (rms) (effective) difference of potential between any two conductors of the circuit concerned. [**70,** 2008]

> FPN: Some systems, such as 3-phase 4-wire, single-phase 3-wire, and 3-wire direct-current, may have various circuits of various voltages.

**Voltage, Nominal.** A nominal value assigned to a circuit or system for the purpose of conveniently designating its voltage class (e.g., 120/240 volts, 480Y/277 volts, 600 volts). The actual voltage at which a circuit operates can vary from the nominal within a range that permits satisfactory operation of equipment. [**70,** 2008]

> FPN: See ANSI C84.1-2006, *Electric Power Systems and Equipment — Voltage Ratings (60 Hz).*

**Voltage to Ground.** For grounded circuits, the voltage between the given conductor and that point or conductor of the circuit that is grounded; for ungrounded circuits, the greatest voltage between the given conductor and any other conductor of the circuit. [**70,** 2008]

Although the term *voltage to ground* includes the word ground, in an ungrounded circuit that term refers to the voltage between conductors.

**Working On (energized electrical conductors or circuit parts).** Coming in contact with energized electrical conductors or circuit parts with the hands, feet, or other body parts, with tools, probes, or with test equipment, regardless of the personal protective equipment a person is wearing. There are two categories of "working on": *Diagnostic (testing)* is taking readings or measurements of electrical equipment with approved test equipment that does not require making any physical change to the equipment; *repair* is any physical alteration of electrical equipment (such as making or tightening connections, removing or replacing components, etc.).

Any work that requires a person to cross the prohibited approach boundary is considered to be working on an *energized electrical conductor or circuit* part and is subject to all associated requirements. Measuring voltage requires that the prohibited approach boundary be breached, suggesting that measuring voltage exposes a worker to an electrical hazard.

The definition of the term *working on* establishes two distinctly different types of tasks that are included within this definition: diagnostic testing and repair. By defining these two types of tasks considered to be "working on," the definition suggests that different procedural approaches are in order depending on the task.

# ARTICLE 110
# General Requirements for Electrical Safety-Related Work Practices

The requirements in Article 110 address how people interact with electrical equipment. The base expectation of Article 110 is that if an installation meets the requirements of the *NEC* and is installed according to the manufacturer's instructions, that installation is safe when operating normally. The requirements of this article are intended to apply when something is wrong and the equipment is not operating normally. A second assumption is that the equipment is maintained adequately. The requirements in this article address only shock (electrocution) and the thermal aspects of an arc flash.

Electrical hazards constitute only one type of hazard to which workers might be exposed. While workers can be exposed to hazards associated with many other sources of energy, *NFPA 70E* only covers exposures to electrical hazards. Although the hazard/risk analysis required by this standard considers only shock/electrocution and thermal burns, the analysis should consider all hazards that could be associated with the work task.

## 110.1 Scope.

Chapter 1 covers electrical safety-related work practices and procedures for employees who are exposed to an electrical hazard in workplaces covered in the scope of this standard. Electric circuits and equipment not included in the scope of this standard might present a hazard to employees not qualified to work near such facilities. Requirements have been included in Chapter 1 to protect unqualified employees from such hazards.

Chapter 1 defines safety-related work practices for employees who are or could be exposed to electrical shock, electrocution, or the thermal hazard associated with arc flash. The practices described in this chapter apply to all workers, regardless of their state of training, their discipline, work task assignment, the business of their employer, or the industrial, commercial or institutional segment. The requirements defined in this chapter can protect both qualified and unqualified workers.

Article 110 identifies work practices and procedures that can reduce or eliminate exposure of people to hazards associated with electrical energy. The scope statement does not limit application to exposed energized electrical conductors or circuit parts. Exposure to circuit conductors or parts of an unknown state also is covered since these conductors could be or could become energized.

Electrical hazards are related to the physical behavior of electrical current and related energy, and they might be widely scattered across a site or facility. Electrical injuries are not related to employer or industrial segment. The scope statement indicates that this standard applies wherever an employee is exposed to an electrical hazard. The work practices contained in this article apply to all work tasks where exposure or potential exposure to exposed energized conductors or circuit parts exist. No industrial segment, including the construction industry, is excluded from the scope.

## 110.2 Purpose.

These practices and procedures are intended to provide for employee safety relative to electrical hazards in the workplace.

FPN: For general categories of electrical hazards, see Annex K.

Persons who work with energized electrical conductors and circuit parts are especially exposed to electrical hazards in the workplace. However, all persons who use equipment connected to an electrical energy source could be exposed to an electrical hazard.

*NFPA 70E* is all about work practices, whereas the *NEC* is primarily about installation. *NFPA 70E* provides work practices that minimize exposure to electrical hazards associated with equipment or installations that *are* or *are not* covered by the *NEC*. The intent is for *NFPA 70E* to be all-inclusive where potential exposure to injury from an electrical energy source exists. The intent of *NFPA 70E* is to provide practices that are needed to protect both qualified and unqualified persons from exposure to hazards associated with electrical energy.

Article 110 covers both work practices and procedures. Its sole purpose is to identify work practices that provide for employee safety from electrical hazards. The requirements in Article 110 historically have proven to reduce exposure to hazards effectively.

### 110.3 Responsibility.

The safety-related work practices contained in Chapter 1 shall be implemented by employees. The employer shall provide the safety-related work practices and shall train the employee who shall then implement them.

Section 110.3 assigns responsibility to each participant in the operation and function of an electrical safety program. Employers must furnish an electrical safety program, and employees must implement the program's requirements. When employers and employees work together closely, the safety program achieves the best result.

Work practices and procedures must include input from, and be embraced by, both the workers (employees) and the line organization (employer). Employers are accountable for generating procedures, and employees are accountable for executing work tasks as defined in the procedures published by the employer. Employers tend to be more familiar with big-picture issues, such as objectives, production schedules, and deadline pressures than employees. At the same time, employees are more familiar with the idiosyncrasies of specific equipment and systems. When employees and employers work together to produce policies, procedures, and practices, the resulting product is much more effective.

### 110.4 Organization.

Chapter 1 of this standard is divided into four articles. Article 100 provides definitions for terms used in one or more of the chapters of this document. Article 110 provides general requirements for electrical safety-related work practices. Article 120 provides requirements for establishing an electrically safe work condition. Article 130 provides requirements for work involving electrical hazards.

Four protective strategies are embedded within this chapter and are considered consecutively based on the importance of the protective strategy. The first and most important protective strategy is creating an electrically safe work condition. The second protective strategy is training, and the third protective strategy is planning. Personal protective equipment (PPE) is the fourth and last protective strategy considered within the chapter.

The provisions identified in Article 110 apply to electrical equipment and circuits generally before an electrically safe work condition exists *(unknown condition)*. The provisions of Article 120 cover establishing control of exposure to electrical hazards by deenergizing the electrical circuit and recognizing the possibility of reenergization (*known to be deenergized*). The provisions contained in Article 130 deal with the condition in which equipment or circuits remain energized while the task is performed *(known to be energized)*.

**CASE STUDY**

Lois and Dick were contractor employees, assigned to a textile manufacturing plant.* They were part of a crew of contractor employees assigned to install a new motor control center (MCC) during a plant turnaround. At the plant, the turnaround was the period of time in which annual maintenance was performed. The new motor control center was not required for the plant to restart, but terminations were necessary while the unit substation could be shut down. With this plan, the textile plant could avoid working on energized bus in the future. This was a good idea that would avoid exposure to shock in the future.

Lois and Dick had the new MCC set into place prior to the shutdown, with cable tray and conduit already installed. Load cables and utilization equipment would be added in the future, as new equipment was needed for the new product.

Statistically, the textile plant had a good safety record. Statistics suggested that personnel safety was highly valued by the organization. The contract between the textile plant and the contractor was standard. The contractor was responsible for testing, as necessary, to ensure the integrity of the installation. The contractor was to follow all legal requirements, including OSHA requirements. However, the contract failed to indicate which portion of the OSHA standards contract employees were expected to follow.

Lois and Dick were familiar with all requirements of the construction standard in 29 CFR 1926. In fact, their last training had relied on the pamphlet obtained from OSHA that included all construction standards. They knew when safety harnesses were needed. They were familiar with shoring requirements. They knew about GFCIs and used one on all electrical hand tools. Their employer was a licensed construction contractor. They were proud of their abilities and their careers.

On Monday of the shutdown week, Lois and Dick pushed the feed cable into the main feeder section of the MCC, terminated the conductors, and marked each conductor with red, blue, and yellow identification tape, according to the plant standard marking system. Next, they proceeded to the secondary switchgear and pushed the conductors into the correct compartment. Robert, the textile plant electrician, told Lois and Dick that the unit substation was deenergized. He showed them the unit substation main switch so they could hang a tag, if needed. Until now, no tags had been necessary.

Of course, 29 CFR 1910, Subpart S, contained the OSHA rules that guided Robert as he executed his maintenance and servicing work. Like Lois and Dick, Robert was highly respected for his knowledge and ability. He was well schooled in the electrical safety-related work practices defined in Subpart S.

After hanging a tag on the main switch, Lois and Dick proceeded to train the new conductors into position and terminate them on the load side of the spare breaker. They considered their work complete and told Robert they would see him later.

The construction supervisor asked Lois if the new cable and MCC had been tested. Lois indicated that the cable had been tested immediately after it was pulled, and she gave the supervisor the record of that test. The supervisor assigned the two electricians to other work for the rest of the afternoon and delivered the cable test record to Robert.

The next morning, the contractor supervisor asked specifically if the MCC had been tested. Lois indicated that the MCC had been tested when it was installed, but no record was kept. Noting that a significant amount of time had passed since the MCC was set into place, the supervisor asked Lois and Dick to test the MCC structure before they released the new unit to the textile plant for use.

Early Thursday morning, Lois and Dick checked out the high-potential (hi-pot) equipment from the shop and went to test the MCC structure. In order to isolate the MCC for the test, Dick removed the conductors from the main feeder section. Even though the conductor ends were touching, there was no need to tape the conductor ends. After all, they would be reterminated as soon as the test was completed. The door of the main compartment was left open so they could connect the hi-pot leads.

Neither Lois nor Dick recognized that their work was no longer "construction work." This last task should have been executed as servicing and maintenance work, because the MCC had been connected to a source of energy.

In the textile plant electrical shop, Robert had completed all his maintenance jobs and was preparing for the plant to start production on Friday morning, as planned. He elected to go to the same substation and heat it up. The secondary switchgear was now energized. Next, Robert proceeded to energize MCC units and other equipment supplied by the substation.

Lois had connected the hi-pot leads and was kneeling next to Dick to operate the machine when Robert energized the new feeder. A violent explosion occurred. Most of the energy went over the heads of both Lois and Dick. However, both were sprayed with droplets of molten copper. Neither Lois nor Dick was physically injured to any significant degree, but both strained several muscles as they attempted to scramble out of the way. The door of the compartment scraped and cut Lois's back as she tried to escape.

The circuit breaker in the secondary switchgear did its job and rapidly opened the circuit. Its proper operation prevented a more significant fireball.

Robert believed that the work was complete. Lois and Dick thought the startup was scheduled for Friday morning, and they did not expect the unit to be energized.

The textile plant electrician and the construction electricians were following different rules. Each facility should have had a set of rules that applied to everyone on site. Following different standards almost always sets a trap that likely will spring at an unexpected time.

---

*This account is based on an actual incident. The names, including the name of the facility, have all been changed to protect those involved. Any similarity to actual names or facilities is strictly coincidental.

Reprinted from Ray A. Jones and Jane G. Jones, *Electrical Safety in the Workplace,* 2000, with permission of Jones and Bartlett Publishers, Sudbury, MA.

### 110.5 Relationships with Contractors (Outside Service Personnel, etc.).

Employers could be on-site or outside employers. An "on-site" employer might be landlord of a facility. On-site or "host" employers might have more than one division functioning at a particular facility location. For instance, a single location might have an exploration division and a downstream division at the same physical location. Although the exploration division might consider the downstream division to be "landlord," both divisions would be considered as the on-site employer from the perspective of consensus standards. An on-site employer can be a site owner or another employer acting as the site landlord.

An "outside" employer is one who is contractually obligated to the on-site employer. Service contractors, such as soda machine vendors, are outside employers. Prime contractors and subcontractors also are outside employers. An outside employer can be a construction contractor, a maintenance contractor, an equipment vendor, or a service provider. When an owner enters into a contractual relationship with a construction or maintenance contractor, the owner becomes the on-site employer and the contractor becomes the outside employer.

Both *NFPA 70E* and OSHA standards are applicable to employers. A specific requirement might be applicable to an on-site employer or an outside employer. Multiple contractors might be working in the same physical area at a facility. Each contractor and the on-site employer have a responsibility for personal safety.

Each employer can be the *creating*, *exposing*, *correcting*, or *controlling* employer. A *creating employer* is one who caused the hazardous condition to exist. An *exposing employer* is one who assigns a work task to workers knowing that a hazardous condition exists or who fails to take reasonable precautions to prevent or avoid the hazardous condition. A *correcting employer* is one who is responsible for correcting the hazardous condition. A *controlling employer* is one who has the authority to correct the hazardous condition. Control among employers or crafts can be established by contract or by exercise of worker control.

**(A) Host Employer Responsibilities.**

(1) The host employer shall inform contract employers of:

a. Known hazards that are covered by this standard, that are related to the contract employer's work, and that might not be recognized by the contract employer or its employees

b. Information about the employer's installation that the contract employer needs to make the assessments required by Chapter 1

(2) The host employer shall report observed contract-employer-related violations of this standard to the contract employer.

The host employer might observe violations and unsafe conditions created by the outside employer. In that instance, the host employer must correct the violation or unsafe condition if imminent risk of injury exists and then report it to the outside employer. If no risk of injury is imminent, the host employer should report it to the outside employer for correction. In any event, the host employer must inform the outside employer that the violation or unsafe condition exists or existed.

A host employer might be a facility owner or a general contractor. Section 110.5(A) discusses the relationship between employers that operate in the same physical facility.

In general, a host employer is required to inform the contractor or subcontractor of all hazards to which employees might be exposed when executing the terms and limits of the contract. In some cases, a subcontractor might be more expert at identifying and understanding hazards associated with the contract. For instance, electrical contractors frequently are hired because of increased understanding of electrical equipment and hazards. However, a host employer is expected to have greater knowledge of general hazards that might exist in a facility. For instance, a chemical plant employer must ensure that an electrical contractor recognizes all chemical hazards at the facility. At the same time, the electrical contractor would be expected to be more familiar with electrical hazards, especially if the host employer does not have its own engineering staff.

A contractor employer must inform the host employer about the contractor's electrical safety program. If the contractor's employees are expected to execute any unique functions, the host employer must be informed.

The host employer should not be expected to correct the safety violations of a contractor's employee unless a violation involves an immediate and serious danger to people. Violations observed by the host employer must be reported to the contractor for correction.

**(B) Contract Employer Responsibilities.**

(1) The contract employer shall ensure that each of his or her employees is instructed in the hazards communicated to the contract employer by the host employer. This instruction is in addition to the basic training required by this standard.

Contractor employees must implement the requirements of the contractor's electrical safety program. Other sections of this standard define specific and general training requirements. If the host employer has communicated specific hazards that might exist in a facility, the contractor must ensure that all contractor employees have been warned about the hazard and provided any information that is necessary to avoid exposure to the hazards.

(2) The contract employer shall ensure that each of his or her employees follows the work practices required by this standard and safety-related work rules required by the host employer.

The contractor employer is responsible for ensuring that all contract employees assigned to work on the facility follow safety rules and requirements required by the host employer. If procedures that comprise the safety program provided by the host employer and the program provided by the outside employer are different, the outside employer should ensure that outside employees are aware of the differences and instructed about which procedures are overriding.

(3) The contract employer shall advise the host employer of:

a. Any unique hazards presented by the contract employer's work,
b. Any unanticipated hazards found during the contract employer's work that the host employer did not mention, and
c. The measures the contractor took to correct any violations reported by the host employer under paragraph (A)(2) of this section and to prevent such violation from recurring in the future.

The outside employer is required to inform the host employer of any conditions that change as a result of the contracted work. For instance, if the contract required change in capacity of a building or facility service such that a new flash hazard analysis might be necessary, the outside contractor must inform the host employer of any potential increase or change in possible hazard exposure. In many instances, the outside employer has the greater expertise and knowledge.

If the host employer identifies unsafe conditions or hazards for correction, the outside employer must inform the host employer how corrections were made. The outside employer must inform the host employer what changes have been made to ensure that the violation will not happen again in the future. It is very important that coordination meetings with precise documentation take place to ensure responsibility is assigned to the proper organization.

### 110.6 Training Requirements.

**(A) Safety Training.** The training requirements contained in this section shall apply to employees who face a risk of electrical hazard that is not reduced to a safe level by the applicable electrical installation requirements. Such employees shall be trained to understand the specific hazards associated with electrical energy. They shall be trained in safety-related work practices and procedural requirements as necessary to provide protection from the electrical hazards associated with their respective job or task assignments. Employees shall be trained to identify and understand the relationship between electrical hazards and possible injury.

FPN: For further information concerning installation requirements, see *NFPA 70®*, *National Electrical Code®*, 2008 edition.

A base expectation of Article 110 is that when operating normally, an installation is safe if it meets all of the following conditions:

- It is installed in accordance with the *NEC*.
- It has been installed in accordance with the manufacturers' instructions.
- It has been installed in accordance with any requirements that are mandated by a product listing.
- It is adequately maintained.

However, if the installation or equipment fails to meet these conditions, employees face an elevated risk of injury. For example, an equipment door or cover left with fasteners unlatched does not meet the requirements of the *NEC*. Equipment that is not adequately maintained does not meet these requirements and causes employees to be exposed to an electrical hazard.

Each employee who is or might be exposed to an elevated risk of injury by exposure to an electrical hazard must be trained to understand the specific hazards to which he or she might be exposed. To increase understanding, the training should include the following:

- What electrical hazards are present in the workplace
- How each electrical hazard affects body tissues
- How to determine the degree of each hazard
- How to avoid exposure to each hazard
- How to minimize risk by body position
- What PPE is needed for the employee to execute his or her work assignment
- How to select and inspect PPE
- What employer-provided procedures, including specific work practices, the employee must implement
- How increased duration of exposure to an electrical hazard results in a higher frequency of injuries
- How to perform a hazard/risk analysis
- How to determine limited, restricted, and prohibited approach boundaries and recognize that these boundaries are related to protection from exposure to electrical shock and electrocution

Although Section 110.6(A) defines specific necessary training, the key term in this section is *understand*. The safety training must include all information and training processes necessary to achieve understanding. The training must develop an understanding of how to avoid all hazard exposure or how to minimize exposure if a hazard remains when the work is to be performed. Workers must understand the characteristics of protective equipment that might be necessary to avoid injury if an incident occurs while exposure exists. Workers must understand that avoiding exposure is the only viable means of avoiding injury in event of an incident while the task is being performed. Workers should understand when, if, and how exposure to each hazard might exist at each step in the work task.

**(B) Type of Training.** The training required by this section shall be classroom or on-the-job type, or a combination of the two. The degree of training provided shall be determined by the risk to the employee.

Classroom training is effective for some objectives, and on-the-job training (OJT) is effective for others. In most instances, effective training makes use of both training processes. The qualification of the instructor is very important. Frequently, new employees learn by emulating a more experienced worker. If the experienced worker understands electrical safety requirements and practices them, the mentoring process is good. However, because a significant amount of new information is available, mentors also might require training.

Some employees have minimal exposure to electrical hazards, especially when equipment is effectively maintained. For instance, arc flash training is not necessary for an office worker who is never exposed to an arc flash. Workers who are not exposed to medium and high voltages might not need to be trained to understand the characteristics of higher-voltage energy.

If any risk of injury from electrical energy exists, employees must be trained to recognize and deal with risk. Sometimes, the risk of injury is of sufficient magnitude that the work task should not be executed.

Training should include formal presentations and actual performance of the work under the supervision of knowledgeable persons. Some work tasks contain minimum exposure, and others expose workers to significant hazards. For instance, operating personnel could be required to operate the handle of a disconnect switch or to push a reset button with doors closed and latched, while maintenance personnel could be required to open doors or remove covers and perform diagnostic tasks. Additionally, the training provided to each of these workers could be different depending on the risk of injury.

**(C) Emergency Procedures.** Employees exposed to shock hazards shall be trained in methods of release of victims from contact with exposed energized electrical conductors or circuit parts. Employees shall be regularly instructed in methods of first aid and emergency procedures, such as approved methods of resuscitation, if their duties warrant such training. Training of employees in approved methods of resuscitation, including cardiopulmonary resuscitation, shall be certified by the employer annually.

Workers who are or might be exposed to shock/electrocution must be trained to respond to emergency conditions that might exist. Employee training must ensure that each employee understand the steps necessary to release a victim who might be in contact with an energized conductor. The employee must be trained to know that the first action in responding to an electrical contact incident must be to remove the source of the electricity and the second action is to request emergency assistance. Workers also must understand how to request assistance.

Employees must understand emergency first-aid procedures, and each worker must know the procedure necessary to reach emergency assistance. Employees who are qualified to work on or near exposed energized electrical conductors or circuit parts should be trained to perform cardiopulmonary resuscitation (CPR) and emergency first aid. CPR and emergency first-aid training should be up-to-date. In some instances, unqualified employees might be expected to provide emergency first aid or CPR. The above applies to workers who are or might be expected to perform emergency response to an electrical hazard and should be qualified to perform that role.

**(D) Employee Training.**

**(1) Qualified Person.** A qualified person shall be trained and knowledgeable of the construction and operation of equipment or a specific work method and be trained to recognize and avoid the electrical hazards that might be present with respect to that equipment or work method.

(a) Such persons shall also be familiar with the proper use of the special precautionary techniques, personal protective equipment, including arc-flash, insulating and shielding materials, and insulated tools and test equipment. A person can be considered qualified with respect to certain equipment and methods but still be unqualified for others.

(b) Such persons permitted to work within the Limited Approach Boundary of exposed energized electrical conductors and circuit parts operating at 50 volts or more shall, at a minimum, be additionally trained in all of the following:

(1) The skills and techniques necessary to distinguish exposed energized electrical conductors and circuit parts from other parts of electrical equipment
(2) The skills and techniques necessary to determine the nominal voltage of exposed energized electrical conductors and circuit parts
(3) The approach distances specified in Table 130.2(C) and the corresponding voltages to which the qualified person will be exposed
(4) The decision-making process necessary to determine the degree and extent of the hazard and the personal protective equipment and job planning necessary to perform the task safely

(c) An employee who is undergoing on-the-job training and who, in the course of such training, has demonstrated an ability to perform duties safely at his or her level of training and who is under the direct supervision of a qualified person shall be considered to be a qualified person for the performance of those duties.

(d) Tasks that are performed less often than once per year shall require retraining before the performance of the work practices involved.

(e) Employees shall be trained to select an appropriate voltage detector and shall demonstrate how to use a device to verify the absence of voltage, including interpreting indications provided by the device. The training shall include information that enables the employee to understand all limitations of each specific voltage detector that may be used.

For a person to be considered "qualified," he or she must have the craft training necessary to be knowledgeable in the construction and operation of the equipment associated with the work task or with the specific work method. The person must also be trained in the selection of PPE including using the PPE in a "dry run" to ensure the PPE does not limit the person's dexterity or vision. If the work task involves a circuit with little association with equipment, a qualified person also must be knowledgeable of the circuit. For instance, when a work task must be performed in a manhole containing cables and conductors only, the worker must be familiar with support requirements and fireproofing methods that might be in the manhole in addition to the unique hazards associated with the confined space and the potential for flammable and toxic atmosphere.

For a person to be considered qualified, he or she must have received the safety training identified in Sections 110.6(A) and 110.6 (C) of this standard, in addition to employee training contained in this section. A worker might be considered qualified for one task and unqualified for another.

Workers must be familiar with steps necessary to determine which parts of the equipment or circuit(s) are conductive and if those parts are energized. Qualified workers must also be familiar with the limited, restricted, and prohibited approach distances for the nominal voltage of the equipment or circuit.

The safety training must describe how to conduct a hazard/risk analysis. Qualified workers must be trained to select PPE based on the existence of a hazard and the degree of the hazard to which they might be exposed, and they must be able to plan and safely execute the work task being considered.

Workers must be trained to select an appropriate voltage-detecting device. The selected voltage-detecting device must be appropriate for the circuit associated with the work. Non-contact devices are appropriate in some circumstances; however, voltmeters that make direct contact with the conductor in question provide the best chance of avoiding an error. Qualified workers must understand all limitations associated with the voltage-detecting device. Qualified workers must be able to execute a visual inspection of the device.

Qualified workers must receive training, as necessary, to ensure that they are familiar with requirements defined in the employer's electrical safety program, with *NFPA 70E*, and with other applicable codes and standards. The training must establish that a qualified worker understands the limitations associated with protective equipment, tools, and test equipment. A qualified worker must recognize and accept his or her personal limitations associated with both skill and knowledge.

An apprentice or other worker undergoing training to become a qualified person is considered qualified if he or she is under the direct supervision of a qualified person.

A worker who has not performed a work task involved with work on or near an exposed energized electrical conductor or circuit part for one or more years is considered to be unqualified for the task until the worker has been retrained.

**(2) Unqualified Persons.** Unqualified persons shall be trained in and be familiar with any of the electrical safety-related practices that might not be addressed specifically by Chapter 1 but are necessary for their safety.

All persons who use electrically operated equipment have some potential exposure to electrical hazards. Employers must ensure that all employees understand where electrical hazards exist. Potential exposure to electrical hazards varies by job assignment. Workers who are not trained to become qualified persons must be trained to understand how they might be injured from shock/electrocution or arc flash. Unqualified workers must understand the limit of their work assignment as it relates to electrical hazards.

**(3) Retraining.** An employee shall receive additional training (or retraining) under any of the following conditions:

(a) If the supervision or annual inspections indicate that the employee is not complying with the safety-related work practices

(b) If new technology, new types of equipment, or changes in procedures necessitate the use of safety-related work practices that are different from those that the employee would normally use

(c) If he or she must employ safety-related work practices that are not normally used during his or her regular job duties

Training must be provided as necessary to ensure that knowledge of the workers, both qualified and unqualified, is up-to-date. Workers who are assigned to a new work position must receive the necessary training associated with the assignment prior to beginning work that could expose him or her to an electrical hazard. Any change to the electrical safety program must trigger the training necessary to rebuild understanding of the current program or procedure requirements.

When a supervisor recognizes that a worker is not implementing requirements defined in the employer's electrical safety program through normal supervision or through annual or other inspections, the worker must receive additional training before being permitted to work on or near exposed energized electrical conductors or circuit parts. If additional or new equipment that provides new or additional hazards or methods of exposure is added, workers must receive more training that develops the necessary understanding.

Workers who are reassigned to work tasks associated with equipment that normally is not used in their day-to-day functions are particularly exposed to potential injury. Such workers must be trained to establish the necessary level of understanding.

**(E) Training Documentation.** The employer shall document that each employee has received the training required by paragraph 110.6(D). This documentation shall be made when the employee demonstrates proficiency in the work practices involved and shall be maintained for the duration of the employee's employment. The documentation shall contain each employee's name and dates of training.

FPN: Employment records that indicate that an employee has received the required training are an acceptable means of meeting this requirement.

Each employer must update the necessary records to indicate that each qualified worker has received the training required by Section 110.6. The record must include the nature and depth of training provided. The documentation must identify the worker and the date that the training was received. If hard copy documentation is a *normal* employer record, then the training documentation could be hard copy. If electronic records are the norm for an employer, then an electronic record is acceptable. However, the record must be available for inspection in some form by a third party.

### 110.7 Electrical Safety Program.

Each employer must develop and implement an electrical safety program that directly addresses all employee exposure to each specific electrical hazard that exists on the work site. The electrical safety program implemented by outside (contractor or vendor) employers must address each electrical hazard to which employees might be exposed. The electrical safety program (and related training) must be appropriate for the conditions that exist.

**(A) General.** The employer shall implement and document an overall electrical safety program that directs activity appropriate for the voltage, energy level, and circuit conditions.

FPN No. 1: Safety-related work practices are just one component of an overall electrical safety program.

An electrical safety program can contain different elements and take multiple forms. However, the electrical safety program must be written, published, and available to all affected employees. It can be part of a more extensive safety program, or it can be an independent program, but in either situation, the program must cover all circuit conditions to which an employee might be exposed. The program should contain all procedures necessary to guide all work activity for the employer.

An electrical safety program includes all the following components needed to provide necessary guidance to employees:

- Updated training, as necessary
- Policies generated by the management structure
- Current procedures that guide worker actions
- Review processes that ensure that procedures are changed when necessary
- PPE necessary to protect workers when they are exposed to an electrical hazard
- Auditing processes to monitor developing knowledge about equipment and requirements
- Controls that establish expectations for workers
- Processes that define and execute enforcement of program requirements

The electrical safety program should contain the necessary unique components and practices that are based on the worker's experience and qualifications as well as any unique aspects of the work environment. For instance, workers who are unfamiliar with a specific workplace or job assignment might need guidance that is different from workers who have been in the same job assignment for several years. Workers assigned to work in an electrically classified hazardous work environment might need different guidance from workers in a commercial enterprise.

The electrical safety program must provide guidance that is based on the type and degree of hazards that exist. When workers are not exposed or potentially exposed to transmission circuits, for instance, knowledge of that work environment is not necessary.

The electrical safety program is more effective as an integral component of an overall safety program. However, injuries that result from exposure to electrical hazards should be recorded and analyzed completely. Injury records should document the energy source that caused the injury.

FPN No. 2: ANSI/AIHA Z10-2005, *American National Standard for Occupational Safety and Health Management Systems*, provides a framework for establishing a comprehensive electrical safety program as a component of an employer's occupational safety and health program.

There are several different documents that can provide effective guidance about developing an electrical safety program. ANSI/AIHA Z10, *American National Standard for Occupational Health and Safety Management Systems*, discusses components that are based on the Plan-Do-Act-Check cycle. This standard emphasizes the need for continual improvement of the program. ANSI/AIHA Z10 is similar to Canada's Z1000 standard covering the same subject.

**CASE STUDY**

Haines Electric* is a small employer in the Columbus area. Brothers Johnny and Jack Haines own the company as equal partners. Johnny is the administrator for the company, and Jack runs the field operation. Haines Electric had been profitable since it was founded 25 years ago, but, in recent years, profitability was diminishing. In fact, in the past year, Haines Electric had lost money for the first time in its history.

The workload had diminished because Haines Electric was being outbid on many jobs. Competitors were receiving many of the contracts that had once gone to Haines Electric. As the administrator, Johnny began searching the books in an attempt to explain why the company's bids were becoming less competitive.

Johnny's search didn't take long. As he looked at all the components that make up the cost segment of Haines' bid, it was obvious that next to the cost of labor, insurance constituted the largest cost. In prior reviews, Johnny had considered insurance simply a fixed cost of doing business in the Columbus area.

Both Johnny and Jack treated their employees as friends. They were kind, considerate, and generous. It now seemed that the brothers would have to make some hard choices because the company simply could not afford to continue to lose money. The Haines brothers would have to make some changes or Haines Electric might have to go out of business.

Out of concern for the economic well-being of their employees, the Haines brothers asked all supervisors to gather in the shop one morning so that they could relay the poor business picture. Most of the supervisors did not understand how the situation could be so bad, when clients were so happy with their work. Johnny explained that the major problem was the high cost of insurance. One supervisor suggested a way to reduce insurance costs would be to reduce injuries, which would reduce expenditures to the insurance company. In turn, reduced insurance expenditures would translate into reduced insurance rates. The idea made sense to Johnny and Jack.

Over the years, employees at Haines Electric had experienced safety incidents and injuries from time to time. In the early days, Haines Electric employees were highly talented and trained. In fact, most of the supervisors were electricians when the company began. However, the type of work the company was doing had changed through the years. At first, the company concentrated solely on new construction, and their employees were particularly talented in construction activities. In recent years, however, manufacturers were trying to get more production from equipment and facilities already installed. As a result, the type of had work shifted from new construction to improvements in existing facilities.

Employees were exposed to safety hazards in many different ways, as compared to the simpler hazards of the early days. The work environment had changed drastically over the years. Hazards do not change with a shift in work environment; however, in most instances, the degree of a hazard is likely to be different, perhaps greater and perhaps less. Not only is it likely that the degree of hazard will change, but the degree of exposure will change as well.

In the supervisors' meeting, the brothers learned that Haines Electric effectively had no electrical safety program. A supervisor suggested the brothers approach one of their clients to ask for help. The selected client had an effective electrical safety program. In fact, the Haines Electric supervisor had been following the electrical safety program as a contractual requirement. The client suggested by the supervisor was happy to share his knowledge and experience with the Haines brothers.

Johnny and Jack asked the supervisor to spend a few days in their office to help shape a Haines Electric safety program. It would emphasize the brothers' concern for their employees' well-being. It would include a policy that clearly stated that unnecessary exposure to electrical hazards is unacceptable. Their employees had to understand that decreased exposure to hazards would result in fewer injuries. The program would include training that emphasized personal responsibility and accountability, as well as employee involvement in generating and managing Haines' electrical safety program. The electrical safety program was only slightly modified from the same program implemented by the manufacturing location that shared the program initially.

Jack was instrumental in selling the program to Haines' employees. He mounted an offensive to make certain that all the program elements were implemented, including deenergization of equipment before beginning work and a special "test before touch" policy.

Safety incidents began to drop, even incidents not associated with electrical energy. Injuries almost disappeared. Insurance rates decreased significantly. Clients even began to award contracts to Haines based on the low number of incidents and injuries instead of the contract price. Haines Electric returned to a profitable position. In fact, the workload increased to the point that the Haines brothers had another strategic decision to make: How large should they allow the company to become?

---

*This account is based on an actual incident. The names, including the name of the facility, have all been changed to protect those involved. Any similarity to actual names or facilities is strictly coincidental.

Reprinted from Ray A. Jones and Jane G. Jones, *Electrical Safety in the Workplace,* 2000, with permission of Jones and Bartlett Publishers, Sudbury, MA.

**(B) Awareness and Self-Discipline.** The electrical safety program shall be designed to provide an awareness of the potential electrical hazards to employees who might from time to time work in an environment influenced by the presence of electrical energy. The program shall be developed to provide the required self-discipline for employees who occasionally must perform work that may involve electrical hazards. The program shall instill safety principles and controls.

Regardless of the content of an electrical program, workers must implement its requirements and guidance to avoid injury. Employers provide PPE, but workers must select, wear, and use the equipment within its limits when warranted. Although the written program might contain effective measures, workers must follow the guidance provided in the program.

Some major components of an effective electrical safety program might not be written as requirements in the program. Awareness includes being aware of electrical hazards and how exposure to them occurs. Awareness also includes being aware of other people in the area, being aware that work conditions might change, being aware of personal conditions, and being focused solely on the work being attempted.

An electrical safety program should be developed from an understanding of both electrical hazards and procedural requirements. Including workers in the process of generating and maintaining requirements can help to develop a program that the workers will support and comply with. The program should be assessed and feedback should be sought to ensure that it is compliant with the law and this standard, and that it meets the safety goals of the organization.

**(C) Electrical Safety Program Principles.** The electrical safety program shall identify the principles upon which it is based.

FPN: For examples of typical electrical safety program principles, see Annex E.

**(D) Electrical Safety Program Controls.** An electrical safety program shall identify the controls by which it is measured and monitored.

FPN: For examples of typical electrical safety program controls, see Annex E.

**(E) Electrical Safety Program Procedures.** An electrical safety program shall identify the procedures for working within the Limited Approach Boundary of energized electrical conductors and circuit parts operating at 50 volts or more or where an electrical hazard exists before work is started.

FPN: For an example of a typical electrical safety program procedure, see Annex E.

The written electrical safety program must contain procedures that describe steps that are required when executing a work task or job. The program must contain procedures that guide workers' actions. Each procedure should identify all required steps. The work process might require actions in which exposure to an electrical hazard exists. If exposure exists, however, the procedure must provide any necessary guidance.

Procedures might need to be changed or modified when conditions change. For instance, a motor control center from a different manufacturer or different models from the same manufacturer might change the method or degree of exposure of the worker. Procedures then, should be reevaluated to ensure that workers will not be at risk. It could be necessary to modify the procedure, but all workers should be aware of the variance and should be made cognizant of any new risks and all procedures to mitigate the risk.

**(F) Hazard/Risk Evaluation Procedure.** An electrical safety program shall identify a hazard/risk evaluation procedure to be used before work is started within the Limited Approach Boundary of energized electrical conductors and circuit parts operating at 50 volts or more or where an electrical hazard exists. The procedure shall identify the hazard/risk process that shall be used by employees to evaluate tasks before work is started.

One procedure in the published program must define the process that employees will use to assess both the hazard and the risk associated with each work task on or near exposed energized electrical conductors or circuit parts operating at 50 volts or more. That hazard/risk evaluation procedure must identify a process that enables employees to determine whether a hazard exists. It must evaluate electrical shock and arc flash. The process must provide sufficient information to enable both the employer and employee to make an informed decision on whether the risk of injury is acceptable or not acceptable.

When the hazard/risk evaluation is complete, the qualified workers should be able to recognize when an electrical hazard exists and how exposure to the hazard might occur.

Determining if an electrical hazard exists is not a simple process. The electrical safety program must contain a procedure that describes how the risk of injury from an electrical hazard should be assessed. The process should assist workers as they determine the following:

- If an electrical hazard exists (Each possible hazard must be considered.)
- The degree or capacity of the hazard (Determine the maximum possible injury that could result.)

- If the worker will be exposed to the electrical hazard (if the task requires a worker to be within the danger zone associated with the possible exposure)
- What work practice can minimize exposure to potential injury
- What PPE can minimize the risk of injury
- If authorization is required before beginning the task
- If a written plan is necessary before beginning the task
- If the risk of injury is acceptable

The hazard/risk evaluation procedure should define what steps are necessary to determine the above information. The process of executing the evaluation could take one or more of several forms, but the process must be documented in a procedure.

When work is performed on or near exposed electrical conductors or circuit parts, the risk of injury is elevated. Specific work practices and PPE might reduce the risk of injury; however, if electrical energy remains in the vicinity of the work task, the risk of injury is greater than if an electrically safe work condition exists. The hazard/risk evaluation procedure must define the process to be followed to determine if the risk of injury is acceptable or unacceptable.

FPN No. 1: The hazard/risk evaluation procedure may include identifying when a second person could be required and the training and equipment that person should have.

This FPN is new in the 2009 edition of the standard. The origin of the FPN was a proposal to require a second person for rescue reasons whenever entering the prohibited approach boundary. Rather than always require a second worker, the committee stated that the requirement should be risk and task based. If the decision is that a second person is warranted, then that person needs to be trained in rescue techniques and CPR.

The worker who might be at risk of injury should conduct the hazard/risk analysis and review it with his or her supervisor. When the risk of injury associated with the work task involves more than one worker, the workers should collaborate on the analysis to ensure that each of the parties understands the risks.

FPN No. 2: For an example of a Hazard/Risk Analysis Evaluation Procedure Flow Chart, see Annex F.

Typically, analyzing a work task to determine if a hazard exists and if a worker might be exposed to a hazard takes the form of a series of questions that become increasingly discerning. Section F-1 of Annex F illustrates the concept of increasingly discerning questions.

FPN No. 3: For an example of a Hazard/Risk Evaluation Procedure, see Annex F.

Evaluating risk requires a person to exercise judgment. Unless an electrically safe work condition exists, some risk of injury from an electrical hazard exists. The risk of injury can depend on circuit conditions or on the degree (capacity) of the hazard. A work practice might reduce the risk of injury; however, some risk of injury always remains when a hazard exists.

Evaluating the risk of injury requires a person to make a determination of the degree of each hazard, the chance of contact with each hazard, any mediating protective equipment, and similar conditions associated with each work task. The worker then must assess if the risk of injury is sufficiently low to be acceptable. If contact with a remaining hazard cannot result in an injury, the work task probably is acceptable. However, if contact with the remaining hazard could result in an injury, the risk is likely to be unacceptable.

The potential for injury probably is somewhere between these two extremes. Workers must determine if the risk of injury is sufficiently low to be acceptable. Of course, as the degree of injury increases, greater assurance that contact with the hazard will be avoided should exist. Annex F provides an example of how to assess risk.

**(G) Job Briefing.**

**(1) General.** Before starting each job, the employee in charge shall conduct a job briefing with the employees involved. The briefing shall cover such subjects as hazards associated with the job, work procedures involved, special precautions, energy source controls, and personal protective equipment requirements.

Supervisors of workers who are assigned a work task involving potential exposure to an electrical hazard must discuss with the employees involved all hazards associated with the task (see Exhibit 110.1). When the supervisor assigns a worker to a specific task, he or she should ensure that the conversation includes a discussion of all hazards associated with the task. The purpose of the discussion is to ensure that workers consider if or how they might be injured when executing the work task.

A job briefing is another name for a discussion of the work task. The briefing must be held prior to beginning each work task that is associated with work on or near exposed, energized electrical conductors or circuit parts. The briefing must include a discussion of electrical hazards and how employees might be exposed to them. At a minimum, the discussion should include the following subjects:

- Electrical hazards associated with the work task
- Procedures that must be followed when executing the work task
- Any special precautions that are required by the working conditions
- Where and how to remove the source of energy
- Emergency response and emergency communications
- Required PPE
- Other work in the immediate physical area
- Other work associated with the same electrical circuits or equipment

*EXHIBIT 110.1* *Reviewing pertinent drawings as part of the job briefing.*

**(2) Repetitive or Similar Tasks.** If the work or operations to be performed during the work day or shift are repetitive and similar, at least one job briefing shall be conducted before the start of the first job of the day or shift. Additional job briefings shall be held if changes that might affect the safety of employees occur during the course of the work.

If the work task is a repetitive task that is performed several times during the day, a single job briefing held before the worker performs the task for the first time is satisfactory. If significant changes that might affect the safety of employees occur during the day, however, a new job briefing is required.

In some instances, a worker's job assignment includes routine tasks. A worker's job assignment can determine what different tasks are performed each shift. In this instance, the supervisor should begin each shift/day with a discussion of hazards to which workers could be exposed during the shift. The job briefing should be supplemented as necessary by discussing any other appropriate issue, such as emergency response.

**(3) Routine Work.** A brief discussion shall be satisfactory if the work involved is routine and if the employee, by virtue of training and experience, can reasonably be expected to recognize and avoid the hazards involved in the job. A more extensive discussion shall be conducted if either of the following apply:

(1) The work is complicated or particularly hazardous.
(2) The employee cannot be expected to recognize and avoid the hazards involved in the job.

The job briefing should be as extensive as necessary to ensure that workers have a complete understanding of their exposure to an electrical hazard. If the task is simple and routine, such as exchanging filters, only a brief discussion is necessary. If, however, the work task is complex or unfamiliar to the worker, a more complete job briefing is necessary. A worker from another employer or another area of the facility should not be expected to be familiar with a new work environment. Thus, the job briefing should be complete and ensure that the particulars of the environment are explained to the "new" worker.

The job briefing provides an opportunity for supervisors or lead persons to review plans that have been developed to cover the work.

Some workers are trained to perform specific work tasks that are typically repetitive in nature. For instance, a worker could be expert at conducting an oil analysis, and the employer has documentation illustrating the specific expertise. In that case, the worker could be expected to perform an oil analysis without a specific job briefing. Yet, if the work task involves risk of injury that is not directly associated with the analysis, a job briefing is necessary. For instance, if executing an oil analysis requires the worker to be exposed to a potential arcing fault from an adjacent circuit, a job briefing is necessary.

FPN: For an example of a job briefing form and planning checklist, see Annex I.

Annex I provides a checklist that can be used to conduct a job briefing. Annex I also provides a checklist that can be used to guide a work plan. A job briefing should address issues specifically associated with the assigned work task. Annex I contains a list of issues that could be included in the discussion or work plan. However, the job briefing or plan should discuss only the issues associated with the specific work task.

**(H) Electrical Safety Auditing.** An electrical safety program shall be audited to help ensure that the principles and procedures of the electrical safety program are being followed. The frequency of audit shall be determined by the employer, based on the complexity of the procedures and the type of work being covered. Where the audit determines that the principles and procedures of the electrical safety program are not being followed, appropriate revisions shall be made.

The electrical safety program must define necessary auditing. Components of the audit can be discrete sections of the procedure or integrated into an overall auditing procedure. Each audit should be documented. The documentation should record the date and results of each audit. The audit must contain at least the following four components:

**1.** The audit must determine if workers are implementing the requirements of the electrical safety program. This component provides information about understanding of the program as well as information about how much supervisors emphasize the program. This component describes the health of the electrical safety program.

**2.** The audit must determine if the program addresses all hazards that might exist on a work site. This component must determine if workers are or might be subject to risk of injury not covered by the electrical safety program.

**3.** The audit must define the process to ensure procedures are revised, as new information becomes available. In the event that an incident or injury occurs, all associated procedures must be reviewed and revised as necessary. Any shortcoming determined in the field or procedure audit must trigger change to procedures that are in place or generation of an entirely new procedure. However, all procedures contained in the electrical safety program must be reviewed on a frequency not greater than three years and revised as appropriate.

**4.** The audit must define how any revision of specific procedures or a general change is communicated to workers.

### 110.8 Working While Exposed to Electrical Hazards.

Under normal conditions, electrical conductors energized at a voltage level less than 50 volts do not present an electrical shock hazard. However, a thermal hazard can exist in circuits that have a significant capacity to deliver energy, even when the voltage level is less than 50. For instance, battery installations can be connected so that arcing resulting from a short circuit could present a significant thermal hazard.

Note that many control circuits operate at a voltage level less than 50 volts. Creating an open circuit or short circuit could result in creating a different type of hazard. A process-upset condition can result in exposure to a chemical hazard or creating an unacceptable environmental condition.

Installations that comply with the *NEC*, that meet the installation instructions of the equipment manufacturer, and that are maintained appropriately are considered safe when operating normally. However, when electrical equipment changes state, such as being switched from energized to deenergized or vice-versa, an overload relay is reset, a door is opened or closed, a circuit breaker is reset, or other conditions where physical movement occurs might result in initiating an arcing fault. Depending on the state and condition of the equipment and the functional circuit protective devices, a worker could be exposed to an arcing fault.

Safety-related work practices must be consistent with the parameters of the hazard. The PPE used by the worker must be consistent with the characteristics of the hazard and exposure. Protective equipment must be selected based on the degree of the hazard. If the risk of injury does not exist because of no exposure, then no PPE is necessary. The boundaries determine if risk of injury exists. Selecting and using PPE is considered a work practice.

Circuits or conductors energized at 50 volts or more must be assumed to present a shock and/or thermal hazard. Employers are required to provide procedures that define required practices to prevent injury to employees. Work that is performed on or near exposed energized electrical conductors or a circuit part is particularly dangerous. Work on or near exposed conductors that are not in an electrically safe work condition expose workers to injury. Only qualified persons may perform work on or near these conductors.

Workers must select and use work practices that are consistent with the degree of the potential hazard. For instance, if the potential hazard includes possible exposure to an arc flash, the worker must select flame-resistant clothing and other PPE that is at least as protective as the potential hazard is dangerous. It is acceptable for the rating of the PPE to exceed the degree of exposure, but unacceptable for the rating of the PPE to be less than the degree of exposure.

**(A) General.** Safety-related work practices shall be used to safeguard employees from injury while they are exposed to electrical hazards from electrical conductors or circuit parts that are or can become energized. The specific safety-related work practices shall be consistent with the nature and extent of the associated electrical hazards.

Employers are required to provide procedures that contain practices that prevent injury to employees. Work that is performed on or near exposed energized electrical conductors or circuit parts is particularly dangerous. Work on or near exposed conductors that are not in an electrically safe work condition expose workers to injury. Only qualified persons may perform work on or near these conductors.

Workers must select and use work practices that are consistent with the degree of the potential hazard. For instance, if the potential hazard includes possible exposure to an arc flash, the worker must select flame-resistant clothing and other PPE that is at least as protective as the potential hazard is dangerous. It is acceptable for the rating of the PPE to exceed the degree of exposure, but unacceptable for the rating of the PPE to be less than the degree of exposure.

**(1) Energized Electrical Conductors and Circuit Parts — Safe Work Condition.** Energized electrical conductors and circuit parts to which an employee might be exposed shall be put into an electrically safe work condition before an employee works within the Limited Approach Boundary of those conductors or parts, unless work on energized components can be justified according to 130.1.

The primary protective strategy must be to establish an electrically safe work condition. After this strategy is executed, all electrical energy has been removed from all conductors and circuit parts to which the worker could be exposed. After the electrically safe work condition has been established, no PPE is necessary, and unqualified workers are permitted to execute the remainder of the work.

The only exception to this requirement is if deenergizing the circuit conductors or equipment cannot be justified as described below. Justification for work on or near exposed energized electrical conductors or circuit parts must be in writing. Of course, some functions require a circuit to be energized. A qualified person can perform work on or near exposed energized conductors or circuit parts under the following conditions:

- Deenergizing the conductors or equipment could result in an increased hazard. For instance, a life support system might be dependent on the continuation of the electrical service.
- Deenergizing the conductors or equipment could require a complete shutdown of a continuous process. For instance, the design of the electrical circuit is such that a continuous processing facility must be taken completely out of production.

If the circuit/equipment voltage is less than 50 volts, only the shock hazard is not present. Other electrical hazards might still be associated with the task. For instance, the capacity of the source of energy might present a thermal or pressure hazard in event of a short circuit condition.

Workers are often reluctant to question a decision by a supervisor that a work task must be conducted while the circuit remains energized. Section 110.8 provides authority for any employee to question the need for a task to be performed while the circuit or equipment remains in an energized condition. Often there is a tendency for workers to err on the side of accepting exposure to electrical hazards, while managers and supervisors tend to be reluctant to accept increased exposure to hazards.

**(2) Energized Electrical Conductors and Circuit Parts — Unsafe Work Condition.** Only qualified persons shall be permitted to work on electrical conductors or circuit parts that have not been put into an electrically safe work condition.

An electrical hazard is considered to be present until an electrically safe work condition exists. Unqualified persons are not permitted to perform any task with potential exposure to an electrical hazard.

The acts of opening a disconnecting means, measuring for absence of voltage, visually verifying a physical break in the power conductors, and installing safety grounds all contain a risk of injury. These acts are necessary to create an electrically safe work condition, and, until they are completed, the worker should be wearing PPE based on the degree of potential hazard. PPE that is rated greater than 40 calories per square centimeter (cal/cm$^2$) could be required until the electrically safe work condition exists but FPN 2 of Section 130.7(A) should be reviewed. Workers who implement the requirements for an electrically safe work condition must be qualified for that work task.

**(B) Working Within the Limited Approach Boundary of Exposed Electrical Conductors or Circuit Parts that Are or Might Become Energized.** Prior to working within the Limited Approach Boundary of exposed electrical conductors and circuit parts operating at 50 volts or more, lockout/tagout devices shall be applied in accordance with 120.1, 120.2, and 120.3. If, for reasons indicated in 130.1, lockout/tagout devices cannot be applied, 130.2(A) through 130.2(D)(2) shall apply to the work.

Locks and tags are both required if they can be installed on circuits with an operating voltage of 50 volts or greater. The installation of locks and tags must be in accordance with a lockout/tagout program defined by the requirements discussed in Article 120 of this standard. In many instances, employers and workers discuss lockout or tagout as if no further action is necessary. When the energy source is electrical, however, lockout is only one step in establishing an electrically safe work condition. When an electrically safe work condition cannot be established, all requirements described in Article 130 must be observed.

Section 110.8(A)(1) defines the only circumstances in which the equipment or circuit might remain energized while a work task in performed. If sufficient justification (as defined in this section) exists and the increased risk of injury is deemed to be acceptable, the requirements defined in Article 130 apply.

**(1) Electrical Hazard Analysis.** If the energized electrical conductors or circuit parts operating at 50 volts or more are not placed in an electrically safe work condition, other safety-related work practices shall be used to protect employees who might be exposed to the electrical hazards involved. Such work practices shall protect each employee from arc flash and from contact with energized electrical conductors or circuit parts operating at 50 volts or more directly with any part of the body or indirectly through some other conductive object. Work practices that are used shall be suitable for the conditions under which the work is to be performed and for the voltage level of the energized electrical conductors or circuit parts. Appropriate safety-related work practices shall be determined before any person is exposed to the electrical hazards involved by using both shock hazard analysis and arc flash hazard analysis.

Qualified workers who are permitted to work on or near exposed energized conductors or circuit parts must select and use work practices that provide protection from shock, arc flash, and other electrical hazards. The work practices that are used must minimize any potential for injury. For instance, body position is one factor that a qualified person should recognize as an element of the analysis that could reduce exposure to electrical shock or arc flash. Every situation should be determined individually.

The hazard/risk analysis must determine whether any conductor will remain energized for the duration of the work task. The analysis must determine the shock approach boundaries and the arc flash protection boundary.

Both a shock analysis and an arc flash analysis are required before any person is permitted to approach the exposed energized electrical conductors or circuit parts. These analyses must answer the following questions:

- Does a shock hazard exist?
- Will the worker be exposed to the shock hazard at any point during the work task?
- What is the degree of the hazard?
- What protective equipment is necessary to minimize the exposure?
- Does an arc flash hazard exist?
- Will the worker be exposed to a thermal hazard at any point during the work task?
- What is the degree of the arc flash hazard?
- What protective equipment is necessary to minimize exposure to the thermal hazard?
- Does a co-occupancy hazard exist?
- What measures will be taken to minimize the impact of other work?
- Will other workers be exposed to an electrical hazard because of the work task?
- Will the worker be exposed to any other electrical hazard while executing the work task?
- What authorization is necessary to justify executing the work task while the exposed conductor(s) is (are) energized?
- What workers are required to be within an approach boundary?
- Are unqualified workers required?
- How will the voltage on the conductor (or nearby conductors) be determined?
- Is the risk of injury acceptable?

Employers must conduct an arc flash hazard analysis and place a label on an external surface of equipment that indicates the degree of the thermal hazard or recommends PPE for protection from the degree of existing thermal hazard. The *NEC* requires a label that warns workers of a potential thermal hazard. Employers must provide sufficient information for an employee to conduct the necessary hazard analysis. Labels are one element of the hazard analysis.

**(a) Shock Hazard Analysis. A shock hazard analysis shall determine the voltage to which personnel will be exposed, boundary requirements, and the personal protective equipment necessary in order to minimize the possibility of electrical shock to personnel.**

FPN: See 130.2 for the requirements of conducting a shock hazard analysis.

The shock hazard analysis is only one part of the hazard/risk analysis. The shock hazard analysis is intended to determine if a risk of electrocution or shock might exist when the work task is being executed. That risk increases as a worker approaches an exposed energized electrical conductor or circuit parts. Requirements that reduce the risk are associated with the limited, restricted, and prohibited approach boundaries. The degree of the risk increases as the voltage increases. The shock hazard analysis must determine if work practices and protective equipment reduce the risk of electrocution to an acceptable level.

The shock hazard analysis determines if a risk of electrocution or shock might exist when the work task is being executed. The chance that a worker might receive a shock or be electrocuted is greater as the distance between a worker and an exposed energized electrical conductor or circuit part is less. Requirements that reduce the risk are associated with the limited, restricted and prohibited approach boundaries. Although, approach boundaries provide broad indications that the chance of injury is increasing or decreasing, the concept applies between the limits of a specific approach boundary. A qualified worker must know that and act accordingly.

As the circuit voltage level increases, the amount of potential current flow increases with the voltage. The degree of a potential injury increases as the voltage increases. The shock hazard analysis must determine necessary work practices and PPE to reduce the risk of injury (electrical shock or electrocution) to an acceptable level.

(b) Arc Flash Hazard Analysis. An arc flash hazard analysis shall determine the Arc Flash Protection Boundary and the personal protective equipment that people within the Arc Flash Protection Boundary shall use.

FPN: See 130.3 for the requirements of conducting an arc flash hazard analysis.

An arc flash hazard analysis must determine if the worker could be exposed to the extreme temperature generated by the electrical current during an arcing fault. The analysis has two purposes. First, the analysis must determine the location of the flash protection boundary. Second, the analysis must determine the rating of protective clothing that must be worn by the worker.

Several different methods are available for establishing the degree of the potential arc flash hazard. No specific method is required, and no preferred method of predicting the degree of the thermal hazard exists. Methods of predicting arc flash intensity are evolving, and research is currently underway. The research will help facilities select an appropriate method to predict the thermal hazard for their electrical safety program. Annex D provides discussion of several methods for calculating potential incident energy.

The flash hazard analysis must provide sufficient information for a worker to select PPE that can protect him or her from injury as the result of the thermal effects of any potential arcing fault. Note that PPE for protection from the thermal hazard does not necessarily provide protection from any other hazard.

If the flash hazard analysis suggests that the intensity of the arc flash could expose a worker to 40 cal/cm$^2$ or more, the work must not be performed unless an electrical safe work condition has been established. If the intensity is greater than 40 cal/cm$^2$, no protective equipment exists that can protect the worker from the intense pressure (arc blast) that also will be produced by the arcing fault. However, when creating an electrical safe work condition, it might be necessary to operate disconnecting means in their designed state (doors closed for example) with significant incident energy. See FPN 2 in Section 130.7(A).

The product of the arc flash hazard analysis must provide sufficient information for a worker to select PPE that will protect him or her from injury as the result of the thermal effects of any potential arcing fault. If the arc flash hazard analysis suggests that the intensity of the arc flash could expose a worker to more than 40 cal/cm$^2$, the work must not be performed unless an electrical safe work condition has been established.

**(2) Energized Electrical Work Permit.** When working on energized electrical conductors or circuit parts that are not placed in an electrically safe work condition (i.e., for the reasons of increased or additional hazards or infeasibility per 130.1), work to be performed shall be considered energized electrical work and shall be performed by written permit only.

FPN: See 130.1(B) for the requirements of an energized electrical work permit.

Electrical workers often accept the increased risk associated with working on or near exposed energized electrical conductors or circuit parts. Also, managers and supervisors often do not recognize the increased risk of electrical injury. Experience suggests that if managers and supervisors are advised that a significant risk of injury exists, they are reluctant to accept the increased risk. If a manager or supervisor is requested to authorize the work on or near exposed energized electrical conductors or circuit parts by signing a permit, he or she will be more critical of the plan to execute the work on or near exposed energized electrical conductors or circuit parts.

All work on or near exposed energized electrical conductors or circuit parts must be authorized by a written document. [See Annex J for a sample energized electrical work permit.] The written authorization must be kept on file until the work is completed. High-level managers should review the number of energized work permits that are authorized over a specific period of time. For instance, if the number of permits is increasing, the high-level manager should question the process that accepts the increased exposure to electrical hazards.

A written permit ensures that all parties have reviewed the work task and agree that an electrically safe work condition cannot be established. [See OSHA 29 CFR 1910.333(a)(1) and Section 130.1.] The parties should agree that justification for accepting the increased risk is satisfactory.

The intent of the written permit is to ensure that the decision to accept the increased risk of injury is shifted to include both the worker and a manager or supervisor. Managers and supervisors must be involved in the decision. Any worker always has the option to refuse to do any task that he or she feels has an unacceptable risk.

Accountability exists regardless of whether workers, supervisors, or managers recognize it. A written permit ensures that workers, supervisors, and managers have a chance to exercise inherent options.

**(3) Unqualified Persons.** Unqualified persons shall not be permitted to enter spaces that are required to be accessible to qualified employees only, unless the electric conductors and equipment involved are in an electrically safe work condition.

Unqualified persons are not trained to recognize when or if an electrical hazard exists, and they must not be closer to an exposed energized electrical conductor or circuit part than the limited approach boundary or the arc flash protection boundary, whichever is greater. An unqualified person is permitted to be within the limited approach boundary, provided he or she has been advised by a qualified person how to avoid contact with the exposed energized electrical conductors or circuit parts and how to select and wear PPE that is necessary for protection from the effects of an arcing fault, shock, or electrocution.

**(4) Safety Interlocks.** Only qualified persons following the requirements for working inside the Restricted Approach Boundary as covered by 130.2(C) shall be permitted to defeat or bypass an electrical safety interlock over which the person has sole control, and then only temporarily while the qualified person is working on the equipment. The safety interlock system shall be returned to its operable condition when the work is completed.

Two types of safety interlocks are often used with electrical equipment and circuits. One type is a mechanical interlock that prevents a door from being opened or a disconnect means from being closed or opened while the equipment is energized. The second type is an electrical circuit contact, such as a limit switch or proximity switch, which prevents a circuit component from being operated unless a specific circuit condition exists.

Interlocks are installed to prevent a specific hazardous condition. Only qualified workers who have been trained to understand how defeating a safety interlock establishes an unsafe condition should be permitted to defeat a safety interlock.

Any safety interlock that is defeated or bypassed creates an unsafe condition, and the worker must be aware of it. The safety interlock must be returned to its normal condition when the work task is completed. Maintaining a bypass log might enable a worker to remember that a safety interlock is in a bypass condition.

**110.9 Use of Equipment.**

**(A) Test Instruments and Equipment.**

**(1) Rating.** Test instruments, equipment, and their accessories shall be rated for circuits and equipment to which they will be connected.

FPN: See ANSI/ISA-61010-1 (82.02.01)/UL 61010-1, *Safety Requirements for Electrical Equipment for Measurement, Control, and Laboratory Use – Part 1: General Requirements*, for rating and design requirements for voltage measurement and test instruments intended for use on electrical systems 1000 volts and below.

Test instruments are safety equipment and should be considered in the same category of protection as voltage-rated gloves or arc-rated face shields. Although they have some characteristics of a tool, all test equipment and accompanying accessories should be purchased without artificial cost restraints. Test equipment must be selected based on the intended use and expected voltage or current rating. Leads and probes are an integral part of the test equipment and must be rated at least as great as the instrument.

See Exhibit 110.2 for an example of a voltage meter. Note that the Cat rating (per ANSI/ISA-61010-1 (82.02.01)/ UL 61010-1) is on the lower left of the device (see insert) and look on the back for an approved testing organization listing. Notice that the leads are appropriate for the use.

*EXHIBIT 110.2 Meter for measuring voltage. (Courtesy of Fluke Corporation)*

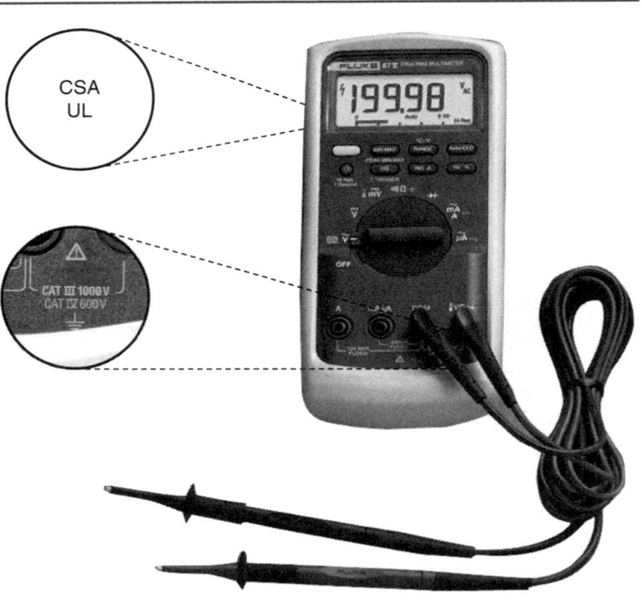


Equipment and instruments that contact an exposed potentially energized electrical contact or conductor might expose a worker to both electrocution and arc flash. These devices must be rated for the expected service. Voltage-detecting devices must have a voltage rating at least as great as the maximum operating voltage of the circuit or conductor. Accordingly, instruments that measure other circuit parameters must be rated accordingly.

The FPN to 110.9(A)(1) explains that one important factor for test instruments used to measure voltage is a static discharge rating for devices used on conductors rated at 1000 volts or less. Any static discharge, such as a lightning discharge, has the potential to damage an instrument that happens to be in contact with a conductor; even if the lightning discharge is remote from the work location.

The rating of the measuring device also must be at least as great as the parameter being measured. Section 110.9 also requires that workers observe any duty rating assigned by the manufacturer.

**(2) Design.** Test instruments, equipment, and their accessories shall be designed for the environment to which they will be exposed, and for the manner in which they will be used.

Section 110.9(A)(2) requires that the design of instruments and similar equipment must be consistent with the conditions of use. For example, voltmeters should have a static rating (category) that is consistent with their use. All components of the instrument must be designed for the application. Design of the meter, probes, and mounting/holding device must be consistent with the environment in which the device is used. ANSI/ISA 82.02.01 describes static discharge ratings for voltmeters and other devices that make direct contact with an electrical circuit.

Instruments that measure voltage, current, and other parameters must be selected for the conditions of use. Contacting an exposed energized electrical conductor with an instrument to measure voltage normally results in a small arc immediately before contact is made or broken. Therefore, devices such as these must not be used when the atmosphere is explosive.

**(3) Visual Inspection.** Test instruments and equipment and all associated test leads, cables, power cords, probes, and connectors shall be visually inspected for external defects and damage before each use. If there is a defect or evidence of damage that might expose an employee to injury, the defective or damaged item shall be removed from service, and no employee shall use it until repairs and tests necessary to render the equipment safe have been made.

A visual inspection must be conducted prior to each use of test instruments and equipment. All test instruments must be inspected for physical damage before each use. Only listed test equipment should be used, and the worker must look for the listing on the device before using it. The inspection must include all leads, probes, and other attachments. If any damage, such as a cracked case, cut or pinched leads, or damaged probe tips is observed, the instrument must be removed from service, repaired, and tested before it is used again.

If a defect is found during the visual inspection, a tag or label indicating that the instrument is defective should be attached to the instrument or equipment and the instrument should be removed from service. No attempt should be made to repair a lead or probe in which a defect is found. Any defective lead or probe should be destroyed and replaced with a new one.

**(4) Operation Verification.** When test instruments are used for the testing for the absence of voltage on conductors or circuit parts operating at 50 volts or more, the operation of the test instrument shall be verified before and after an absence of voltage test is performed.

When a voltage-testing device is used to test for absence of voltage, an indication of zero volts might mean that no voltage is present when the test was performed, or it could mean that the instrument has failed. Workers must be able to determine if an electrocution hazard exists, and the worker must be aware of that condition. The voltmeter or device must be verified as operating normally before testing for absence of voltage. After the voltage test has been conducted, the voltmeter or device again must be verified as operating normally to ensure that a failure did not occur during the testing for absence of voltage.

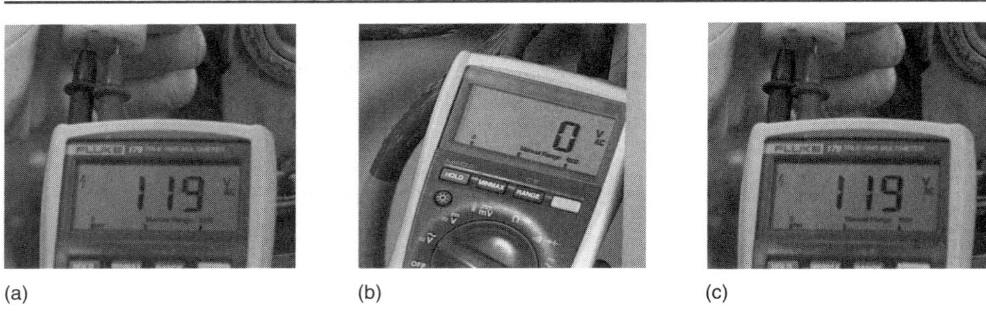

***EXHIBIT 110.3*** *Steps for testing for an absence of voltage: (a) first, verify operation of the tester, then (b) confirm there is no voltage, and then (c) again confirm operation of the tester.*

The operation must be verified before and after use in every instance in which the voltage is 50 volts or more. The shock or electrocution hazard is significantly reduced if the circuit voltage is less than 50 volts. The 50-volt limit does not rely on the frequency of the circuit and applies to direct current as well as high frequency circuits. See Exhibit 110.3 for an illustration of the procedure to test for absence of voltage.

**(B) Portable Electric Equipment.** This section applies to the use of cord-and-plug-connected equipment, including cord sets (extension cords).

**(1) Handling.** Portable equipment shall be handled in a manner that will not cause damage. Flexible electric cords connected to equipment shall not be used for raising or lowering the equipment. Flexible cords shall not be fastened with staples or hung in such a fashion as could damage the outer jacket or insulation.

All electrical equipment that receives its energy through a cord/plug/receptacle and is easily moved from one position to another is considered to be portable electric equipment. For instance, a drill motor, which is connected by a cord and plug, is portable and covered by the requirements of this section. Although a battery-operated drill motor is not covered by the requirements of this section, this section covers the battery charger that provides energy to the battery unless it is permanently wired in a fixed position.

Moving or lifting the flexible equipment by the cord often damages the cord. Flexible cords are constructed to resist damage from use; however, the damage sometimes occurs if the portable cord is placed in a fixed position and held in place by staples or similar devices that tend to crush or abrade the cord. When a portable cord is installed through a doorway, it is subjected to crushing when the door closes. When necessary to install a portable cord through a doorway, the door should be blocked from closing to prevent the cord from being pinched if the door should close.

Workers could be tempted to move cord-connected portable equipment by holding the cord and to pull the cord cap from a receptacle by grasping the cord instead of the cord cap. Practices such as these often damage the cord and expose conductors, so these practices must be discouraged.

Flexible cords connected to portable equipment and extension cords must be protected from potential damage when in use. When installing supports for flexible cords, workers must avoid support mechanisms that might damage the insulation.

**(2) Grounding-Type Equipment.**

(a) A flexible cord used with grounding-type utilization equipment shall contain an equipment grounding conductor.

Including a grounding conductor in the supply conductor is one primary protective measure to prevent exposure to shock and electrocution. Double insulation is another protective measure. Either of these protective measures provides significant protection from shock or electrocution. Ground-fault circuit interrupters (GFCIs) are another primary protective measure that is discussed in a later section. Unless the portable equipment is rated as double insulated, the cord supplying the energy must include a grounding conductor. The plug must be a grounding-type attachment plug and the receptacle must be a grounding-type receptacle.

The grounding conductor is an integral part of the safety system built into equipment by manufacturers for tools and devices that are not double insulated. The integrity of the grounding conductor is paramount to minimizing the chance that a worker becomes a part of the electrical circuit providing power to the tool. Each person must visually inspect the tool before it is used. Any defect, such as a cord cap with a missing or damaged ground pin, is sufficient indication to cause the tool to be removed from use. If a defect is found in the visual inspection, the tool must be removed from service until it is repaired and the integrity of the flexible cord reaffirmed.

Grounding conductors are not required on tools that are rated by the manufacturer as double insulated. These tools are required to be inspected for indication of damage.

(b) Attachment plugs and receptacles shall not be connected or altered in a manner that would interrupt continuity of the equipment grounding conductor.

Additionally, these devices shall not be altered in order to allow use in a manner that was not intended by the manufacturer.

Since the grounding conductor is a primary protective measure, the integrity of the grounding conductor is important. All components of the circuit must provide the same integrity as the grounding conductor. Plugs and receptacles are constructed, tested, and listed for specific use. Altering the plug (or receptacle) in a manner not intended by the manufacturer destroys the integrity of the grounding conductor. Workers sometimes remove the grounding pin on a plug or twist the hot and neutral pins to enable the plug to mate with a different receptacle. Either of these actions is prohibited by this requirement.

(c) Adapters that interrupt the continuity of the equipment grounding conductor shall not be used.

Workers sometimes field-fabricate an adapter by installing a plug of one construction on one end and with a cord cap of a different construction on the other to enable portable equipment or cord to mate with a receptacle of different configuration. This practice must be avoided.

**(3) Visual Inspection of Portable Cord-and-Plug-Connected Equipment and Flexible Cord Sets.**

By their nature, cords and plugs installed on portable electric equipment are subjected to use and subsequent damage. When the cord or plug is damaged, the worker might be exposed to shock or electrocution by a fault in the equipment, a damaged grounding conductor, or by damage to the insulation on the circuit conductors. Although a visual inspection might not identify all possible problems, a thorough visual inspection provides significant assurance that the integrity of the cord will function as intended. When performing the inspection, the worker should visually observe each end of the cord or cord set and ensure that all pins are in place and unmodified. The worker then should begin at one end and run his or her hand along the complete surface of the cord to the other end. The surface should be smooth with no indentations, cuts, or abrasions. Any indentation, cut, or abrasion should be inspected further to ensure that the full insulating quality of the cord is complete. Indentations from crushing or pinching should trigger further inspection to ensure that conductivity of the grounding conductor is complete.

(a) Frequency of Inspection. Before each use, portable cord-and-plug-connected equipment shall be visually inspected for external defects (such as loose parts or deformed and missing pins) and for evidence of possible internal damage (such as a pinched or crushed outer jacket).

Cords and plugs can be damaged while in storage or while being moved from one position to another. The cord and plug must be visually inspected to identify any damage before being energized for use. See Exhibits 110.4, 110.5, and 110.6.

*Exception: Cord-and-plug-connected equipment and flexible cord sets (extension cords) that remain connected once they are put in place and are not exposed to damage shall not be required to be visually inspected until they are relocated.*

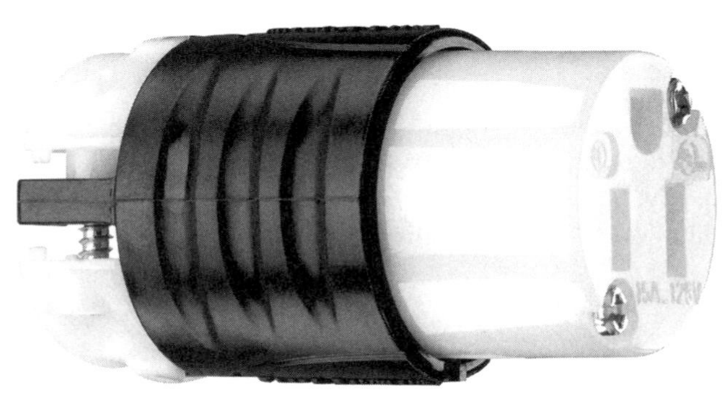

***EXHIBIT 110.4*** *A cord cap. (Courtesy of Pass & Seymour/Legrand®)*

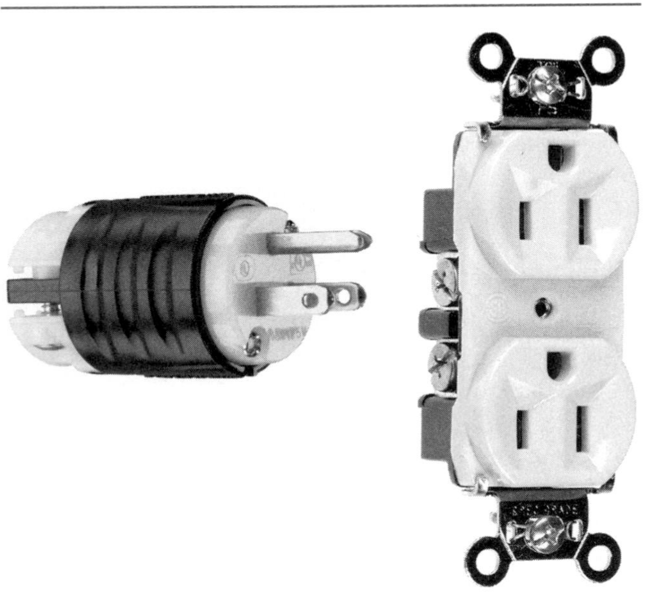

***EXHIBIT 110.5*** *A three-pronged plug (left), which goes into a three-pronged receptacle (right). (Courtesy of Pass & Seymour/Legrand®)*

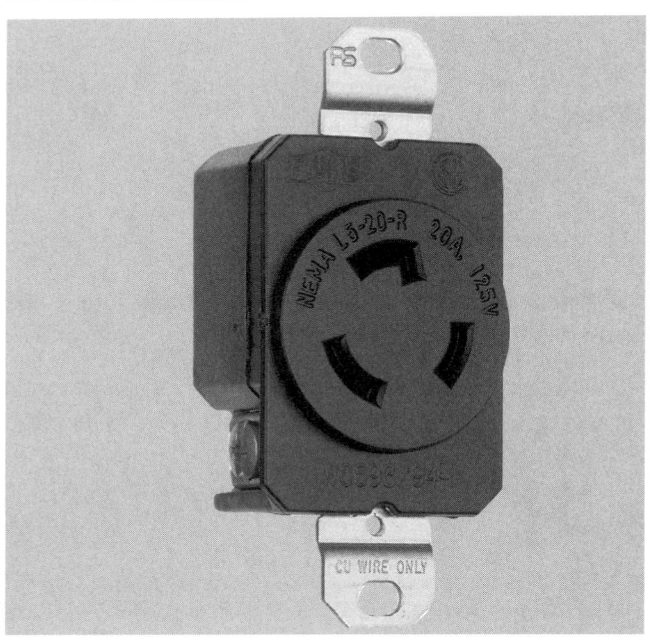


***EXHIBIT 110.6*** *A Twistlock plug. (Courtesy of Pass & Seymour/Legrand®)*

Cords that supply equipment that is not used as portable equipment such as a water cooler (*NEC* 2008 Section 422.52 requires GFCI on water coolers), drill press, or computer terminal should be inspected when installed. Although no additional inspection is required until the equipment is moved to another location, periodic inspection of the grounding conductor integrity is recommended.

(b) Defective Equipment. If there is a defect or evidence of damage that might expose an employee to injury, the defective or damaged item shall be removed from service, and no employee shall use it until repairs and tests necessary to render the equipment safe have been made.

When a defect is identified in portable electric equipment or a cord set, the equipment or cord set must be removed from service. A tag or other method that identifies the equipment or cord set as defective should be attached. The tag or label should warn workers that the equipment or cord set should not be used. The warning should remain with the equipment or cord set until repairs have been completed and the operability of the equipment or cord set has been verified by testing.

The integrity of the grounding conductor of any portable or cord-connected tools and devices must be verified by tests prior to returning the tool or device to service. The integrity of the insulating system for double-insulated tools must be verified by tests after repairs have been made.

(c) Proper Mating. When an attachment plug is to be connected to a receptacle, the relationship of the plug and receptacle contacts shall first be checked to ensure that they are of mating configurations.

When cord-and-plug-connected equipment is placed into service, the pin configuration of the plug must match the configuration of the receptacle. Pins must not be damaged or removed. Attachment devices that permit the installation of devices with pin configurations that do not match must ensure the integrity of all connecting pins of the cord cap.

Moisture can provide a conducting path from the hot conductor in a cord cap to the surface of the device. A person inserting a wet cord cap into an energized receptacle is exposed to a shock hazard. The person handling the wet cord cap must be wearing PPE that is rated at least as great as the circuit voltage.

If the cord-connected equipment is subject to moisture ingress, a listed GFCI must protect the cord. Some work environments could contain another conductive compound. A GFCI must protect any cord-connected equipment or device installed in these areas.

(d) Conductive Work Locations. Portable electric equipment used in highly conductive work locations (such as those inundated with water or other conductive liquids) or in job locations where employees are likely to contact water or conductive liquids shall be approved for those locations. In job locations where employees are likely to contact or be drenched with water or conductive liquids, ground-fault circuit-interrupter protection for personnel shall also be used.

Any conductive path from a worker's hand to an energized conductor exposes that worker to shock or electrocution. A worker standing in water or other conductive material must wear adequate insulating gloves when removing the plug from an energized receptacle. If the environment contains highly conductive compounds, gloves with adequate voltage rating are required. If the receptacle or plug is wet from water or other conductive material, adequately rated gloves are required.

Interior work locations that are wash-down areas should be considered conductive unless the possibility of moisture does not exist. The best option is to use a GFCI installed at the supply end of the cord set or portable cord-connected equipment if the equipment is plugged directly into a permanently installed receptacle.

As noted previously, if the cord-connected equipment is used outdoors, it is subject to moisture ingress and a listed ground-fault circuit interrupter must protect the cord. Some work environments could contain another conductive compound. A GFCI must protect any cord-connected equipment or device installed in these areas.

**(4) Connecting Attachment Plugs.**

(a) Employees' hands shall not be wet when plugging and unplugging flexible cords and cord-and-plug-connected equipment if energized equipment is involved.

Moisture can provide a conducting path from the hot conductor in a cord cap to the surface of the device. A person inserting a wet cord cap into an energized receptacle is exposed to a shock hazard. The person handling the wet cord cap must be wearing PPE that is rated at least as great as the circuit voltage.

Any conductive path from a worker's hand to an energized conductor exposes that worker to shock or electrocution. If the worker is standing in water or covered by another conductive material, he or she must wear adequate insulating gloves when removing the plug from an energized receptacle.

(b) Energized plug and receptacle connections shall be handled only with insulating protective equipment if the condition of the connection could provide a conductive path to the employee's hand (if, for example, a cord connector is wet from being immersed in water).

Any conductive path from a worker's hand to an energized conductor exposes that worker to shock or electrocution. A worker standing in water or other conductive material must wear adequate insulating gloves when removing the plug from an energized receptacle. If the environment contains highly conductive compounds, gloves with adequate voltage rating are required. If the receptacle or plug is wet from water or other conductive material, adequately rated gloves are required.

(c) Locking-type connectors shall be secured after connection.

Some attachment plugs are intended to be held in place mechanically by twisting the plug after the conducting pins have been fully inserted into place. Other forms of mechanical interlocks also are available. Any interlocking mechanism provided by the manufacturer should be secure after the connection is complete.

Twistlock plugs and some other portable cord-connecting devices are designed to provide a connection that is secure from accidental withdrawal. Devices that provide the additional security must be inserted so that the design intent of the device is complete. For instance, twistlock plugs must be turned to the "secure" position.

**(C) GFCI Protection Devices.** GFCI protection devices shall be tested per manufacturer's instructions.

All manufacturers of listed GFCI devices provide installation and testing instructions.

**(D) Overcurrent Protection Modification.** Overcurrent protection of circuits and conductors shall not be modified, even on a temporary basis, beyond that permitted by applicable portions of electrical codes and standards dealing with overcurrent protection.

Overcurrent devices are important components of a safety system to prevent conductors and devices from current flow that exceeds the conductor's ability to safely conduct current.

Exceeding the conductor or equipment rating can result in overheating. One result of overheating is ignition of the insulation or other nearby flammable material. Device and component overheating can result in massive component failure.

Overcurrent devices, which limit the amount of current to less than the ampacity of the circuit, can be substituted. However, the interrupting rating of the smaller overcurrent device must be at least as great as the original.

Overcurrent devices with a current-limiting rating reduce the amount of energy that might be let through. Limiting the amount of energy available on the downstream side of the device reduces the flash protection boundary.

Note that all overcurrent devices must comply with the requirements of the *NEC*.

FPN: For further information concerning electrical codes and standards dealing with overcurrent protection, refer to Article 240 of *NFPA 70, National Electrical Code*.

### CASE STUDY

It was about 3:00 P.M. on October 23. The sun was shining brightly when Alfred went to check on the work in the Poly building at the Gulf Coast MESP plant near Houston.* Alfred was a construction supervisor for SSC Construction. He was a valuable site representative for the company. He always paid particular attention to requirements that the customer included in the contract. Alfred knew that SSC was very serious about its lockout procedure. He also knew that the MESP lockout procedure required workers to remove all lockout devices at the end of the day, regardless of whether or not the work was complete.

The job in the Poly building was to replace a 45-kVA transformer with a new 75-kVA unit. No new wiring was required on the primary — both the wire and the disconnect size were large enough to supply the larger unit. New secondary wiring was already in place. The final task was to make the transformer connections in the transformer. The disconnect switch was the energy-isolating device, and it was locked at the beginning of each day. The MESP procedure included a requirement for a blue tag, representing transfer of ownership, to be installed on the disconnect switch to indicate permission for maintenance or contractors to install their lock.

The MESP procedure required each MESP employee to install a personal lock. Following OSHA's construction standards, Alfred had written an SSC procedure that permitted the supervisor to install locks to cover the construction employees.

Although Alfred listened carefully to MESP people, he frequently didn't adequately care for his own workers. On this day, quitting time for SSC electricians was 3:30 P.M., since the workday had started at 5:00 A.M. When Alfred reached the work area, he saw that the electrician was just finishing the primary connections. Alfred told the electrician to complete the primary connection and clean up. The new secondary conductors could be completed the next day. The position of the new larger secondary conductors made it very difficult to replace the front cover on the transformer. The electrician knew that MESP was adamant about lockout, so he did not worry about replacing the transformer cover.

As required by the MESP lockout procedure, Alfred proceeded to the panelboard and removed the SSC lock from the disconnect switch. At the time, Alfred thought it curious that the blue tag was missing, but the blue tag was not really his responsibility. He had no way of knowing that the blue tag had accidentally been torn from the switch earlier that day.

During the next shift, lights were needed in the instrument control room of the Poly building. The second-shift electrician went to the panelboard to energize the lighting panel for the instrument control room. Seeing the switch turned off and seeing neither lockout locks nor blue tags, the shift electrician moved the switch handle to "on." However, the lights failed to come on. After a short investigation, the shift electrician reported back to the shift supervisor that the transformer was disconnected. The lights would not work that night. The work in the instrument control room would have to wait until another day. The shift electrician moved on to the work indicated on his next work ticket.

On Sunday, Matthew, the SSC electrician, entered the plant early in the morning. He was a new employee for SSC and wanted to do the best job he could. He knew what his work assignment would be that day, and he proceeded directly to the Poly building. Matthew did not question whether the circuit was locked out or not. He knew that MESP was adamant about lockout, so he prepared to begin work. Alfred arrived just before the work was started. Not really caring about the response, Alfred asked Matthew about the World Series game. Alfred was thinking about where to go next as Matthew replied that he had not seen the game.

Alfred was leaving the area when Matthew began to manhandle the new grounding conductor into position. The end of the new grounding conductor contacted the "A-phase" secondary tap on the transformer. The noise associated with the arc flash resounded throughout the Poly building. Matthew was wearing safety glasses, a hard hat, denim jeans, and a cotton shirt with long sleeves. His clothing did not ignite. He was treated at the local hospital for some burns on his hands and released. Again, an electrocution had been avoided only by luck.

---

*This account is based on an actual incident. The names, including the name of the facility, have all been changed to protect those involved. Any similarity to actual names or facilities is strictly coincidental.

Reprinted from Ray A. Jones and Jane G. Jones, *Electrical Safety in the Workplace,* 2000, with permission of Jones and Bartlett Publishers, Sudbury, MA.

# ARTICLE 120
## Establishing an Electrically Safe Work Condition

The most effective way to prevent an electrical injury is to completely remove the source of electrical energy and eliminate the possibility of its reappearance. To do that, workers must identify all possible sources of electricity and locate the disconnecting means of each source. Electricity sometimes can appear from the load side. The worker must identify all potential sources of energy, eliminate them, visually verify (if possible) that they are eliminated, lock them out, test to verify absence of voltage, and ground the conductors or parts, if necessary. Only then can a worker establish an electrically safe work condition, which is a lifesaving process.

The process of establishing an electrically safe work condition can expose a worker to an electrical hazard. When switches are moved from one position to another, equipment failure can initiate an internal arcing fault. Although unlikely, where equipment is ungrounded, for some reason, an internal failure could result in the enclosure becoming energized. Any internal arcing fault subjects the equipment to high temperature and a pressure wave. Depending on the integrity of the equipment, circuit, and overcurrent device, a worker could be exposed to arc flash, arc blast, shock, electrocution, and flying parts and pieces. Equipment that meets the requirements of the *NEC* and that is well maintained is unlikely to result in unexpected exposure to an electrical hazard when the equipment is operating normally. However, the worker should conduct a hazard/risk analysis of the work task before executing it.

Equipment rated as arc resistant by a manufacturer provides assurance to the worker. When the switch operating handle is moved from one position to another, an internal arcing fault within arc-resistant equipment would not expose the worker to effects from the fault as long as the equipment doors are closed and latched. Arc-resistant equipment directs the pressure and products of combustion away from a worker standing near the front of the equipment. In most cases, the pressure relief is to the top of the equipment. Workers must recognize, however, that when they are troubleshooting a problem and the door is less than fully latched or a cover is removed, the arc-resistant nature of the equipment no longer exists. Arc-resistant equipment has no effect on exposure to shock or electrocution with the exception that no penetration for ventilation purposes exists on the front of the equipment.

Arc-rated equipment has no bearing on the need to conduct a hazard/risk analysis or any other requirement defined in *NFPA 70E*. Arc-resistant equipment offers unique protection for a person who operates the switch handle from one position to another when the equipment doors are closed and latched. If fasteners are not completely closed on arc-resistant equipment, the equipment is no longer arc resistant.

### 120.1 Process of Achieving an Electrically Safe Work Condition.

An electrically safe work condition shall be achieved when performed in accordance with the procedures of 120.2 and verified by the following process:

An electrically safe work condition does not exist until *all* of the six steps in Section 120.1 have been completed. Until then, workers could contact an exposed live part, and they must wear PPE.

If an electrically safe work condition exists, no electrical energy is in immediate vicinity of the work task(s). All danger of injury from an electrical hazard has been removed, and neither protective equipment nor special safety training is required. Unqualified workers may be used to perform work on equipment after an electrically safe work condition has been established. However, workers must be capable of executing the technical aspects of the work task.

(1) Determine all possible sources of electrical supply to the specific equipment. Check applicable up-to-date drawings, diagrams, and identification tags.

Establishing an electrically safe work condition ensures that all sources or potential sources of electrical energy have been identified. A worker should use all possible sources of information to identify and locate all sources of electrical energy. In many cases, some electrical energy sources might be labeled well and operated frequently. A sneak circuit sometimes exists that can be identified only by reviewing diagrammatic-type drawings. Circuits containing transformers or potential transformers must be checked and links or disconnects opened to confirm no possible back feeds are possible. Checking for back feeds is especially important when connecting temporary power in situations when equipment is taken out of service for repair or maintenance.

Diagrammatic-type drawings must be maintained in an up-to-date condition to provide accurate information. Panel schedules that are complete and up-to-date might provide the same information that is gleaned from a diagram. For instance, lighting circuits with separate neutrals can be identified from an up-to-date panel schedule.

Although experienced workers can be a source of information, written documentation is necessary to ensure that the information is available to all workers.

(2) After properly interrupting the load current, open the disconnecting device(s) for each source.

The rating of some disconnecting equipment is insufficient to interrupt the load current demanded by the utilization equipment. These disconnecting means must not be operated unless the load has been removed by another action. Unless the contacts in the disconnecting means have a rating at least as great as the current being conducted, the contacts can be destroyed and initiate a significant failure. In some cases, such a failure might escalate to an arcing fault within the equipment.

Even when equipment is load rated, interrupting full-load current reduces the life of the disconnecting means. Driven equipment should be stopped to reduce the amount of current in the circuit before the disconnecting means is operated. See Exhibit 120.1, which shows the opening of the disconnect from the "on" to the "off" position.

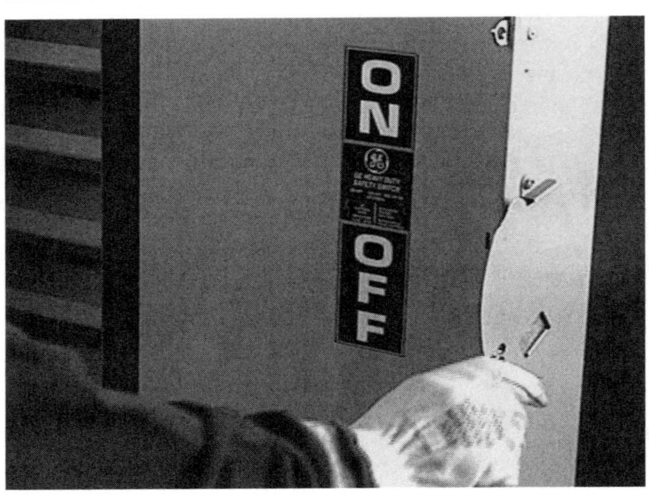


***EXHIBIT 120.1*** *Opening of disconnect.*

(3) **Wherever possible, visually verify that all blades of the disconnecting devices are fully open or that drawout-type circuit breakers are withdrawn to the fully disconnected position.**

A disconnect switch sometimes fails to open all phase conductors when the handle is operated. Ensuring that operating the handle of the device actually establishes a physical break in *all* conductors is critical. After operating the handle of the disconnecting device, the worker should open the door or cover or view the switch blades through a viewing window if available and observe the physical opening in each blade of the disconnect switch. See Exhibit 120.2. Opening a door or removing a cover could expose a worker to electrical hazards. Therefore, the worker must be protected from those hazards by PPE.

The physical opening of the contacts is sometimes difficult or impossible to observe directly. In those instances, the worker should verify the physical opening by measuring voltage on the load side of the device after the handle has been operated. The voltage test must include measuring voltage to ground from each conductor and between each conductor and each of the other conductors (phase-to-ground and phase-to-phase). Taking these voltage measurements should be considered working on or near exposed live parts, regardless of the position of the operating handle.

(4) **Apply lockout/tagout devices in accordance with a documented and established policy.**

Employers must establish and implement a lockout/tagout procedure. Employees must implement all aspects of the published lockout/ragout procedure.

OSHA requires employers to generate and implement a lockout/tagout procedure and train employees on the content of the procedure. Employees then should implement the requirements of the procedure provided by the employer. Note however, that exposure to electrical hazards is excluded from OSHA's requirements defined in 29 CFR 1910.147. Exposure

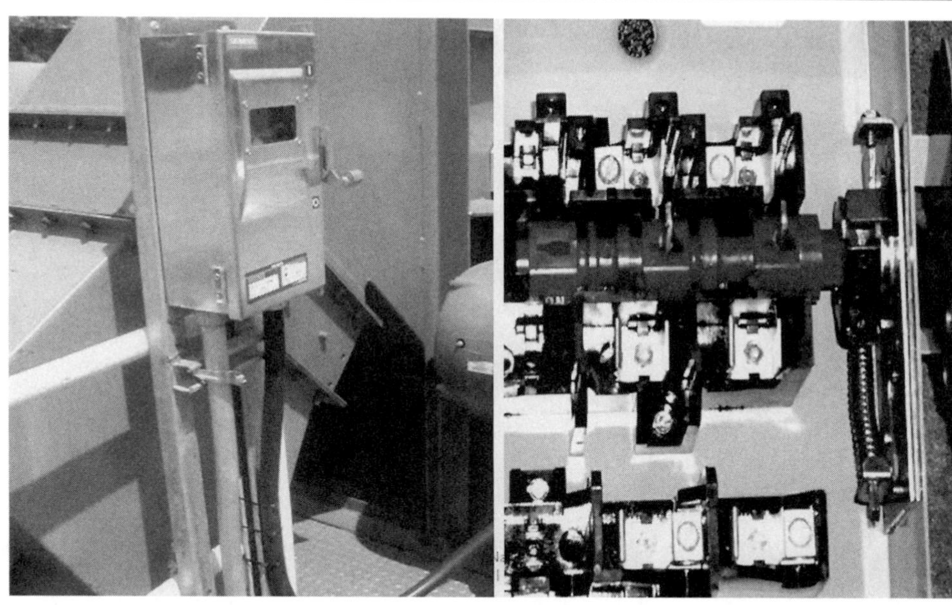

***EXHIBIT 120.2*** *Verifying opening of contacts.*

to electrical hazards is covered in the electrical section of the general industry rules (29 CFR 1910.333 and 29 CFR 1910.269) and in the safety and health regulations for construction standards (in 29 CFR 1926.417).

Only qualified workers are to be authorized to implement electrical lockout/tagout. Exhibit 120.3 shows a worker attaching a lock and tag to the open switch.

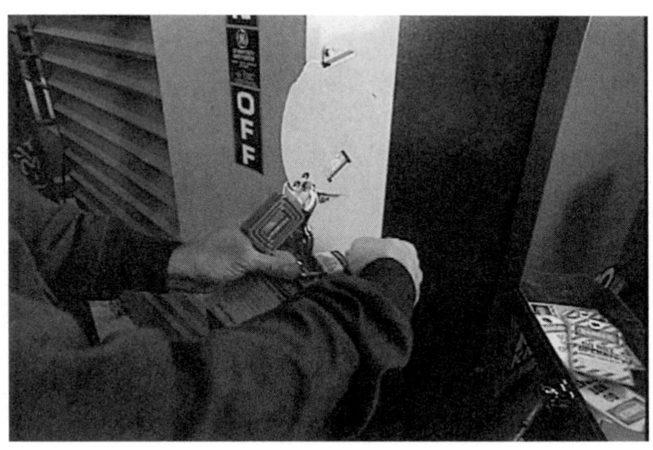


***EXHIBIT 120.3*** *Applying a lock and tag to a switch.*

(5) Use an adequately rated voltage detector to test each phase conductor or circuit part to verify they are deenergized. Test each phase conductor or circuit part both phase-to-phase and phase-to-ground. Before and after each test, determine that the voltage detector is operating satisfactorily.

The testing procedure is shown in Exhibit 120.4. First, the voltage tester is used to test a known source to verify that the tester is operating properly, as shown in Exhibit 120.4(a). Then the voltage tester is used to confirm that the equipment to be deenergized has zero voltage, as shown in Exhibit 120.4(b). Completing the procedure, the voltage tester is used again to test a known source to confirm that the voltage tester has not failed during the testing process, as shown in Exhibit 120.4(c).

FPN: See ANSI/ISA-61010-1 (82.02.01)/ UL 61010-1, *Safety Requirements for Electrical Equipment for Measurement, Control, and Laboratory Use – Part 1: General Requirements*, for rating and design requirements for voltage measurement and test instruments intended for use on electrical systems 1000 V and below.

In each instance, the qualified worker must determine that no voltage exists on each conductor to which he or she could be exposed. That determination must be made by measuring the voltage to ground and the voltage to all other conductors from any potential power source using a voltage detector rated for the maximum voltage available from any potential source of energy. The voltage-detecting device must be functionally tested before taking the measurement and then again after the measurement is taken to ensure that the device is working satisfactorily. See the three steps of Exhibit 120.4.

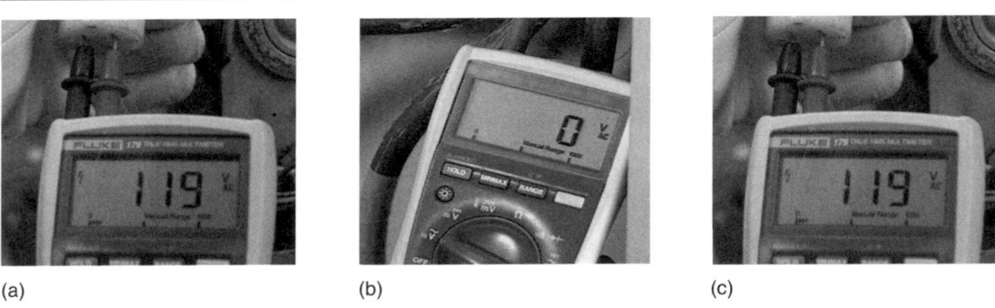


(a) (b) (c)

***EXHIBIT 120.4*** *Test procedure.*

A measurement must be made from each conductor to ground and between each conductor to each other conductor from a potential source of energy. Note that the voltage detector might indicate no voltage to ground in an ungrounded circuit.

The voltage-detecting device should be selected on the basis of the service. The rating of the device must be at least as great as the expected voltage. The device must have an adequate static discharge rating as discussed in ANSI/ISA 61010-1. See Exhibit 110.2 for an example of what to consider in a voltage measuring device.

**(6) Where the possibility of induced voltages or stored electrical energy exists, ground the phase conductors or circuit parts before touching them. Where it could be reasonably anticipated that the conductors or circuit parts being deenergized could contact other exposed energized conductors or circuit parts, apply ground connecting devices rated for the available fault duty.**

Electrical conductors sometimes break and fall onto another conductor installed at a lower elevation. For instance, more than one outside overhead line might be installed on a single pole or support structure. Physical damage to a higher conductor could cause it to fall onto a lower installed conductor and result in reenergizing the lower conductor.

A set of safety grounds is necessary to protect workers from potentially hazardous voltage, such as in the following accidental instances:

- A long conductor installed in proximity to the deenergized conductor could induce a hazardous voltage onto the otherwise deenergized conductor (magnetic coupling).
- Another worker could inadvertently connect a conductor that is not locked out and add a source of energy to a circuit under repair.

- Other situations that could reintroduce voltage to a conductor under repair could be identified by the hazard/risk analysis.

The hazard/risk analysis can identify other situations that could reintroduce voltage to a conductor under repair. Safety grounds provide protection only if the rating is sufficiently great to conduct any potentially available energy. Inadequately rated safety grounds introduce a hazard that otherwise would not exist. The rating must be established by the manufacturer or another generally accepted rating process. See Exhibit 120.5 for an example of an approved grounding assembly.

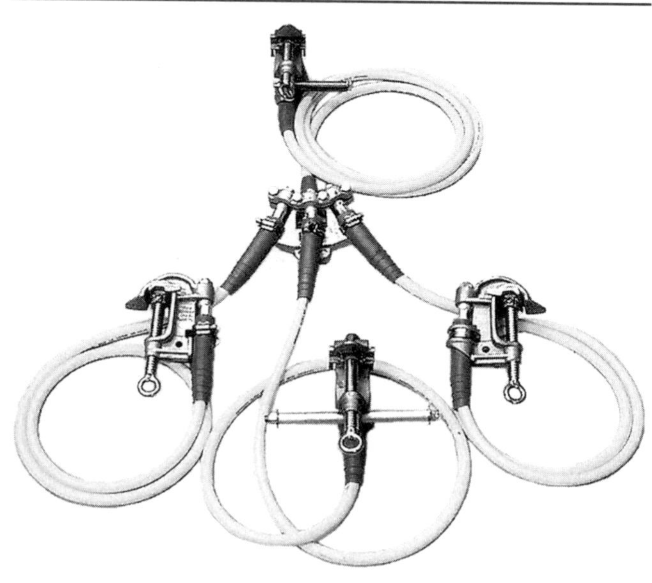

*EXHIBIT 120.5 An approved grounding assembly. (Courtesy of Salisbury Electrical Safety, LLC)*

**120.2 Deenergized Electrical Conductors or Circuit Parts That Have Lockout/Tagout Devices Applied.**

Each employer shall identify, document, and implement lockout/tagout procedures conforming to Article 120 to safeguard employees from exposure to electrical hazards. The lockout/tagout procedure shall be appropriate for the experience and training of the employees and conditions as they exist in the workplace.

The most effective way to prevent an electrical injury is to completely remove the source of electrical energy and eliminate the possibility of its reappearance. To do that, workers must identify all possible sources of electricity and locate the disconnecting means for each source. Electricity sometimes can appear from the load side. The worker must identify all potential sources of energy, eliminate them, visually verify (if possible) that they are eliminated, lock them out, test to verify absence of voltage, and ground the conductors or parts, if warranted. Only then does a worker establish an electrically safe work condition, which is a lifesaving process.

**(A) General.** All electrical circuit conductors and circuit parts shall be considered energized until the source(s) of energy is (are) removed, at which time they shall be considered deenergized. All electrical circuit conductors and circuit parts shall not be considered to be in an electrically safe work condition until all of the applicable requirements of Article 120 have been met.

FPN: See 120.1 for the six-step procedure to verify an electrically safe work condition.

Electrical conductors and circuit parts that have been disconnected, but not under lockout/tagout, tested, and grounded (where appropriate) shall not be considered to be in an electrically safe work condition, and safe work practices appropriate for the circuit voltage and energy level shall be used. Lockout/tagout requirements shall apply to fixed, permanently installed equipment, to temporarily installed equipment, and to portable equipment.

Lockout/tagout is only one step in the process of establishing an electrically safe working condition. An electrically safe work condition does not exist until *all* of the six steps have been completed. Until then, workers could contact an exposed conductor that is energized or exposed to a thermal hazard so appropriate PPE must be worn. Installing locks and tags does not ensure that electrical hazards have been removed. Workers must select and use work practices that are identical to working on or near exposed live parts until an electrically safe work condition has been established.

The lockout/tagout procedure must apply to all exposed live parts, regardless of whether they are temporary, permanent, or portable. When implementing the procedural requirements, workers must select and use work practices (including PPE) that are appropriate for the voltage and energy level of the circuit as if it were known to be energized.

A flash protection boundary exists until an electrically safe work condition is established. Workers within that boundary must wear arc-rated, flame-resistant protective clothing as well as protection for the worker's head, face, and hands as determined by the hazard/risk analysis. If it is necessary for a worker to penetrate the restricted approach boundary, the worker must be wearing shock-protective equipment that is rated at least as high as the full-circuit voltage.

If an electrically safe work condition exists, no electrical energy is in proximity of the work task(s). All danger of injury from an electrical hazard has been removed, and neither protective equipment nor special safety training is required. However, hazards associated with other energy sources are not changed by the electrically safe work condition.

**(B) Principles of Lockout/Tagout Execution.**

**(1) Employee Involvement.** Each person who could be exposed directly or indirectly to a source of electrical energy shall be involved in the lockout/tagout process.

Any worker who could be exposed to an electrical hazard when executing the work task must be involved in the lockout/tagout process. Temporary workers and contract workers also must understand how the lockout/tagout procedure modifies their exposure to electrical hazards and must participate in the process. When multiple employers are involved in the work process, such as when a contractor is involved, Section 110.5 requires each employer to share information about hazards and procedures. All workers who might be exposed to electrical hazards must be involved in the lockout/tagout process.

FPN: An example of direct exposure is the qualified electrician who works on the motor starter control, the power circuits, or the motor. An example of indirect exposure is the person who works on the coupling between the motor and compressor.

In most cases, electrical energy is converted to another form before it is used. Motors, electrical space heaters, lamps in lighting fixtures, and similar equipment convert electrical energy to another form. Workers who are exposed to the converted form of electrical energy are exposed *indirectly* to electrical hazards. Employees who are exposed to shock or arc flash are exposed *directly* to an electrical hazard. If workers are not exposed to an electrical hazard, the lockout/tagout procedure must comply with the provisions of OSHA 29 CFR 1910.147, "The Control of Hazardous Energy (Lockout/Tagout)."

**(2) Training.** All persons who could be exposed shall be trained to understand the established procedure to control the energy and their responsibility in executing the procedure.

New (or reassigned) employees shall be trained (or retrained) to understand the lockout/tagout procedure as related to their new assignment.

Each employer must provide training for all workers who might be involved in the process of establishing an electrically safe work condition. The training must develop understanding of how to avoid exposure to electrical hazards. The training should describe each step in the process and its importance.

When employees are reassigned to an area, each employee must understand the process of establishing an electrically safe work condition associated with his or her new assignment.

As employers, contractors must provide training for each employee who might be exposed to an electrical hazard. Contractor employees must understand that each facility might have unique characteristics. Contractors and facility owners must exchange information about creating an electrically safe work condition and ensure that their respective employees understand all issues that are important to the other employer.

Each person who is associated with executing the job must be trained to understand his or her role in the lockout/tagout process. Each person must understand and accept individual responsibility for the integrity of the lockout/tagout process. Employees who have been reassigned (either permanently or temporarily) must be retrained to understand their new role in the lockout/tagout procedure.

**(3) Plan.** A plan shall be developed on the basis of the existing electrical equipment and system and shall utilize up-to-date diagrammatic drawing representation(s).

A plan must be developed for each lockout/tagout. The planner might be the person in charge, the supervisor, or the worker. The plan must identify the location, both physically and electrically, that requires a lockout or tagout device. The plan must be based on up-to-date diagrammatic information.

The diagrammatic information can be a single-line diagram or a three-line diagram. The drawing can be hand marked to illustrate recent circuit modifications, provided the information accurately depicts the configuration of the circuit. The information on the drawing must match the information on the equipment labels. The plan should include identifying information that is found on both the drawing and the equipment label.

Single-line diagrams provide essential information when the worker is creating an electrically safe work condition, provided the drawing illustrates current information. If the diagram is inaccurate, however, workers could be injured as a result of one or more sources of energy that are not removed. Accurate information might be provided by a completed panel schedule or similar representation of circuit information. The purpose of the information is to clearly indicate all sources of electrical energy that are or might be available at any point in the electrical circuit. Regardless of the manner of recording this information, the information must be current. Information recorded on the drawing (or other record) must match the information on the equipment or device label.

**(4) Control of Energy.** All sources of electrical energy shall be controlled in such a way as to minimize employee exposure to electrical hazards.

All sources of energy must be removed by operating the disconnecting means. After the disconnecting means is opened, all employees who might be exposed to the electrical hazard should be involved in the process of installing devices that provide adequate assurance that the energy cannot be reapplied. In most instances, locks provide that assurance, whereas tags might become illegible or otherwise ineffective.

**(5) Identification.** The lockout/tagout device shall be unique and readily identifiable as a lockout/tagout device.

Devices used for control of energy or control of exposure to electrical energy must not be used for any other purpose. The device must be physically different from any other lock or similar devices that are used by either the facility owner or by any contractor working at the site.

Workers and supervisors must be able to recognize a lockout/tagout device by sight. There must be no possibility of confusing lockout/tagout devices with locks or tags used for other purposes. For instance, information tags and process control locks must not be confused with tags or locks used for lockout/tagout. Exhibit 120.6 shows an example of a lockout station containing locks, tags, and devices.

***EXHIBIT 120.6*** *A lockout station. (Courtesy of Salisbury Electrical Safety, LLC)*

**(6) Voltage.** Voltage shall be removed and absence of voltage verified.

When the source of energy has been removed by operating all disconnecting means, workers must verify that no voltage is present in the vicinity of the work location. No source of electrical energy can be present within the Restricted Approach Boundary if the worker(s) is (are) qualified employees. If the workers are not qualified persons, no source of energy can be present within the Limited Approach Boundary.

When verifying absence of voltage, workers must be qualified to perform the test for absence of voltage and wearing the necessary protective equipment. The testing device can take several forms; however, the integrity of the testing device must be verified before and after the voltage test. (See Exhibit 120.4.)

**(7) Coordination.** The established electrical lockout/tagout procedure shall be coordinated with all of the employer's procedures associated with lockout/tagout of other energy sources.

Some projects involve contractors and employers other than the facility or equipment owner. Since each employer is required to implement a lockout/tagout procedure, different requirements can exist. To ensure that the requirements of each procedure are observed, each employer might need to add requirements to one or more procedures.

All employers must attend a coordination meeting and review and discuss each procedural requirement that is more or less restrictive than other employers' lockout/tagout procedure. The result of the meeting should be that all lockout/tagout procedures in effect on the project are coordinated with each other. The meeting must include the facility owner and all contractors who might be working in the physical area associated with the project. Minutes of this meeting should be kept and distributed to all attendees to ensure proper understanding.

Each employer must audit at least one work task where lockout/tagout has been applied.

The audit must determine if all requirements of the lockout/tagout procedure were observed. The audit also must determine if the requirements contained in the published procedure are sufficient to ensure that the electrical energy is satisfactorily controlled.

**(C) Responsibility.**

Although employees, employers, and facility owners could be assigned different responsibilities by this and other standards, the most effective lockout processes are the result of collaboration between and among all participants in the process.

**(1) Procedures.** The employer shall establish lockout/tagout procedures for the organization, provide training to employees, provide equipment necessary to execute the details of the procedure, audit execution of the procedures to ensure employee understanding/compliance, and audit the procedure for improvement opportunity and completeness.

Each employer must generate, write, and publish a lockout/tagout procedure that applies to all work involving exposure, or potential exposure, to an electrical hazard. If a tool or equipment is required to implement the details of the procedure, the employer must purchase and supply the tool(s) or equipment in sufficient quantity.

Each employer must train every employee to understand his or her role in implementing the details of the procedure. The training must ensure that each employee understands the enforcement aspects of respecting an installed lockout or tagout device. Each employer is responsible for initiating audits, as necessary, to ensure that the lockout/tagout procedure is effective and that no changes are warranted. The audit must determine if the requirements of the published procedure are being implemented.

**(2) Form of Control.** Three forms of hazardous electrical energy control shall be permitted: individual employee control, simple lockout/tagout, and complex lockout/tagout. *[See 120.2(D).]* For the individual employee control and the simple lockout/tagout, the qualified person shall be in charge. For the complex lockout/tagout, the person in charge shall have overall responsibility.

The basic idea in controlling exposure to electrical energy is to ensure that all possible sources of electrical energy are disconnected and cannot reappear unexpectedly. In some instances, workers might not understand the details of an installed electrical circuit, either the circuitry or the location of electrical equipment. Consequently, the workers depend on the integrity of the information provided to them. Each employer is expected to assign to one person (person in charge) the responsibility of determining the integrity and completeness of each lockout/tagout. The person in charge need not be the same individual for all sites or for all lockouts or tagouts. Instead, the person in charge could change with each application of the procedure.

When a failure occurs in the lockout or tagout process, the failure is likely to be the result of inadequate communication between or among participants in the process of establishing an electrically safe work condition. To avoid the chance of inadequate communication, a single person should be charged with the responsibility of ensuring that the integrity of the lockout process is completed and that all workers are removed from the risk of injury before releasing the equipment or circuit for use after the work task is completed. The lockout/tagout procedure should be as simple as possible to enable workers to understand the reasons for each requirement. For instance, if a worker is troubleshooting an individually mounted disconnect switch that is directly in front of that worker, nobody else could operate the switch handle without the worker's knowledge. However, if the worker moves from the position directly in front of the switch, the handle could be operated without his or her knowledge. In this illustration, the worker would act as the person in charge. This would be considered an individual employee control.

Sometimes a single disconnect switch provides energy for a device such as a motor. A worker who intends to remove the motor should operate the disconnecting means for the motor. Since the potential hazard would not be continually in view, the worker should install a lockout device on the disconnect switch. Again, the worker would act as the person in charge and this would be considered a simple lockout/tagout.

In other instances, implementing the requirements of the lockout/tagout procedure can be complex. Adding any additional factor, such as more workers or multiple sources of energy, increases the complexity of the lockout/tagout. As the complexity of the lockout/tagout increases, the need to understand the electrical circuit increases. A single person in charge can improve the chance of accurately locating and controlling all potential sources of electrical energy. This would be considered a complex lockout/tagout. Although tagout is a recognized method of energy control, the integrity of the energy control is much greater if lockout devices are used.

FPN: For an example of a lockout/tagout procedure, see Annex G.

**(3) Audit Procedures.** An audit shall be conducted at least annually by a qualified person and shall cover at least one lockout/tagout in progress and the procedure details. The audit shall be designed to correct deficiencies in the procedure or in employee understanding.

In 120.2(C)(3), an annual audit of lockout/tagout procedures is required to locate and correct any deficiencies in the procedure or in employee understanding of those procedures. Each employer must audit at least one work task where lockout/tagout has been applied. The audit must determine if all requirements of the lockout/tagout procedure were observed. The audit also must determine if the requirements contained in the published procedure are sufficient to ensure that the electrical energy is satisfactorily controlled. The objective of the audit should be to identify and correct any weaknesses (or potential weaknesses) in the procedure, in employee training, or in enforcement of the requirements.

**(D) Hazardous Electrical Energy Control Procedure.**

**(1) Individual Qualified Employee Control Procedure.** The individual qualified employee control procedure shall be permitted when equipment with exposed conductors and circuit parts is deenergized for minor maintenance, servicing, adjusting, cleaning, inspection, operating conditions, and the like. The work shall be permitted to be performed without the placement of lockout/tagout devices on the disconnecting means, provided the disconnecting means is adjacent to the conductor, circuit parts, and equipment on which the work is performed, the disconnecting means is clearly visible to the individual qualified employee involved in the work, and the work does not extend beyond one shift.

When a worker is working on or within the enclosure of a disconnecting means with a single source of energy, such as replacing a motor fuse, he or she does not need to install a lockout/tagout device. The worker is physically positioned to ensure that he or she is in control of the operating mechanism of the switch, provided the worker does not leave the front of the switch. However, if he or she turns away from the switch or leaves the area, the lockout/tagout must shift from an individual qualified employee control procedure to a simple lockout/tagout, and a lockout/tagout device becomes necessary.

Some equipment, such as a disconnect switch, cannot be opened with a lock installed. If the work task is to measure voltage or change a fuse in this disconnecting means, a lockout device cannot be installed. The purpose of the individual qualified employee control procedure is to permit tasks such as these without a lockout/tagout device. It is important to note that the individual qualified employee control procedure suggests that the exposed conductor be continuously within sight and within arm's reach.

**(2) Simple Lockout/Tagout Procedure.** All lockout/tagout procedures that are not under individual qualified employee control according to 120.2(D)(1) or complex lockout/tagout according to 120.2(D)(3) shall be considered to be simple lockout/tagout procedures. All lockout/tagout procedures that involve only a qualified person(s) deenergizing one set of conductors or circuit part source for the sole purpose of performing work within the Limited Approach Boundary electrical equipment shall be considered to be a simple lockout/tagout. Simple lockout/tagout plans shall not be required to be written for each application. Each worker shall be responsible for his or her own lockout/tagout.

Some work tasks involve only a single source of electrical energy and a single disconnecting means. If the disconnecting means is capable of accepting a lockout/tagout device, the lockout/tagout can be considered to be a simple lockout/tagout procedure and no written lockout/tagout plan is necessary. A simple lockout/tagout must be a planned activity, although a written plan is not always necessary. Exhibit 120.7 shows a motor disconnect that has been locked open and tagged.

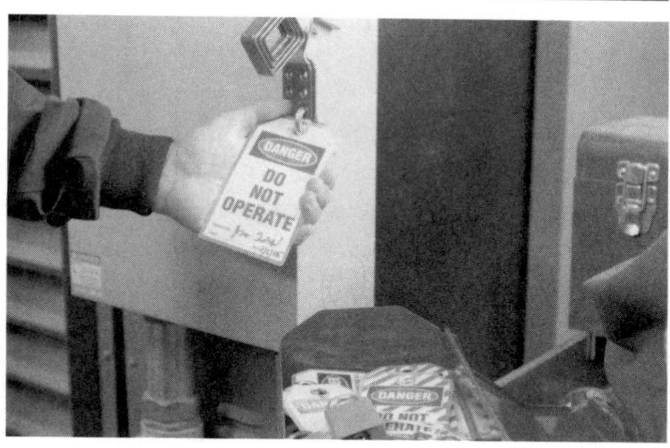


*EXHIBIT 120.7 A disconnect switch that is locked open and tagged.*

**(3) Complex Lockout/Tagout Procedure.**

(a) A complex lockout/tagout plan shall be permitted where one or more of the following exist:

(1) Multiple energy sources
(2) Multiple crews
(3) Multiple crafts
(4) Multiple locations
(5) Multiple employers
(6) Different disconnecting means
(7) Particular sequences
(8) A job or task that continues for more than one work period

The conditions listed in 120.2(D)(3) are known to increase the complexity and difficulty of a lockout/tagout. If one or more of these conditions exist, the lockout/tagout is defined as complex, and a person in charge is necessary to avoid difficulty in communication.

(b) All complex lockout/tagout procedures shall require a written plan of execution that identifies the person in charge.

Each complex lockout/tagout must be under the direct control of a single person in charge who is identified in a written plan. The person in charge must be assigned and must accept the responsibility of ensuring that an electrically safe work condition is established before any work task associated with the job can begin. The person in charge also must accept the responsibility of ensuring that all people who are assigned to the job are accounted for before the electrically safe work condition is removed.

As the degree of complexity increases, eliminating the risk of injury becomes more difficult and requires greater formality in the method of communication between and among workers and supervisors. A written plan of execution that details the process of controlling exposure to electrical energy is necessary. After the written plan has been generated, reviewed, and accepted, all persons associated with the complex lockout/tagout should be advised of the plan's content, their role in executing the plan, the name of the person in charge, and all interaction with any other employer, crew, or craft.

A written plan must identify each step required to install lockout and tagout devices. To clearly establish the authority of the person in charge, the plan must be reviewed with or by all workers. The plan must identify the following:

- The disconnecting means
- Who will install lockout/tagout devices
- How the absence of voltage will be verified
- How employees will be accounted for before, during, and after the work is complete

**(c) The complex lockout/tagout procedure shall vest primary responsibility in an authorized employee for a set number of employees working under the protection of a group lockout or tagout device (such as an operation lock). The person in charge shall be held accountable for safe execution of the complex lockout/tagout.**

The person in charge must be both a qualified person and an authorized employee. Additional employees often are assigned to an area to provide sufficient manpower to complete all necessary work tasks while a shutdown is under way. Contract employees or other temporary employees might be unfamiliar with the location of electrical circuits and disconnecting means. The person in charge is responsible for ensuring that no employee, including each temporary employee, is unnecessarily exposed to an electrical hazard.

The person in charge must understand that he or she is accountable for generating, implementing, and monitoring the implementation of the plan.

**(d) Each authorized employee shall affix a personal lockout or tagout device to the group lockout device, group lockbox, or comparable mechanism when he or she begins work, and shall remove those devices when he or she stops working on the machine or equipment being serviced or maintained.**

Some projects involve employers other than the facility owner, such as contractors. Since each employer is required to implement a lockout/tagout procedure, different requirements might exist. To ensure that the requirements of each procedure are observed, the requirements of one or more procedures might need to have additional requirement(s). The person in charge must ensure that the basic concerns of each employer are addressed in the written lockout/tagout plan.

To adhere to the basic principle of each worker being in control of the hazardous energy, each employee involved in the work task must affix his or her personal lockout device to the group lockout device or to each individual lockout point. The person in charge is responsible to ensure adherence to this principle. See Exhibits 120.8, 120.9, and 120.10 for examples of complex lockout/tagout devices.

**EXHIBIT 120.8** *Complex lockout/tagout devices.*

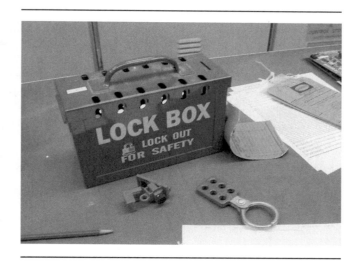

**EXHIBIT 120.9** *Lockbox.*

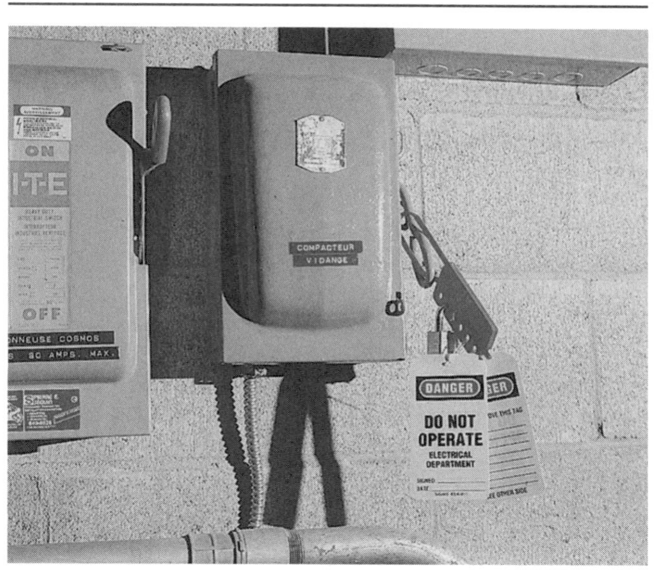


**EXHIBIT 120.10** *Disconnect switch with locks from two individuals. (Courtesy of Salisbury Electrical Safety, LLC)*

(e) The complex lockout/tagout procedure shall address all the concerns of employees who might be exposed. All complex lockout/tagout plans shall identify the method to account for all persons who might be exposed to electrical hazards in the course of the lockout/tagout.

The lockout/tagout procedure and plan must identify the method that will be used to account for all employees involved with the work. The person in charge is responsible for ensuring that all workers are protected by the complex lockout/tagout. However, each worker must understand the process by which the person in charge will implement this responsibility.

**(4) Coordination.**

(a) The established electrical lockout/tagout procedure shall be coordinated with all other employer's procedures for control of exposure to electrical energy sources such that all employer's procedural requirements are adequately addressed on a site basis.

Some projects involve contractors and employers other than the facility owner. Each employer is required to implement a lockout/tagout procedure, so different requirements can exist. To ensure that the requirements of each procedure are observed, each employer might need to add requirements to one or more procedures.

When the lockout/tagout procedure for different employers must be implemented on the same worksite, the procedures must be coordinated with each other and employees of each employer must be instructed about any unique aspect. Employees must understand the change and the necessity for the change.

(b) The procedure for control of exposure to electrical hazards shall be coordinated with other procedures for control of other hazardous energy sources such that they are based on similar/identical concepts.

Other standards require an employer to implement a "control of hazardous energy procedure" that covers all sources of energy *except* electrical energy. The electrical lockout/tagout procedure must be coordinated with any other control of hazardous energy procedure to ensure that the requirements of each procedure have a similar basis and similar requirements. For instance, the general control of hazardous energy procedure and the electrical lockout/tagout procedure must have similar or identical requirements for locks and tags.

The electrical lockout/tagout procedure could be integrated into an overall control of hazardous energy procedure; however, that procedure must address all the issues identified in this standard.

(c) The electrical lockout/tagout procedure shall always include voltage testing requirements where there might be direct exposure to electrical energy hazards.

The published lockout/tagout procedure must contain the requirements necessary to ensure that all employees know whether they are exposed to an electrical hazard. The procedure should identify acceptable voltage-testing devices and contain a requirement to ensure that the voltage-testing device is functioning properly, both before and after each use. Employee training must ensure that each qualified employee is familiar with the requirements for testing voltage. See Exhibits 110.2 and 110.3.

Employees who use a voltage-detecting device must understand how to use the device, how to protect themselves from any associated hazard, and how to interpret all possible indications provided by the voltage-detecting device.

(d) Electrical lockout/tagout devices shall be permitted to be similar to lockout/tagout devices for control of other hazardous energy sources, such as pneumatic, hydraulic, thermal, and mechanical, provided such devices are used only for control of hazardous energy and for no other purpose.

Devices used for control of hazardous energy must be easily recognizable. Lockout/tagout devices that are used to control exposure to electrical energy should have the same physical characteristics as devices used for control of other forms of energy so that workers are not confused.

**(5) Training and Retraining.** Each employer shall provide training as required to ensure employees' understanding of the lockout/tagout procedure content and their duty in executing such procedures.

Employees who work at different job assignments must understand how their personal safety is impacted by the lockout/tagout. Each employee must accept as their duty the principle of helping their coworkers avoid injury. Workers that are temporarily reassigned to increase the number of workers available for a specific job or work task must be trained to understand their role in implementing and maintaining the lockout/tagout. Each worker must understand that adhering to the requirements of the published procedure can affect the safety of other workers involved in the work process.

**(E) Equipment.**

**(1) Lock Application.** Energy isolation devices for machinery or equipment installed after January 2, 1990, shall be capable of accepting a lockout device.

OSHA requires that energy-isolating devices installed after January 2, 1990, be capable of receiving a lock. Disconnecting means that are used as energy-isolating devices must be capable of being locked in the open position. An energy-isolating device might be "load-rated" and be used to break an operating electrical circuit without significant damage. On the other hand, if an energy-isolating device is not load-rated, the device might destroy itself if used to break the circuit of operating equipment. In either case, the disconnecting means must be capable being locked in the open position. Any disconnecting means that cannot be locked in the open (or disconnected) position must not be used as an energy-isolating device.

**(2) Lockout/Tagout Device.** Each employer shall supply, and employees shall use, lockout/tagout devices and equipment necessary to execute the requirements of 120.2(E). Locks and tags used for control of exposure to electrical energy hazards shall be unique, shall be readily identifiable as lockout/tagout devices, and shall be used for no other purpose.

Employers must provide the necessary equipment, including locks, tags, chains, and so on, for employees to use to control exposure to electrical hazards.

Lockout/tagout devices for control of exposure to electrical energy can be identical to lockout/tagout devices used for the control of hazardous energy from other energy sources. They must be unique and readily identifiable, such as having a unique color or some other easily recognizable characteristic. Lockout/tagout devices used for *control of exposure* to electrical energy or for *control of energy* of other kinds must not be used for any other purpose.

**(3) Lockout Device.**

(a) A lockout device shall include a lock (either keyed or combination).

(b) The lockout device shall include a method of identifying the individual who installed the lockout device.

(c) A lockout device shall be permitted to be only a lock, provided the lock is readily identifiable as a lockout device, in addition to having a means of identifying the person who installed the lock.

A lockout device might include a tag, chain, cable tie, or other component; however, lockout devices are intended to mean "a lock." The lock may be used in conjunction with other components, but the basic lockout device is a lock. The lock can be operated with either a key or a combination. That key (or combination) must prevent unauthorized removal of the lock, and the lock's installer must be in control of the key (or combination).

The lockout device must include information that identifies the person who installed the lock and be installed in a manner that prevents operation of the energy-isolating device. The lockout device also must contain information suggesting disciplinary action for unauthorized removal.

(d) Lockout devices shall be attached to prevent operation of the disconnecting means without resorting to undue force or the use of tools.

(e) Where a tag is used in conjunction with a lockout device, the tag shall contain a statement prohibiting unauthorized operation of the disconnecting means or unauthorized removal of the device.

(f) Lockout devices shall be suitable for the environment and for the duration of the lockout.

Lockout devices, including all necessary components, must have sufficient durability to withstand the environment in which they are used. Any marking installed by the worker must withstand the environment for the duration of the lockout also.

(g) Whether keyed or combination locks are used, the key or combination shall remain in the possession of the individual installing the lock or the person in charge, when provided by the established procedure.

One basic premise of lockout requirements is that the person who might be exposed to an electrical hazard is in control of the electrical energy. To accomplish that purpose, the key to the lock must remain in the possession of the person who installed it.

**(4) Tagout Device.**

(a) A tagout device shall include a tag together with an attachment means.

(b) The tagout device shall be readily identifiable as a tagout device and suitable for the environment and duration of the tagout.

(c) A tagout device attachment means shall be capable of withstanding at least 224.4 N (50 lb) of force exerted at a right angle to the disconnecting means surface. The tag attachment means shall be nonreusable, attachable by hand, self-locking, and nonreleasable, equal to an all-environmental tolerant nylon cable tie.

(d) Tags shall contain a statement prohibiting unauthorized operation of the disconnecting means or removal of the tag.

*Exception to (a), (b), and (c): A "hold card tagging tool" on an overhead conductor in conjunction with a hotline tool to install the tagout device safely on a disconnect that is isolated from the worker(s).*

The term *tagout device* is intended to mean a tag and other equipment necessary to attach the complete assembly to an energy-isolating device. The tagout device must be unique and easily recognizable as a tagout device. The device must be attached to the energy-isolating device with a component that ensures that the tag stays in place. The tagout device must contain information suggesting disciplinary action for unauthorized removal.

Tagout is deemed to be an acceptable method of indicating that a worker is working on a circuit and that the energy-isolating device must not be operated. In an industrial or commercial environment, lockout is a more positive method to ensure that an energy-isolating device will not be operated while workers could be exposed. Tagout must not be used if it is possible to install a lockout device. As of January 2, 1990, all new energy-isolating devices must be capable of accepting a lock.

For many years, utility systems have successfully relied on a "hold card" to provide warning that operating a disconnecting means would place workers in danger. The work environment for utility workers means that the location of the disconnecting means might be several miles away from the work site. Utility workers are trained to respect the system associated with the hold card, which results in a positive and effective system of energy control. Employees of utility systems must be covered by a written policy that describes how the hold card system functions. When contractor employees perform utility maintenance or construction, the contractor must provide a program that is at least as effective as the program of the utility authorizing the contractor's work.

**(5) Electrical Circuit Interlocks.** Up-to-date diagrammatic drawings shall be consulted to ensure that no electrical circuit interlock operation can result in reenergizing the circuit being worked on.

Any drawing that is used as a reference to locate or identify equipment or circuits that impact the safety of workers must be current. In many instances, a diagrammatic drawing, such as a single-line diagram, is used to determine the possibility for a circuit to be reenergized from backfeeds, either directly or from a "sneak" circuit. The drawing must be up-to-date, and it must accurately depict the installation.

As an installed circuit or system changes, record single-line diagrams, schematic diagrams or similar drawings should be marked to record the change in the system. The record drawing then must be the document used for reference purposes until the file is updated with the new information.

**(6) Control Devices.** Locks/tags shall be installed only on circuit disconnecting means. Control devices, such as pushbuttons or selector switches, shall not be used as the primary isolating device.

A control device cannot be directly installed in the circuit providing energy to a utilization device, such as a motor. Normally, the "pushbutton" operates a contactor or relay that then closes the energy delivery circuit. A computer or other electronic signal can serve as an input device to accomplish the same purpose. Although a control device can be fitted with a mechanism intended to accept a lock, the control device cannot ensure that the energy delivery circuit remains in a safe condition.

OSHA 29 CFR 1926.417 is often interpreted that tags can be installed on control devices. However, 29 CFR 1926.417 also contains a statement saying that the circuit must be rendered inoperative. Control devices do not render a circuit inoperative; therefore, installing a tag (or a lock) on a control device simply does not meet the objective of avoiding injury.

If the device being locked or tagged does not create a break in the conductors providing energy to the equipment or circuit, the device must not be used as a lockout or tagout point.

**(F) Procedures.** The employer shall maintain a copy of the procedures required by this section and shall make the procedures available to all employees.

Each employer must generate, document, and publish a lockout/tagout procedure that addresses control of exposure to electrical energy for the organization. The procedure must be in writing and made available to all employees. Although an employer is responsible for providing the procedure, employees should be involved in gathering the information to produce the procedure. The lockout/tagout procedure must define the sequential steps necessary to complete the task. The procedure required by this section must be implemented as one step in the process of establishing an electrically safe work condition. The lockout/tagout procedure can be included in an overall lockout/tagout procedure for an employer or site, or it can be a standalone procedure. In either instance, all the requirements of this section must be addressed.

**(1) Planning.** The procedure shall require planning, including 120.2(F)(1)(a) through 120.2(F)(2)(n).

The procedure must include a requirement for planning, and it must indicate if the plan must be in writing and if any authorization is necessary.

The procedure must contain a requirement for a plan to exist whenever a lockout or tagout is implemented. That plan must consider the information in Section 120.2(F)(1)(a) through (F)(1)(c).

(a) Locating Sources. Up-to-date single-line drawings shall be considered a primary reference source for such information. When up-to-date drawings are not available, the employer shall be responsible for ensuring that an equally effective means of locating sources of energy is employed.

Locating all possible sources of electrical energy accurately and completely is crucial. Section 120.2(F)(1)(a) is intended to apply to those times when up-to-date drawings are not available. For instance, small businesses might not be able to maintain drawings easily. Those employers must provide an alternative means of locating all possible sources of electrical energy accurately and completely. If changes to the electrical system are unlikely, the employer could prepare a document that identifies all possible sources of energy for each system component and make the list available at each workstation.

(b) Exposed Persons. The plan shall identify persons who might be exposed to an electrical hazard and the personal protective equipment required during the execution of the job or task.

The plan must identify who could be exposed to an electrical hazard while the work task is being executed. The plan must recognize that implementing a lockout or tagout could expose one or more workers to an electrical hazard. The plan must identify what PPE workers must use while workers implement the lockout or tagout. On occasion, operating a disconnecting means or opening a door to verify absence of voltage results in a failure that creates a faulted condition. The plan must address this possibility.

(c) Person In Charge. The plan shall identify the person in charge and his or her responsibility in the lockout/tagout.

A person in charge must be identified by name or position for each complex lockout/tagout in the plan. Each person involved in the work task must be told the name of the person in charge as well as his or her responsibilities.

(d) Individual Qualified Employee Control. Individual qualified employee control shall be in accordance with 120.2(D)(1).

The procedure must identify the conditions under which individual qualified employee control may be used.

(e) Simple Lockout/Tagout. Simple lockout/tagout procedure shall be in accordance with 120.2(D)(2).

The procedure must identity the conditions that describe a simple lockout/tagout and indicate who is authorized to implement a simple lockout/tagout.

(f) Complex Lockout/Tagout. Complex lockout/tagout procedure shall be in accordance with 120.2(D)(3).

The procedure must identify the conditions that require the lockout/tagout to be considered complex. The procedure must describe the specific steps associated with implementing a complex lockout and define a requirement for a person in charge.

**(2) Elements of Control.** The procedure shall identify elements of control.

(a) Deenergizing Equipment (Shutdown). The procedure shall establish the person who performs the switching and where and how to deenergize the load.

The procedure must identify who is authorized to operate a disconnecting means to deenergize the equipment. The procedure must indicate if or when arc-rated FR clothing is necessary. The procedure must describe the conditions under which the worker must remove the load current from the circuit before operating the disconnecting means.

If a different employer owns the equipment, such as if the work is being executed by a contractor, the plan must describe any limit of authority for workers. For instance, if contract employees are authorized to operate the disconnecting means or if contract employees are not authorized to open a door to verify absence of voltage, the procedure must describe those limits of authority.

(b) Stored Energy. The procedure shall include requirements for releasing stored electric or mechanical energy that might endanger personnel. All capacitors shall be discharged, and high capacitance elements shall also be short-circuited and grounded before the associated equipment is touched or worked on. Springs shall be released or physical restraint shall be applied when necessary to immobilize mechanical equipment and pneumatic and hydraulic pressure reservoirs. Other sources of stored energy shall be blocked or otherwise relieved.

Stored energy can exist in many different forms. Electrical energy can be stored in capacitors or other capacitive elements, such as a long cable. Stored energy, in all its forms, must be relieved or blocked. Electrical equipment frequently contains pneumatic pressure components, which are sources of stored energy that must be relieved.

In some instances, electrical switches and other disconnecting means contain springs. The plan must describe if or how the kinetic energy must be blocked.

(c) Disconnecting Means. The procedure shall identify how to verify that the circuit is deenergized (open).

In some cases, verifying that the disconnecting means is open is a simple matter of opening a door and observing the position of the contacts in the disconnecting means. In other cases, the contacts are not readily visible. When direct observation of the contacts is not possible, workers must be instructed about how to determine that all power conductors are disconnected from the source of energy.

(d) Responsibility. The procedure shall identify the person who is responsible to verify that the lockout/tagout procedure is implemented and who is responsible to ensure that the task is completed prior to removing locks/tags. A mechanism to accomplish lockout/tagout for multiple (complex) jobs/tasks where required, including the person responsible for coordination, shall be included.

One person must be assigned the responsibility of making certain that the requirements of the lockout/tagout procedure are implemented. The procedure also should define the person responsible for verifying that the job or task is complete before the locks and tags are removed.

The procedure must define the process for executing a complex lockout/tagout. If more than one employer has work on the site, or if multiple procedures apply for any reason, the procedure must define the process for achieving coordination between or among the procedures.

Workers must have a complete understanding of their role in implementing the lockout/tagout. Significant injuries can occur if workers exceed the limit of their responsibility. Fulfilling responsibility without exceeding the associated authority is key to successful application of each lockout/tagout.

(e) Verification. The procedure shall verify that equipment cannot be restarted. The equipment operating controls, such as pushbuttons, selector switches, and electrical interlocks, shall be operated or otherwise it shall be verified that the equipment cannot be restarted.

Verifying that the correct disconnecting means have been operated to remove the source of energy is an important step in lockout/tagout. In 29 CFR 1910.147, OSHA requires that the control device be operated to ensure that the equipment cannot operate after locks or tags have been applied. However, operating control devices, such as pushbuttons, is not adequate to verify absence of voltage. Additional measures are necessary.

(f) Testing. The procedure shall establish the following:

(1) What voltage detector will be used, the required personal protective equipment, and who will use it to verify proper operation of the voltage detector before and after use
(2) A requirement to define the boundary of the work area
(3) A requirement to test before touching every exposed conductor or circuit part(s) within the defined boundary of the work area
(4) A requirement to retest for absence of voltage when circuit conditions change or when the job location has been left unattended
(5) Where there is no accessible exposed point to take voltage measurements, planning considerations shall include methods of verification.

Many injuries result from inadequate testing for absence of voltage. The procedure or plan must specifically identify the following:

- The testing device to be used
- Who will use the testing device
- The boundary of the safe zone established by the lockout/tagout

The procedure and/or plan also must define the following:

- A requirement for testing every conductor every time before a person touches them
- A requirement to retest for absence of voltage each time the worker leaves the work area for any reason and for any length of time
- The mechanism to be used to determine that the conductor is, in fact, deenergized if no points are exposed to test for voltage

If the work task is associated with a potential exposure to an electrical hazard but has no exposed point for measuring voltage, the procedure or plan must identify the method for verifying that no voltage exists before the work task is completed. If the work task is to cut an existing insulated conductor or cable, for instance, the procedure or plan must provide instructions for the worker.

(g) Grounding. Grounding requirements for the circuit shall be established, including whether the grounds shall be installed for the duration of the task or temporarily are established by the procedure. Grounding needs or requirements shall be permitted to be covered in other work rules and might not be part of the lockout/tagout procedure.

The procedure or plan must consider all possible ways that a voltage could reappear in the vicinity of the point of the work. The plan must define any requirement for any temporary grounds or safety grounds. If other work rules consider and address conductor grounding, the lockout/tagout procedure or plan is not required to address the same issue.

An employer's procedure could establish a simple rule that all work tasks associated with any voltage source more than 250 volts requires safety grounds. Establishing general rules aids a worker who is attempting to determine if a safety ground is necessary.

When safety grounds are used, workers must verify that the rating of the device equals or exceeds the available fault current at the point where the ground set is installed. Workers must ensure that the attaching hardware is also adequate.

(h) Shift Change. A method shall be identified in the procedure to transfer responsibility for lockout/tagout to another person or to the person in charge when the job or task extends beyond one shift.

If a second crew continues work at the end of one shift, all responsibilities must transfer to different people. The person in charge of the first shift must transfer his or her responsibility to a person in the oncoming shift. Workers who installed locks and tags should remove their locks and tags, and the oncoming workers must install their own personal locks and tags.

The procedure or plan must define the method to be followed in transferring responsibility from one person in charge to another. The procedure must define any requirement for workers to remove locks and tags or to replace them should they be required to continue work on the task when they return to work again. See the Case Study on p. 79.

(i) Coordination. The procedure shall establish how coordination is accomplished with other jobs or tasks in progress, including related jobs or tasks at remote locations, including the person responsible for coordination.

When more than one task or job is being executed in the same physical or electrical circuit area, the actions of one employee might have an impact on the actions of another employee. One person must be assigned responsibility for ensuring that work tasks by different trades or contractors are coordinated to minimize the possibility of one person's actions having a negative impact on another.

(j) Accountability for Personnel. A method shall be identified in the procedure to account for all persons who could be exposed to hazardous energy during the lockout/tagout.

The procedure or plan must define the method that will be used to account for all workers when the work task is complete or at the end of the shift. The person in charge is usually assigned to account for everyone involved in the work task. When contract workers are involved with the work task in conjunction with workers from another employer, a single person in charge is necessary.

(k) Lockout/Tagout Application. The procedure shall clearly identify when and where lockout applies, in addition to when and where tagout applies, and shall address the following:

The procedure must define when lockout is acceptable and when tagout is acceptable. If tagout is permitted, the employer should be able to justify to a third party that a lockout device could not be installed. All disconnecting means installed since January 2, 1990, are required to accept a lockout device.

If tagout is permitted, the procedure or plan must define clearly and unambiguously individual responsibility and accountability for each person potentially exposed to an electrical hazard.

Electrical utilities have successfully used a version of tagout for many years. The work environment that exists in transmission and distribution lines is significantly different from that on an industrial facility. Disconnecting means are generally located several miles from the point of work. Typically, utility crews communicate directly with a dispatcher who installs a "hold" card on the disconnecting means. Lockout/tagout requirements for utilities are discussed in OSHA 29 CFR 1910.269, Subpart R.

(1) Lockout shall be defined as installing a lockout device on all sources of hazardous energy such that operation of the disconnecting means is prohibited and forcible removal of the lock is required to operate the disconnecting means.

When a disconnecting means has a lock installed correctly, the disconnecting means cannot be operated. Forcible removal of a lock must be prohibited by the employer's procedure. Unauthorized removal of the lock should result in dismissal of the worker.

(2) Tagout shall be defined as installing a tagout device on all sources of hazardous energy, such that operation of the disconnecting means is prohibited. The tagout device shall be installed in the same position available for the lockout device.

When tagout is used, the tagout device must be easily visible to workers. The device must be easily recognizable and installed in the same position that a lockout device would otherwise be installed. Tagout devices must be suitable for the environment in which they are used.

(3) Where it is not possible to attach a lock to existing disconnecting means, the disconnecting means shall not be used as the only means to put the circuit in an electrically safe work condition.

Any disconnecting means that does not accept a lockout device must not be the only means of controlling the energy source. When tagout is the authorized method of controlling exposure to electrical energy, the tagout must be supplemented by at least one additional safety measure.

(4) The use of tagout procedures without a lock shall be permitted only in cases where equipment design precludes the installation of a lock on an energy isolation device(s). When tagout is employed, at least one additional safety measure shall be employed. In such cases, the procedure shall clearly establish responsibilities and accountability for each person who might be exposed to electrical hazards.

Tagout must not be selected as the energy control measure unless the disconnecting means does not accept a lock. Any equipment installed after January 2, 1990, must accept a lockout device. If tagout is the selected energy control measure, the employer's procedure must clearly identify the acceptable measures that might be used as the additional safety measure.

FPN: Examples of additional safety measures include the removal of an isolating circuit element such as fuses, blocking of the controlling switch, or opening an extra disconnecting device to reduce the likelihood of inadvertent energization.

(l) Removal of Lockout/Tagout Devices. The procedure shall identify the details for removing locks or tags when the installing individual is unavailable. When locks or tags are removed by someone other than the installer, the employer shall attempt to locate that person prior to removing the lock or tag. When the lock or tag is removed because the installer is unavailable, the installer shall be informed prior to returning to work.

The person who installed the lockout/tagout devices must remove them. In an emergency situation, however, if that person is not available, a supervisory member of the line organization may remove the lockout/tagout device provided an attempt to locate the person fails and removing the lockout/tagout device cannot expose a person to an electrical hazard. The procedure must describe all steps that must be taken before declaring that an emergency need to remove the lockout device exists. If a member of the line organization removes the lockout/tagout device, the person who installed the device must be informed that the lockout/tagout device has been removed before he or she returns to work.

(m) Release for Return to Service. The procedure shall identify steps to be taken when the job or task requiring lockout/tagout is completed. Before electric circuits or equipment are reenergized, appropriate tests and visual inspections shall be conducted to verify that all tools, mechanical restraints and electrical jumpers, shorts, and grounds have been removed, so that the circuits and equipment are in a condition to be safely energized. Where appropriate, the employees responsible for operating the machines or process shall be notified when circuits and equipment are ready to be energized, and such employees shall provide assistance as necessary to safely energize the circuits and equipment. The procedure shall contain a statement requiring the area to be inspected to ensure that nonessential items have been removed. One such step shall ensure that all personnel are clear of exposure to dangerous conditions resulting from reenergizing the service and that blocked mechanical equipment or grounded equipment is cleared and prepared for return to service.

The procedure must define all steps necessary to ensure that the work task is complete and that no one can be exposed unexpectedly to an electrical hazard when the electrical service is restored. The person in charge must ensure that no person is exposed or potentially exposed to an electrical energy source before removing the lockout devices from the disconnecting means.

The person in charge might choose to require workers to sign in and sign out of an area while a lockout exists. However, requiring workers to sign in and out does not replace the need for the person in charge to visually inspect the area.

(n) Temporary Release for Testing/Positioning. The procedure shall clearly identify the steps and qualified persons' responsibilities when the job or task requiring lockout/tagout is to be interrupted temporarily for testing or positioning of equipment; then the steps shall be identical to the steps for return to service.

In some instances, it might be desirable to restore electrical energy temporarily to reposition a mechanical device, such as a large agitator, to facilitate an additional mechanical work task. Restoring electrical energy to reposition equipment is very hazardous and should be avoided. Restoring electrical energy to facilitate testing electrical circuits should be avoided also. The procedure must provide specific details of all necessary actions should this function become necessary.

FPN: See 110.9 and 130.4 for requirements when using test instruments and equipment.

### 120.3 Temporary Protective Grounding Equipment.

Several different names are used to refer to temporary protective grounding equipment. Safety grounds, grounding sets, grounding devices, and similar terms all refer to equipment intended to provide intentional grounding of an electrical circuit. Sometimes temporary protective grounds take the form of electrical conductors spliced or otherwise joined together with connection devices for connecting to an exposed electrical conductor. A manufacturer normally provides this type of device with an established fault-duty rating. See Exhibit 120.5 for a typical set of safety grounds.

Grounding devices sometimes are sold by the equipment manufacturer typically for switchgear. They normally are built on a frame for inserting into a space or cubicle that normally holds a circuit breaker or fusible switch. These devices also have a fault-duty rating and are constructed by the manufacturer of the circuit breaker or switch.

**(A) Placement.** Temporary protective grounds shall be placed at such locations and arranged in such a manner as to prevent each employee from being exposed to hazardous differences in electrical potential.

When protective grounds (safety grounds) are necessary to avoid possibly reenergizing a conductor that would expose a worker to an unexpected electrical hazard, the grounds must be installed on the conductor at a point between the worker and the source of energy. If the unexpected source of electricity could be from both directions in the electrical circuit, safety grounds must be installed on both sides of the worker. The safety grounds should be installed in a manner that establishes a zone of equipotential where each employee is working.

**(B) Capacity.** Temporary protective grounds shall be capable of conducting the maximum fault current that could flow at the point of grounding for the time necessary to clear the fault.

A set of safety grounds that is exposed to a fault current is subjected to significant mechanical forces. The physical and electrical integrity of the conductors that comprise the ground set must be sufficient to withstand both the mechanical and electrical forces associated with the fault. When applied, care must be taken to properly locate the conductors to ensure the movement of the conductors due to mechanical forces does not cause harm to the worker. The ground set must have a rating established by the manufacturer and must be applied within that rating. To establish the necessary rating, available fault current must be determined by a system analysis. The rating of the ground set must be at least as great as the available fault current. Safety grounds should be tested in accordance with ASTM F 855, *Standard Specifications for Temporary Protective Grounds to Be Used on De-energized Electric Power Lines and Equipment.*

Safety grounds could be fabricated in the field. However, employers who choose this option must be able to verify that the rating of the safety grounds exceeds the available fault current. This option is not recommended.

**(C) Equipment Approval.** Temporary protective grounding equipment shall meet the requirements of ASTM F 855, *Standard Specification for Temporary Protective Grounds to be Used on De-energized Electric Power Lines and Equipment.*

**(D) Impedance.** Temporary protective grounds shall have an impedance low enough to cause immediate operation of protective devices in case of accidental energizing of the electric conductors or circuit parts.

The objective of safety grounds is to provide a path to earth so that the overcurrent protective device can operate. The impedance of the grounding path must be low enough to permit a significant fault current to flow through the overcurrent device.

The impedance of the ground-fault return path through earth should be verified on a frequency determined through use. If the impedance of the ground-fault return path is high, the overcurrent device might not operate or might not operate rapidly enough to limit damage to equipment.

# ARTICLE 130
# Work Involving Electrical Hazards

Deciding to work on or near exposed energized electrical conductors should be a last resort and made only after all other possibilities for establishing an electrically safe work condition have been exhausted. For an employer, accepting the risk of having an employee working on or near an exposed energized electrical conductor has significant implications, such as if the employee is injured or killed. For a worker, accepting the risk of working on or near an exposed energized electrical conductor significantly increases his or her chance for injury, severe burns, or electrocution, as well as the chance for damaged equipment and lost production or use. This is an extremely serious decision that must be supported by a written work permit that is reviewed by several people.

Workers must not work on or near exposed energized electrical conductors unless they are trained and qualified to recognize and avoid contact. The employee must determine where a difference of potential 50 volts or more exists between exposed parts within arm's reach of the work task. The work must be planned, and the plan must be shared with all employees who might be involved in or associated with any job task involving work on or near exposed energized electrical conductors.

Employers are required to supply, and employees are required to wear, personal protective equipment (PPE) that is selected to protect from all hazards associated with the overall job. The PPE to be used must be inspected before each use to ensure the integrity of the equipment and that it has been maintained in usable condition.

When performing maintenance, a worker might be required to accept some exposure to electrical hazards. Any worker who performs maintenance on electrical equipment must be qualified for the particular maintenance task being contemplated. He or she must wear PPE that is rated at least as great as the degree of the potential hazard.

The act of creating an electrically safe work condition can expose a worker to either electrocution and/or arc flash. Until the electrically safe work condition has been established, the worker must wear PPE suitable for the maximum degree of all associated hazards.

As suggested by FPN 2 of 130.3(A) and again in the FPN of Annex D, Section D.7, when a thermal hazard exceeds 40 cal/cm$^2$, the work should not be performed until after an electrically safe work condition exists. In some instances, creating an electrically safe work condition might require some exposure to a thermal hazard exceeding 40 cal/cm$^2$. Employers should implement processes, such as remote switching, to reduce the risk to an acceptable level.

### 130.1 Justification for Work.

**(A) General.** Energized electrical conductors and circuit parts to which an employee might be exposed shall be put into an electrically safe work condition before an employee works within the Limited Approach Boundary of those conductors or parts.

The language used in this section suggests that justification is necessary for working on or near an exposed energized electrical conductor or circuit part. Workers might be exposed to both electrocution and severe burns when performing such tasks. Any work task that exposes a worker to injury must be justified. Work performed within the Limited Approach Boundary and/or within the Arc Flash Protection Boundary must be justified if the risk of injury to the worker is elevated because the work has to be performed while energized.

The basic rule for justifying work on or near any exposed energized electrical conductor is that equipment at 50 volts or more must be placed in an electrically safe work condition before an employee works on the equipment. Employees are exposed to an electrical hazard at all times when they are within the Limited Approach Boundary. Therefore, the employer must be able to demonstrate that the work cannot be performed with the equipment or circuit in an electrically safe work condition. Only the following two explanations for not creating an electrically safe work condition are acceptable:

- If deenergizing the electrical circuit would result in an increased, additional, or greater hazard, the task may be performed with the circuit energized. An example of an increased hazard might be that loss of electrical power could result in an environmental spill or a runaway process. An example of an *additional* hazard could be the loss of electrical power to life-support equipment.

- If deenergizing the electrical circuit is infeasible due to equipment design or operational limitations, the task may be performed with the circuit energized. An example of infeasible due to equipment design might be that removing the source of voltage for a single instrument circuit would require a complete shutdown of a continuous process.

In some instances, an employer or employee incorrectly assumes the terms *infeasible* and *inconvenient* are interchangeable. The words *infeasible* and *inconvenient* should not be confused because the difference between the two words is significant. The word *inconvenient* cannot serve to justify work on or near exposed energized electrical conductors. Most shutdowns are inconvenient, but do not justify exposing workers to potential harm. If work is performed on or near an exposed energized electrical conductor, the employer must be able to document that the work task meets the criteria for one of the satisfactory reasons for executing the work with the circuit energized. A shutdown could be infeasible if, for example, it involves a continuous process where interruption of a part of the process creates a dangerous condition. Another example would be the shutdown of power to an operating suite in a hospital, where surgical procedures are in progress.

**(1) Greater Hazard.** Energized work shall be permitted where the employer can demonstrate that deenergizing introduces additional or increased hazards.

Removing the source of electrical energy from unique equipment can result in a greater hazard existing at a different location in the facility. In other instances, deenergizing the equipment could result in creating additional hazards at a different location. The additional or greater hazard could be from an electrical energy source or from a different type of energy source, such as a chemical or environmental hazard. See FPN 1 below for some examples, but note that FPNs are not mandatory text.

**(2) Infeasibility.** Energized work shall be permitted where the employer can demonstrate that the task to be performed is infeasible in a deenergized state due to equipment design or operational limitations.

*Infeasible* is defined in Merriam-Webster's *Collegiate Dictionary,* tenth edition, as "not capable of being done or carried out." The dictionary suggests that the phrase *not possible* is a synonym. If it is not possible to deenergize the exposed energized electrical conductors or circuit part by performing the work at a different time, then the work task is *infeasible* to perform in an electrically safe work condition.

If the work task is to measure load current, for instance, the task can only be performed with the circuit energized and operating. In general, diagnostic analysis requires the circuit or equipment to be energized.

Although 130.1(A)(2) permits energized work due to issues with equipment design or operational limitations, the employer and the employee should be very selective in determining when energized work is permitted. For example, a lack of illumination is an operational limitation, but it does not establish infeasibility. A worker can always bring in lighting from a portable source or another circuit. See FPN 2 below for some examples, but note that FPNs are not mandatory text.

**(3) Less Than 50 Volts.** Energized electrical conductors and circuit parts that operate at less than 50 volts to ground shall not be required to be deenergized where the capacity of the source and any overcurrent protection between the energy source and the worker are considered and it is determined that there will be no increased exposure to electrical burns or to explosion due to electric arcs.

If the voltage of the circuit or equipment is less than 50 volts, the risk of electrocution is reduced to an acceptable level. However, the risk associated with an arcing fault might be significant. To determine if the risk of injury from an arcing fault, the worker must perform a hazard/risk analysis that considers the capacity of the system to create a thermal hazard at the point where the work task is to be performed.

The electrical energy source might be alternating current, at any frequency, or direct current. Batteries and other storage devices can provide sufficient energy to create an arcing fault and resulting hazardous condition.

Control circuits that are energized at less than 50 volts might be hazardous. For instance, disconnecting or shorting a control circuit could create a process condition that results in releasing another type of energy. Even if the capacity of the electrical energy source is limited, the integrity of the circuit might be critical.

FPN No. 1: Examples of increased or additional hazards include, but are not limited to, interruption of life support equipment, deactivation of emergency alarm systems, and shutdown of hazardous location ventilation equipment.

FPN No. 2: Examples of work that might be performed within the Limited Approach Boundary of exposed energized electrical conductors or circuit parts because of infeasibility due to equipment design or operational limitations include performing diagnostics and testing (e.g., start-up or troubleshooting) of electric circuits that can only be performed with the circuit energized and work on circuits that form an integral part of a continuous process that would otherwise need to be completely shut down in order to permit work on one circuit or piece of equipment.

FPN No. 3: The occurrence of arcing fault inside an enclosure produces a variety of physical phenomena very different from a bolted fault. For example, the arc energy resulting from an arc developed in air will cause a sudden pressure increase and localized overheating. Equipment and design practices are available to minimize the energy levels and the number of at-risk procedures that require an employee to be exposed to high level energy sources. Proven designs such as arc-resistant switchgear, remote racking (insertion or removal), remote opening and closing of switching devices, high-resistance grounding of low-voltage and 5 kV (nominal) systems, current limitation, and specification of covered bus within equipment are techniques available to reduce the hazard of the system.

### (B) Energized Electrical Work Permit.

The energized work permit is intended to ensure that the increased risk (and increased possibility of injuries) associated with exposure to an exposed energized electrical conductor receives adequate consideration. Equipment owners and managers do not always understand that working on or near exposed energized electrical conductors exposes workers unnecessarily to electrical hazards. Workers, especially employees of a contractor, often accept exposure to an electrical hazard when the equipment owner seems to be unwilling to slow or stop production.

The energized electrical work permit provides workers, managers, and equipment owners with an opportunity to recognize the increased exposure to electrocution or a thermal hazard and make a decision based on the conditions that exist in the field. The energized electrical work permit correlates with OSHA's requirement to deenergize the equipment or circuit except in specific conditions. The energized electrical work permit ensures that all affected personnel understand that exposure to an electrical hazard is increased when the equipment must remain energized while the work task is performed and there is also increased cost of executing the work task.

Except that mentioned in 130.1(B)(3), all work to be done on or near an exposed energized electrical conductor or circuit parts must be authorized by a written permit. By signing the permit, the person authorizing the work is accepting responsibility for the exposure. That individual, therefore, often tends to find alternative ways to accomplish such work without exposing workers to electrical hazards. Generally, when an energized work permit is required, work on or near exposed energized electrical conductors is reduced.

**(1) Where Required.** When working on energized electrical conductors or circuit parts that are not placed in an electrically safe work condition (i.e., for the reasons of increased or additional hazards or infeasibility per 130.1), work to be performed shall be considered energized electrical work and shall be performed by written permit only.

Permits that cover routine work tasks to be performed by trained and qualified employees can be written to cover a long period of time. For instance, a worker might be trained and qualified to replace a fuse that involves an exposed energized electrical conductor or circuit part. If the worker is trained to understand the electrical hazards associated with exchanging the fuse and is wearing any necessary PPE, a permit might be issued that covers an extended period, such as three months.

**(2) Elements of Work Permit.** The energized electrical work permit shall include, but not be limited to, the following items:

(1) A description of the circuit and equipment to be worked on and their location
(2) Justification for why the work must be performed in an energized condition (130.1)
(3) A description of the safe work practices to be employed [110.8(B)]
(4) Results of the shock hazard analysis [110.8(B)(1)(a)]
(5) Determination of shock protection boundaries [130.2(B) and Table 130.2(C)]
(6) Results of the arc flash hazard analysis (130.3)
(7) The arc flash protection boundary [130.3(A)]
(8) The necessary personal protective equipment to safely perform the assigned task [130.3(B), 130.7(C)(9), and Table 130.7(C)(9)]
(9) Means employed to restrict the access of unqualified persons from the work area [110.8(A)(2)]
(10) Evidence of completion of a job briefing, including a discussion of any job-specific hazards [110.7(G)]
(11) Energized work approval (authorizing or responsible management, safety officer, or owner, etc.) signature(s)

When the energized work permit is created, the items listed in 130.1(B)(2) must be considered, and the permit should provide evidence of their consideration. Including these items in the work permit shows that the items have been addressed and that consideration of electrical hazards is of primary importance.

**(3) Exemptions to Work Permit.** Work performed within the Limited Approach Boundary of energized electrical conductors or circuit parts by qualified persons related to tasks such as testing, troubleshooting, voltage measuring, etc., shall be permitted to be performed without an energized electrical work permit, provided appropriate safe work practices and personal protective equipment in accordance with Chapter 1 are provided and used. If the purpose of crossing the Limited Approach Boundary is only for visual inspection and the Restricted Approach Boundary will not be crossed, then an energized electrical work permit shall not be required.

In general, diagnostic work such as testing, troubleshooting, and voltage measuring require the equipment or circuit to be energized. Although the specific actions associated with these and similar work tasks are not routine, employees should be specifically chosen based on their knowledge and skill. They must be qualified to perform the specific actions necessary to accomplish the necessary function. Although an energized electrical work permit is not necessary for diagnostic work, the electrical safety program should identify the specific workers who are authorized to perform these functions. If the work task requires a worker to remove or change a conductor or to add a new conductor, the work task is not diagnostic in nature and must be covered by an energized electrical work permit. As noted in 130.1(B)(3), if the work task is solely for visual inspection, no energized electrical work permit is necessary. However, all other requirements of Chapter 1 must be implemented.

FPN: For an example of an acceptable energized electrical work permit, see Annex J.

### 130.2 Approach Boundaries to Energized Electrical Conductors or Circuit Parts.

*NFPA 70E* identifies two types of approach boundaries: Shock Protection Boundaries and Arc Flash Protection Boundary. Each type is determined by the kind of hazard associated with it. *NFPA 70E* considers potential contact with an exposed energized electrical conductor, with the accompanying potential for shock and electrocution, as a source of injury from current flowing through a worker's body and call them the shock protection boundaries. *NFPA 70E* considers exposure to a potential arcing fault as a different source of injury in which direct contact with an exposed energized electrical conductor or circuit part does not occur. Safe approach boundaries are established on the basis of potential exposure to injury. Although the boundaries are related to the source of energy, they must be considered in completely different analyses. See Exhibit 130.1 for Shock Protection Boundaries (Prohibited Approach, Restricted Approach, and Limited Approach) and Arc Flash Protection Boundary.

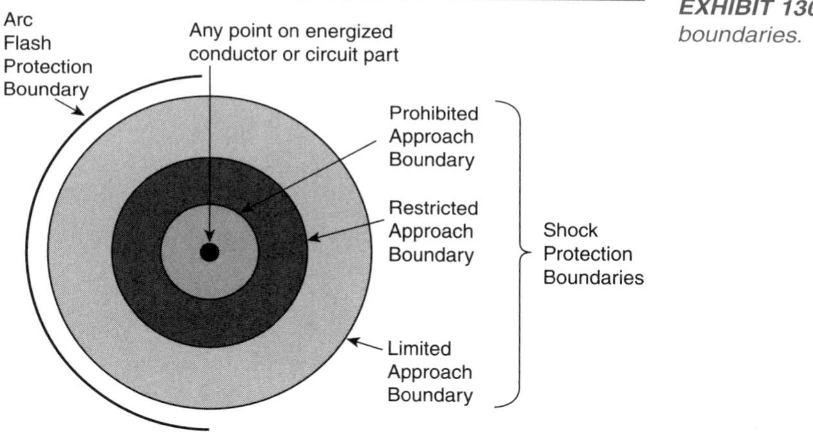


***EXHIBIT 130.1*** *Approach boundaries.*

**(A) Shock Hazard Analysis.** A shock hazard analysis shall determine the voltage to which personnel will be exposed, boundary requirements, and the personal protective equipment necessary in order to minimize the possibility of electric shock to personnel.

As the approach distance to an exposed energized electrical conductor decreases, the risk of direct contact with the conductor increases. Workers must determine the closest approach to a conductor that is necessary to perform the work task. The Shock Approach Boundary is determined based on that necessary approach distance. When the worker conducts the shock hazard analysis, he or she must determine the voltage of all conductors in the vicinity of the worker's body or tool that will be used.

The risk of injury increases as the approach distance decreases, so the worker must select safe work practices that provide minimum risk of injury. Qualified persons are expected to have the knowledge and skill to be able to approach an exposed energized electrical conductor closer than an unqualified worker. Therefore, an unqualified worker must remain at a greater distance than a qualified worker. The shock approach boundaries provide a trigger for added protection for all workers.

**(B) Shock Protection Boundaries.** The shock protection boundaries identified as Limited, Restricted, and Prohibited Approach Boundaries are applicable to the situation in which approaching personnel are exposed to energized electrical conductors or circuit parts. See Table 130.2(C) for the distances associated with various system voltages.

The electrical safety program must include a procedure that provides workers with guidance related to required protection when the approach distance is less than the appropriate Shock Protection Boundary. The protection boundaries are as follows:

- Limited Approach Boundary
- Restricted Approach Boundary
- Prohibited Approach Boundary

The dimension associated with each of these boundaries depends on the maximum voltage to which a worker might be exposed. The dimensions are given in Table 130.2(C).

The Limited Approach Boundary is the closest approach distance for an *unqualified* worker unless additional protective measures are used. The Restricted Approach Boundary is the closest approach distance for a *qualified* worker unless additional protective measures are used. The Prohibited Approach Boundary must not be crossed unless the work task is guided by the measures identified in Section 130.1.

Shock approach boundaries are related to direct contact only and do not consider exposure to arc flash.

FPN: In certain instances, the Arc Flash Protection Boundary might be a greater distance from the exposed energized electrical conductors or circuit parts than the Limited Approach Boundary. The Shock Protection Boundaries and the Arc Flash Hazard Boundary are independent of each other.

**(C) Approach to Exposed Energized Electrical Conductors or Circuit Parts Operating at 50 Volts or More.** No qualified person shall approach or take any conductive object closer to exposed energized electrical conductors or circuit parts operating at 50 volts or more than the Restricted Approach Boundary set forth in Table 130.2(C), unless any of the following apply:

(1) The qualified person is insulated or guarded from the energized electrical conductors or circuit parts operating at 50 volts or more and no uninsulated part of the qualified person's body crosses the Prohibited Approach Boundary set forth in Table 130.2(C). Insulating gloves or insulating gloves and sleeves are considered insulation only with regard to the energized parts upon which work is being performed. If there is a need for an uninsulated part of the qualified person's body to cross the Prohibited Approach Boundary, a combination of Sections 130.2(C)(1), 130.2(C)(2), and 130.2(C)(3) shall be used to protect the uninsulated body parts.

(2) The energized electrical conductors or circuit part operating at 50 volts or more are insulated from the qualified person and from any other conductive object at a different potential.

(3) The qualified person is insulated from any other conductive object as during live-line bare-hand work.

The Restricted Approach Boundary is the closest approach distance for a *qualified* person. If necessary to cross the Restricted Approach Boundary, the qualified person must take

***TABLE 130.2(C)*** *Approach Boundaries to Energized Electrical Conductors or Circuit Parts for Shock Protection (All dimensions are distance from energized electrical conductor or circuit part to employee.)*

| *(1)* | *(2)* | *(3)* | *(4)* | *(5)* |
|---|---|---|---|---|
| | ***Limited Approach Boundary***[1] | | ***Restricted Approach Boundary***[1]***; Includes Inadvertent Movement Adder*** | |
| ***Nominal System Voltage Range, Phase to Phase***[2] | ***Exposed Movable Conductor***[3] | ***Exposed Fixed Circuit Part*** | | ***Prohibited Approach Boundary***[1] |
| Less than 50 | Not specified | Not specified | Not specified | Not specified |
| 50 to 300 | 3.05 m (10 ft 0 in.) | 1.07 m (3 ft 6 in.) | Avoid contact | Avoid contact |
| 301 to 750 | 3.05 m (10 ft 0 in.) | 1.07 m (3 ft 6 in.) | 304.8 mm (1 ft 0 in.) | 25.4 mm (0 ft 1 in.) |
| 751 to 15 kV | 3.05 m (10 ft 0 in.) | 1.53 m (5 ft 0 in.) | 660.4 mm (2 ft 2 in.) | 177.8 mm (0 ft 7 in.) |
| 15.1 kV to 36 kV | 3.05 m (10 ft 0 in.) | 1.83 m (6 ft 0 in.) | 787.4 mm (2 ft 7 in.) | 254 mm (0 ft 10 in.) |
| 36.1 kV to 46 kV | 3.05 m (10 ft 0 in.) | 2.44 m (8 ft 0 in.) | 838.2 mm (2 ft 9 in.) | 431.8 mm (1 ft 5 in.) |
| 46.1 kV to 72.5 kV | 3.05 m (10 ft 0 in.) | 2.44 m (8 ft 0 in.) | .990 m (3 ft 3 in.) | 660 mm (2 ft 2 in.) |
| 72.6 kV to 121 kV | 3.25 m (10 ft 8 in.) | 2.44 m (8 ft 0 in.) | 1.016 m (3 ft 4 in.) | 838 mm (2 ft 9 in.) |
| 138 kV to 145 kV | 3.36 m (11 ft 0 in.) | 3.05 m (10 ft 0 in.) | 1.168 m (3 ft 10 in.) | .965 m (3 ft 4 in.) |
| 161 kV to 169 kV | 3.56 m (11 ft 8 in.) | 3.56 m (11 ft 8 in.) | 1.29 m (4 ft 3 in.) | 1.14 m (3 ft 9 in.) |
| 230 kV to 242 kV | 3.97 m (13 ft 0 in.) | 3.97 m (13 ft 0 in.) | 1.721 m (5 ft 8 in.) | 1.57 m (5 ft 2 in.) |
| 345 kV to 362 kV | 4.68 m (15 ft 4 in.) | 4.68 m (15 ft 4 in.) | 2.794 m (9 ft 2 in.) | 2.641 m (8 ft 8 in.) |
| 500 kV to 550 kV | 5.8 m (19 ft 0 in.) | 5.8 m (19 ft 0 in.) | 3.61 m (11 ft 10 in.) | 3.45 m (11 ft 4 in.) |
| 765 kV to 800 kV | 7.24 m (23 ft 9 in.) | 7.24 m (23 ft 9 in.) | 4.851 m (15 ft 11 in.) | 4.7 m (15 ft 5 in.) |

Note: For Arc Flash Protection Boundary, see 130.3(A).

[1] See definition in Article 100 and text in 130.2(D)(2) and Annex C for elaboration.

[2] For single-phase systems, select the range that is equal to the system's maximum phase-to-ground voltage multiplied by 1.732.

[3] A condition in which the distance between the conductor and a person is not under the control of the person. The term is normally applied to overhead line conductors supported by poles.

additional precautionary measures. If a qualified person must approach an exposed energized electrical conductor closer than the Restricted Approach Boundary, insulating materials with a defined voltage rating must be placed between the person and the conductor. The insulating material can take several forms. The insulating material can be installed so that the conductor is insulated from possible contact. The worker can be insulated by wearing appropriately rated PPE, or the worker can be insulated from ground as in live-line bare-hand work.

In some cases, using more than one protective scheme is desirable. For instance, appropriately rated rubber blankets might be installed to cover or partially cover one or more conductors, and the worker also might wear appropriate voltage-rated PPE.

Insulating the qualified person from ground as the sole protective measure requires special training and qualification. At best, this alternative is hazardous and is not recommended except when the work is located on an overhead conductor at an elevated position.

Section 130.2(C) defines requirements that must be observed before a qualified person is permitted to cross the Restricted Approach Boundary. This section recognizes that a tool or other object is considered to be an extension of the person's body. If a worker is holding an object in his or her hand, the requirements of this section apply to both the worker and the object. The effect of these requirements is that the worker (or extended body part) is prevented from being exposed to any difference of potential more than 50 volts.

The purpose of columns 2 and 3 of Table 130.2(C) is to recognize that an electrical conductor can move. If the conductor is fixed into position, the distance between the worker and the conductor is under control of the employee. If that distance can vary because the conductor can move, such as a bare overhead conductor or a conductor installed on racks in a manhole, or if the distance can vary because the platform (articulating arm) on which the employee is standing can move, then the distance is not under the worker's control, and column 2 of Table 130.2(C) applies. That column would also apply if a large size conductor is being disconnected from a fixed part, such as a connection point on an equipment bus, and it is anticipated that the act of disconnection could cause the conductor to swing away from the connection point.

**(D) Approach by Unqualified Persons.** Unqualified persons shall not be permitted to enter spaces that are required to be accessible to qualified employees only, unless the electric conductors and equipment involved are in an electrically safe work condition.

By definition, unqualified persons have not been trained to recognize and react to risk of shock or electrocution. The Limited Approach Boundary is the approach limit for unqualified persons. Only qualified persons should be permitted to be within the space defined by the Limited Approach Boundary. If the conductors are placed in an electrically safe work condition, approach boundaries no longer exist, and unqualified workers can approach the conductor without risk of injury.

**(1) Working At or Close to the Limited Approach Boundary.** Where one or more unqualified persons are working at or close to the Limited Approach Boundary, the designated person in charge of the work space where the electrical hazard exists shall advise the unqualified person(s) of the electrical hazard and warn him or her to stay outside of the Limited Approach Boundary.

Workers who are not associated with an electrical work task or tasks could be exposed to an electrical hazard. For instance, a painter could be working in the same room with an exposed energized electrical conductor or circuit part. The electrical supervisor and the painter's supervisor must establish a communication path to ensure that the painter is advised of the location of the exposed energized electrical conductor and how to avoid exposure to any associated electrical hazard(s). Signs or barricades might be necessary.

**(2) Entering the Limited Approach Boundary.** Where there is a need for an unqualified person(s) to cross the Limited Approach Boundary, a qualified person shall advise him or her of the possible hazards and continuously escort the unqualified person(s) while inside the Limited Approach Boundary. Under no circumstance shall the escorted unqualified person(s) be permitted to cross the Restricted Approach Boundary.

In some instances, an unqualified person might be required to perform one or more tasks within the Limited Approach Boundary. If this becomes necessary, a qualified person must ensure that the unqualified person is advised about the location of all exposed energized electrical conductors. The unqualified person must be advised that the risk of shock or electrocution exists, and a qualified person must escort the unqualified person at all times. The unqualified person must not cross the Restricted Approach Boundary under any circumstances.

### 130.3 Arc Flash Hazard Analysis.

An arc flash hazard analysis shall determine the Arc Flash Protection Boundary and the personal protective equipment that people within the Arc Flash Protection Boundary shall use.

The arc flash hazard analysis shall be updated when a major modification or renovation takes place. It shall be reviewed periodically, not to exceed five years, to account for changes

**in the electrical distribution system that could affect the results of the arc flash hazard analysis.**

**The arc flash hazard analysis shall take into consideration the design of the overcurrent protective device and its opening time, including its condition of maintenance.**

The arc flash hazard analysis consists of one or more processes that determine the following information:

- If a thermal hazard exists
- Parameters of the Arc Flash Protection Boundary
- What PPE is necessary to mitigate the thermal hazard

An arc flash hazard analysis is a review of an electrical circuit to determine its capacity to deliver sufficient thermal energy to cause a burn from an arcing fault. One purpose of the analysis is to determine the distance from the potential arcing fault point that would expose a person to a second-degree burn (Arc Flash Protection Boundary). Any body part that is within the Arc Flash Protection Boundary must be protected from the thermal effects of that fault. The analysis must determine that point. The Arc Flash Protection Boundary can be determined by individual calculations at a specific point in a circuit. Another purpose of the arc flash hazard analysis, which is defined in Section 100, is to determine the PPE. One way to determine the PPE is to determine the incident energy available, and another way is to use Tables 130.7(C)(9), 130.7(C)(10), and 130.7(C)(11) in the standard or the methods in Annex D.

The Arc Flash Protection Boundary is intended to trigger the need for PPE that might protect the worker from potential thermal injury. After the Arc Flash Protection Boundary has been determined, the worker must be able to select PPE that would minimize the possibility of a second-degree burn. Any body part that is closer to a potential arcing fault than the Arc Flash Protection Boundary must be protected from thermal injury.

Determining the degree of possible exposure is complex and involves the interaction of as many as 14 variables. Examples of the variables include the type of enclosure, the spacing and orientation of the arcing parts, and the voltage level of the circuit. An arcing fault is a transformation of electrical energy to other forms of energy, such as heat and pressure. Heat energy is the hazard addressed in the arc flash hazard analysis. The release of thermal energy in an arcing fault is dependent on many variables. Although the variables are known, how they interact is difficult to predict.

Methods to calculate the degree of possible thermal exposure currently are evolving. Several have been developed and are available now, and all estimate a result that is stated in terms of calories per square centimeter. Flame-resistant (FR) protective clothing and equipment also are rated in calories per square centimeter (cal/cm$^2$). Workers must wear FR clothing that has a rating equal to or higher than the potential arc intensity.

The process used to perform the arc flash hazard analysis must be defined in the employer's electrical safety program. Each employer is responsible for ensuring that qualified employees are trained to perform an arc flash hazard analysis. The process must be documented in a procedure published as a part of the electrical safety program.

Contractors are responsible to ensure that their employees recognize the conditions that might exist in the field. If a label that provides the Arc Flash Protection Boundary is installed on equipment, for instance, the procedure must provide guidance regarding whether the information is acceptable or not. If no label exists on the equipment, the procedure must provide appropriate guidance for employees. Employers are responsible for the safety of their employees, regardless of the conditions.

As employers, owners must provide sufficient information to protect their own employees. Owners should perform an arc flash hazard analysis to generate the necessary information. The arc flash hazard analysis must be reviewed at any time major changes occur to the electrical system that might have a bearing on the amount of available fault current or overcurrent protection. However, a review must be conducted at intervals not exceeding five years.

The arc flash hazard analysis must also take into consideration the design of the overcurrent protective device and its opening time, including its condition of maintenance. See Section 21.10 of NFPA 70B for maintenance of protective devices.

*Exception No. 1: An arc flash hazard analysis shall not be required where all of the following conditions exist:*

*(1) The circuit is rated 240 volts or less.*

*(2) The circuit is supplied by one transformer.*

*(3) The transformer supplying the circuit is rated less than 125 kVA.*

A detailed arc flash hazard analysis is not necessary when all of the conditions exist that are described in Section 130.3, Exception No. 1. Although conditions vary among installations, the conditions described in this exception limit the potential arc flash hazard. However, workers should wear heavy leather gloves to protect their hands from any arc that might occur due to the fact that a worker's hands are much closer to the potential arc than the torso.

*Exception No. 2: The requirements of 130.7(C)(9), 130.7(C)(10), and 130.7(C)(11) shall be permitted to be used in lieu of a detailed incident energy analysis.*

FPN No. 1: Improper or inadequate maintenance can result in increased opening time of the overcurrent protective device, thus increasing the incident energy.

FPN No. 2: For additional direction for performing maintenance on overcurrent protective devices, see Chapter 2, Safety-Related Maintenance Requirements.

Section 130.3, Exception No. 2 identifies alternative methods to determine the necessary PPE. However, each alternative method has limitations that should determine when the alternative method should be used and when it is not permitted to be used.

Although associated with many variables, determining the risk of injury from an arcing fault depends primarily on three major parameters:

1. Available fault current
2. Duration of the arcing fault
3. Distance between the worker and the arcing fault

The duration of the fault current depends on the clearing time of the overcurrent device at the amount of current flowing in the fault circuit. Overcurrent devices and fault-current return conductors must be maintained adequately for the overcurrent devices to operate in accordance with their design parameters.

### (A) Arc Flash Protection Boundary.

Arcing faults rarely occur in open air. In the majority of instances, arcing faults occur within an enclosure. An arcing fault is most likely to occur when movement occurs within the enclosure. For instance, movement occurs when a switch contact is closed or opened, when a contactor operates, or when a door is opened or cover removed. Many arcing faults are associated with movement initiated by a person, such as when a person reaches inside a cubicle with a conductive object in his or her hand.

Section 130.3(A) also recognizes that other methods can be used to determine the Arc Flash Protection Boundary and the degree of protection, and it permits their use under engineering supervision. Annex D illustrates some methods of determining the Arc Flash Protection Boundary and the rating of protective equipment.

**(1) Voltage Levels Between 50 Volts and 600 Volts.** In those cases where detailed arc flash hazard analysis calculations are not performed for systems that are between 50 volts and 600 volts, the Arc Flash Protection Boundary shall be 4.0 ft, based on the product of clearing time of 2 cycles (0.033 sec) and the available bolted fault current of 50 kA or any combination not exceeding 100 kA cycles (1667 ampere seconds). When the product of clearing times and bolted fault current exceeds 100 kA cycles, the Arc Flash Protection Boundary shall be calculated.

A default Arc Flash Protection Boundary of 4 ft can be used, provided the system capacity does not exceed 100 kA cycles. The available bolted-fault current must not exceed 50 kA, with duration of 2 cycles. (The product of 50 kA and 2 cycles is 100 kA cycles.) If the product of the available bolted-fault current (in thousands of amperes) and the clearing time (in cycles) of the overcurrent device is greater than 100, the default Arc Flash Protection Boundary must not be used. If the default conditions are exceeded, the Arc Flash Protection Boundary must be calculated.

It is important to note that some information about the installation must be known to use the default Arc Flash Protection Boundary. For instance, the worker must know both the clearing time of the overcurrent device and the available fault current. These values differ by location in the facility power system. They need to be known for each location in the facility where work will be performed when PPE is being chosen.

**(2) Voltage Levels Above 600 Volts.** At voltage levels above 600 volts, the Arc Flash Protection Boundary shall be the distance at which the incident energy equals 5 J/cm$^2$ (1.2 cal/cm$^2$). For situations where fault-clearing time is equal to or less than 0.1 sec, the Arc Flash Protection Boundary shall be the distance at which the incident energy level equals 6.24 J/cm$^2$ (1.5 cal/cm$^2$).

When the circuit voltage exceeds 600 volts, the Arc Flash Protection Boundary must be calculated. The Arc Flash Protection Boundary is the distance at which the incident energy equals 1.2 cal/cm$^2$. If the duration of the arcing fault can be limited to less than 6 cycles, the injury limit can be increased to 1.5 cal/cm$^2$. A thermal burn occurs because the temperature of cells in skin tissue is raised to a level that damages their structure. Although extremely fast, the transfer of heat from the source (arcing fault) into the surface of the skin does not happen instantaneously. If the duration of the arcing fault is limited to less than 6 cycles, the injury limit may be increased to 1.5 cal/cm$^2$. The information needed to perform this calculation normally is available from a facility coordination study.

FPN: For information on estimating the Arc Flash Protection Boundary, see Annex D.

Annex D illustrates several different methods of estimating the Arc Flash Protection Boundary and incident energy. Annex D is intended to illustrate several methods but not place a limit on other methods that could be used to generate the information. Information related to predicting the degree of the thermal hazard associated with an arcing fault is evolving.

**(B) Protective Clothing and Other Personal Protective Equipment (PPE) for Application with an Arc Flash Hazard Analysis.** Where it has been determined that work will be performed within the Arc Flash Protection Boundary identified by 130.3(A), one of the following methods shall be used for the selection of protective clothing and other personal protective equipment:

If a worker's body or part of a worker's body must be within the Arc Flash Protection Boundary to perform a task, an analysis must be performed to determine the amount of energy that may be incident upon the worker's body or clothing. The worker then must select and use PPE that is rated at least as great as the predicted incident energy that might be available from the energy source.

In choosing protective clothing, purchasers must compare the degree of arc flash hazard with the ratings of clothing construction. Therefore, the degree of the potential arc flash hazard must be determined in calories per square centimeter or converted to that measure. If the clothing rating is equal to or greater than the degree of arc flash hazard, the clothing can protect the worker from a second-degree burn in most exposures. Although most second-degree burns would be prevented, protection is not absolute. Section 130.3(B) requires that body parts closer to the potential arc source should have additional protection.

The term *incident energy* is intended to be the amount of *thermal* energy that could be received (incident upon) by either clothing or exposed skin. FR clothing must be rated at least as great as the incident energy potentially available in the arcing fault.

Because understanding of the thermal hazard is evolving, an employer might choose to define protective clothing by several different methods. PPE requirements also could be defined by calculating the available incident energy by one of several different methods. Appropriate PPE also could be determined on the basis of Tables 103.7(C)(9), 103.7(C)(10), and 103.7(C)(11). There are several ways to do the analysis and determine the PPE. The employer's procedure that defines the flash hazard analysis must define the process selected by the employer.

The distance between an arcing fault and a worker is a very important consideration. The intensity of the incident energy decreases exponentially as that distance increases. If a worker could increase the distance of his or her body from the exposed energized electrical conductors, he or she certainly should do so. An example of increasing the distance is to use a remote racking mechanism for switchgear.

Parts of a worker's body that are closer to the potential arc (such as the hands) must have increased protection. No method has been identified to suggest the amount of thermal protection needed, but a worker's hands probably are exposed to much higher temperatures than other body parts.

Section 130.3(B) requires that FR clothing be worn within the Flash Protection Boundary, which is generally considered to be 1.5 cal/cm$^2$. As described in Table 130.7(C)(11), the Hazard/Risk Category 0 does not contain FR clothing. Other than in Table 130.7(C)(9), Table 130.7(C)(10), and Table 130.7(C)(11), Hazard/Risk Category O is not mentioned in *NFPA 70E*. If an arc flash hazard analysis is performed in lieu of using Table 130.7(C)(9), there is no option of using Hazard/Risk Category 0. FPN 1 of 130.7(C)(9) can be interpreted to mean that for tasks that the committee thought do not require FR clothing, they would permit that task to be performed using Hazard/Risk Category 0.

**(1) Incident Energy Analysis. The incident energy analysis shall determine, and the employer shall document, the incident energy exposure of the worker (in calories per square centimeter). The incident energy exposure level shall be based on the working distance of the employee's face and chest areas from a prospective arc source for the specific task to be performed. Arc-Rated FR clothing and other personal protective equipment (PPE) shall be used by the employee based on the incident energy exposure associated with the specific task. Recognizing that incident energy increases as the distance from the arc flash decreases, additional PPE shall be used for any parts of the body that are closer than the distance at which the incident energy was determined.**

An incident energy analysis must be used to predict the amount of incident energy that might be available from the energy source at the point in the circuit where the work task is to be performed. Analytical methods predict incident energy at a specific distance between the worker and the potential arc source. If the distance between the worker and the potential arc source is different from the dimension used by the analysis method, the thermal energy might be greater or smaller. The selected PPE should be determined by the thermal energy that will be impressed at the working distance.

Any body parts, such as the hands and arms, might require greater protection than the worker's chest and torso. These parts should be protected accordingly.

Protective equipment and clothing that is intended for protection from an arcing fault must be rated by the manufacturer for use in an environment influenced by an electrical arc. There are two types of FR-protective apparel are available, but only one is arc rated. As an example, FR clothing for race car drivers or pilots is not suitable for arc flash protection. Arc rated FR clothing is the only FR clothing marked with the cal/cm$^2$ rating.

FPN: For information on estimating the incident energy, see Annex D.

(2) Hazard/Risk Categories. The requirements of 130.7(C)(9), 130.7(C)(10), and 130.7(C)(11) shall be permitted to be used for the selection and use of personal and other protective equipment.

Sections 130.7(C)(9), 130.7(C)(10), and 130.7(C)(11) identity protective equipment by categories.

Table 130.7(C)(11) illustrates Categories 0, 1, 2, 3, and 4 and assigns maximum protection afforded by each hazard/risk category. To clarify the word maximum, if a task in Table 130.7(C)(9) requires H/R Cat 3 PPE, then PPE with a rating no less than 25 cal/cm$^2$ needs to be used. See Table 130.7(C)(10). Category 0 carries no maximum protective characteristic since the material is ignitable. However, Category 0 clothing must not include any clothing that is considered meltable.

Table 130.7(C)(9) provides a list of work tasks intended to suggest protective equipment based on standard configurations and conditions. The employer could choose to use this table to determine the amount of protection that is necessary. When this table is used, however, the currents and clearing times in the notes to the table still have to be checked to confirm that use of the table is permitted. Also, using the table is an alternative method to determine protective equipment. It is suggested that an incident energy calculation be performed if Table 130.7(C)(9) suggests H/R Cat 4 PPE for medium voltage tasks since some stronger power systems could result in incident energies above 40 cal/cm$^2$. If a task is not in the table or the working distance is closer than the notes in the table, then the table cannot be used and it is necessary to do an incident energy analysis for that task. If an incident energy analysis is performed and the result is higher than using the tables, the table value cannot be substituted. As an example, an incident energy analysis results in an incident energy value of 76 cal/cm$^2$ for racking in a 4 kV breaker. For that same task, Table 130.7(C)(9) only requires H/R Cat 4 (40 cal/cm$^2$) PPE. The higher value should be used to protect the worker.

Determining the thermal intensity produced in an arcing fault is difficult and complex. Many small businesses have an electrical system that does not exceed 600 volts. Small businesses are less likely to have access to engineering supervision needed to execute complex calculations, and their electrical installations usually have limited capacity and relatively simple circuitry. These tables can help this type of user.

**(C) Equipment Labeling.** Equipment shall be field marked with a label containing the available incident energy or required level of PPE.

After the arc flash hazard analysis has been completed, equipment must be labeled to provide sufficient information for a worker to select the necessary arc-rated protective equipment. PPE needs to be worn when working within the Arc Flash Protection Boundary. Only the equipment where the incident energy level is greater than 1.2 cal/cm$^2$ needs to be marked, but that decision depends on a facility's safety program. Incident energy is determined by the parameters of the circuit components as installed. Therefore, electrical equipment manufacturers cannot predict the amount of incident energy. That is why the labels cannot be added to the equipment prior to shipment. The label must be installed in the field and based on an analysis of field conditions. Also see *NEC* Section 110.16 for arc flash warning label requirements.

The basic purpose of the label is to provide warning that an arc flash hazard exists in the equipment to which the label is attached. Other than the basic information to be included on the label, *NFPA 70E* does not describe how the label should look. ANSI Z 535, *American National Standard for Product Safety Signs and Labels*, provides a description for warning labels. To ensure that signs and labels on a facility have the best chance of communicating the appropriate hazard, the label required by Section 130.3(C) should comply with the requirements of ANSI Z 535.

One of two forms of information are required on the label: either the incident energy or the level of PPE. Only cal/cm$^2$ incident energy available or cal/cm$^2$ requirement for the PPE are acceptable to meet this requirement. The organization safety plan must include which form of information is on the label and how that form of information should be interpreted by the worker. Since incident energy level varies at the distance from the arc, it is also suggested that the working distance also be on the label. Then a worker knows that the marked incident energy is at a certain working distance and if their task requires a different working distance, they need to reconsider the PPE level. *NFPA 70E* does not limit any additional information that can be included on the label. Additional information, such as the available fault current, can be provided on the label. However, the label must provide sufficient information for a worker to determine the equipment that is necessary for protection.

It is the employer's responsibility to ensure that employees have the necessary information to select arc-rated protective equipment. Although owners might be responsible for providing the necessary label on equipment containing an arc flash hazard, employers are responsible to ensure that their workers are protected as required by the work task.

A hazard/risk analysis is necessary, regardless of labeling. Exhibit 130.2 shows examples of four warning labels that meet the requirements of Section 110.16 of the *NEC* and 130.3(C) of *NFPA 70E*. Exhibits 130.2(a) and (b) show warning signs that have orange background and white lettering. Exhibits 130.2(c) and (d) show danger signs with red background and white lettering. These are a few of the many possible labels. The information on the label depends on the facility's safety program.

***EXHIBIT 130.2*** *Sample arc flash warning labels. (Adapted from Clarion Safety Systems)*

(a) Orange background with black lettering

WARNING
**Arc Flash Hazard**
Wear 12 cal/cm$^2$ PPE
18" working distance

(b) Orange background with black lettering

WARNING
**Arc Flash Hazard**
15.5 cal/cm$^2$ incident energy
at 18" working distance

(c) Red background with white lettering

DANGER
**Arc Flash Hazard**
Incident energy level is 56 cal/cm$^2$
Do not work energized

(d) Red background with white lettering

DANGER
**Arc Flash Hazard**
Wear 16 cal/cm$^2$ PPE at
18" working distance

**130.4 Test Instruments and Equipment Use.**

Only qualified persons shall perform testing work within the Limited Approach Boundary of energized electrical conductors or circuit parts operating at 50 volts or more.

A Limited Approach Boundary exists only if the circuit voltage is 50 volts or greater. Only qualified persons are permitted to be within the Limited Approach Boundary. To be qualified to perform testing, such as measuring voltage or load current, the worker must be trained to understand the electrical hazards associated with the work task and select the necessary PPE.

Workers must be trained to understand that when they are performing work tasks involving testing they are exposed to shock and electrocution. Each qualified person must be trained to understand how to use the specific meter (see Exhibit 110.2) and to understand and interpret its indication(s). The meter must be in good working condition, appropriate for the task, and inspected before use.

All employees who are qualified persons must be trained to test for the absence of voltage. Each qualified person must be able to operate every meter that he or she could be expected to use and to interpret any possible meter indication. No voltage-testing device should be available for use until each qualified person has been trained to use it. Workers must understand all limitations of the testing instrument.

**130.5 Work Within the Limited Approach Boundary of Uninsulated Overhead Lines.**

Electrical hazards do not change because a conductor is elevated. However, the method of exposure is significantly different. Exposure to electrical hazards associated with overhead conductors or cables is inherently different from exposure to conductors that are solidly supported and accessible to members of the general public. In most cases, overhead conductors are guarded by being located at an elevated position and generally not available for incidental contact. When working on an overhead conductor, workers are likely to be supported by some type of elevating or articulating platform. In some instances, workers may be supported by a semi-permanent platform such as a scaffold or by a permanently installed platform or deck. When using such support methods, workers are unlikely to have significant opportunity to move about and escape any release of energy or avoid direct contact. Reacting to an emergency situation is more difficult, which emphasizes the need for an emergency recovery plan.

**(A) Uninsulated and Energized.** Where work is performed in locations containing uninsulated energized overhead lines that are not guarded or isolated, precautions shall be taken to prevent employees from contacting such lines directly with any unguarded parts of their body or indirectly through conductive materials, tools, or equipment. Where the work to be performed is such that contact with uninsulated energized overhead lines is possible, the lines shall be deenergized and visibly grounded at the point of work, or suitably guarded.

Electrical conductors that are not fully insulated for the circuit voltage have the same potential for shock and electrocution as conductors that are completely bare. Some overhead conductors have a covering as protection from environmental degradation.

Exhibit 130.3 illustrates the danger of operating equipment in close proximity of overhead power lines and the need to treat these lines as energized, unless they have been deenergized and visibly grounded.

No work task should be performed unless the worker is protected from unintentional contact with any overhead lines. Workers carrying long objects must exercise caution and avoid penetrating the space defined by the Limited Approach Boundary. When long objects are moved, workers should be assigned to both ends of the object to maintain control of each end. Any object that is not fully insulated for the circuit voltage is considered to be conductive. To

*EXHIBIT 130.3 Dangerous use of equipment that is capable of reaching overhead lines.*

be insulated for the voltage, an object must have a rating established by the manufacturer and rated according to a standard testing method.

Unqualified workers must not approach an overhead line that is not in an electrically safe work condition. Qualified workers must observe and comply with the approach boundaries identified in Table 130.2(C), Approach Boundaries to Energized Electrical Conductors or Circuit Parts for Shock Protection.

**(B) Deenergizing or Guarding.** If the lines are to be deenergized, arrangements shall be made with the person or organization that operates or controls the lines to deenergize them and visibly ground them at the point of work. If arrangements are made to use protective measures, such as guarding, isolating, or insulation, these precautions shall prevent each employee from contacting such lines directly with any part of his or her body or indirectly through conductive materials, tools, or equipment.

In many cases, responsibility for operation and maintenance of outside overhead lines is assigned to a crew or a person. The operation and maintenance of transmission and distribution lines might be the responsibility of a utility or other similar group. The person responsible for operation and maintenance of the affected conductors must be consulted and directly involved in deenergizing and grounding the overhead conductors.

Suitable guards could be installed to prevent accidental contact with the overhead lines. However, the guards must be of sufficient strength to control the approach or any possible movement of the person or object to eliminate the chance of unintentional contact. In most instances, line hose *is not satisfactory* to prevent unintentional contact.

For a guard to be adequate, it must be capable of preventing contact with the energized conductor by any part of a worker's body or being damaged by a tool or equipment that is in use.

Creating an electrically safe work condition by deenergizing the conductors and installing safety grounds provides the best option to avoid exposure to electrical hazards. Safety grounds, as shown in Exhibit 120.5, are required for overhead conductors unless exceptional precautions are made to avoid risk of injury. Safety grounds must be installed in a manner that provides a zone of equipotential for the work area. When other conductors exist in the immediate vicinity of the work area, the additional conductors must be guarded from potential contact.

**(C) Employer and Employee Responsibility.** The employer and employee shall be responsible for ensuring that guards or protective measures are satisfactory for the conditions. Employees shall comply with established work methods and the use of protective equipment.

Employers are responsible for providing the electrical safety program, and employees are responsible for implementing the requirements of the procedures defined in the program. Both employers and employees are responsible for ensuring that any installed guards are adequate for the conditions. That shared responsibility involves procedures and/or work methods provided by the employer and implemented by the employee. The employer and employee must work together to ensure that effective procedures exist, are applied stringently, and are reviewed frequently.

**(D) Approach Distances for Unqualified Persons.** When unqualified persons are working on the ground or in an elevated position near overhead lines, the location shall be such that the employee and the longest conductive object the employee might contact do not come closer to any unguarded, energized overhead power line than the Limited Approach Boundary in Table 130.2(C), Column 2.

Section 130.5(D) underscores the importance of the Limited Approach Boundary. Unqualified workers must observe the specified dimension. See Table 130.2(C).

The approach distance for unqualified persons remains the same, regardless of the installation method for the conductor(s). The Limited Approach Boundary given in Table 130.2(C) defines that distance. Table 130.2(C) has two columns, Columns 2 and 3, that define different approach distances. The limited approach distance depends on whether the distance between the conductor and the worker is under the worker's control or not under the worker's control. If the supporting platform can move, as would be the case for an articulating platform, Column 2 of Table 130.2(C) applies.

If the conductor is supported on a messenger or similar support method, the conductor can move as the wind blows, making the distance between the worker and the conductor *not* under the control of the worker, and Column 2 still applies. If the overhead conductor is fixed into position, as is the case with solid bus conductors, and the worker is standing on a fixed platform or scaffold, the worker has control of the distance between himself or herself and the conductor, and Column 3 applies.

FPN: Objects that are not insulated for the voltage involved should be considered to be conductive.

Some conductors have a covering over the conductor material that is intended to serve as protection from the effects of the environment. Many electricians refer to such conductors as *weatherproof conductors*. This covering is *not* insulating material and generally has no established voltage rating. Weatherproof conductors must be considered to be uninsulated.

**(E) Vehicular and Mechanical Equipment.**

The requirements of Section 130.5(E) apply to any type of mechanical equipment unless the conductor(s) is (are) in an electrically safe work condition. For instance, these requirements apply to operating a man lift or any other articulating equipment.

**(1) Elevated Equipment.** Where any vehicle or mechanical equipment structure will be elevated near energized overhead lines, they shall be operated so that the Limited Approach

Boundary distance of Table 130.2(C), Column 2, is maintained. However, under any of the following conditions, the clearances shall be permitted to be reduced:

The Limited Approach Boundary given in Column 2 of Table 130.2(C) defines the closest dimension that any vehicle or mechanical equipment structure can be elevated to an exposed energized overhead conductor unless conditions defined in this section permit closer approach. Distances are difficult to estimate when standing on the ground or sitting in the seat of a crane or other mobile equipment.

(1) If the vehicle is in transit with its structure lowered, the Limited Approach Boundary to overhead lines in Table 130.2(C), Column 2, shall be permitted to be reduced by 1.83 m (6 ft). If insulated barriers, rated for the voltages involved, are installed and they are not part of an attachment to the vehicle, the clearance shall be permitted to be reduced to the design working dimensions of the insulating barrier.

When a vehicle or equipment is moving from one location to another, the risk of contact is reduced since the elevating structure will not be moved while under or near an overhead conductor. If the conductor is adequately insulated or guarded, the risk of contact is reduced. Any elevating structure, such as a boom or dump truck bed, must be in the resting position. If not in a resting position, the Limited Approach Boundary cannot be reduced.

(2) If the equipment is an aerial lift insulated for the voltage involved, and if the work is performed by a qualified person, the clearance (between the uninsulated portion of the aerial lift and the power line) shall be permitted to be reduced to the Restricted Approach Boundary given in Table 130.2(C), Column 4.

When qualified workers are supported by an aerial lifting device, such as a truck boom that is fully insulated from contact with earth, the minimum unprotected approach distance is defined as the Restricted Approach Boundary given in Column 4 of Table 130.2(C). The definition of a qualified person means that the worker has received training in the operation of the aerial lifting device in addition to all other required training.

**(2) Equipment Contact.** Employees standing on the ground shall not contact the vehicle or mechanical equipment or any of its attachments, unless either of the following conditions apply:

(1) The employee is using protective equipment rated for the voltage.
(2) The equipment is located so that no uninsulated part of its structure (that portion of the structure that provides a conductive path to employees on the ground) can come closer to the line than permitted in 130.5(E)(1).

Electrocutions sometimes occur to workers who are in contact with mobile equipment when the equipment touches an exposed energized overhead conductor. Although the contact could be several feet away from the worker, he or she provides the conductive path to earth if the worker is touching the equipment with unprotected hands or another body part. Hand lines and tag lines sometimes serve as the point of contact for a worker. This is a dangerous practice, and these lines should not be used. Hand lines and tag lines serve no useful purpose.

Workers who are outside the equipment that is in contact with an energized conductor have greater exposure to electrocution than workers who are inside the equipment cab. Workers must not enter the physical area surrounding equipment operating in proximity to overhead conductors. Such workers are exposed to touch potential that is effectively equal to the operating circuit voltage if they are in contact with the equipment when it touches an energized conductor. Workers are exposed to electrocution by step potential if they are standing on the ground near the equipment when it makes contact with an overhead line.

A barricade should be erected around the physical area to surround equipment that could contact an overhead line. Signs should be installed to warn people to stay out of the area. The barricaded area should not permit approach closer than the Limited Approach Boundary.

**(3) Equipment Grounding.** If any vehicle or mechanical equipment capable of having parts of its structure elevated near energized overhead lines is intentionally grounded, employees working on the ground near the point of grounding shall not stand at the grounding location whenever there is a possibility of overhead line contact. Additional precautions, such as the use of barricades, dielectric overshoe footwear, or insulation, shall be taken to protect employees from hazardous ground potentials (step and touch potential).

FPN: Upon contact of the elevated structure with the energized lines, hazardous ground potentials can develop within a few feet or more outward from the grounded point.

Some safety programs require vehicular and other mobile equipment to be grounded with a temporary grounding conductor connected to an existing earth ground or a temporary ground rod. In event of the equipment contacting an exposed energized overhead conductor, the grounding conductor becomes an integral path for any fault current to flow. The grounding conductor, then, expands the touch and step potential hazard to include the grounding conductor and the ground rod (or other earth ground connection point). If such a grounding conductor is installed, the worker must not be within the Limited Approach Boundary of any portion of the ground circuit.

Barricades and warning signs should be erected that include the grounding conductor and ground rod. The barricades and warning signs should prevent any person from entering the space defined by the Limited Approach Boundary.

## 130.6 Other Precautions for Personnel Activities.

In the absence of an electrically safe work condition, workers are at risk of injury from an electrical hazard. Regardless of the selected work practice and PPE, a risk of injury exists. Work practices and PPE can reduce the risk, but some degree of risk remains, and supervisors and workers must rely on personal judgment to reduce the risk to its lowest level. Section 130.6 discusses some of the possible precautions that are necessary.

### (A) Alertness.

**(1) When Hazardous.** Employees shall be instructed to be alert at all times when they are working within the Limited Approach Boundary of energized electrical conductors or circuit parts operating at 50 volts or more and in work situations where electrical hazards might exist.

Supervisors must ensure that workers are reminded to be alert at all times when inside the space defined by the Limited Approach Boundary. Employees have an inherent duty to observe these precautions.

**(2) When Impaired.** Employees shall not be permitted to work within the Limited Approach Boundary of energized electrical conductors or circuit parts operating at 50 volts or more, or where other electrical hazards exist, while their alertness is recognizably impaired due to illness, fatigue, or other reasons.

Supervisors represent the employer and have a duty to observe the physical and mental condition of their employees. Any recognizable sign of impairment is justifiable reason to avoid assigning an employee to a work task that is within the Limited Approach Boundary of exposed energized or potentially energized conductors.

When the job line-up instruction is given, in addition to alertness, the supervisor must evaluate whether any employee is impaired for any reason. In addition to illness and fatigue, impairment could be the result of substance abuse or legal drugs. An employee's ability to act or react could be affected by conditions surrounding his or her personal life. The supervisor should look for signs that the employee's ability to think or act could be impaired for any reason.

**(3) Changes in Scope.** Employees shall be instructed to be alert for changes in the job or task that may lead the person outside of the electrically safe work condition or expose the person to additional hazards that were not part of the original plan.

When the job line-up is given, workers must be instructed about the details of the expected work task. The instruction must define the limits of the work task and employees should be instructed to be aware of any potential change in the original plan. If the work task cannot be executed as planned, the work task must be stopped until a new job line-up has been provided. A change in the condition or work environment that was not a part of the original plan might result in significant exposure to an electrical hazard.

Contractor or temporary employees must be keenly aware of the work plan and implement each element of the plan. If any change in the circuit or equipment condition changes during the execution of the work, contractor or temporary employees should not attempt to troubleshoot the problem. The work task should be stopped until their supervisor has provided a new job line-up.

A change in scope or work environment is a major or contributing factor in many incidents that result in injuries. Qualified workers must react to any change in scope or circuit condition.

**(B) Blind Reaching.** Employees shall be instructed not to reach blindly into areas that might contain exposed energized electrical conductors or circuit parts where an electrical hazard exists.

"Reaching blindly" is when a worker reaches to feel any point that is not directly visible. Using a mirror as an aid to see behind a device or object is still considered to be reaching blindly. If an exposed energized electrical conductor exists, employees must not reach into the area unless they have direct visual observation. If the hazard/risk analysis indicates that an electrical hazard exists, reaching blindly must be avoided.

If a worker's ability to observe an area is restricted by insufficient lighting, reaching into the poorly lighted area is blind reaching just as if reaching behind a physical obstruction. Through immediate supervision, the employer should ensure that adequate lighting is provided for each work task associated with an exposed energized electrical conductor or circuit part. Blind reaching must be avoided.

**(C) Illumination.**

**(1) General.** Employees shall not enter spaces containing electrical hazards unless illumination is provided that enables the employees to perform the work safely.

The electrical safety program should provide guidance for employees regarding the ability to physically discern all aspects of the equipment or installation. If area lighting has been removed for any reason, workers should consider installing temporary lighting. When the worker performs the required hazard/risk analysis, illumination should be one condition that is evaluated.

Additional illumination could be required due to darkness of a face shield. Prior to start of work, the work area should be viewed wearing the face shield to see if additional illumination is necessary.

**(2) Obstructed View of Work Area.** Where lack of illumination or an obstruction precludes observation of the work to be performed, employees shall not perform any task within the Limited Approach Boundary of energized electrical conductors or circuit parts operating at 50 volts or more or where an electrical hazard exists.

An obstruction could be a barrier installed by the equipment manufacturer to protect specific points within the equipment or a barrier that has been installed to isolate an energized component. Workers must be able to see the point that is intended to be touched with hands, tools, or equipment.

**(D) Conductive Articles Being Worn.** Conductive articles of jewelry and clothing (such as watchbands, bracelets, rings, key chains, necklaces, metalized aprons, cloth with conductive thread, metal headgear, or metal frame glasses) shall not be worn where they present an electrical contact hazard with exposed energized electrical conductors or circuit parts.

Workers must be aware if any of their jewelry or clothing could present an electrical hazard. Articles of jewelry that are conductive must be removed or effectively insulated at all times if a worker might contact an exposed energized electrical conductor or circuit part. Clothing that has metal or conductive threads or fibers must not be worn if contact with an exposed energized electrical conductor or circuit part is possible. Metalized aprons or face shields must not be worn within the Limited Approach Boundary. Eyeglasses containing exposed conductive components must be restrained and covered with appropriate PPE so that it is impossible for them to fall into or touch an exposed energized conductor or circuit part. Conductive body piercing jewelry must be removed before entering the Limited Approach Boundary.

**(E) Conductive Materials, Tools, and Equipment Being Handled.**

**(1) General.** Conductive materials, tools, and equipment that are in contact with any part of an employee's body shall be handled in a manner that prevents accidental contact with energized electrical conductors or circuit parts. Such materials and equipment include, but are not limited to, long conductive objects, such as ducts, pipes and tubes, conductive hose and rope, metal-lined rules and scales, steel tapes, pulling lines, metal scaffold parts, structural members, bull floats, and chains.

A conductive object being held by an employee extends the reach of the employee. In the context of 130.6(E)(1), the word *conductive* means any object that does not have an assigned rating. Manufacturers (and sometimes testing laboratories) test materials and equipment to determine the effective insulation level and resultant voltage rating. All objects that do not have a voltage rating must be considered to be conductive.

Long objects are difficult to control and should be handled by two workers, one on each end of the object. This work practice enables the object to be moved without crossing the Limited Approach Boundary.

Metal "fish lines" or fish lines with a metal "nosing" should not be used for a work task associated with an exposed energized electrical conductor or circuit part. Nonconductive pulling and fishing equipment should be used.

**(2) Approach to Energized Electrical Conductors and Circuit Parts.** Means shall be employed to ensure that conductive materials approach exposed energized electrical conductors or circuit parts no closer than that permitted by 130.2.

When long conductive objects are being handled in the vicinity of exposed energized electrical conductors or circuit parts, each end of the object should be under control of different workers. For instance, unless they have an established voltage rating, ladders are conductive objects and should be handled by assigning one employee to each end of the ladder.

**(F) Confined or Enclosed Work Spaces.** When an employee works in a confined or enclosed space (such as a manhole or vault) that contains exposed energized electrical conductors or circuit parts operating at 50 volts or more or where an electrical hazard exists, the employer shall provide, and the employee shall use, protective shields, protective barriers, or insulating materials as necessary to avoid inadvertent contact with these parts and the effects of the electrical hazards. Doors, hinged panels, and the like shall be secured to prevent their swinging into an employee and causing the employee to contact exposed energized electrical conductors or circuit parts rating at 50 volts or more or where an electrical hazard exists.

Manholes, hand holes, vaults, and large sections of equipment could enable an employee to enter an area that has exposed conductors that could be energized. Only authorized qualified employees should be permitted to enter these areas.

Hazard/risk analysis must consider all hazards that might be associated with the work task. Toxic gases or low oxygen content within the confined space could expose a worker to a life-threatening situation. OSHA has specific requirements for work performed in confined spaces that are not related to an electrical hazard. The oxygen content and flammability of the atmosphere in the confined space should be analyzed before any person enters it along with all other requirements defined in the appropriate OSHA standard. If it is determined that a task cannot be delayed, a hazard/risk analysis must be performed and documented. If the analysis determines that the risks could be reduced to an acceptable level by installing barriers, shields, or other isolating devices, the task can be performed, provided all hazards identified in the hazard/risk analysis are mitigated. An energized electrical work permit must cover the work task if the task could expose a worker to injury from an electrical hazard.

Doors, hinged panels, and similar covers must be held open by a secure means to avoid the possibility that the door or cover could swing and surprise the worker while he or she is exposed to a shock hazard.

**(G) Housekeeping Duties.** Where energized electrical conductors or circuit parts present an electrical contact hazard, employees shall not perform housekeeping duties inside the Limited Approach Boundary where there is a possibility of contact, unless adequate safeguards (such as insulating equipment or barriers) are provided to prevent contact. Electrically conductive cleaning materials (including conductive solids such as steel wool, metalized cloth, and silicone carbide, as well as conductive liquid solutions) shall not be used inside the Limited Approach Boundary unless procedures to prevent electrical contact are followed.

If any housekeeping work task could expose the worker to an exposed energized electrical conductor or circuit part, the worker must be a qualified person. Unqualified persons must remain outside the Limited Approach Boundary. Many housekeeping and cleaning materials are conductive, and therefore, must not be used inside the Limited Approach Boundary.

Work tasks sometimes generate trash and require housekeeping at the conclusion or while the work task is being executed. Unless an electrically safe work condition exists, a qualified person must perform the clean up, since no unqualified person should be permitted within the Limited Approach Boundary.

**(H) Occasional Use of Flammable Materials.** Where flammable materials are present only occasionally, electric equipment capable of igniting them may not be used, unless measures are taken to prevent hazardous conditions from developing. Such materials include, but are not limited to, flammable gases, vapors, or liquids; combustible dust; and ignitible fibers or flyings.

The propellant used to spread the cleaning compound in some spray cans could be flammable. Many cleaning materials are flammable. Liquids sometime vaporize readily and flow into nooks and crannies of electrical equipment. Any electrical arc caused by making or breaking a circuit stands the chance of initiating a small explosion that could damage equipment. Any cleaning liquid that remains would probably ignite and burn.

If it is necessary to use flammable liquids or gases for any purpose in the immediate vicinity of an exposed electrical conductor or circuit part, an electrically safe work condition must be created before the material is used.

FPN: Electrical installation requirements for locations where flammable materials are present on a regular basis are contained in *NFPA 70, National Electrical Code.*

**(I) Anticipating Failure.** When there is evidence that electric equipment could fail and injure employees, the electric equipment shall be deenergized unless the employer can demonstrate that deenergizing introduces additional or increased hazards or is infeasible because of equipment design or operational limitation. Until the equipment is deenergized or repaired, employees shall be protected from hazards associated with the impending failure of the equipment.

Electrical equipment frequently offers indications that failure is impending, and workers should be capable of recognizing these indications. If an enclosure feels hot to the touch, for instance, the chance of equipment failure is increased. If unusual noises or sounds are heard, the chance of equipment failure is increased. Unusual smells also are an indication of impending failure. Workers must be aware that noises, sparks, hot surfaces, and unusual smells suggest that equipment is in the process of failing. If any of these indications is observed or recognized, the worker should not approach the equipment.

If these or other indications of impending failure exist, the equipment should be deenergized from a remote location. A circuit breaker or switch that supplies voltage to the suspect equipment should be opened. Disconnecting means located in equipment that has an indication of impending failure should not be operated unless the worker is protected from the effects of equipment failure.

**(J) Routine Opening and Closing of Circuits.** Load-rated switches, circuit breakers, or other devices specifically designed as disconnecting means shall be used for the opening, reversing, or closing of circuits under load conditions. Cable connectors not of the load-break type, fuses, terminal lugs, and cable splice connections shall not be permitted to be used for such purposes, except in an emergency.

Only equipment that has been rated to serve as load-break equipment can be used for routine control of electrical equipment or circuits. The manufacturer establishes the load-break rating after testing to ensure that the equipment can safely interrupt the full-load current expected in the circuit. Unless the rating equals or exceeds the full-load current of the circuit, the disconnecting device must not be operated unless the circuit has been deenergized.

With the exception of fuse-cutout devices operated with hot sticks, fuses are not designed to be removed from an energized circuit. Cable connectors are not designed to be opened under load. Removing a fuse or opening a cable connector that is conducting current can initiate an arc that could escalate into an arcing fault. These and similar devices must not be disconnected or removed unless the load current has been interrupted by another means. Disconnect switches that are not load rated can initiate an arcing fault if operated under load.

**(K) Reclosing Circuits After Protective Device Operation.** After a circuit is deenergized by a circuit protective device, the circuit shall not be manually reenergized until it has been determined that the equipment and circuit can be safely energized. The repetitive manual reclosing of circuit breakers or reenergizing circuits through replaced fuses shall be prohibited. When it is determined from the design of the circuit and the overcurrent devices involved that the automatic operation of a device was caused by an overload rather than a fault condition, examination of the circuit or connected equipment shall not be required before the circuit is reenergized.

Overcurrent protective devices operate if the rated current flow is exceeded. If an overcurrent device (fuse, circuit breaker, relay, overload, and similar devices) operates, a reasonable assumption is that the rated current has been exceeded. One possible reason for an overcurrent device to operate is that a short-circuit condition exists. Reclosing a circuit into a short circuit could have a disastrous result and must be avoided. Circuit breakers must not be reset and reclosed. Fuses must not be replaced and the switch closed again into a faulted condition. These devices should not be closed unless it has been determined that they are safe to be closed.

A second possible cause for a specific overcurrent device to operate is that the device has been exposed to an overloaded condition, such as a motor overload. A motor could experience an overloaded condition, due to a temporary mechanical condition in the driven equipment. If the design of the circuit suggests a mechanical overload, the circuit can be returned to its operating position. For instance, a motor overload relay can be reset.

### 130.7 Personal and Other Protective Equipment.

PPE generally is designed to afford protection from a single hazard and for a specific body part. For instance, voltage-rated gloves are designed for protection from electrocution. An arc-rated flame-resistant shirt is designed to protect the upper torso from a thermal hazard. Although some specific items of PPE might offer protection from more than one hazard, PPE is not designed for that purpose.

Hazards that are identified by the hazard/risk analysis should be considered separately. Protective equipment must be selected for each discrete hazard and then merged together into an overall system of protection for all hazards that the hazard/risk analysis identifies.

**(A) General.** Employees working in areas where electrical hazards are present shall be provided with, and shall use, protective equipment that is designed and constructed for the specific part of the body to be protected and for the work to be performed.

When selecting PPE, the worker must identify each hazard as well as the degree of each hazard. Next, the worker must determine which part of his or her body will be within the boundary associated with the hazard. If a worker's face is within the Arc Flash Protection Boundary, for instance, a thermally rated face shield is satisfactory. If the worker's entire head is within the Arc Flash Protection Boundary, protection for the back of the worker's head is required in addition to protection for the worker's face.

If a worker must approach an exposed energized electrical conductor or circuit part within the Arc Flash Protection Boundary or any of the Shock Protection Boundaries, some protective equipment is necessary.

FPN No. 1: The PPE requirements of 130.7 are intended to protect a person from arc flash and shock hazards. While some situations could result in burns to the skin, even with the protection selected, burn injury should be reduced and survivable. Due to the explosive effect of some arc events, physical trauma injuries could occur. The PPE requirements of 130.7 do not address protection against physical trauma other than exposure to the thermal effects of an arc flash.

During an arcing fault, electrical energy is converted to several other forms of energy. *NFPA 70E* does not define equipment for protection from the other forms of energy that are generated by an arcing fault. For instance, a significant pressure wave is generated in each arcing fault. The amount of force that could result from the pressure wave is not currently predictable. With the exception of PPE, the other strategies promoted by *NFPA 70E* offer some protection from the other hazards. However, only possible electrocution and thermal burns are addressed by PPE requirements.

FPN No. 2: When incident energy exceeds 40 cal/cm² at the working distance, greater emphasis may be necessary with respect to de-energizing before working within the Limited Approach Boundary of the exposed electrical conductors or circuit parts.

In addition to the thermal energy, an arcing fault generates a significant pressure wave. The degree of the pressure wave and possible results from it are not currently predictable with reasonable reliability. If the thermal energy exceeds 40 cal/cm², the accompanying pressure wave might injure any worker who is near.

**(B) Care of Equipment.** Protective equipment shall be maintained in a safe, reliable condition. The protective equipment shall be visually inspected before each use. Protective equipment shall be stored in a manner to prevent damage from physically damaging conditions and from moisture, dust, or other deteriorating agents.

FPN: Specific requirements for periodic testing of electrical protective equipment are given in 130.7(C)(8) and 130.7(F).

To ensure that PPE is effective when needed, it must be inspected and maintained at specific intervals and stored appropriately. The electrical safety program should define these intervals and inspection methods. If inspection intervals and test methodology are defined in a national consensus standard, the electrical safety program inspection methods and intervals must be at least as frequent as those stated in the consensus standard. Tables 130.7(C)(6)(c) and 130.7(C)(8) list national consensus standards that cover electrical PPE.

Flame-resistant clothing must be cleaned and maintained as defined by the clothing manufacturer. Such clothing slowly loses its flame-resistant characteristics when cleaned. There are very specific laundering instructions provided by the FR clothing manufacturers. If workers launder their PPE, they must follow those instructions, which generally require different wash/rinse cycles than are used for household washing. Not all bleaching additives are acceptable. If a laundry service is used, the laundry facility must be aware of the FR clothing manufacturer's laundering instructions. Manufacturer's instructions must be implemented with regard to restoring the flame-resistant characteristics. When FR clothing is shared by workers, the electrical safety program should consider health aspects of shared PPE when determining the cleaning frequency.

**(C) Personal Protective Equipment.**

The final barrier that has a chance of preventing an injury from an electrical hazard is PPE. The hazard/risk analysis determines if, when, and how a worker could be at risk of injury from an electrical hazard. If an employee is or could be exposed to an electrical hazard, he or she must use PPE that protects from injury.

**(1) General.** When an employee is working within the Arc Flash Protection Boundary he or she shall wear protective clothing and other personal protective equipment in accordance with 130.3. All parts of the body inside the Arc Flash Protection Boundary shall be protected.

Any worker, qualified or unqualified, who is within the Arc Flash Protection Boundary must be protected from the thermal effects of any potential arcing fault. If only a part of the worker's body would be within the Arc Flash Protection Boundary and would be exposed to injury from the thermal hazard, the portion of the body that is exposed must be protected from injury. If the worker's entire body is within the Arc Flash Protection Boundary, all parts of the body, including the back of the head and torso, must be protected from thermal injury. The PPE used for protection from thermal injury from an arcing fault must be arc-rated, flame-resistant equipment.

**(2) Movement and Visibility.** When flame-resistant (FR) clothing is worn to protect an employee, it shall cover all ignitible clothing and shall allow for movement and visibility.

When protective apparel and equipment are selected and sized appropriately, the worker's movement is not restricted. All parts of the worker's ignitable clothing must be protected from the significant incident energy to avoid igniting the covered apparel.

Ignitable clothing provides some additional protection and can be used in conjunction with arc-rated clothing for thermal insulation by some protective systems. See the FPN in Section 130.7(C)(11) to determine the rating of a protective system. Any clothing that will melt is not considered to provide any thermal insulation and cannot be considered for use as a component of arc-rated protective clothing.

The protective equipment must not unduly restrict the ability of the worker to see in the necessary direction. If needed, normal area lighting should be supplemented by temporary lighting for some face shields. Rehearsing a work task on similar equipment in an electrically safe work condition and wearing PPE is one way to determine if PPE is non-restricting.

When women wear PPE designed to fit men, the clothing could increase the risk of an incident as a result of the fit. Several manufacturers of arc-rated FR equipment provide clothing that is specifically designed for women. See Exhibit 130.12 later in this chapter.

**(3) Head, Face, Neck, and Chin (Head Area) Protection.** Employees shall wear nonconductive head protection wherever there is a danger of head injury from electric shock or burns due to contact with energized electrical conductors or circuit parts or from flying objects resulting from electrical explosion. Employees shall wear nonconductive protective equipment for the face, neck, and chin whenever there is a danger of injury from exposure to electric arcs or flashes or from flying objects resulting from electrical explosion. If employees use hairnets and/or beard nets, these items must be non-melting and flame resistant.

FPN: See 130.7(C)(13)(b) for arc flash protective requirements.

When an arcing fault occurs, electrical energy is converted into other forms of energy. Energy is released that covers the electromagnetic spectrum. Some of the electrical energy is converted into visible light, some to infrared energy, some to ultraviolet energy, and some to each of the remaining bands of the spectrum.

The significant current flow heats the conductor material, usually copper or aluminum, which melts and is vaporized. When the conductor material is vaporized, it expands. The air in the immediate vicinity of the arc current is heated very rapidly. The heated air also expands. The expanding conductor material and heated air creates a pressure wave that is similar to an explosion and swells outward from the fault current. Loose parts and materials are propelled outward from the arc.

Workers must wear electrically rated (Class E or G) hard hats to protect their heads from flying parts and pieces. All exterior apparel must be nonconductive. Exterior apparel must have an arc rating at least as great as the predicted incident energy. Exterior apparel must not ignite nor melt.

**(4) Eye Protection.** Employees shall wear protective equipment for the eyes whenever there is danger of injury from electric arcs, flashes, or from flying objects resulting from electrical explosion.

Electromagnetic energy is radiated from an arc. The pressure wave associated with the rapid heating expels parts of molten material from the arc. The worker must wear PPE to protect his or her eyes from damage. Eyeglasses or spectacles that meet the requirements of ANSI Z87.1, *Occupational and Educational Personal Eye and Face Protection Devices,* offer protection from impact and also filter a significant portion of the damaging ultraviolet energy.

If the worker's head is within the Arc Flash Protection Boundary, the worker's eyes must be protected from the thermal hazard as well. If the worker is wearing a face shield, the face shield must have an arc rating at least as great as the predicted incident energy. If the worker is wearing a hood for thermal protection, the viewing window must protect the worker's eyes from the thermal hazard.

Eyeglasses or spectacles that have exposed conductive components could fall into an exposed energized electrical conductor and initiate an arcing fault.

**(5) Body Protection.** Employees shall wear FR clothing wherever there is possible exposure to an electric arc flash above the threshold incident-energy level for a second-degree burn [5 J/cm$^2$ (1.2 cal/cm$^2$)].

FPN: Such clothing can be provided as an arc flash suit jacket and arc flash suit pants, shirts and pants, or as coveralls, or as a combination of jacket and pants, or, for increased protection, as coveralls with jacket and pants. Various weight fabrics are available. Generally, the higher degree of protection is provided by heavier weight fabrics and/or by layering combinations of one or more layers of FR clothing. In some cases, one or more layers of FR clothing are worn over flammable, non-melting clothing.

Employees who are within the Arc Flash Protection Boundary must wear FR clothing for protection from the thermal effects of an arcing fault. The requirement is that all ignitable clothing must be protected by at least one layer of clothing that has an established incident-energy rating. Clothing could be shirt and pants, coveralls, or any other assembly that provides protection for ignitable clothing or exposed skin.

Some FR clothing is rated for use with incident energy exposures up to 100 cal/cm$^2$. The *NFPA 70E* Technical Committee has determined that although it could be possible to protect a person from such extreme thermal exposure, the clothing would be unlikely to protect the worker from the effects of the accompanying pressure wave. If the arc flash hazard analysis indicates an exposure of more than 40 cal/cm$^2$, the task must not be performed until an electrically safe work condition exists. FR clothing with a very high incident energy rating might be needed to perform the steps necessary to establish an electrically safe work condition. However, this is the only task that should be accomplished with the equipment energized.

Several manufacturers make FR clothing using different materials. Generally, as the weight of the material increases, the degree of protection also increases. If FR clothing is worn in layers, some air is trapped between the layers, providing extra thermal insulation. Layering FR clothing increases the amount of thermal insulation afforded by the overall material.

FR clothing could be worn over flammable or non-melting materials, provided that it would keep the flammable clothing below its ignition point.

Ordinary clothing is categorized by the weight of the material per unit of area. Although FR clothing also can be categorized by the same method, it is categorized in calories per square centimeter. If FR clothing rated only by weight is on hand, the manufacturer should be consulted to determine the protective characteristics of the material.

Although arc-rated FR clothing provides protection from the thermal energy associated with an arcing fault, the apparel should be considered to provide no protection from shock.

**(6) Hand and Arm Protection.** Hand and arm protection shall be provided in accordance with (a), (b), and (c) below.

**(a) Shock Protection.** Employees shall wear rubber insulating gloves with leather protectors where there is a danger of hand injury from electric shock due to contact with energized electrical conductors or circuit parts. Employees shall wear rubber insulating gloves with leather protectors and rubber insulating sleeves where there is a danger of hand and arm injury from electric shock due to contact with energized electrical conductors or circuit parts. Rubber insulating gloves shall be rated for the voltage for which the gloves will be exposed.

*Exception: Where it is necessary to use rubber insulating gloves without leather protectors, the requirements of ASTM F 496, Standard Specification for In-Service Care of Insulating Gloves and Sleeves, shall be met.*

FPN: Table 130.7(C)(9) provides further information on tasks where rubber insulating gloves are required.

If an electrically safe work condition has not been established, the qualified worker must wear protection for his or her hands and arms. The protection must eliminate the possibility that the worker might be electrocuted. With the exception of rubber insulating gloves with leather protectors, shock-protective equipment might not provide protection from the associated thermal hazard. See Exhibit 130.4 for hand protection and rating labels.

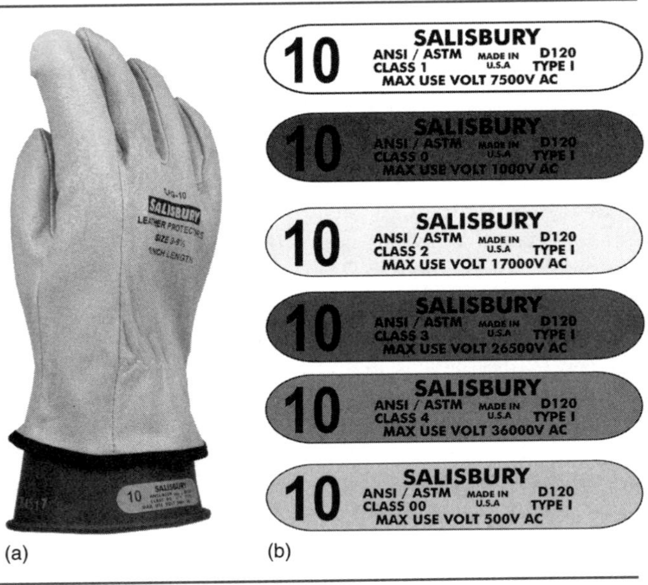


***EXHIBIT 130.4*** *Hand protection consisting of (a) rubber insulating gloves and leather protectors with an appropriate rating label and (b) rating labels. (Courtesy of Salisbury Electrical Safety, LLC)*

**(b) Arc Flash Protection.** Hand and arm protection shall be worn where there is possible exposure to arc flash burn. The apparel described in 130.7(C)(13)(c) shall be required for protection of hands from burns. Arm protection shall be accomplished by the apparel described in 130.7(C)(5).

If an electrically safe work condition has not been established, the worker must wear PPE that protects his or her hands and arms from the thermal hazard. The same clothing worn for body protection must provide flash protection for the worker's arms. Clothing selected and worn to protect the upper torso from thermal exposure must have long sleeves, and the sleeves must not be shortened or rolled up. Apparel that provides thermal protection for the worker's arms should be an integral part of the apparel that protects the upper torso.

Should an arcing fault occur, the worker's hands are likely to be much closer to the electrical arc than his or her torso. Although not rated for thermal protection, voltage-rated gloves with leather protectors provide significant thermal protection. Where the worker's hands are within the Arc Flash Protection Boundary, rubber insulating gloves should not be worn without leather protectors. FR-rated gloves are also available. PPE that provides thermal protection offers no acceptable protection from shock or electrocution.

**(c) Maintenance and Use.** Electrical protective equipment shall be maintained in a safe, reliable condition. Insulating equipment shall be inspected for damage before each day's use and immediately following any incident that can reasonably be suspected of having caused damage. Insulating gloves shall be given an air test, along with the inspection. Electrical protective equipment shall be subjected to periodic electrical tests. Test voltages and the maximum intervals between tests shall be in accordance with Table 130.7(C)(6)(c).

***TABLE 130.7(C)(6)(c)*** *Rubber Insulating Equipment, Maximum Test Intervals*

| *Rubber Insulating Equipment* | *When to Test* | *Governing Standard* for Test Voltage* |
|---|---|---|
| Blankets | Before first issue; every 12 months thereafter† | ASTM F 479 |
| Covers | If insulating value is suspect | ASTM F 478 |
| Gloves | Before first issue; every 6 months thereafter† | ASTM F 496 |
| Line hose | If insulating value is suspect | ASTM F 478 |
| Sleeves | Before first issue; every 12 months thereafter† | ASTM F 496 |

*ASTM F 478, *Standard Specification for In-Service Care of Insulating Line Hose and Covers*; ASTM F 479, *Standard Specification for In-Service Care of Insulating Blankets*; ASTM F 496, *Standard Specification for In-Service Care of Insulating Gloves and Sleeves.*

†If the insulating equipment has been electrically tested but not issued for service, it may not be placed into service unless it has been electrically tested within the previous 12 months.

Manufacturers of arc-rated, flame-resistant clothing and other arc-rated protective equipment provide instructions for cleaning and care of their products. The electrical safety program must describe the maintenance and cleaning process or processes that ensure the integrity of the apparel. Workers should inspect their arc-rated clothing and ensure that the apparel is not soiled with a flammable contaminant. Any patches or labels attached to arc-rated, flame-resistant clothing also must be arc-rated, flame-resistant material.

Shock-protective equipment must be physically inspected immediately prior to using it. The integrity of shock-protective equipment is paramount to avoid electrocution. The insulating integrity of this equipment must be verified by tests conducted in accordance with the manufacturer's instructions and the appropriate standards as shown in Table 130.7(C)(6)(c). Table 130.7(C)(6)(c) describes the maximum testing intervals for various products and identifies the appropriate testing standard.

FPN: See OSHA 1910.137 and ASTM F 496, *Standard Specification for In-Service Care of Insulating Gloves and Sleeves.*

**(7) Foot Protection.** Where insulated footwear is used as protection against step and touch potential, dielectric overshoes shall be required. Insulated soles shall not be used as primary electrical protection.

If the hazard/risk analysis indicates that the worker's feet and legs could be exposed to an arc flash, FR clothing worn to protect the lower torso must protect the worker's legs from exposure. Heavy-duty leather work shoes also must cover the worker's feet.

The integrity of the insulating quality of shoes with insulated soles cannot be established easily after the worker has been wearing them in a work environment. Therefore, shoes with insulated soles must not serve as the primary protection from touch and step potential. If such protection is warranted, the worker must wear dielectric overshoes (boots).

**(8) Standards for Personal Protective Equipment (PPE).** Personal protective equipment (PPE) shall conform to the standards given in Table 130.7(C)(8).

***TABLE 130.7(C)(8)*** *Standards on Protective Equipment*

| ***Subject*** | ***Number and Title*** |
|---|---|
| Head protection | ANSI Z89.1, *Requirements for Protective Headwear for Industrial Workers*, 2003 |
| Eye and face protection | ANSI Z87.1, *Practice for Occupational and Educational Eye and Face Protection*, 2003 |
| Gloves | ASTM D 120, *Standard Specification for Rubber Insulating Gloves*, 2002a (R 2006) |
| Sleeves | ASTM D 1051, *Standard Specification for Rubber Insulating Sleeves*, 2007 |
| Gloves and sleeves | ASTM F 496, *Standard Specification for In-Service Care of Insulating Gloves and Sleeves*, 2006 |
| Leather protectors | ASTM F 696, *Standard Specification for Leather Protectors for Rubber Insulating Gloves and Mittens*, 2006 |
| Footwear | ASTM F 1117, *Standard Specification for Dielectric Overshoe Footwear*, 2003<br>ASTM F 2412, *Standard Test Methods for Foot Protection*, 2005<br>ASTM F 2413, *Standard Specification for Performance Requirements for Foot Protection*, 2005 |
| Visual inspection | ASTM F 1236, *Standard Guide for Visual Inspection of Electrical Protective Rubber Products*, 1996 (R 2007) |
| Apparel | ASTM F 1506, *Standard Performance Specification for Flame Resistant Textile Materials for Wearing Apparel for Use by Electrical Workers Exposed to Momentary Electric Arc and Related Thermal Hazards*, 2002a |
| Raingear | ASTM F 1891, *Standard Specification for Arc and Flame Resistant Rainwear*, 2006 |
| Face protective products | ASTM F 2178, *Standard Test Method for Determining the Arc Rating and Standard Specification for Face Protective Products*, 2006 |
| Fall protection | ASTM F 887, *Standard Specifications for Personal Climbing Equipment*, 2005 |

FPN: Non-FR or flammable fabrics are not covered by a standard in Table 130.7(C)(8). See 130.7(C)(14) and 130.7(C)(15).

The standards listed in Table 130.7(C)(8) describe tests necessary for equipment to be considered acceptable. The table does not cover equipment that is not ordinarily considered PPE. For instance, voltmeters provide information that enables workers to protect themselves from electrical shock, but voltmeters are not covered in Table 130.7(C)(8).

**(9) Selection of Personal Protective Equipment When Required for Various Tasks.** Where selected in lieu of the incident energy analysis of 130.3(B)(1), Table 130.7(C)(9) shall be used to determine the hazard/risk category and requirements for use of rubber insulating gloves and insulated and insulating hand tools for a task. The assumed maximum short-circuit current capacities and maximum fault clearing times for various tasks are listed in the notes to Table 130.7(C)(9). For tasks not listed, or for power systems with greater than the assumed

***TABLE 130.7(C)(9)*** *Hazard/Risk Category Classifications and Use of Rubber Insulating Gloves and Insulated and Insulating Hand Tools*

| ***Tasks Performed on Energized Equipment*** | ***Hazard/Risk Category*** | ***Rubber Insulating Gloves*** | ***Insulated and Insulating Hand Tools*** |
|---|---|---|---|
| **Panelboards or Other Equipment Rated 240 V and Below — Note 1** | | | |
| Perform infrared thermography and other non-contact inspections outside the restricted approach boundary | 0 | N | N |
| Circuit breaker (CB) or fused switch operation with covers on | 0 | N | N |
| CB or fused switch operation with covers off | 0 | N | N |
| Work on energized electrical conductors and circuit parts, including voltage testing | 1 | Y | Y |
| Remove/install CBs or fused switches | 1 | Y | Y |
| See Exhibit 130.5 | | | |
| Removal of bolted covers (to expose bare, energized electrical conductors and circuit parts) | 1 | N | N |
| Opening hinged covers (to expose bare, energized electrical conductors and circuit parts) | 0 | N | N |
| Work on energized electrical conductors and circuit parts of utilization equipment fed directly by a branch circuit of the panelboard | 1 | Y | Y |
| **Panelboards or Switchboards Rated >240 V and up to 600 V (with molded case or insulated case circuit breakers) — Note 1** | | | |
| Perform infrared thermography and other non-contact inspections outside the Restricted Approach Boundary | 1 | N | N |
| CB or fused switch operation with covers on | 0 | N | N |
| CB or fused switch operation with covers off | 1 | Y | N |
| Work on energized electrical conductors and circuit parts, including voltage testing | 2* | Y | Y |
| Work on energized electrical conductors and circuit parts of utilization equipment fed directly by a branch circuit of the panelboard or switchboard | 2* | Y | Y |
| **600 V Class Motor Control Centers (MCCs) — Note 2 (except as indicated)** | | | |
| Perform infrared thermography and other non-contact inspections outside the restricted approach boundary | 1 | N | N |
| CB or fused switch or starter operation with enclosure doors closed | 0 | N | N |
| Reading a panel meter while operating a meter switch | 0 | N | N |

*(continues)*

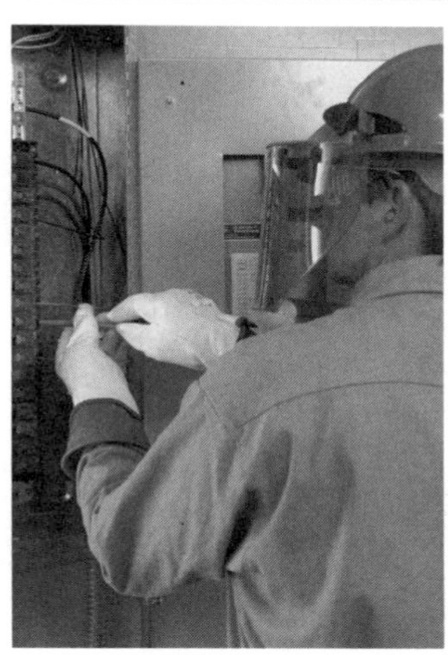

***EXHIBIT 130.5*** *H/R Cat 1 PPE to remove or install CB in a panelboard.*

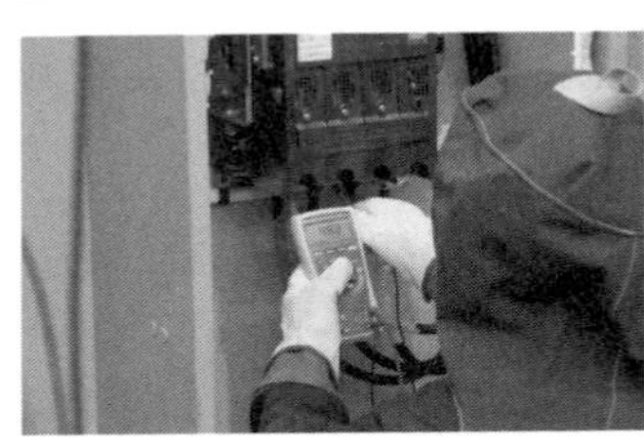

**EXHIBIT 130.6** *Measuring voltage in 480 V motor control center using H/R Cat 2* PPE.*

**EXHIBIT 130.7** *Current measurement in a 480 V motor control center also requires H/R Cat 2* PPE.*

**EXHIBIT 130.8** *Insertion of a 480 V drawout breaker with switchgear doors open using H/R Cat 4 PPE.*

*TABLE 130.7(C)(9) Continued*

| *Tasks Performed on Energized Equipment* | *Hazard/Risk Category* | *Rubber Insulating Gloves* | *Insulated and Insulating Hand Tools* |
|---|---|---|---|
| CB or fused switch or starter operation with enclosure doors open | 1 | N | N |
| Work on energized electrical conductors and circuit parts, including voltage testing | 2* | Y | Y |
| See Exhibits 130.6 and 130.7 | | | |
| Work on control circuits with energized electrical conductors and circuit parts 120 V or below, exposed | 0 | Y | Y |
| Work on control circuits with energized electrical conductors and circuit parts >120 V, exposed | 2* | Y | Y |
| Insertion or removal of individual starter “buckets” from MCC — Note 3 | 4 | Y | N |
| Application of safety grounds, after voltage test | 2* | Y | N |
| Removal of bolted covers (to expose bare, energized electrical conductors and circuit parts) — Note 3 | 4 | N | N |
| Opening hinged covers (to expose bare, energized electrical conductors and circuit parts) — Note 3 | 1 | N | N |
| Work on energized electrical conductors and circuit parts of utilization equipment fed directly by a branch circuit of the motor control center | 2* | Y | Y |
| **600 V Class Switchgear (with power circuit breakers or fused switches) — Note 4** | | | |
| Perform infrared thermography and other non-contact inspections outside the restricted approach boundary | 2 | N | N |
| CB or fused switch operation with enclosure doors closed | 0 | N | N |
| Reading a panel meter while operating a meter switch | 0 | N | N |
| CB or fused switch operation with enclosure doors open | 1 | N | N |
| Work on energized electrical conductors and circuit parts, including voltage testing | 2* | Y | Y |
| Work on control circuits with energized electrical conductors and circuit parts 120 V or below, exposed | 0 | Y | Y |
| Work on control circuits with energized electrical conductors and circuit parts >120 V, exposed | 2* | Y | Y |
| Insertion or removal (racking) of CBs from cubicles, doors open or closed | 4 | N | N |
| See Exhibit 130.8 | | | |

*TABLE 130.7(C)(9) Continued*

| *Tasks Performed on Energized Equipment* | *Hazard/Risk Category* | *Rubber Insulating Gloves* | *Insulated and Insulating Hand Tools* |
|---|---|---|---|
| Application of safety grounds, after voltage test | 2* | Y | N |
| Removal of bolted covers (to expose bare, energized electrical conductors and circuit parts) | 4 | N | N |
| Opening hinged covers (to expose bare, energized electrical conductors and circuit parts) | 2 | N | N |
| **Other 600 V Class (277 V through 600 V, nominal) Equipment — Note 2 (except as indicated)** | | | |
| Lighting or small power transformers (600 V, maximum) | | | |
| Removal of bolted covers (to expose bare, energized electrical conductors and circuit parts) | 2* | N | N |
| Opening hinged covers (to expose bare, energized electrical conductors and circuit parts) | 1 | N | N |
| Work on energized electrical conductors and circuit parts, including voltage testing | 2* | Y | Y |
| Application of safety grounds, after voltage test | 2* | Y | N |
| Revenue meters (kW-hour, at primary voltage and current) | — | — | — |
| Insertion or removal | 2* | Y | N |
| Cable trough or tray cover removal or installation | 1 | N | N |
| Miscellaneous equipment cover removal or installation | 1 | N | N |
| Work on energized electrical conductors and circuit parts, including voltage testing | 2* | Y | Y |
| Application of safety grounds, after voltage test | 2* | Y | N |
| Insertion or removal of plug-in devices into or from busways | 2* | Y | N |
| **NEMA E2 (fused contactor) Motor Starters, 2.3 kV Through 7.2 kV** | | | |
| Perform infrared thermography and other non-contact inspections outside the restricted approach boundary | 3 | N | N |
| Contactor operation with enclosure doors closed | 0 | N | N |
| Reading a panel meter while operating a meter switch | 0 | N | N |
| Contactor operation with enclosure doors open | 2* | N | N |
| Work on energized electrical conductors and circuit parts, including voltage testing | 4 | Y | Y |
| Work on control circuits with energized electrical conductors and circuit parts 120 V or below, exposed | 0 | Y | Y |
| Work on control circuits with energized electrical conductors and circuit parts >120 V, exposed | 3 | Y | Y |

*(continues)*

*TABLE 130.7(C)(9) Continued*

| *Tasks Performed on Energized Equipment* | *Hazard/Risk Category* | *Rubber Insulating Gloves* | *Insulated and Insulating Hand Tools* |
|---|---|---|---|
| Insertion or removal (racking) of starters from cubicles, doors open or closed | 4 | N | N |
| Application of safety grounds, after voltage test | 3 | Y | N |
| Removal of bolted covers (to expose bare, energized electrical conductors and circuit parts) | 4 | N | N |
| Opening hinged covers (to expose bare, energized electrical conductors and circuit parts) | 3 | N | N |
| Insertion or removal (racking) of starters from cubicles of arc-resistant construction, tested in accordance with IEEE C37.20.7, doors closed only | 0 | N | N |
| **Metal Clad Switchgear, 1 kV Through 38 kV** | | | |
| Perform infrared thermography and other non-contact inspections outside the restricted approach boundary | 3 | N | N |
| CB operation with enclosure doors closed | 2 | N | N |
| Reading a panel meter while operating a meter switch | 0 | N | N |
| CB operation with enclosure doors open | 4 | N | N |
| Work on energized electrical conductors and circuit parts, including voltage testing | 4 | Y | Y |
| Work on control circuits with energized electrical conductors and circuit parts 120 V or below, exposed | 2 | Y | Y |
| Work on control circuits with energized electrical conductors and circuit parts >120 V, exposed | 4 | Y | Y |
| Insertion or removal (racking) of CBs from cubicles, doors open or closed | 4 | N | N |
| Application of safety grounds, after voltage test | 4 | Y | N |
| Removal of bolted covers (to expose bare, energized electrical conductors and circuit parts) | 4 | N | N |
| Opening hinged covers (to expose bare, energized electrical conductors and circuit parts) | 3 | N | N |
| Opening voltage transformer or control power transformer compartments | 4 | N | N |
| **Arc-Resistant Switchgear Type 1 or 2 (for clearing times of <0.5 sec with a perspective fault current not to exceed the arc resistant rating of the equipment)** | | | |
| CB operation with enclosure door closed | 0 | N | N |
| Insertion or removal (racking) of CBs from cubicles, doors closed | 0 | N | N |
| Insertion or removal of CBs from cubicles with door open | 4 | N | N |

*TABLE 130.7(C)(9) Continued*

| *Tasks Performed on Energized Equipment* | *Hazard/Risk Category* | *Rubber Insulating Gloves* | *Insulated and Insulating Hand Tools* |
|---|---|---|---|
| Work on control circuits with energized electrical conductors and circuit parts 120 V or below, exposed | 2 | Y | Y |
| Insertion or removal (racking) of ground and test device with door closed | 0 | N | N |
| Insertion or removal (racking) of voltage transformers on or off the bus door closed | 0 | N | N |
| **Other Equipment 1 kV Through 38 kV** | | | |
| Metal-enclosed interrupter switchgear, fused or unfused | — | — | — |
| Switch operation of arc-resistant-type construction, tested in accordance with IEEE C37.20.7, doors closed only | 0 | N | N |
| Switch operation, doors closed | 2 | N | N |
| Work on energized electrical conductors and circuit parts, including voltage testing | 4 | Y | Y |
| Removal of bolted covers (to expose bare, energized electrical conductors and circuit parts) | 4 | N | N |
| Opening hinged covers (to expose bare, energized electrical conductors and circuit parts) | 3 | N | N |
| Outdoor disconnect switch operation (hookstick operated) | 3 | Y | Y |
| Outdoor disconnect switch operation (gang-operated, from grade) | 2 | Y | N |
| Insulated cable examination, in manhole or other confined space | 4 | Y | N |
| Insulated cable examination, in open area | 2 | Y | N |

General Notes (applicable to the entire table):

(a) Rubber insulating gloves are gloves rated for the maximum line-to-line voltage upon which work will be done.

(b) Insulated and insulating hand tools are tools rated and tested for the maximum line-to-line voltage upon which work will be done, and are manufactured and tested in accordance with ASTM F 1505, *Standard Specification for Insulated and Insulating Hand Tools.*

(c) Y = yes (required), N = no (not required).

(d) For systems rated less than 1000 volts, the fault currents and upstream protective device clearing times are based on an 18 in. working distance.

(e) For systems rated 1 kV and greater, the Hazard/Risk Categories are based on a 36 in. working distance.

(f) For equipment protected by upstream current limiting fuses with arcing fault current in their current limiting range ($^1/_2$ cycle fault clearing time or less), the hazard/risk category required may be reduced by one number.

Specific Notes (as referenced in the table):

1. Maximum of 25 kA short circuit current available; maximum of 0.03 sec (2 cycle) fault clearing time.
2. Maximum of 65 kA short circuit current available; maximum of 0.03 sec (2 cycle) fault clearing time.
3. Maximum of 42 kA short circuit current available; maximum of 0.33 sec (20 cycle) fault clearing time.
4. Maximum of 35 kA short circuit current available; maximum of up to 0.5 sec (30 cycle) fault clearing time.

**maximum short-circuit current capacity or with longer than the assumed maximum fault clearing times, an arc flash hazard analysis shall be required in accordance with 130.3.**

Table 130.7(C)(9) lists common work tasks, and this table can be used to determine a hazard/risk category. A hazard analysis, including an arc flash hazard analysis, has been performed for each common task listed in the table. The hazard analysis is based on parameters that commonly are found in industrial workplaces and are identified as Specific Notes at the bottom of the table. If the circuit conditions exceed the requirements in the notes that follow the table, an incident energy analysis must be performed. If Table 130.7(C)(9) requires H/R Category 4 PPE for medium voltage tasks, an incident energy analysis is suggested since some stronger power systems can result in incident energies above 40 cal/cm$^2$.

The protective equipment identified in Table 130.7(C)(9) considers both electrical circuit parameters and the physical attributes of the equipment and work task. Work tasks that are not listed in the table must be subjected to a hazard/risk analysis that considers both shock and arc flash hazards and includes the incident energy analysis required by 130.3B(1).

The content of Table 130.9(C)(9) is based on experience, so the protective equipment contained in the table might be different from protective equipment determined by other methods. Also, the table does not attempt to address risk associated with the work task. Workers and their supervisors must assess risk based on the specific conditions that exist on a work site.

FPN No. 1: The work tasks and protective equipment identified in Table 130.7(C)(9) were identified by a task group and the protective clothing and equipment selected was based on the collective experience of the task group. The protective clothing and equipment is generally based on determination of estimated exposure levels.

In several cases where the risk of an arc flash incident is considered low, very low, or extremely low by the task group, the hazard/risk category number has been reduced by 1, 2, or 3 numbers, respectively. The collective experience of the task group is that in most cases closed doors do not provide enough protection to eliminate the need for PPE for instances where the state of the equipment is known to readily change (e.g., doors open or closed, rack in or rack out). The premise used by the Task Group is considered to be reasonable, based on the consensus judgment of the full NFPA 70E Technical Committee.

FPN No. 2: Both larger and smaller available short-circuit currents could result in higher available arc flash energies. If the available short-circuit current increases without a decrease in the opening time of the overcurrent protective device, the arc flash energy will increase. If the available short-circuit current decreases, resulting in a longer opening time for the overcurrent protective device, arc flash energies could also increase.

Dual-function overcurrent devices, such as current-limiting fuses or circuit breakers, do not limit the current when the fault current is less than the trigger point of the current-limiting element of the device. Until the fault current reaches the lower end of the current-limiting characteristic, very high incident energy exposures can exist because the clearing time is long. Fault current normally is dependent on the impedance of the ground-fault current return path. The integrity of the ground-fault current return path is crucial to ensure that any fault current would be sufficient to reach the current-limiting range of the device.

FPN No. 3: Energized electrical conductors or circuit parts that operate at less than 50 volts may need to be de-energized to satisfy an "electrically safe work condition." Consideration should be given to the capacity of the source, any overcurrent protection between the energy source and the worker, and whether the work task related to the source operating at less than 50 volts increases exposure to electrical burns or to explosion from an electric arc.

The generally accepted lowest voltage that can result in a significant shock is 50 volts. That voltage level is selected by several standards as the lower limit of the danger level. However, employers and employees should be aware that electrical burns also could be received at lower voltages.

FPN No. 4: See 130.1(B)(2)(6) for requirements on documenting the available short-circuit current and fault clearing time.

Notes to Table 130.7(C)(9) define the available short-circuit current and fault-clearing time that were assumed to determine the hazard category for the table. The required PPE for work tasks that have higher short-circuit current and/or higher fault-clearing times must be determined by a different process. Note that the clearing times in the Specific Notes are well within the current-limiting range of current-limiting devices. However, if the fault current is below the current-limiting range of the overcurrent device, the clearing times could be longer than in the Specific Notes and the incident energy should be calculated. Also see 130.3, Exception 1 when evaluating the tasks in the panelboards and other equipment rated 240 V and below. Exhibits 130.5 through 130.8 show PPE for the specific task marked in Table 130.7(C)(9).

**(10) Protective Clothing and Personal Protective Equipment Matrix.** Once the Hazard/Risk Category has been identified from Table 130.7(C)(9) (including associated notes)

***TABLE 130.7(C)(10)*** *Protective Clothing and Personal Protective Equipment (PPE)*

| *Hazard/Risk Category* | *Protective Clothing and PPE* |
|---|---|
| **Hazard/Risk Category 0** | |
| Protective Clothing, Nonmelting (according to ASTM F 1506-00) or Untreated Natural Fiber | Shirt (long sleeve)<br>Pants (long) |
| FR Protective Equipment | Safety glasses or safety goggles (SR)<br>Hearing protection (ear canal inserts)<br>Leather gloves (AN) (Note 2) |
| **Hazard/Risk Category 1** | |
| FR Clothing, Minimum Arc Rating of 4 (Note 1) | Arc-rated long-sleeve shirt (Note 3)<br>Arc-rated pants (Note 3)<br>Arc-rated coverall (Note 4)<br>Arc-rated face shield or arc flash suit hood (Note 7)<br>Arc-rated jacket, parka, or rainwear (AN) |
| FR Protective Equipment | Hard hat<br>Safety glasses or safety goggles (SR)<br>Hearing protection (ear canal inserts)<br>Leather gloves (Note 2)<br>Leather work shoes (AN) |
| **Hazard/Risk Category 2** | |
| FR Clothing, Minimum Arc Rating of 8 (Note 1) | Arc-rated long-sleeve shirt (Note 5)<br>Arc-rated pants (Note 5)<br>Arc-rated coverall (Note 6)<br>Arc-rated face shield or arc flash suit hood (Note 7)<br>Arc rated jacket, parka, or rainwear (AN) |
| FR Protective Equipment | Hard hat<br>Safety glasses or safety goggles (SR)<br>Hearing protection (ear canal inserts)<br>Leather gloves (Note 2)<br>Leather work shoes |
| **Hazard/Risk Category 2*** | |
| FR Clothing, Minimum Arc Rating of 8 (Note 1) | Arc-rated long-sleeve shirt (Note 5)<br>Arc-rated pants (Note 5)<br>Arc-rated coverall (Note 6)<br>Arc-rated arc flash suit hood (Note 10)<br>Arc-rated jacket, parka, or rainwear (AN) |

*(continues)*

*TABLE 130.7(C)(10) Continued*

| *Hazard/Risk Category* | *Protective Clothing and PPE* |
|---|---|
| FR Protective Equipment | Hard hat<br>Safety glasses or safety goggles (SR)<br>Hearing protection (ear canal inserts)<br>Leather gloves (Note 2)<br>Leather work shoes |
| **Hazard/Risk Category 3** | |
| FR Clothing, Minimum Arc Rating of 25 (Note 1) | Arc-rated long-sleeve shirt (AR) (Note 8)<br>Arc-rated pants (AR) (Note 8)<br>Arc-rated coverall (AR) (Note 8)<br>Arc-rated arc flash suit jacket (AR) (Note 8)<br>Arc-rated arc flash suit pants (AR) (Note 8)<br>Arc-rated arc flash suit hood (Note 8)<br>Arc-rated jacket, parka, or rainwear (AN) |
| FR Protective Equipment | Hard hat<br>FR hard hat liner (AR)<br>Safety glasses or safety goggles (SR)<br>Hearing protection (ear canal inserts)<br>Arc-rated gloves (Note 2)<br>Leather work shoes |
| **Hazard/Risk Category 4** | |
| FR Clothing, Minimum Arc Rating of 40 (Note 1) | Arc-rated long-sleeve shirt (AR) (Note 9)<br>Arc-rated pants (AR) (Note 9)<br>Arc-rated coverall (AR) (Note 9)<br>Arc-rated arc flash suit jacket (AR) (Note 9)<br>Arc-rated arc flash suit pants (AR) (Note 9)<br>Arc-rated arc flash suit hood (Note 9)<br>Arc-rated jacket, parka, or rainwear (AN) |
| FR Protective Equipment | Hard hat<br>FR hard hat liner (AR)<br>Safety glasses or safety goggles (SR)<br>Hearing protection (ear canal inserts)<br>Arc-rated gloves (Note 2)<br>Leather work shoes |

AN = As needed (optional)

AR = As required

SR = Selection required

Notes:

1. See Table 130.7(C)(11). Arc rating for a garment or system of garments is expressed in cal/cm$^2$.

2. If rubber insulating gloves with leather protectors are required by Table 130.7(C)(9), additional leather or arc-rated gloves are not required. The combination of rubber insulating gloves with leather protectors satisfies the arc flash protection requirement.

3. The FR shirt and pants used for Hazard/ Risk Category 1 shall have a minimum arc rating of 4.

4. Alternate is to use FR coveralls (minimum arc rating of 4) instead of FR shirt and FR pants.

5. FR shirt and FR pants used for Hazard/ Risk Category 2 shall have a minimum arc rating of 8.

6. Alternate is to use FR coveralls (minimum arc rating of 8) instead of FR shirt and FR pants.

7. A face shield with a minimum arc rating of 4 for Hazard/Risk Category 1 or a minimum arc rating of 8 for Hazard/Risk Category 2, with wrap-around guarding to protect not only the face, but also the forehead, ears, and neck (or, alternatively, an arc-rated arc flash suit hood), is required.

8. An alternate is to use a total FR clothing system and hood, which shall have a minimum arc rating of 25 for Hazard/Risk Category 3.

9. The total clothing system consisting of FR shirt and pants and/or FR coveralls and/or arc flash coat and pants and hood shall have a minimum arc rating of 40 for Hazard/Risk Category 4.

10. Alternate is to use a face shield with a minimum arc rating of 8 and a balaclava (sock hood) with a minimum arc rating of 8 and which covers the face, head and neck except for the eye and nose areas

See Exhibit 130.9 for both cases noted in Note 10.

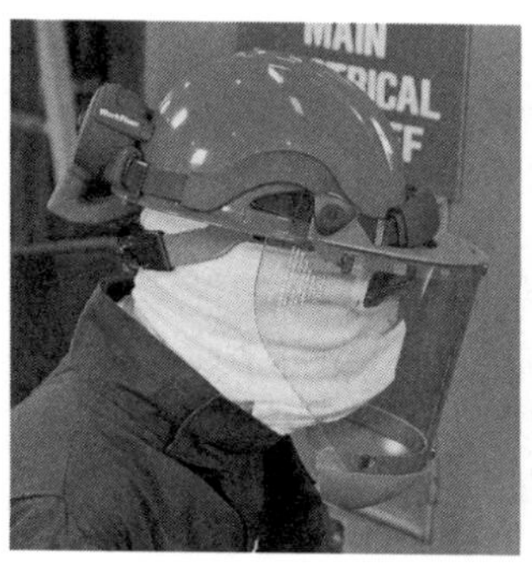


(a)

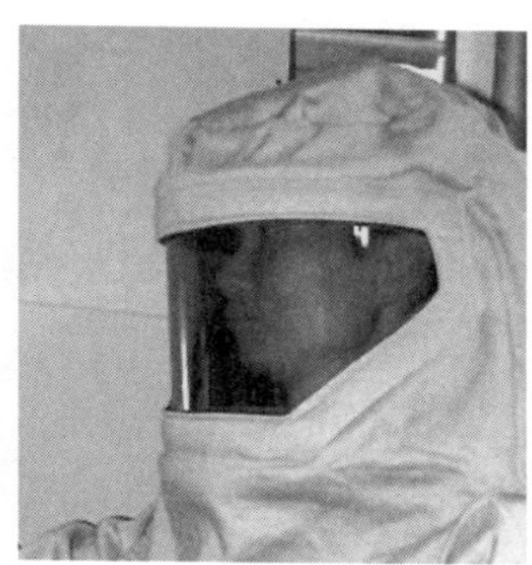

(b)

*EXHIBIT 130.9 A face shield with (a) a balaclava sock hood or (b) a flash suit hood. (Courtesy of Salisbury Electrical Safety, LLC)*

and the requirements of 130.7(C)(9), Table 130.7(C)(10) shall be used to determine the required PPE for the task. Table 130.7(C)(10) lists the requirements for protective clothing and other protective equipment based on Hazard/Risk Category numbers 0 through 4. This clothing and equipment shall be used when working within the Arc Flash Protection Boundary.

The protective clothing matrix shown in Table 130.7(C)(10) is intended to provide helpful information. The content of the table is *illustrative only*. FR clothing is available in many different constructions. Using FR clothing in layers is one way to achieve a higher level of protection. Table 130.7(C)(10) suggests acceptable combinations of clothing items to achieve a desired hazard/risk category. Other combinations are also be possible. Section 130.7 requires that any body part within the Arc Flash Protection Boundary be protected from the thermal effects of an arcing fault. Defining a hazard/risk category is a viable way to translate incident energy exposure to selection of FR clothing.

See Exhibit 130.9 for two solutions to the Hazard Risk Category 2* head protection. If an incident energy analysis is done rather that using Table 130.7(C)(9) and the result is 7.9 cal/cm$^2$, the same PPE as shown in Exhibit 130.9 needs to be worn because the back of the head needs protection. Exhibits 130.9(a) and (b) are both acceptable for H/R Cat 2* per Note 10 of Table 130.7(C)(10).

FPN No. 1: See Annex H for a suggested simplified approach to ensure adequate PPE for electrical workers within facilities with large and diverse electrical systems.

To simplify administration of FR clothing, an employer might implement a procedure that has two or three different assemblies of arc flash protective equipment. An employer could define a general requirement for workers to wear a minimum level of protective clothing. The procedure could define a secondary level of protection that is easily recognized by employees, and still a third level of protection that is required in special situations. The keys to this type of requirement are that the protection is adequate for the greatest exposure for each level of protective clothing and that each worker can recognize when it is necessary to wear a higher level of protection.

FPN No. 2: The PPE requirements of this section are intended to protect a person from arc flash and shock hazards. While some situations could result in burns to the skin, even with the protection described in Table 130.7(C)(10), burn injury should be reduced and survivable. Due to the explosive effect of some arc events, physical trauma injuries could occur. The PPE requirements of this section do not address protection against physical trauma other than exposure to the thermal effects of an arc flash.

The nature of thermal energy enables the wearing of FR clothing to reduce the chance of thermal injury. However, because the nature of an arcing fault is unpredictable, determining the degree of each hazard associated with an arcing fault is difficult, and the effectiveness of the determination might not be complete. The effects of arc blast also are not predictable. Equipment for protection from the pressure wave or other conditions associated with an arcing fault is not available. Any work performed on or near exposed energized electrical conductors exposes a worker to an elevated risk of injury.

**(11) Protective Clothing Characteristics.** Table 130.7(C)(11) lists examples of protective clothing systems and typical characteristics, including the degree of protection, for various clothing. The protective clothing selected for the corresponding Hazard/Risk Category number determined from Table 130.7(C)(9) (including associated notes) and the requirements of 130.7(C)(9) shall have an arc rating of at least the value listed in the last column of Table 130.7(C)(11).

***TABLE 130.7(C)(11)*** *Protective Clothing Characteristics*

| ***Hazard/Risk Category*** | ***Clothing Description*** | ***Required Minimum Arc Rating of PPE [J/cm²(cal/cm²)]*** |
|---|---|---|
| 0 | Nonmelting, flammable materials (i.e., untreated cotton, wool, rayon, or silk, or blends of these materials) with a fabric weight at least 4.5 oz/yd² | N/A |
| 1 | Arc-rated FR shirt and FR pants or FR coverall | 16.74 (4) |
| 2 | Arc-rated FR shirt and FR pants or FR coverall | 33.47 (8) |
| 3 | Arc-rated FR shirt and pants or FR coverall, and arc flash suit selected so that the system arc rating meets the required minimum | 104.6 (25) |
| 4 | Arc-rated FR shirt and pants or FR coverall, and arc flash suit selected so that the system arc rating meets the required minimum | 167.36 (40) |

Note: Arc rating is defined in Article 100 and can be either ATPV or $E_{BT}$. ATPV is defined in ASTM F 1959, *Standard Test Method for Determining the Arc Thermal Performance Value of Materials for Clothing*, as the incident energy on a material or a multilayer system of materials that results in a 50% probability that sufficient heat transfer through the tested specimen is predicted to cause the onset of a second-degree skin burn injury based on the Stoll curve, cal/cm². $E_{BT}$ is defined in ASTM F 1959 as the incident energy on a material or material system that results in a 50% probability of breakopen. Arc rating is reported as either ATPV or $E_{BT}$, whichever is the lower value.

HRC 0 is not FR clothing. The use of HRC 0 is only for those tasks listed in Table 130.7(C)(9) that permit HRC 0. If Table 130.7(C)(9) is not used or a task is not listed in the table, an incident energy analysis must be used. Section 130.7(C)(5) requires FR clothing to be used for all incident energy levels above 1.2 cal/cm². HRC 0 requires a fabric weight of at least 4.5 oz/yd². As non-FR materials get laundered and worn, the fabric weight could get lower than that required, and it is difficult to confirm that the material is still suitable for HRC 0 use. Even if Table 130.7(C)(9) is used, there is nothing that prohibits the use of FR clothing rated 4 cal/cm² when performing a task from that table that requires HRC 0 PPE.

FPN: The arc rating for a particular clothing system can be obtained from the FR clothing manufacturer.

Table 130.7(C)(11) provides general information that can help a worker understand the process for selecting clothing based on a hazard/risk category designation. This table describes the protective nature of clothing that meets a specific hazard/risk category. The table can be used with other methods of determining the necessary protective clothing.

Table 130.7(C)(11) is to serve as an illustration only. The table does not describe any required combination or construction of a protective system. The manufactured system can differ from the content of the table. The clothing manufacturer must be consulted. See the FPN in Section 130.7(C)(11).

When FR clothing meeting the requirements of ASTM F 1959, *Standard Test Method for Determining the Arc Thermal Performance Value of Materials for Clothing,* is exposed to an electrical arc to establish its rating, the rating is the failure mode in which the material chars and breaks open. Charring causes the fibers of the material to close the mesh in the material weave. The charred fabric could break open and expose the material (or skin) underneath. The arc rating of FR clothing is either the arc-thermal performance value (ATPV — ASTM F 1959) or the energy level ($E_{BT}$ — ASTM F 1959) that caused breakopen to occur when the fabric was tested.

PPE rated in cal/cm$^2$ is suitable for that incident energy level. The rating of layered PPE can only be determined by test. The rating of a clothing system recommended by a FR clothing manufacturer is only determined by testing those specific layers together. See the FPN in Section 130.7(C)(11).

**(12) Factors in Selection of Protective Clothing.** Clothing and equipment that provide worker protection from shock and arc flash hazards shall be utilized. Clothing and equipment required for the degree of exposure shall be permitted to be worn alone or integrated with flammable, nonmelting apparel. If FR clothing is required, it shall cover associated parts of the body as well as all flammable apparel while allowing movement and visibility. All personal protective equipment shall be maintained in a sanitary and functionally effective condition. Personal protective equipment items will normally be used in conjunction with one another as a system to provide the appropriate level of protection.

The role of protective clothing is to mitigate the potential for thermal injury. To do that, the protective nature of the FR clothing must be rated for use in an environment that is or could be influenced by the presence of electrical energy. The nature of an arcing fault is that thermal energy is generated very rapidly. The protective clothing must be capable of providing the necessary protection in that environment.

To avoid injury from a thermal hazard, the worker's clothing must ensure that the worker's skin surface does not receive more than 1.2 cal/cm$^2$. The protective clothing must prevent any flammable underclothing from igniting and burning. Additionally, the protective clothing must ensure that any underclothing with limited meltable fabric does not melt onto the worker's skin. See the exception in 130.7(C)(14) for limited meltable fabric explanation.

Materials that have an established arc flash rating can be used either alone or in combination with materials without an established rating. Materials from one manufacturer also can be used with materials from a different manufacturer, provided both sets of clothing have an established rating. The worker's clothing should be viewed as a system that avoids injury from the thermal hazard. The system can consist of multiple clothing items but the worker must understand that the system includes all garments being worn. Arc-rated FR-protective equipment must contain a label or other mark that describes the maximum incident energy rating.

Clothing must be maintained in a clean and sanitary condition and stored in a way that has no deleterious effect on the protective characteristics of the clothing.

FPN: Protective clothing includes shirts, pants, coveralls, jackets, and parkas worn routinely by workers who, under normal working conditions, are exposed to momentary electric arc and related thermal hazards. Flame-resistant rainwear worn in inclement weather is included in this category of clothing.

(a) Layering. Nonmelting, flammable fiber garments shall be permitted to be used as underlayers in conjunction with FR garments in a layered system for added protection. If nonmelting, flammable fiber garments are used as underlayers, the system arc rating shall be sufficient to prevent breakopen of the innermost FR layer at the expected arc exposure incident energy level to prevent ignition of flammable underlayers.

FPN: A typical layering system might include cotton underwear, a cotton shirt and trouser, and a FR coverall. Specific tasks might call for additional FR layers to achieve the required protection level.

Layering increases the overall protective characteristics of FR clothing. If ignitable fabrics are used as underlayers, the arc flash rating of the clothing system must not permit the outer layer to break open and directly expose the ignitable material to the arcing fault. The outer layers must limit the temperature rise of the ignitable underlayers to no more than 1.2 cal/cm$^2$.

Air is a good thermal insulator. Wearing multiple layers of clothing traps air in between the clothing layers. The layering effect tends to improve the thermal insulating efficiency of the overall protective system. However, although the air between layers improves the thermal insulating efficiency, the increase in efficiency does not increase the arc rating of the protective clothing system. The clothing manufacturer should be consulted to determine the arc rating of the overall system of protective clothing.

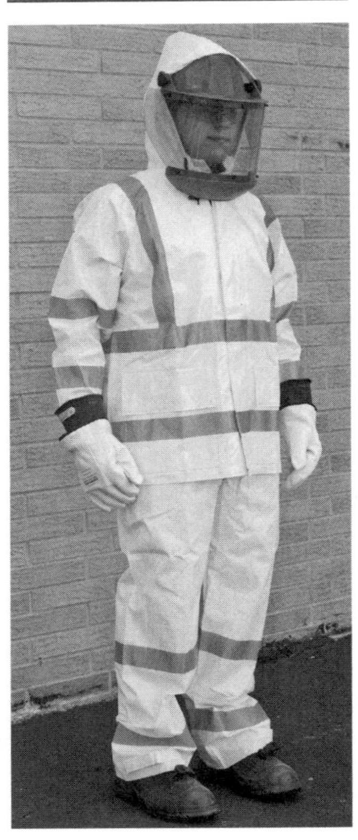

***EXHIBIT 130.10*** *FR rainwear. (Courtesy of Salisbury Electrical Safety, LLC)*

(b) Outer Layers. Garments worn as outer layers over FR clothing, such as jackets or rainwear, shall also be made from FR material.

When the protective clothing system consists of arc-rated, FR-protective clothing and clothing that is flammable, the worker must ensure that all flammable material is not exposed to the arcing fault. The flammable clothing decreases the amount of thermal energy that is conducted through the clothing to the worker's skin. However, all flammable materials must be protected from ignition. Note that arc-rated, FR-material rainwear is available in the marketplace. See Exhibit 130.10 for an example of this type of clothing.

(c) Underlayers. Meltable fibers such as acetate, nylon, polyester, polypropylene, and spandex shall not be permitted in fabric underlayers (underwear) next to the skin.

*Exception: An incidental amount of elastic used on nonmelting fabric underwear or socks shall be permitted.*

FPN No. 1: FR garments (e.g., shirts, trousers, and coveralls) worn as underlayers that neither ignite nor melt and drip in the course of an exposure to electric arc and related thermal hazards generally provide a higher system arc rating than nonmelting, flammable fiber underlayers.

FPN No. 2: FR underwear or undergarments used as underlayers generally provide a higher system arc rating than nonmelting, flammable fiber underwear or undergarments used as underlayers.

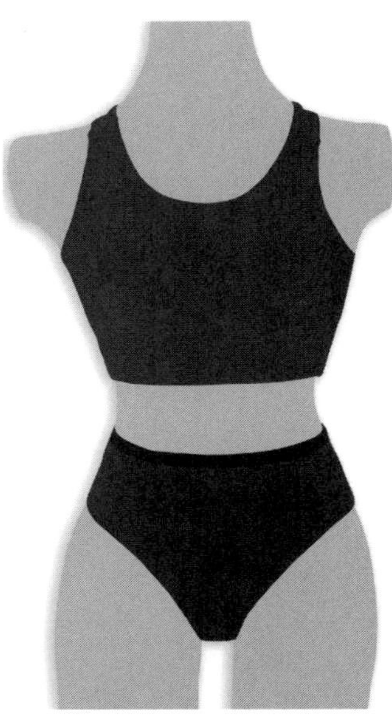

***EXHIBIT 130.11*** *FR underlayers for women. (Courtesy of DRIFIRE, LLC)*

Significant injuries occur when fabrics melt onto a worker's skin. Clothing made from materials that melt, such as acetate, nylon, polyester, polypropylene, or spandex, must not be worn next to a worker's skin. However, an incidental quantity of these fabrics that is used in the elastic bands in underwear is permitted. Exhibit 130.11 shows an example of FR underlayers that are available for women.

(d) Coverage. Clothing shall cover potentially exposed areas as completely as possible. Shirt sleeves shall be fastened at the wrists, and shirts and jackets shall be closed at the neck.

When required by the hazard/risk analysis, FR clothing must completely cover all body areas within the Arc Flash Boundary. Shirt sleeves must be fastened at the wrists, and the top button of shirts and/or jackets must be fastened to minimize the chance that heated air could reach below the FR clothing. Shirt sleeves should fit under the gauntlet of the protective gloves to minimize the chance that thermal energy could enter under the shirt sleeves.

(e) Fit. Tight-fitting clothing shall be avoided. Loose-fitting clothing provides additional thermal insulation because of air spaces. FR apparel shall fit properly such that it does not interfere with the work task.

The fit of FR clothing is important to the safety of the worker. When the surface of FR clothing is heated, heat is conducted through the material. If the FR clothing is touching skin, the heat energy that is conducted through the FR clothing could result in a burn. To minimize this chance, FR clothing must fit loosely to provide additional thermal insulation. However, FR clothing must not be so loose that it interferes with the worker's movements. It is also important to remove all contents of pockets so the clothing fit is loose in the pocket areas. FR clothing specifically sized for women is also available, as shown in Exhibit 130.12.

***EXHIBIT 130.12*** *FR clothing specifically sized for women. (Courtesy of Workrite Uniform Company)*

(f) Interference. The garment selected shall result in the least interference with the task but still provide the necessary protection. The work method, location, and task could influence the protective equipment selected.

The plan for the work task must define the protective garments to be worn by the worker. As the plan is developed, the location and position of the worker must be considered to provide the best chance that the PPE does not interfere with the worker's movements as he or she executes the work task. A dry run of the task wearing PPE will identify whether there are PPE restrictions and if enough illumination is available for the task.

**(13) Arc Flash Protective Equipment.**

(a) Arc Flash Suits. Arc flash suit design shall permit easy and rapid removal by the wearer. The entire arc flash suit, including the hood's face shield, shall have an arc rating that is suitable for the arc flash exposure. When exterior air is supplied into the hood, the air hoses and pump housing shall be either covered by FR materials or constructed of nonmelting and nonflammable materials.

Should an emergency situation develop, a worker might need to remove the PPE rapidly, so the design of the PPE must enable the worker to do so. In addition, the rating of the flash suit must at least be equal to the potential incident energy exposure.

When a worker wears a hood with a viewing window, the concentration of oxygen under the hood can decrease if the hood fits tightly. Air can be supplied to the worker inside the hood. However, the air hoses and pump (if necessary) must be protected from ignition as necessary. Any air hoses should be on the opposite side of the worker from the potential thermal exposure.

Face shields and viewing windows must have an arc rating similar to an arc rating for FR clothing. Face shields and viewing windows also must meet the impact requirements of ANSI Z87.1, *Occupational and Educational Personal Eye and Face Protection Devices*. Safety glasses, which are called *spectacles* in ANSI Z87.1, must be worn under the face shield or viewing window. Goggles should not be worn because they are permitted to have an ignitable and meltable component.

(b) Face Protection. Face shields shall have an arc rating suitable for the arc flash exposure. Face shields without an arc rating shall not be used. Eye protection (safety glasses or goggles) shall always be worn under face shields or hoods.

FPN: Face shields made with energy-absorbing formulations that can provide higher levels of protection from the radiant energy of an arc flash are available, but these shields are tinted and can reduce visual acuity and color perception. Additional illumination of the task area might be necessary when these types of arc-protective face shields are used.

(c) Hand Protection.

(1) Leather or FR gloves shall be worn where required for arc flash protection.
(2) Where insulating rubber gloves are used for shock protection, leather protectors shall be worn over the rubber gloves.

FPN: Insulating rubber gloves and gloves made from layers of flame-resistant material provide hand protection against the arc flash hazard. Heavy-duty leather (e.g., greater than 12 oz/yd$^2$) gloves provide protection suitable up to Hazard/Risk Category 2. The leather protectors worn over insulating rubber gloves provide additional arc flash protection for the hands. During high arc flash exposures leather can shrink and cause a decrease in protection.

The leather protectors worn over insulating rubber gloves provide additional arc flash protection for the hands. During high arc flash exposures leather can shrink and cause a decrease in protection.

The hands normally are the most exposed part of a worker's body. Arc flash analyses generally are based on exposure at a distance of 18 in. or 36 in. Because the worker's hands are much closer, the thermal exposure is much greater to the hands. Additional thermal protection is warranted; however, no method exists for determining the degree of exposure for a worker's hands.

Gloves made from FR materials are available. The worker must wear voltage-rated gloves with heavy-duty leather over-protectors. The combination of rubber gloves and leather over-protectors provides significant protection. When wearing layers of hand protection, rehearsal of the work task will determine if there is sufficient dexterity for the task.

Experience has shown that leather gloves stitched with cotton thread tend to offer significant protection if the overcurrent protection functions as intended by the *NEC* requirements. The cotton stitching can disintegrate; however, the leather still offers protection for the hands underneath.

(d) Foot Protection. Heavy-duty leather work shoes provide some arc flash protection to the feet and shall be used in all tasks in Hazard/Risk Category 2 and higher and in all exposures greater than 4 cal/cm$^2$.

Shoes with an arc rating are not available. However, experience has shown that heavy-duty leather work shoes offer significant protection for the feet. Normally, the worker's feet are less exposed than his or her hands or head. However, workers should not wear shoes made from lightweight material. In most cases, heavy-duty leather workshoes with integral steel toes are satisfactory.

Only heavy-duty leather work shoes must be worn if the hazard/risk analysis indicates that arc flash protection is necessary while the task is executed.

**(14) Clothing Material Characteristics.** FR clothing shall meet the requirements described in 130.7(C)(14) and 130.7(C)(15).

Section 130.7(C)(14) describes requirements that must be met by FR clothing. FR clothing that has an arc rating based on testing defined by ASTM F 1506, *Standard Performance Specification for Flame Resistant Textile Materials for Wearing Apparel for Use by Electrical Workers Exposed to Momentary Electric Arc and Related Thermal Hazards,* and ASTM F 1959, *Standard Test Method for Determining the Arc Thermal Performace Value of Materials for Clothing,* adheres to these requirements.

FPN No. 1: FR materials, such as flame-retardant treated cotton, meta-aramid, para-aramid, and poly-benzimidazole (PBI) fibers, provide thermal protection. These materials can ignite but will not continue to burn after the ignition source is removed. FR fabrics can reduce burn injuries during an arc flash exposure by providing a thermal barrier between the arc flash and the wearer.

FPN No. 2: Non-FR cotton, polyester-cotton blends, nylon, nylon-cotton blends, silk, rayon, and wool fabrics are flammable. These fabrics could ignite and continue to burn on the body, resulting in serious burn injuries.

FPN No. 3: Rayon is a cellulose-based (wood pulp) synthetic fiber that is a flammable but nonmelting material.

Clothing made from flammable synthetic materials that melt at temperatures below 315°C (600°F), such as acetate, acrylic, nylon, polyester, polyethylene, polypropylene, and spandex, either alone or in blends, shall not be used.

FPN: These materials melt as a result of arc flash exposure conditions, form intimate contact with the skin, and aggravate the burn injury.

*Exception: Fiber blends that contain materials that melt, such as acetate, acrylic, nylon, polyester, polyethylene, polypropylene, and spandex shall be permitted if such blends in fabrics meet the requirements of ASTM F 1506, Standard Performance Specification for Textile Material for Wearing Apparel for Use by Electrical Workers Exposed to Momentary Electric Arc and Related Thermal Hazards, and if such blends in fabrics do not exhibit evidence of a melting and sticking hazard during arc testing according to ASTM F 1959, Standard Test Method for Determining the Arc Thermal Performance Value of Materials for Clothing [see also 130.7(C)(15)].*

Small amounts of flammable or meltable materials can be used in the elastic bands of underwear. See 130.7(C)(12)(c), Exception.

**(15) Clothing and Other Apparel Not Permitted.** Clothing and other apparel (such as hard hat liners and hair nets) made from materials that do not meet the requirements of 130.7(C)(14) regarding melting, or made from materials that do not meet the flammability requirements shall not be permitted to be worn.

Apparel made from materials that are not arc-rated FR material must not be worn. For instance, hairnets could melt on a worker's hair and head (although arc-rated hairnets are available). Ear warmers and head covers must not be worn unless they are arc-rated FR material.

FPN: Some flame-resistant fabrics, such as non-FR modacrylic and nondurable flame-retardant treatments of cotton, are not recommended for industrial electrical or utility applications.

*Exception No. 1: Nonmelting, flammable (non-FR) materials shall be permitted to be used as underlayers to FR clothing, as described in 130.7(C)(14), and also shall be permitted to be used for Hazard/Risk Category 0 as described in Table 130.7(C)(10).*

*Exception No. 2: Where the work to be performed inside the Arc Flash Protection Boundary exposes the worker to multiple hazards, such as airborne contaminants, under special permission by the authority having jurisdiction and where it can be shown that the level of protection is adequate to address the arc flash hazard, non-FR Personnel Protective Equipment shall be permitted.*

**(16) Care and Maintenance of FR Clothing and FR Arc Flash Suits.**

(a) Inspection. FR apparel shall be inspected before each use. Work clothing or arc flash suits that are contaminated, or damaged to the extent their protective qualities are impaired, shall not be used. Protective items that become contaminated with grease, oil, or flammable liquids or combustible materials shall not be used.

The qualified worker must inspect his or her FR clothing before wearing it. The qualified worker should be trained to understand that if any flammable substance is on the surface of the FR clothing, the rating of the FR clothing is voided. The clothing must be free from tears, cuts, or rips.

(b) Manufacturer's Instructions. The garment manufacturer's instructions for care and maintenance of FR apparel shall be followed.

Flame-resistant clothing must be cleaned and maintained as defined by the clothing manufacturer. Such clothing slowly loses its flame-resistant characteristics when cleaned. There are very specific laundering instructions provided by the FR clothing manufacturers. If workers launder their PPE, they must follow those instructions, which generally require different wash/rinse cycles than are used for household washing. Not all bleaching additives are acceptable. If a laundry service is used, the laundry facility must be aware of the FR clothing manufacturer's laundering instructions. Manufacturer's instructions must be implemented with regard to restoring the flame-resistant characteristics. When FR clothing is shared by workers, the electrical safety program should consider health aspects of shared PPE when determining the cleaning frequency.

(c) Storage. FR apparel shall be stored in a manner that prevents physical damage; damage from moisture, dust, or other deteriorating agents; or contamination from flammable or combustible materials.

Arc-rated FR clothing is intended to protect workers from the thermal hazard associated with an arcing fault. For the protective clothing to perform as intended, it must be protected when in use and in storage. Contaminants such as grease and oil must be avoided. Exposure to flammable materials also must be avoided. Contamination reduces the thermal protection provided by the clothing.

(d) Cleaning, Repairing, and Affixing Items. When FR clothing is cleaned, manufacturer's instructions shall be followed to avoid loss of protection. When FR clothing is repaired, the same FR materials used to manufacture the FR clothing shall be used to provide repairs. When trim, name tags, and/or logos are affixed to FR clothing, guidance in ASTM F 1506, *Standard Performance Specification for Textile Material for Wearing Apparel for Use by Electrical Workers Exposed to Momentary Electric Arc and Related Thermal Hazards*, shall be followed [see Table 130.7(C)(8)].

Manufacturers provide cleaning instructions for their products used for thermal protection. Those instructions should be followed closely.

**(D) Other Protective Equipment.**

**(1) Insulated Tools and Equipment.** Employees shall use insulated tools and/or handling equipment when working inside the Limited Approach Boundary of exposed energized electrical conductors or circuit parts where tools or handling equipment might make accidental contact. Table 130.7(C)(9) provides further information for tasks that require insulated and insulating hand tools. Insulated tools shall be protected from damage to the insulating material.

FPN: See 130.2(B), Shock Protection Boundaries.

(a) Requirements for Insulated Tools. The following requirements shall apply to insulated tools:

(1) Insulated tools shall be rated for the voltages on which they are used.
(2) Insulated tools shall be designed and constructed for the environment to which they are exposed and the manner in which they are used.
(3) Insulated tools and equipment shall be inspected prior to each use. The inspection shall look for damage to the insulation or damage that may limit the tool from performing its intended function or could increase the potential for an incident (e.g., damaged tip on a screwdriver).

When working inside the Limited Approach Boundary, workers must select and use work practices, including using insulated tools that provide maximum protection from a release of energy. If contact with the exposed energized electrical conductor or circuit part is likely, the worker must use insulated tools only. An unqualified worker within the Limited Approach Boundary is considered to be likely to contact an exposed energized electrical conductor. Also, a qualified worker performing a work task within the Restricted Approach Boundary is considered likely to contact an exposed energized electrical conductor. The term *insulated* means that the tool manufacturer has assigned a voltage rating to the insulating material. If the task requires the worker to cross or work in the vicinity of the Prohibited Approach Boundary, insulated tools are required.

Qualified workers are expected to be competent to inspect an insulated tool for potential damage. Qualified workers also must be able to determine whether the voltage rating remains intact.

Exhibit 130.13(a) shows a marked insulated tool and Exhibit 130.13(b) shows a covered tool, which is not insulated. It is important for workers to look for the markings on the tool before using it.

*__EXHIBIT 130.13__ Example of (a) insulated tool and (b) non-insulated tool. (Courtesy of Ideal Industries, Inc.)*

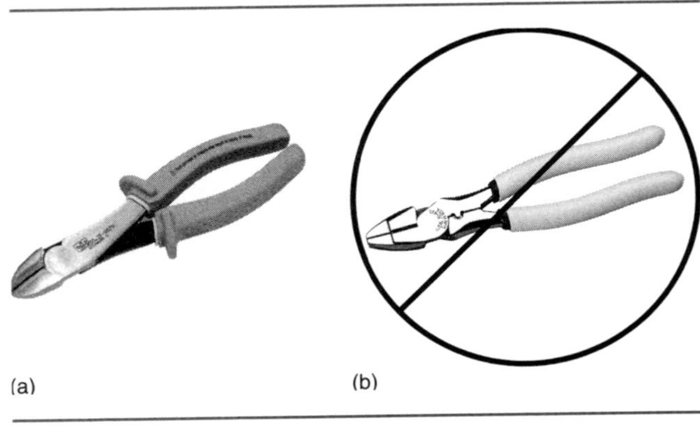


**(b) Fuse or Fuse Holding Equipment.** Fuse or fuse holder handling equipment, insulated for the circuit voltage, shall be used to remove or install a fuse if the fuse terminals are energized.

Fuses should not be removed or installed in a fuse holder that is energized. If fuses are removed in an emergency situation, workers must use fuse-handling equipment that is rated for the voltage.

Authorized and qualified workers can remove and install fuses routinely with the fuse terminals energized using a live-line tool (hot stick).

**(c) Ropes and Handlines.** Ropes and handlines used within the Limited Approach Boundary of exposed energized electrical conductors or circuit parts operating at 50 volts or more, or used where an electrical hazard exists, shall be nonconductive.

If the workers feel that ropes and handlines are necessary to control the lift, they should reevaluate the lift to determine if it could be performed from a different position, making ropes or handlines unnecessary.

**(d) Fiberglass-Reinforced Plastic Rods.** Fiberglass-reinforced plastic rod and tube used for live line tools shall meet the requirements of applicable portions of electrical codes and standards dealing with electrical installation requirements.

FPN: For further information concerning electrical codes and standards dealing with installation requirements, refer to ASTM F 711, *Standard Specification for Fiberglass-Reinforced Plastic (FRP) Rod and Tube Used in Live Line Tools.*

**(e) Portable Ladders.** Portable ladders shall have nonconductive side rails if they are used where the employee or ladder could contact exposed energized electrical conductors or circuit parts operating at 50 volts or more or where an electrical hazard exists. Nonconductive ladders shall meet the requirements of ANSI standards for ladders listed in Table 130.7(F).

(f) Protective Shields. Protective shields, protective barriers, or insulating materials shall be used to protect each employee from shock, burns, or other electrically related injuries while that employee is working within the Limited Approach Boundary of energized conductors or circuit parts that might be accidentally contacted or where dangerous electric heating or arcing might occur. When normally enclosed energized conductors or circuit parts are exposed for maintenance or repair, they shall be guarded to protect unqualified persons from contact with the energized conductors or circuit parts.

(g) Rubber Insulating Equipment. Rubber insulating equipment used for protection from accidental contact with energized conductors or circuit parts shall meet the requirements of the ASTM standards listed in Table 130.7(F).

(h) Voltage-Rated Plastic Guard Equipment. Plastic guard equipment for protection of employees from accidental contact with energized conductors or circuit parts, or for protection of employees or energized equipment or material from contact with ground, shall meet the requirements of the ASTM standards listed in Table 130.7(F).

(i) Physical or Mechanical Barriers. Physical or mechanical (field-fabricated) barriers shall be installed no closer than the Restricted Approach Boundary distance given in Table 130.2(C). While the barrier is being installed, the Restricted Approach Boundary distance specified in Table 130.2(C) shall be maintained, or the energized conductors or circuit parts shall be placed in an electrically safe work condition.

**(E) Alerting Techniques.**

People who are not involved in the work task can be exposed to an electrical hazard when the work task is being executed. To avoid unnecessary exposure to electrical hazards, people must be provided with a warning that an electrical hazard exists.

**(1) Safety Signs and Tags.** Safety signs, safety symbols, or accident prevention tags shall be used where necessary to warn employees about electrical hazards that might endanger them. Such signs and tags shall meet the requirements of ANSI Z535, *Series of Standards for Safety Signs and Tags*, given in Table 130.7(F).

For a warning sign to be useful, the sign must be readable and communicate the needed warning.

FPN: Safety signs, tags, and barricades used to identify energized "look-alike" equipment can be employed as an additional preventive measure.

**(2) Barricades.** Barricades shall be used in conjunction with safety signs where it is necessary to prevent or limit employee access to work areas containing energized conductors or circuit parts. Conductive barricades shall not be used where it might cause an electrical hazard. Barricades shall be placed no closer than the Limited Approach Boundary given in Table 130.2(C).

Like signs, barricades are not intended to prevent approach to an area. Instead, a barricade is intended to act as a warning device. When installed, the barricade should enclose the area containing the electrical hazard. The barricade must not be closer to the exposed energized electrical conductor or circuit part than the Limited Approach Boundary. The barricade should be placed so as not impede the exit of workers within the boundary.

**(3) Attendants.** If signs and barricades do not provide sufficient warning and protection from electrical hazards, an attendant shall be stationed to warn and protect employees. The primary duty and responsibility of an attendant providing manual signaling and alerting shall be to keep unqualified employees outside a work area where the unqualified employee might be exposed to electrical hazards. An attendant shall remain in the area as long as there is a potential for employees to be exposed to the electrical hazards.

When an attendant is necessary to deliver the warning, the attendant should have no other duty.

**(4) Look-Alike Equipment.** Where work performed on equipment that is deenergized and placed in an electrically safe condition exists in a work area with other energized equipment that is similar in size, shape, and construction, one of the altering methods in 130.7(E)(1), (2), or (3) shall be employed to prevent the employee from entering look-alike equipment.

When an installation has multiple similar processes, it is likely that similar electrical equipment exists in the same physical area, with different labels as the only visible difference. Workers must be particularly aware that equipment similar to the one under maintenance exists. Workers should consider installing some temporary identifying mark to reduce the chance of opening the wrong equipment.

**(F) Standards for Other Protective Equipment.** Other protective equipment required in 130.7(D) shall conform to the standards given in Table 130.7(F).

Table 130.7(F) identifies standards that define requirements for specific equipment. All equipment listed in the table impacts safe work practices that a qualified worker should implement. Equipment identified in this table must comply with the latest edition of the referenced standard.

***TABLE 130.7(F)*** *Standards on Other Protective Equipment*

| ***Subject*** | ***Number and Title*** |
|---|---|
| Ladders | ANSI A14.1, *Safety Requirements for Portable Wood Ladders,* 2000<br>ANSI A14.3, *Safety Requirements for Fixed Ladders,* 2002<br>ANSI A14.4, *Safety Requirements for Job-Made Ladders,* 2002<br>ANSI A14.5, *Safety Requirement for Portable Reinforced Plastic Ladders,* 2000 |
| Safety signs and tags | ANSI Z535, *Series of Standards for Safety Signs and Tags,* 2006 |
| Blankets | ASTM D 1048, *Standard Specification for Rubber Insulating Blankets,* 2005 |
| Covers | ASTM D 1049, *Standard Specification for Rubber Covers,* 1998 (R 2002) |
| Line hoses | ASTM D 1050, *Standard Specification for Rubber Insulating Line Hoses,* 2005 |
| Line hoses and covers | ASTM F 478, *Standard Specification for In-Service Care of Insulating Line Hose and Covers,* 1999 (R 2007) |
| Blankets | ASTM F 479, *Standard Specification for In-Service Care of Insulating Blankets,* 2006 |
| Fiberglass tools/ladders | ASTM F 711, *Standard Specification for Fiberglass-Reinforced Plastic (FRP) Rod and Tube Used in Line Tools,* 2002 (R 2007) |
| Plastic guards | ASTM F 712, *Standard Test Methods and Specifications for Electrically Insulating Plastic Guard Equipment for Protection of Workers,* 2006 |
| Temporary grounding | ASTM F 855, *Standard Specification for Temporary Protective Grounds to Be Used on De-energized Electric Power Lines and Equipment,* 2004 |
| Insulated hand tools | ASTM F 1505, *Standard Specification for Insulated and Insulating Hand Tools,* 2007 |

# Safety-Related Maintenance Requirements

# CHAPTER 2

### Summary of Changes

**Article 200**

- **200.1(3) FPN:** Added reference to ANSI/NETA MTS, *Standard for Maintenance Testing Specification.*

**Article 205**

- **205.3:** Added a new section requiring maintenance of overcurrent protective devices.

**Article 210**

- **210.5:** Added an FPN discussing the relationship of protective device maintenance and incident energy.

**Article 225**

- **225.1:** Revised to indicate that fuseholders for current-limiting fuses must not be modified.

**Article 250**

- **250.1:** Added bypass jumpers and insulated and insulating hand tools to the list of items requiring maintenance.
- **250.3(B):** Added a new FPN referring to ASTM F 2249, *Standard Specification for In-Service Test Methods for Temporary Grounding Jumper Assemblies Used on De-Energized Power Lines and Equipment.*
- **250.3(C):** Added a section covering storage of grounding and testing devices.

### Contents

205.5 Grounding and Bonding
205.6 Guarding of Energized Conductors and Circuit Parts
205.7 Safety Equipment
205.8 Clear Spaces
205.9 Identification of Components
205.10 Warning Signs
205.11 Identification of Circuits
205.12 Single and Multiple Conductors and Cables
205.13 Flexible Cords and Cables
210 Substations, Switchgear Assemblies, Switchboards, Panelboards, Motor Control Centers, and Disconnect Switches
210.1 Enclosures
210.2 Area Enclosures
210.3 Conductors
210.4 Insulation Integrity
210.5 Protective Devices
215 Premises Wiring
215.1 Covers for Wiring System Components
215.2 Open Wiring Protection
215.3 Raceways and Cable Trays
220 Controller Equipment
220.1 Scope
220.2 Protection and Control Circuitry
225 Fuses and Circuit Breakers
225.1 Fuses
225.2 Molded-Case Circuit Breakers
225.3 Circuit Breaker Testing
230 Rotating Equipment
230.1 Terminal Boxes
230.2 Guards, Barriers, and Access Plates
235 Hazardous (Classified) Locations
235.1 Scope
235.2 Maintenance Requirements for Hazardous (Classified) Locations
240 Batteries and Battery Rooms
240.1 Ventilation
240.2 Eye and Body Wash Apparatus
240.3 Cell Flame Arrestors and Cell Ventilation
245 Portable Electric Tools and Equipment
245.1 Maintenance Requirements for Portable Electric Tools and Equipment
250 Personal Safety and Protective Equipment
250.1 Maintenance Requirements for Personal Safety and Protective Equipment
250.2 Inspection and Testing of Protective Equipment and Protective Tools
(A) Visual
(B) Testing
250.3 Safety Grounding Equipment
(A) Visual
(B) Testing
(C) Grounding and Testing Devices

An electrical work environment consists of three interrelated components: installation, safe work practices, and maintenance. Whether the work environment is safe depends on the quality with which these elements are implemented. For instance, safe work practices, which are defined in Chapter 1, are effective when the installation is code compliant and the equipment is maintained appropriately. Any deficiency in the installation or maintenance has the potential to adversely impact safe work practices.

Chapter 2 addresses maintenance of the electrical equipment, which provides for the reliability and predictability necessary for the safe operation of electrical equipment. Historically, electrical equipment has been very reliable, and, in many instances, that reliability is taken for granted. *NFPA 70E* relies on the fact that if equipment is installed according to *NFPA 70®*, *National Electrical Code®* (*NEC®*), and the manufacturer's instructions, it is considered to be safe for operation. However, a minor deviation in the operating range of devices can change the hazard exposure to a worker significantly.

NFPA 70B, *Recommended Practice for Electrical Equipment Maintenance*, is a companion document for *NFPA 70E*. NFPA 70B is a recommended practice that serves to reduce hazards to life and property that can result from failure or malfunction of industrial-type electrical systems and equipment. It provides guidance on maintenance practices and on setting up a preventive maintenance program. *NFPA 70E* addresses the work practices that should be used during maintenance work.

# ARTICLE 200
# Introduction

A comprehensive electrical equipment maintenance program provides two benefits to a company: increased reliability of the electrical systems, which avoids electrical outages and malfunctions, and decreased exposure of workers to electrical hazards. Equipment that functions normally and within its design parameters requires fewer interactions. Historically, one-third of all electrical incidents are caused by a combination of unsafe equipment and unsafe conditions.

The primary reason for unsafe equipment is lack of maintenance or inadequate maintenance. Adequately maintained and operated equipment minimizes employees' exposure to electrical hazards. In workplaces where maintenance practices are required, inadequate maintenance could have an immediate and dramatic impact on personal safety.

Equipment maintenance frequency and adequacy impact the cost of operating a manufacturing process and, consequently, the cost of products. Where maintenance is optimized, minimum product cost can be realized. Optimum maintenance depends on the overall physical environment that exists in the manufacturing process. Owners are expected to determine what maintenance tasks are necessary and when the maintenance tasks should be performed to achieve the optimum operating condition. When a maintenance task is necessary to provide a safe working environment, however, the task is required.

## 200.1 Scope

Chapter 2 addresses the following requirements:

(1) Chapter 2 covers practical safety-related maintenance requirements for electrical equipment and installations in workplaces as included in 90.2. These requirements identify only that maintenance directly associated with employee safety.

When electrical equipment is adequately maintained, the risk of equipment failure is reduced. It is not the intent of Chapter 2 to define impractical requirements. Nor is it the intent to require specific maintenance procedures or periods. Employers must determine a maintenance strategy and then implement the necessary components of the selected strategy. However, some maintenance is necessary to maximize the impact of the electrical safety program. For information on preventive maintenance programs, see NFPA 70B.

(2) Chapter 2 does not prescribe specific maintenance methods or testing procedures. It is left to the employer to choose from the various maintenance methods available to satisfy the requirements of Chapter 2.

NFPA 70B provides information on commissioning and on an effective preventive maintenance program. Commissioning, which is also referred to as acceptance testing, integrated system testing, operational tune-up, and start-up testing, is the process by which baseline test results verify the proper operation and sequence of operation of electrical equipment. This process also develops baseline criteria by which future trend analysis can help to identify equipment deterioration. This trend information is useful in predicting when equipment failure or an out of tolerance condition will occur. This information also can allow for convenient scheduling of outages.

When implemented correctly, a realistic commissioning plan minimizes startup and long-term problems, reduces operational costs, and minimizes future maintenance requirements. It is not unusual for electrical systems to have problems during startup and installation. Often it takes experienced engineers and technicians to identify operational problems and provide solutions to fine-tune the system to operate as it was designed.

Routine maintenance tests are tests that are performed at regular intervals over the service life of equipment. These tests normally are performed concurrently with preventive maintenance on the equipment.

NFPA 70B references manufacturers' recommendations and specifications of the International Electrical Testing Association (NETA). The data from the manufacturer are essential to ensure that equipment is installed and used in accordance with any product limitations. The NETA specifications are useful information that can be used for acceptance testing and periodic maintenance testing to ensure system reliability. Acceptance testing verifies that the equipment functions as intended by the design specifications and provides a benchmark for comparing future test results. The comparison can identify any change in reliability for the equipment being tested. Maintenance tests enable a company to identify potential failures before they occur. A shutdown, then, can be scheduled, and repairs can be made with minimum exposure to workers and that eliminates equipment damage.

(3) For the purpose of Chapter 2, maintenance shall be defined as preserving or restoring the condition of electrical equipment and installations, or parts of either, for the safety of employees who work on, near, or with such equipment. Repair or replacement of individual portions or parts of equipment shall be permitted without requiring modification or replacement of other portions or parts that are in a safe condition.

FPN: Refer to NFPA 70B, *Recommended Practice for Electrical Equipment Maintenance*, and ANSI/NETA MTS-2007, *Standard for Maintenance Testing Specification*, for guidance on maintenance frequency, methods, and tests.

Maintenance often is the most neglected component of a strategy to provide a safe work environment. NFPA 70B provides prescribed maintenance methods and intervals to maximize the reliability of electrical equipment and systems. It describes electrical maintenance subjects and issues surrounding maintenance of electrical equipment. NFPA 70B provides workers with solutions, techniques, and testing intervals for adequate maintenance of electrical equipment.

All electrical equipment has a predictable life cycle. Knowing the usable life of equipment is crucial if a facility is to predict the reliability and safe operation of its equipment. Repair parts for old equipment might be unavailable when a defect is discovered. Recognizing these two factors allows for planning, budgeting, and replacement of equipment in an orderly and safe manner.

# ARTICLE 205
# General Maintenance Requirements

Maintenance tasks, although routine, are key to equipment reliability. Rigid scheduling and documentation of maintenance tasks can help to ensure equipment safety as well as reliability.

## 205.1 Qualified Persons

**Employees who perform maintenance on electrical equipment and installations shall be qualified persons as required in Chapter 1 and shall be trained in, and familiar with, the specific maintenance procedures and tests required.**

A worker qualified for electrical equipment maintenance work must be familiar with the operation, maintenance, and history of the equipment as well as the safety training defined in Chapter 1. Familiarity with equipment and maintenance tools directly affects a worker's ability to recognize and avoid electrical hazards. A worker who is trained for a task and who does not perform that task for more than one year should no longer be considered qualified for the task. Employees should be provided with supplemental training or instruction to ensure that they continue to be qualified for the task.

## 205.2 Single Line Diagram

**A single line diagram, where provided for the electrical system, shall be maintained.**

Single line (S/L) diagrams, which are covered in Section 205.2, are the best source of information for workers to locate the electrical hazards that might be encountered in their daily work routines. Maintaining these drawings in an up-to-date condition provides the following valuable information:

- All the sources of power to a specific piece of equipment
- The interrupting capacity of devices at each point in the system
- All possible paths of potential backfeed

An up-to-date single line diagram enables an electrically safe work condition to be implemented. It provides the correct rating for overcurrent devices and enables calculation of available fault current.

Overcurrent protective devices that are not rated appropriately for the available fault current could malfunction or explode in a short-circuit situation. Shrapnel from this type of incident creates another hazard for the worker.

Single line diagrams are critical for determining electrical hazards at a work site, so all qualified workers must have the ability to read and understand single line diagrams of the systems they work on.

## 205.3 General Maintenance Requirements

**Overcurrent protective devices shall be maintained in accordance with the manufacturers' instructions or industry consensus standards.**

Manufacturers provide maintenance instructions for overcurrent devices when they are purchased. Standards such as NFPA 70B and ANSI/NETA MTS, *Standard for Maintenance*

*Testing Specification*, provide testing and maintenance instructions for some overcurrent devices. Maintenance of overcurrent devices must comply with the requirements defined by the manufacturer or by a consensus standard such as these.

### 205.4 Spaces About Electrical Equipment

All working space and clearances required by electrical codes and standards shall be maintained.

Adequate working space allows workers to perform their tasks in an unencumbered manner. The ability to use tools and equipment within adequate clearances provides the necessary safety factor to prevent an inadvertent movement, which could result in an electrical incident and injury.

Working spaces must be kept clear of temporary obstructions, such as stored equipment. Electrical workers and equipment operators sometimes choose to store materials and tools in an open space in front of electrical equipment. When items are stored in the space, emergency access to the equipment is restricted.

FPN: For further information concerning spaces about electrical equipment, see Article 110, Parts II and III, of *NFPA 70, National Electrical Code*.

### 205.5 Grounding and Bonding

Equipment, raceway, cable tray, and enclosure bonding and grounding shall be maintained to ensure electrical continuity.

Grounding and bonding provide the electrical continuity to enable fault current to return to the energy source. Adequate grounding and bonding of electrical equipment provides personnel and equipment safety benefits by enabling the overcurrent device to operate. In a short-circuit or overload condition, the overcurrent device relies on a clear and effective grounding path to operate within its designed range. Without effective grounding and bonding, the tripping time might be extended, thus increasing the amount of incident energy to which a worker could be exposed.

### 205.6 Guarding of Energized Conductors and Circuit Parts

Enclosures shall be maintained to guard against accidental contact with energized conductors and circuit parts and other electrical hazards.

The base premise for worker safety is that no energized electrical conductors are exposed. In some installations, exposed energized electrical conductors are guarded by a fence or similar barrier. If the gate is kept locked, access is restricted only to authorized and qualified personnel who have the key to the lock.

Unused openings in enclosures must be covered in a manner that prevents inadvertent contact with exposed energized electrical conductor parts. Covering these openings prevents or minimizes the chance of introducing contamination, including conductive dust that could result in an arc flash incident.

### 205.7 Safety Equipment

Locks, interlocks, and other safety equipment shall be maintained in proper working condition to accomplish the control purpose.

Locks and interlocks assist in providing safety for workers and equipment. Locks and interlocks ensure that only authorized persons have access to areas that contain exposed energized electrical conductors or circuit parts. Maintaining these locks and interlocks in good working condition helps to minimize exposure to electrical hazards.

Some equipment arrangements use an interlocking system of keys to control the flow of electrical power through the system and to control the sequence of switch operations. The system's keying mechanism must work smoothly and without incident to accomplish this sequence. This keyed system is typically used to redirect power upon interruption from an incoming feeder or to provide system maintenance while minimizing the amount of energized equipment. Duplicate keys for these interlocked key systems must be destroyed or rigidly controlled.

A worker who does not perform the procedure to transfer power on a regular basis might be unable to remember the steps of the procedure. Therefore, instructions should be readily available to authorized workers so they can execute the required steps for operating the equipment.

### 205.8 Clear Spaces

Access to working space and escape passages shall be kept clear and unobstructed.

Good housekeeping is an important characteristic of a safe work environment. Maintaining adequate access to equipment is essential if a worker is to operate equipment in a safe and timely manner. Material storage that blocks access or prevents safe work practices must be avoided at all times.

Emergency access to an area in which a worker is or might be exposed to an electrical hazard must be maintained. In the event of an emergency, the worker might need to escape, or assistance might be necessary.

### 205.9 Identification of Components

Identification of components, where required, and safety-related instructions (operating or maintenance), if posted, shall be securely attached and maintained in legible condition.

Up-to-date operating or maintenance instructions and necessary warnings are vital to ensure worker safety. Affected workers must be aware that instructions exist and know where they are located, and the instructions must be readily accessible.

### 205.10 Warning Signs

Warning signs, where required, shall be visible, securely attached, and maintained in legible condition.

Warning signs are required to inform both qualified and unqualified workers of potential hazards that they might encounter. For example, Section 130.3(C) requires a warning label for equipment that has a potential arc flash hazard. The label must provide a worker with sufficient information to select appropriate personal protective equipment (PPE). The label must be located so it is clearly visible to qualified persons before examination, adjustment, servicing, or maintenance of the equipment. The signs required by installation standards and the United States Occupational Safety and Health Administration (OSHA) regulations must be legible for worker's use.

### 205.11 Identification of Circuits

Circuit or voltage identification shall be securely affixed and maintained in updated and legible condition.

Sections 110.22, 230.2(E), 230.70(B), and 408.4 of the *NEC* require that circuits be identified and the identification securely affixed to the equipment. Mislabeled equipment sets a trap for workers who assume that they have deenergized the appropriate circuit feeding the equipment. However, circuit identification does not change the requirement to verify the absence of voltage when establishing an electrically safe work condition. The necessity of performing a hazard/risk analysis also remains, regardless of labels or other warnings.

### 205.12 Single and Multiple Conductors and Cables

Electrical cables and single and multiple conductors shall be maintained free of damage, shorts, and ground that would present a hazard to employees.

Single- and multi-conductor cables have a protective shield to protect conductors from physical damage under certain conditions. Age of equipment, environment, and work practices all are factors that could affect the integrity of that protection. Conductor insulation and outer coverings can become dry and brittle over time, and movement around these cables could create a shock and flash hazard. For example, if a company modifies an electrical installation by adding cables to a cable tray, workers crawling along the cable tray to install the new cables or to move existing cable could damage insulation on existing cables and create a shock or flash hazard.

### 205.13 Flexible Cords and Cables

Flexible cords and cables shall be maintained to avoid strain and damage.

Incorrect termination of flexible cords and cables at an enclosure or device are a common problem. Tension placed on the cable easily can damage the outer covering, allowing conductors to be exposed and subjecting the worker to a hazard.

A damaged or missing ground prong is the most common problem with extension cords. The grounding path provided by the ground prong is necessary to prevent electrical shock or electrocution. Before each use, extension cords should be inspected to ensure that the ground prong has not been damaged.

**(1) Damaged Cords and Cables.** Cords and cables shall not have worn, frayed, or damaged areas that present an electrical hazard to employees.

**(2) Strain Relief.** Strain relief of cords and cables shall be maintained to prevent pull from being transmitted directly to joints or terminals.

## ARTICLE 210 Substations, Switchgear Assemblies, Switchboards, Panelboards, Motor Control Centers, and Disconnect Switches

Inspection and appropriate maintenance of the components addressed by Article 210 are imperative, since undetected deterioration can cause severe electrical safety hazards to workers.

### 210.1 Enclosures

Enclosures shall be kept free of material that would create a hazard.

Materials or tools left behind by a worker after performing a work task are a common cause of a fault. Housekeeping duties are critical actions that must be performed before a work task is completed. Workers who are qualified persons must remove all extra materials and all tools before reporting that the work task is complete.

### 210.2 Area Enclosures

Fences, physical protection, enclosures, or other protective means, where required to guard against unauthorized access or accidental contact with exposed energized conductors and circuit parts, shall be maintained.

Fences, their gates, and other enclosures should be inspected regularly to ensure that they continue to guard against entry of unauthorized personnel or animals. The gates or doors, especially where equipped with panic hardware, should be checked regularly for security and appropriate operation. Keys to the locking devices for these areas must be carefully controlled to prevent unauthorized personnel from entering areas and, conversely, to enable timely access in the case of an emergency. Repairing any defects or damage to these area enclosures must be completed promptly; otherwise, the ability of the equipment to prevent entry by personnel might be compromised.

### 210.3 Conductors

Current-carrying conductors (buses, switches, disconnects, joints, and terminations) and bracing shall be maintained to:

(1) Conduct rated current without overheating

Discoloration of copper conductors or terminals is evidence of overheating. One cost-effective method of investigating overheating problems is to perform an infrared scan of the equipment. This scan must be performed while the system is operating. However, performing an infrared scan must be considered a hazardous task. Appropriate PPE, as determined by a hazard/risk analysis, must be selected and worn while the scan is being executed. If evidence of overheating is found, the equipment should be deenergized and the problem investigated and repaired in accordance with manufacturers' specifications.

(2) Withstand available fault current

Short circuits or fault currents represent a significant amount of destructive energy that could be released into electrical systems under abnormal conditions. During normal system operation, electrical energy is controlled and does useful work. However, under fault conditions, short-circuit current can cause serious damage to electrical systems and equipment and create the potential for serious injury to personnel. It is the magnetic force produced by large current flow that causes the damage.

### 210.4 Insulation Integrity

Insulation integrity shall be maintained to support the voltage impressed.

Maintenance testing must be performed periodically to ensure that the insulation is capable of performing its intended function. The integrity of the electrical insulation could deteriorate due to environmental conditions, operating conditions, or aging.

### 210.5 Protective Devices

Protective devices shall be maintained to adequately withstand or interrupt available fault current.

> FPN: Failure to properly maintain protective devices can have an adverse effect on the arc flash hazard analysis incident energy values.

Protective devices are designed to operate within a prescribed range and disconnect the power to equipment or sections of equipment in a timely manner. These devices are intended to minimize the damage to equipment and injury to personnel. When a protective device is incorrectly sized or maintained, any arcing fault could result in injury and equipment damage.

When a protective device fails to operate as intended, employees can be exposed to the incident energy generated in the arcing fault. Protective devices must be maintained as necessary to safely interrupt any current that might be applied to the circuit. If the amount of available fault current increases for any reason, such as a change in upstream components, each protective device must be analyzed to determine if the intended interrupting capability remains.

# ARTICLE 215
# Premises Wiring

Article 215 addresses the maintenance needs of all wiring in a facility.

### 215.1 Covers for Wiring System Components

Covers for wiring system components shall be in place with all associated hardware, and there shall be no unprotected openings.

All unused openings in an electrical enclosure must be closed to afford protection substantially equivalent to the wall of the equipment. All fasteners must be completely installed on all covers. Doors must be closed and latched. The objective of the requirement in Section 215.1 is to protect workers from contact with any exposed energized electrical conductor or circuit part with a body part or conductive object. An opening in an enclosure subjects workers to the introduction of contamination or conductive dust in equipment, which could result in an arcing fault.

### 215.2 Open Wiring Protection

Open wiring protection, such as location or barriers, shall be maintained to prevent accidental contact.

All barriers that are installed to segregate classes of wiring must be maintained to ensure that the intended function continues. Conductors that are protected by location must continue to be protected. If new means of access, such as a ladder, is installed, the protection provided by location might be breached.

### 215.3 Raceways and Cable Trays

Raceways and cable trays shall be maintained to provide physical protection and support for conductors.

Conduit, duct, cable tray, and other conductor methods must be maintained to ensure that support for electrical conductors is adequate. Using flexible conduit as a step frequently destroys its integrity. It could break the raceway at its connectors, or pull it apart at the connectors, which could destroy the ground-fault return capability of the initial installation.

# ARTICLE 220
# Controller Equipment

Article 220 addresses maintenance of control equipment and defines which types of equipment are considered to be controllers.

### 220.1 Scope

**This article shall apply to controllers, including electrical equipment that governs the starting, stopping, direction of motion, acceleration, speed, and protection of rotating equipment and other power utilization apparatus in the workplace.**

A controller can be a remote-controlled magnetic contactor, switch, circuit breaker, or device that normally is used to start and stop motors and other apparatus. In the case of motors, the controller must be capable of interrupting the locked-rotor current of the motor. Stop-and-start stations and similar control circuit components that do not open the power conductors to the motor are not considered to be controllers.

### 220.2 Protection and Control Circuitry

**Protection and control circuitry used to guard against accidental contact with energized conductors and circuit parts and to prevent other electrical or mechanical hazards shall be maintained.**

All protective components that are used to prevent or minimize exposure to an electrical hazard must be maintained to ensure the continued integrity of the installed components.

## ARTICLE 225
## Fuses and Circuit Breakers

Article 225 covers maintenance of fuses and circuit breakers. Precise operation of fuses and circuit breakers in an electrical system is imperative. Therefore, their adequate maintenance is essential to maintaining a safe work environment in a facility.

### 225.1 Fuses

**Fuses shall be maintained free of breaks or cracks in fuse cases, ferrules, and insulators. Fuse clips shall be maintained to provide adequate contact with fuses. Fuseholders for current-limiting fuses shall not be modified to allow the insertion of fuses that are not current-limiting.**

Fuse terminals and fuse clips should be examined for discoloration caused by heat from poor contact or corrosion. Early detection of overheating is possible through the use of infrared examination.

Fuses should have an interrupting rating equal to or greater than the maximum fault current available at their point of application. The interrupting rating of fuses, ranging from 10,000 amperes to 300,000 amperes, should be clearly visible on the fuse label.

Many different types of fuses are used in power distribution systems and utilization equipment. Fuses differ by performance, characteristics, and physical size. Their ratings must be verified to ensure that the circuit design is maintained for the life of the equipment. When replacing fuses, a worker should never alter the fuseholder or force it to accept fuses that do not readily fit. Stocking an adequate supply of spare fuses with appropriate ratings minimizes replacement problems.

Fuses from different manufacturers should not be mixed in the same circuit. Manufacturers provide information about their fuses. Although fuses from multiple manufacturers might have the same rating, their operating characteristics are probably different, making coordinating unlikely.

### 225.2 Molded-Case Circuit Breakers

**Molded-case circuit breakers shall be maintained free of cracks in cases and cracked or broken operating handles.**

Maintenance of molded-case circuit breakers generally can be divided into two categories: mechanical and electrical. Mechanical maintenance consists of inspection for good housekeeping, maintenance of appropriate mechanical mounting and electrical connections, and manual operation of the circuit breakers.

Molded-case circuit breakers should be kept free of external contamination so that internal heat can be dissipated normally. A clean circuit breaker enclosure reduces potential arcing conditions between energized conductors and between energized conductors and ground. The structural strength of the case is important in withstanding the stresses imposed during fault-current interruptions. Therefore, an inspection should be made for cracks in the case, and replacement made if necessary.

Excessive heat in a circuit breaker can cause a malfunction in the form of nuisance tripping and, possibly, an eventual failure. Loose connections are the most common cause of excessive heat. Periodic maintenance checks should involve checking for loose connections or evidence of overheating. All connections should be maintained in accordance with manufacturers' recommendations.

Molded-case circuit breakers can be in service for extended periods and yet never be called on to perform their overload- or short-circuit-tripping functions. Manual operation of the circuit breaker helps keep the contacts clean, but it does not exercise the tripping mechanism. Although manual operations exercise the breaker mechanisms, none of the mechanical linkages in the tripping mechanisms are moved with this exercise. Even though manual operation does not completely check molded-case circuit breakers, it should be completed as it is the best-case testing scenario available.

Electrical maintenance is the second category. Electrical maintenance of a molded case circuit breaker verifies that the circuit breaker will trip at its desired set point. See NFPA 70B for more information.

### 225.3 Circuit Breaker Testing

Circuit breakers that interrupt faults approaching their interrupting ratings shall be inspected and tested in accordance with the manufacturer's instructions.

A fault current that is near interrupting levels is a significant amount of current. High level currents can cause damage, even where catastrophic failure does not occur. Testing of the device will ensure that the high current did not damage the circuit breaker and that it will operate at its set point if called upon again.

Circuit breakers that do not operate within their prescribed rating and range can result in catastrophic failure. One effect of the failure might be a significant increase in danger to personnel. For example, if the circuit breaker is cracked or has a higher current induced through a fault condition, it could explode, striking the worker with shrapnel traveling in excess of 700 miles per hour.

If the circuit breaker does not trip within its prescribed range, it can result in an increase in incident energy, exposing the worker to increased risk. For example, a typical arc flash situation where the worker is 18 inches from a 20-kA short-circuit and 5-cycle tripping time results in an incident energy exposure of 6.44 cal/cm$^2$. If the tripping time is increased to 30 cycles, due to the circuit breaker being out of calibration or improper maintenance, the incident energy is 38.64 cal/cm$^2$.

Circuit breakers should have an initial acceptance test and subsequent maintenance testing at recommended intervals. NFPA 70B and ANSI/NETA MTS are documents that can assist a company in understanding the specific tests and testing intervals that are required to ensure reliability.

# ARTICLE 230
# Rotating Equipment

Movement of rotating equipment and motors present safety hazards to workers, so adequate maintenance of their guards is extremely important.

## 230.1 Terminal Boxes

Terminal chambers, enclosures, and terminal boxes shall be maintained to guard against accidental contact with energized conductors and circuit parts and other electrical hazards.

No motor terminal box must exert pressure on the conductors that are spliced within it. Hardware intended to hold the terminal box in place must be complete and firmly torqued into position.

## 230.2 Guards, Barriers, and Access Plates

Guards, barriers, and access plates shall be maintained to prevent employees from contacting moving or energized parts.

In most instances, a guard or barrier is necessary to prevent an employee from contacting rotating or moving parts. The integrity of the guard or barrier is important to ensure that a worker cannot become entangled in or by the moving part. The guard or barrier must be inspected periodically and repaired as necessary to restore its original integrity.

***CASE STUDY***

Joe Schwartz was an electrician for Brown Enterprises.* He had been an electrician for 29 years, most of the time with Brown. He was very vocal about how work should be performed. His long experience established him as the informal leader of the younger crew.

In an attempt to be more competitive, Brown Enterprises had recently begun to shift to multiple-craft supervision. Joe's current supervisor, Al, was experienced in the mechanical crafts and was familiar with the mechanical aspects of the electrical craft. Al was unaware of how much the rest of the crew respected and embraced Joe's ideas.

A new OSHA rule had just been promulgated, and the site safety manager was aware that wholesale revision of the electrical procedures was in order. The new rule received a great deal of attention in the management ranks. The main point of discussion with the new rule was that a hazard called arc flash was working its way into the regulatory process.

The safety manager asked Al to review the current electrical safety procedures, paying particular attention to what needed to be done to protect people from this "new" hazard. In order to fulfill the request, Al sought help from Brown Enterprises' engineering staff. Because the engineering staff had been keeping abreast of IEEE's developing knowledge of arc flash, an engineer agreed to write a procedure for Al.

Al was confident that the engineer was familiar with how to protect people. He accepted the draft procedure and relayed it to the safety manager for action, without discussing the procedure with his electricians. The safety manager reviewed the procedure and placed it into the system for issue. No attempt was made to gain the "buy-in" of the electrical workers. The gist of the procedure was that switchman's jackets and switchman's hoods were to be worn at all times when a door was opened on any electrical equipment.

When the procedure "hit the street," Joe was very vocal in his denunciation of the procedure. In all his 29 years as an electrician, he had never seen so much overkill. He had seen electrical explosions, but nothing that warranted such an outfit. He simply was not going to wear the prescribed gear. Anyway, he complained he wouldn't even be able to take an ammeter reading while wearing the hood. He doubted that he would be able to see the voltmeter. He refused to follow the new procedure.

On the day that the protective clothing was delivered to the site, Joe was called to troubleshoot a balky motor. He grabbed his tool belt and thought briefly about the new procedure. Again, he made a conscious choice to ignore the procedure.

When Joe arrived at the work site, he noted that the overload button in the front of the motor control center unit was missing. He put on his leather gloves and reached for his screwdriver to release the door fasteners. After the door fasteners were released, he moved the switch handle to the "off" position. As soon as the switch handle began to move, a broken spring on the handle flew off and made contact between phases A and B. The resulting arc flash and blast blew the door open. A fireball and molten metal flew out of the now open door. Joe's glove prevented significant burns on his hand, but the fabric in the sleeve of his shirt melted onto his arm. Joe went to the company's medical facility and was transported to a nearby hospital, where the burn was treated.

Although the PPE identified in the procedure would have prevented injury, other PPE would have been sufficient for Joe's exposure as well. Had Joe participated in the production of the procedure, he probably would have been following the procedural requirements and avoided injury.

---

* This account is based on an actual incident. The names, including the name of the facility, have all been changed to protect those involved. Any similarity to actual names or facilities is strictly coincidental.

Reprinted from Ray A. Jones and Jane G. Jones, *Electrical Safety in the Workplace,* 2000, with permission of Jones and Bartlett Publishers, Sudbury, MA.

# ARTICLE 235
# Hazardous (Classified) Locations

Maintenance of equipment in hazardous areas in a facility presents a special safety problem. Personnel who are assigned to such maintenance must be trained to understand the explosive nature of the materials within the areas and how equipment maintenance is important to a safe environment.

Employees must understand that troubleshooting or maintenance of equipment or circuits presents a special problem if an explosive atmosphere exists. Most instruments or other devices used for troubleshooting or maintenance use a battery or other energy storage device to operate. They must also know that a spark is likely to occur when the testing device contacts a conductor. If an explosive atmosphere exists at the time, an explosion is possible. Before using devices such as these in an electrically hazardous (classified) area, the area should be tested to ensure that an explosive atmosphere does not exist.

### 235.1 Scope

This article covers maintenance requirements in those areas identified as hazardous (classified) locations.

FPN: These locations need special types of equipment and installation to ensure safe performance under conditions of proper use and maintenance. It is important that inspection authorities and users exercise more than ordinary care with regard to installation and maintenance. The maintenance for specific equipment and materials is covered elsewhere in Chapter 2 and is applicable to hazardous (classified) locations. Other maintenance will ensure that the form of construction and of installation that makes the equipment, and materials suitable for the particular location are not compromised.

After maintenance is performed on equipment in a hazardous (classified) area, the integrity of the protective scheme that prevents an explosion must be restored. Purging and pressurization must be re-initiated if that scheme is the selected protective scheme. Explosionproof equipment must be completely resealed with all fasteners in place. The integrity of the flame path must not be damaged.

The maintenance needed for specific hazardous (classified) locations depends on the classification of the specific location. The design principles and equipment characteristics, for example, use of positive pressure ventilation, explosionproof, nonincendive, intrinsically safe, and purged and pressurized equipment, that were applied in the installation to meet the requirements of the area classification must also be known. With this information, the employer and the inspection authority are able to determine whether the installation as maintained has retained the condition necessary for a safe workplace.

## 235.2 Maintenance Requirements for Hazardous (Classified) Locations

Equipment and installations in these locations shall be maintained such that the following apply:

(1) No energized parts are exposed.

*Exception to (1): Intrinsically safe and nonincendive circuits.*

(2) There are no breaks in conduit systems, fittings, or enclosures from damage, corrosion, or other causes.
(3) All bonding jumpers are securely fastened and intact.
(4) All fittings, boxes, and enclosures with bolted covers have all bolts installed and bolted tight.
(5) All threaded conduit shall be wrenchtight and enclosure covers shall be tightened in accordance with the manufacturer's instructions.
(6) There are no open entries into fittings, boxes, or enclosures that would compromise the protection characteristics.
(7) All close-up plugs, breathers, seals, and drains are securely in place.
(8) Marking of luminaires (lighting fixtures) for maximum lamp wattage and temperature rating is legible and not exceeded.
(9) Required markings are secure and legible.

Equipment maintenance should be performed only by qualified personnel trained in safe maintenance practices and the special considerations necessary to maintain electrical equipment for use in hazardous (classified) locations. These individuals should be familiar with requirements for safe electrical installations. They should be trained to identify and eliminate ignition sources, such as high surface temperatures, stored electrical energy, and the buildup of static charges and to identify the need for special tools, equipment, tests, and protective clothing.

# ARTICLE 240
# Batteries and Battery Rooms

There are special safety hazards associated with explosive gases present in battery rooms; therefore, maintenance of these areas requires adequate ventilation and special safety equipment.

### 240.1 Ventilation

**Ventilation systems, forced or natural, shall be maintained to prevent buildup of explosive mixtures. This maintenance shall include a functional test of any associated detection and alarm systems.**

Ventilation of the battery room is required to prevent the buildup of hydrogen. Chapter 52 of NFPA 1, *Fire Code,* requires ventilation to be provided for rooms and cabinets in accordance with the mechanical code and one of the following:

1. A ventilation system designed to limit the maximum concentration of hydrogen to 1.0 percent of the total volume of the room during the worst-case event of simultaneous "boost" charging of all the batteries, in accordance with nationally recognized standards
2. Continuous ventilation that is provided at a rate of not less than 1 $ft^3/min/ft^2$ (5.1 $L/sec/m^2$) of floor area of the room or cabinet

NFPA 111, *Standard on Stored Electrical Energy Emergency and Standby Power Systems,* provides information on the installation requirements for battery rooms.

### 240.2 Eye and Body Wash Apparatus

**Eye and body wash apparatus shall be maintained in operable condition.**

Eye and body wash apparatus and areas must be maintained in a sanitary condition. The function and operation of the apparatus should be tested periodically.

### 240.3 Cell Flame Arresters and Cell Ventilation

**Battery cell ventilation openings shall be unobstructed, and cell flame arresters shall be maintained.**

Ventilation of cells permits gases that are generated within the cell to escape. Ventilation prevents the cell from rupturing. Battery installations should be inspected periodically to ensure adequate ventilation of cells within each battery.

# ARTICLE 245
# Portable Electric Tools and Equipment

Most hazardous conditions associated with portable electric tools and equipment result from improper handling or storage. Maintenance and inspection of such equipment must be included in the facility's electrical safety program as a matter of course.

### 245.1 Maintenance Requirements for Portable Electric Tools and Equipment

**Attachment plugs, receptacles, cover plates, and cord connectors shall be maintained such that the following apply:**

(1) There are no breaks, damage, or cracks exposing energized conductors and circuit parts.
(2) There are no missing cover plates.
(3) Terminations have no stray strands or loose terminals.
(4) There are no missing, loose, altered, or damaged blades, pins, or contacts.
(5) Polarity is correct.

Periodic electrical testing of portable electric tools can uncover operating defects. Immediate correction of these defects ensures continued safe operation and prevents breakdown and more costly repairs. A visual inspection is recommended when a tool is issued as well as after each use just before the tool is returned to the storage area.

Employees should be trained to recognize visible defects such as cut, frayed, spliced, or broken cords; cracked or broken attachment plugs; and missing or deformed grounding prongs. Such defects should be reported immediately and the tool should be removed from service until it is repaired.

Employees should be instructed to report all shocks immediately, no matter how minor, and to cease using the tool. Tools that cause shocks must be removed from service, examined, and repaired before further use. Tools that trip GFCI devices must also be removed from service and a record of the GFCI tripping should be passed on to the next shift.

# ARTICLE 250
# Personal Safety and Protective Equipment

PPE is a worker's final chance to avoid injury in the event of an incident. Therefore, workers should invest a great deal of interest and energy in maintaining this special equipment.

### 250.1 Maintenance Requirements for Personal Safety and Protective Equipment

Personal safety and protective equipment such as the following shall be maintained in a safe working condition:

(1) Grounding equipment
(2) Hot sticks
(3) Rubber gloves, sleeves, and leather protectors
(4) Voltage test indicators
(5) Blanket and similar insulating equipment
(6) Insulating mats and similar insulating equipment
(7) Protective barriers
(8) External circuit breaker rack-out devices
(9) Portable lighting units
(10) Safety grounding equipment
(11) Dielectric footwear
(12) Protective clothing
(13) Bypass jumpers
(14) Insulated and insulating hand tools

To ensure reliability, all equipment must be maintained in accordance with the manufacturers' instructions or listings. Protective equipment also must be maintained in a sanitary condition.

## 250.2 Inspection and Testing of Protective Equipment and Protective Tools

**(A) Visual.** Safety and protective equipment and protective tools shall be visually inspected for damage and defects before initial use and at intervals thereafter, as service conditions require, but in no case shall the interval exceed 1 year, unless specified otherwise by the respective ASTM standards.

Since PPE is the last chance to avoid an injury, each component must be visually inspected immediately before use to verify that no visual defects exist in the equipment. In some instances, the PPE must have a date stamp that indicates when the equipment must be tested. The visual inspection must verify that the equipment is not past due for retesting. Specific ASTM standards describe what aspects of the equipment should be included in the visual inspection. See Table 130.7(C)(8) for more information.

**(B) Testing.** The insulation of protective equipment and protective tools, such as items (1) through (14) of 250.1, shall be verified by the appropriate test and visual inspection to ascertain that insulating capability has been retained before initial use, and at intervals thereafter, as service conditions and applicable standards and instructions require, but in no case shall the interval exceed 3 years, unless specified otherwise by the respective ASTM standards.

See Table 130.7(C)(8) for ASTM standards that describe testing requirements.

## 250.3 Safety Grounding Equipment

Electricians refer to personal protective grounding equipment by several names. *Safety grounds* and *ground sets* are terms that refer to the same equipment. Safety grounding equipment is normally constructed with insulated conductors and terminated in devices intended for connection to a bare conductor or part.

Safety grounding equipment should be assigned an identifying mark for record purposes. The identifying mark, then, can be recorded when the equipment is installed on a circuit that will be reenergized after a work task has been performed.

**(A) Visual.** Personal protective ground cable sets shall be inspected for cuts in the protective sheath and damage to the conductors. Clamps and connector strain relief devices shall be checked for tightness. These inspections shall be made at intervals thereafter as service conditions require, but in no case shall the interval exceed 1 year.

Safety grounding equipment should be visually inspected before each use. However, when an annual inspection is made, a record of the inspection should be made and maintained until the next inspection is performed, when a new record should be established.

**(B) Testing.** Prior to being returned to service, safety grounds that have been repaired or modified shall be tested.

Safety grounding equipment must be capable of conducting any available fault current long enough for the overcurrent protection to clear the fault. When a manufacturer determines a rating for specific devices, a destructive test is normally performed. However, destructive testing is not an option for equipment that will be used again. Guidance for maintenance testing of safety grounding equipment is provided in ASTM F 2249, *Standard Specification for In-Service Test Methods for Temporary Grounding Jumper Assemblies Used on De-Energized Electric Power Lines and Equipment.*

FPN: Guidance for inspecting and testing safety grounds is provided in ASTM F 2249, *Standard Specification for In-Service Test Methods for Temporary Grounding Jumper Assemblies Used on De-Energized Electric Power Lines and Equipment.*

**(C) Grounding and Testing Devices.** Grounding and testing devices shall be stored in a clean and dry area. Grounding and testing devices shall be properly inspected and tested before each use.

Grounding and testing devices are constructed on a frame intended to be inserted into a compartment from which a circuit breaker or disconnect has been removed. Grounding and testing devices can be inserted only into specific spaces. Like safety ground equipment, grounding and testing devices must be visually inspected for defects before each use. Also like safety ground equipment, the integrity of each grounding and testing device should be verified at intervals not exceeding one year. The IEEE standard referenced in the following FPN provides information on integrity tests for grounding and testing devices.

FPN: Guidance for testing of grounding and testing devices is provided in Section 9.5 of IEEE C37.20.6-1997, *Standard for 4.76 kV to 38 kV-Rated Ground and Test Devices Used in Enclosures.*

# Safety Requirements for Special Equipment

## CHAPTER 3

### Summary of Changes

**Article 300**

- **300.3:** Revised to clarify that Article 300 applies to research and development laboratories.
- **Article 310.1 FPN 2:** Revised to add a reference to *NEC* Article 668.
- **310.5(D)(2):** Revised to limit frames of eyeglasses to non-conductive material and to add non-melting face shields to the list of PPE.
- **310.5(D)(3):** Revised to indicate that bonding is permitted to equalize voltage.
- **310.6:** Added an FPN indicating the order of preference for handheld equipment.

**Article 320**

- **320.1:** Added an FPN that lists documents where further information on batteries can be found.
- **320.2:** Revised definitions of *secondary battery, secondary cell,* and *valve-regulated lead acid (VLRA) battery* to align with IEEE definitions.
- **320.3(D):** Added an FPN that references the 2008 *NEC.*
- **320.4(B)(2):** Revised to add a requirement for minimum space between batteries.
- **320.4(C)(2)(c):** Revised the FPN to clarify the requirement for containing possible electrolyte spill.
- **320.5(A)(4):** Revised to require spill containment for some installations.
- **320.6(A)(2):** Revised with additional requirements for ventilation.

**Article 340**

- **340.4:** Revised to update issue date of referenced publications.
- **340.5:** Revised title of the section to specifically address the effect of electrical energy on the human body.

**Article 350**

- Added new article describing exceptions for work practices required by Chapters 1, 2, and 3 and defining terms that are applicable to research and development laboratories. Research and development laboratories are those facilities that test, discover, and develop equipment, devices, and processes, including governmental, manufacturing, and educational facilities.

## Contents

(H) Power
(I) Location of General-Purpose Outlets
320.6 Battery Enclosure Requirements
(A) Enclosure Construction
(B) Battery Takeoff Terminals and Outgoing Busbars and Cables
(C) Battery Compartment Circuits
320.7 Protection
(A) General
(B) Switching and Control Equipment
(C) Ground-Fault Protection
(D) Main Isolating Switch
(E) Section Isolating Equipment
(F) Warning Signs
320.8 Personnel Protective Equipment
320.9 Tools and Equipment
330 Safety-Related Work Practices for Use of Lasers
330.1 Scope
330.2 Definitions
330.3 Safety Training
(A) Personnel to Be Trained
(B) Scope of Training
(C) Proof of Qualification
330.4 Safeguarding of Employees in the Laser Operating Area
(A) Eye Protection
(B) Warning Signs
(C) Master Control
(D)
(E)
(F)
(G)
(H)
330.5 Employee Responsibility
340 Safety-Related Work Practices: Power Electronic Equipment
340.1 Scope
340.2 Definition
340.3 Application
340.4 Reference Standards
340.5 Hazardous Effects of Electricity on the Human Body
340.6 Hazards Associated with Power Electronic Equipment
340.7 Specific Measures for Personnel Safety
(A) Employer Responsibility
(B) Employee Responsibility
350 Safety-Related Work Requirements: Research and Development Laboratories
350.1 Scope
350.2 Definitions
350.3 Applications of Other Articles
350.5 Specific Measures and Controls for Personnel Safety
350.6 Listing Requirements

Chapters 1 and 2 define requirements that apply generally to all workplaces, and Chapter 3 covers safety requirements for special equipment. Some facilities use electrical energy in unique ways and, therefore, have special circumstances and differ from most general industries. In some cases, the electrical energy is an integral part of the manufacturing process. In others, the electrical energy is converted to a form that exposes workers to unique hazards. When electrical energy is used as a process variable, the safe work practices defined in Chapters 1 and 2 can become unsafe or produce unsafe conditions. Chapter 3 is intended to modify the requirements of Chapters 1 and 2 as necessary for use in these situations.

Some workplaces require equipment that is unique. For example, research and development facilities, as well as other laboratory facilities, frequently use equipment that is unique and expose workers to hazards in special ways. General safe work practices might not mitigate that exposure adequately. Chapter 3 enables an employer to comply with appropriate requirements from Chapters 1 and 2 and modify requirements that are not appropriate for the specific conditions.

# ARTICLE 300 Introduction

Article 300 names the special equipment discussed in Chapter 3 and identifies the purpose of the chapter as supplementing and modifying the safety related work practices in Chapters 1 and 2 for work with special equipment. Article 300 also requires the employer to provide safety related work practices and training.

### 300.1 Scope

Chapter 3 covers special electrical equipment in the workplace and modifies the general requirements of Chapter 1.

The scope of this chapter aligns with the scope of *NFPA 70E*. Workplaces that are included in the scope of the standard and also contain special equipment and processes are the subject covered in Chapter 3.

### 300.2 Responsibility

The employer shall provide safety-related work practices and employee training. The employee shall follow those work practices.

Employers and employees are assigned the same responsibility for Chapter 3 requirements as defined for Chapters 1 and 2. The employer must define the electrical safety program, and employees must implement the requirements defined in the program. When employers and employees work together to accomplish both needs, the electrical safety program is more effective.

### 300.3 Organization

Chapter 3 of this standard is divided into articles. Article 300 applies generally. Article 310 applies to electrolytic cells. Article 320 applies to batteries and battery rooms. Article 330 applies to lasers. Article 340 applies to power electronic equipment. Article 350 applies to R&D laboratories.

Each article included in Chapter 3 addresses a single unique workplace. Requirements defined in one article apply only to that workplace and modify requirements of Chapters 1 and

2. Each article in Chapter 3 stands alone and is not intended to apply to other articles in Chapter 3.

FPN: The *NFPA 70E* Technical Committee might develop additional chapters for other types of special equipment in the future.

# ARTICLE 310
# Safety-Related Work Practices for Electrolytic Cells

Article 310 identifies the supplementary or replacement safe work practices workers should use in electrolytic cell line working zones and the special hazards of working with ungrounded dc systems (see Exhibit 310.1).

***EXHIBIT 310.1*** *An electrolytic cell line. (Photo by David Pace and Michael Petry, courtesy of Olin Corporation)*

Each individual cell of an electrolytic cell line is a battery and cannot be deenergized without removing the electrolyte in the vessel. A cell line is a series of individual cells that are connected together electrically. Generally, the process requires a significant amount of direct current and is ungrounded.

Establishing an electrically safe work condition is not a viable method for avoiding injury. Training provided to employees who work in the vicinity of the cells and interconnecting bus must establish an understanding of the hazards associated with an unintentional grounded condition of either an individual cell or the interconnecting bus.

Current flowing in the interconnecting bus generates a significant magnetic field. Workers must understand that the magnetic field might interfere with certain medical devices.

Grounded portable tools and equipment must not be used in the area containing the cells or interconnecting bus.

## 310.1 Scope

The requirements of this chapter shall apply to the electrical safety–related work practices used in the types of electrolytic cell areas.

FPN No. 1: See Annex L for a typical application of safeguards in the cell line working zone.

FPN No. 2: For further information about electrolytic cells, see *NFPA 70, National Electrical Code*, Article 668.

### 310.2 Definitions

For the purposes of this chapter, the following definitions shall apply.

**Battery Effect.** A voltage that exists on the cell line after the power supply is disconnected.

FPN: Electrolytic cells could exhibit characteristics similar to an electrical storage battery, and thus a hazardous voltage could exist after the power supply is disconnected from the cell line.

**Safeguarding.** Safeguards for personnel include the consistent administrative enforcement of safe work practices. Safeguards include training in safe work practices, cell line design, safety equipment, personal protective equipment, operating procedures, and work checklists.

### 310.3 Safety Training

**(A) General.** The training requirements of this chapter shall apply to employees who are exposed to the risk of electrical hazard in the cell line working zone defined in 110.6 and shall supplement or modify the requirements of 110.8, 120.1, 130.1, and 130.5.

**(B) Training Requirements.** Employees shall be trained to understand the specific hazards associated with electrical energy on the cell line. Employees shall be trained in safety-related work practices and procedural requirements to provide protection from the electrical hazards associated with their respective job or task assignment.

### 310.4 Employee Training

**(A) Qualified Persons.**

**(1) Training.** Qualified persons shall be trained and knowledgeable in the operation of cell line working zone equipment and specific work methods and shall be trained to avoid the electrical hazards that are present. Such persons shall be familiar with the proper use of precautionary techniques and personal protective equipment. Training for a qualified person shall include the following:

(1) The skills and techniques to avoid dangerous contact with hazardous voltages between energized surfaces and between energized surfaces and ground. Skills and techniques might include temporarily insulating or guarding parts to permit the employee to work on energized parts.
(2) The method of determining the cell line working zone area boundaries.

**(2) Qualified Persons.** Qualified persons shall be permitted to work within the cell line working zone.

In electrolytic cell working zones, individual cells normally act as batteries. Direct current voltage is supplied by the cells as well as by rectifying equipment. Because the dc voltage normally is ungrounded, hand tools that might contact the dc bus work must not be grounded. Employees who work within the area of the dc bus must be trained to understand the unique hazards associated with ungrounded dc voltage.

Hazard/risk analyses used for ac circuits might not be appropriate for use with dc circuits. Each employee must be trained to understand how he or she might be exposed to a thermal hazard associated with an arc flash. Employees must understand how to select personal protective equipment (PPE) for use when they are exposed to an arcing fault.

**(B) Unqualified Persons.**

**(1) Training.** Unqualified persons shall be trained to recognize electrical hazards to which they may be exposed and the proper methods of avoiding the hazards.

**(2) In Cell Line Working Zone.** When there is a need for an unqualified person to enter the cell line working zone to perform a specific task, that person shall be advised by the designated qualified person in charge of the possible hazards to ensure the unqualified person is safeguarded.

### 310.5 Safeguarding of Employees in the Cell Line Working Zone

**(A) General.** Operation and maintenance of electrolytic cell lines may require contact by employees with exposed energized surfaces such as buses, electrolytic cells, and their attachments. The approach distances referred to in Table 130.2(C) shall not apply to work performed by qualified persons in the cell line working zone. Safeguards such as safety-related work practices and other safeguards shall be used to protect employees from injury while working in the cell line working zone. These safeguards shall be consistent with the nature and extent of the related electrical hazards. Safeguards might be different for energized cell lines and deenergized cell lines. Hazardous battery effect voltages shall be dissipated to consider a cell line deenergized.

FPN No. 1: Exposed energized surfaces might not establish a hazardous condition. A hazardous electrical condition is related to current flow through the body causing shock and arc flash burns and arc blasts. Shock is a function of many factors, including resistance through the body and through skin, of return paths, of paths in parallel with the body, and of system voltages. Arc flash burns and arc blasts are a function of the current available at the point involved and the time of arc exposure.

FPN No. 2: A cell line or group of cell lines operated as a unit for the production of a particular metal, gas, or chemical compound might differ from other cell lines producing the same product because of variations in the particular raw materials used, output capacity, use of proprietary methods or process practices, or other modifying factors. Detailed standard electrical safety-related work practice requirements could become overly restrictive without accomplishing the stated purpose of Chapter 1 of this standard.

Employers must institute an electrical safety program that addresses the issues identified in Chapters 1 and 2. However, the work practices can be modified as necessary to recognize the different types of exposure to electrical hazards. For instance, because each cell acts like a battery, the employer must define actions that are necessary if a worker must contact the dc bus structure. Those procedures must be consistent with the risk associated with the work task.

**(B) Signs.** Permanent signs shall clearly designate electrolytic cell areas.

**(C) Electrical Arc Flash Hazard Analysis.** The requirements of 130.3, Arc Flash Hazard Analysis, shall not apply to electrolytic cell line work zones.

Hazard/risk analyses used for ac circuits might not be appropriate for use with dc circuits. Each employee must be trained to understand how he or she might be exposed to a thermal hazard associated with an arc flash. Employees must understand how to select PPE for use when they are exposed to an arcing fault.

**(1) Arc Flash Hazard Analysis Procedure.** Each task performed in the electrolytic cell line working zone shall be analyzed for the risk of arc flash hazard injury. If there is risk of personal injury, appropriate measures shall be taken to protect persons exposed to the arc flash hazards. These measures shall include one or more of the following:

(1) Provide appropriate personal protective equipment *[see 310.5(D)(2)]* to prevent injury from the arc flash hazard.
(2) Alter work procedures to eliminate the possibility of the arc flash hazard.
(3) Schedule the task so that work can be performed when the cell line is deenergized.

**(2) Routine Tasks.** Arc flash hazard risk analysis shall be done for all routine tasks performed in the cell line work zone. The results of the arc flash hazard analysis shall be used in training employees in job procedures that minimize the possibility of arc flash hazards. The training shall be included in the requirements of 310.3.

**(3) Nonroutine Tasks.** Before a nonroutine task is performed in the cell line working zone, an arc flash hazard risk analysis shall be done. If an arc flash hazard is a possibility during nonroutine work, appropriate instructions shall be given to employees involved on how to minimize the possibility of a hazardous arc flash.

**(4) Arc Flash Hazards.** If the possibility of an arc flash hazard exists for either routine or nonroutine tasks, employees shall use appropriate safeguards.

**(D) Safeguards.** Safeguards shall include one or a combination of the following means.

**(1) Insulation.** Insulation shall be suitable for the specific conditions, and its components shall be permitted to include glass, porcelain, epoxy coating, rubber, fiberglass, plastic, and when dry, such materials as concrete, tile, brick, and wood. Insulation shall be permitted to be applied to energized or grounded surfaces.

**(2) Personal Protective Equipment.** Personal protective equipment shall provide protection from hazardous electrical conditions. Personal protective equipment shall include one or more of the following as determined by authorized management:

(1) Shoes, boots, or overshoes for wet service
(2) Gloves for wet service
(3) Sleeves for wet service
(4) Shoes for dry service
(5) Gloves for dry service
(6) Sleeves for dry service
(7) Electrically insulated head protection
(8) Protective clothing
(9) Eye protection with nonconductive frames
(10) Faceshield (polycarbonate or similar nonmelting type)

a. Standards for Personal Protective Equipment. Personal and other protective equipment shall be appropriate for conditions, as determined by authorized management, and shall not be required to meet the equipment standards in 130.7(C)(8) through 130.7(F) and in Table 130.7(C)(8) and Table 130.7(F).

b. Testing of Personal Protective Equipment. Personal protective equipment shall be verified with regularity and by methods that are consistent with the exposure of employees to hazardous electrical conditions.

**(3) Barriers.** Barriers shall be devices that prevent contact with energized or grounded surfaces that could present a hazardous electrical condition.

**(4) Voltage Equalization.** Voltage equalization shall be permitted by bonding a conductive surface to an electrically energized surface, either directly or through a resistance, so that there is insufficient voltage to create an electrical hazard.

**(5) Isolation.** Isolation shall be the placement of equipment or items in locations such that employees are unable to simultaneously contact exposed conductive surfaces that could present a hazardous electrical condition.

**(6) Safe Work Practices.** Employees shall be trained in safe work practices. The training shall include why the work practices in a cell line working zone are different from similar work situations in other areas of the plant. Employees shall comply with established safe work practices and the safe use of protective equipment.

(a) Attitude Awareness. Safe work practice training shall include attitude awareness instruction. Simultaneous contact with energized parts and ground can cause serious electrical shock. Of special importance is the need to be aware of body position where contact may be made with energized parts of the electrolytic cell line and grounded surfaces.

(b) Bypassing of Safety Equipment. Safe work practice training shall include techniques to prevent bypassing the protection of safety equipment. Clothing may bypass protective equipment if the clothing is wet. Trouser legs should be kept at appropriate length, and shirt sleeves should be a good fit so as not to drape while reaching. Jewelry and other metal accessories that may bypass protective equipment shall not be worn while working in the cell line working zone.

**(7) Tools.** Tools and other devices used in the energized cell line work zone shall be selected to prevent bridging between surfaces at hazardous potential difference.

Significant magnetic forces normally exist in areas in which electrolytic cells are present. Tools that contain magnetic materials or materials that are affected by magnetic fields should not be used in the cell area.

FPN: Tools and other devices of magnetic material could be difficult to handle in energized cells' areas due to their strong dc magnetic fields.

**(8) Portable Cutout Type Switches.** Portable cell cutout switches that are connected shall be considered as energized and as an extension of the cell line working zone. Appropriate procedures shall be used to ensure proper cutout switch connection and operation.

**(9) Cranes and Hoists.** Cranes and hoists shall meet the requirements of 668.32 of *NFPA 70, National Electrical Code*. Insulation required for safeguarding employees, such as insulated crane hooks, shall be periodically tested.

**(10) Attachments.** Attachments that extend the cell line electrical hazards beyond the cell line working zone shall utilize one or more of the following:

(1) Temporary or permanent extension of the cell line working zone
(2) Barriers
(3) Insulating breaks
(4) Isolation

**(11) Pacemakers and Metallic Implants.** Employees with implanted pacemakers, ferromagnetic medical devices, or other electronic devices vital to life shall not be permitted in cell areas unless written permission is obtained from the employee's physician.

Employers must take steps to ensure that workers who wear pacemakers and similar medical devices are not exposed to the magnetic fields that normally exist in the cell area.

FPN: The American Conference of Government Industrial Hygienists (ACGIH) recommends that persons with implanted pacemakers should not be exposed to magnetic flux densities above 10 gauss.

**(12) Testing.** Equipment safeguards for employee protection shall be tested to ensure they are in a safe working condition.

### 310.6 Portable Tools and Equipment

FPN: The order of preference for the energy source for portable handheld equipment is considered to be: (1) battery powered, (2) pneumatic, (3) a portable generator, (4) a nongrounded–type receptacle connected to an ungrounded source.

Portable tools must not be grounded when used in the cell area. Although a grounding conductor normally decreases exposure to an electrical hazard, in cell areas any grounded conductor increases exposure to an electrical hazard. All equipment and tools, including pneumatic tools, must be free from any grounding circuit. Pneumatic tools must be fitted with nonconductive hoses.

**(A) Portable Electrical Equipment.** The grounding requirements of 110.9(B)(2) shall not be permitted within an energized cell line working zone. Portable electrical equipment shall meet the requirements of 668.20 of *NFPA 70, National Electrical Code*. Power supplies for portable electric equipment shall meet the requirements of 668.21 of *NFPA 70, National Electrical Code*.

**(B) Auxiliary Nonelectric Connections.** Auxiliary nonelectric connections such as air, water, and gas hoses shall meet the requirements of 668.31 of *NFPA 70, National Electrical Code*. Pneumatic-powered tools and equipment shall be supplied with nonconductive air hoses in the cell line working zone.

**(C) Welding Machines.** Welding machine frames shall be considered at cell potential when within the cell line working zone. Safety-related work practices shall require that the cell line not be grounded through the welding machine or its power supply. Welding machines located outside the cell line working zone shall be barricaded to prevent employees from touching the welding machine and ground simultaneously where the welding cables are in the cell line working zone.

**(D) Portable Test Equipment.** Test equipment in the cell line working zone shall be suitable for use in areas of large magnetic fields and orientation.

FPN: Test equipment that is not suitable for use in such magnetic fields could result in an incorrect response. When such test equipment is removed from the cell line working zone, its performance might return to normal, giving the false impression that the results were correct.

# ARTICLE 320 Safety Requirements Related to Batteries and Battery Rooms

Article 320 identifies work practices associated with installation and maintenance of batteries containing many cells, such as those used with uninterruptible power supplies (UPS) and unit substation dc power supplies.

Article 320 has been modified for the 2009 edition to achieve better alignment with several IEEE standards covering the same subject.

Working with batteries exposes a worker to both shock and arc flash. A person's body might react to contact with dc voltage differently than from contact with ac voltage. *NFPA 70E* takes a conservative position and considers the risk of shock or electrocution to be the same for both ac and dc exposure.

In addition to the same electrical hazards, batteries also expose a worker to hazards associated with the chemical electrolyte used in the battery. The worker also must understand that battery charging might generate significant quantities of hydrogen and other flammable gases. When selecting work practices and PPE, the worker must consider exposure to these hazards as well.

## 320.1 Scope

The requirements of this article shall apply to the safety requirements related to installations of stationary storage batteries and battery rooms with a stored capacity exceeding 1 kWh or a nominal voltage that exceeds 50 volts but does not exceed 650 volts.

FPN: For further information, refer to the following documents:

(1) NFPA 1, *Fire Code*, 2009

(2) *NFPA 70, National Electrical Code*, Article 480, Storage Batteries, 2008

(3) IEEE Std. 450, *IEEE Recommended Practice for Maintenance, Testing, and Replacement of Vented Lead-Acid Batteries for Stationary Applications*, 2002

(4) IEEE Std. 484, *Recommended Practice for Installation Design and Installation of Vented Lead-Acid Batteries for Stationary Applications*, 2002

(5) IEEE 485, *IEEE Recommended Practice for Sizing Lead-Acid Storage Batteries for Stationary Applications*, 1997

(6) IEEE Std. 937, *Recommended Practice for Installation and Maintenance of Lead-Acid Batteries for Photovoltaic Systems*, 2007

(7) IEEE 1106, *IEEE Recommended Practice for Installation, Maintenance, Testing, and Replacement of Vented Nickel-Cadmium Batteries for Stationary Applications*, 2005

(8) IEEE 1184, *IEEE Guide for Batteries for Uninterruptible Power Supply Systems*, 2006

(9) IEEE Std. 1187, *Recommended Practice for Installation Design and Installation of Valve-Regulated Lead-Acid Storage Batteries for Stationary Applications*, 2002

(10) IEEE 1188, *IEEE Recommended Practice for Maintenance, Testing, and Replacement of Valve Regulated Lead-Acid (VRLA) Batteries for Stationary Applications*, 2005

(11) IEEE 1189, *IEEE Guide for Selection of Valve-Regulated Lead-Acid (VRLA) Batteries for Stationary Applications*, 1996

(12) IEEE 1375, *IEEE Guide for the Protection of Stationary Battery Systems*, 1998 (R 2003)

(13) OSHA 29 CFR 1926.441, "Batteries and battery charging"

(14) OSHA 29 CFR 1910.305(j)(7), "Storage batteries"

### 320.2 Definitions

For the purposes of this chapter, the following definitions shall apply.

The definitions listed in Section 320.2 are intended to apply to Article 320 only. However, the definitions are consistent with *The Authoritative Dictionary of IEEE Standards Terms.*

**Accessories.** Items supplied with the battery to facilitate the continued operation of the battery.

**Authorized Person.** The person in charge of the premises, or other person appointed or selected by the person in charge of the premises, to perform certain duties associated with the battery installation on the premises.

**Battery.** An electrochemical system capable of storing under chemical form the electric energy received and which can give it back by reconversion.

**Battery Enclosure.** An enclosure containing batteries that is suitable for use in an area other than a battery room or an area restricted to authorized personnel.

**Battery Room.** Room specifically intended for the installation of batteries that have no other protective enclosure.

**Capacity.** The quantity of electricity (electric charge) usually expressed in ampere-hour (A-h) that a fully charged battery can deliver under specified conditions.

**Cell.** An assembly of electrodes and electrolyte that constitutes the basic unit of the battery.

**Charging.** An operation during which a battery receives electric energy that is converted to chemical energy from an external circuit. The quantity of electric energy then is known as the charge and is usually measured in ampere-hour.

**Constant Current Charge.** A charge during which the current is maintained at a constant value.

**Constant Voltage Charge.** A charge during which the voltage across the battery terminals is maintained at a constant value.

**Container.** A container for the plate pack and electrolyte of a cell of a material impervious to attack by the electrolyte.

**Discharging.** An operation during which a battery delivers current to an external circuit by the conversion of chemical energy to electric energy.

**Electrolyte.** A solid, liquid, or aqueous salt solution that permits ionic conduction between positive and negative electrodes of a cell.

**Electrolyte Density.** Density of the electrolyte, measured in kilograms per cubic meter at a specific temperature (density of pure water = 1000 kilograms per cubic meter at 4°Celsius).

FPN: The density of an electrolyte was formerly indicated by its specific gravity. Specific gravity is the ratio of the density of the electrolyte to the density of pure water. S.G. = (electrolyte density in kilograms per cubic meter)/1000.

**Flame-Arrested Vent Plug.** A vent plug design that provides protection against internal explosion when the cell or battery is exposed to a naked flame or external spark.

**Gassing.** The formation of gas produced by electrolyte.

**Intercell and Interrow Connection.** Connections made between rows of cells or at the positive and negative terminals of the battery that might include lead-plated terminal plates, cables with lead, plated lugs, and lead-plated rigid copper connectors, and for nickel-cadmium cells, nickel-plated copper intercell connections.

**Intercell Connector Safety Cover.** Insulated cover to shroud the terminals and intercell connectors from inadvertent contact by personnel or accidental short circuiting.

**Nominal Voltage.** An approximate value of voltage used to identify a type of battery.

**Pilot Cell.** A selected cell of a battery that is considered to be representative of the average state of the battery or part thereof.

**Prospective Fault Current.** The highest level of fault current that can occur at a point on a circuit. This is the fault current that can flow in the event of a zero impedance short-circuit and if no protection devices operate.

**Rate.** The current expressed in amperes at which a battery is discharged.

**Secondary Battery.** Two or more rechargeable cells electrically connected and used as a source of energy.

**Secondary Cell.** A rechargeable assembly of electrodes and electrolytes that constitutes the basic unit of a battery.

**Stepped Stand.** Containers placed in rows and these rows are placed at different levels to form a stepped arrangement.

**Terminal Post.** A part provided for the connection of a cell or a battery to external conductors.

**Tiered Stand.** Where rows of containers are placed above containers of the same or another battery.

**Valve-Regulated Lead Acid (VRLA) Battery.** A battery that has no provision for the addition of water or electrolyte or for external measurement of electrolyte specific gravity.

**Vented Battery.** A battery in which the products of electrolysis and evaporation are allowed to escape freely to the atmosphere.

**Vent Plug.** A part closing the filling hole that is also employed to permit the escape of gas.

**VRLA.** Valve-regulated lead acid storage battery.

### 320.3 Battery Connections

Batteries are sources of energy. Therefore, isolating the source of voltage from a cell is not possible. However, limiting the number of cells that are connected together is possible. Workers are exposed to shock and arc flash when performing tasks associated with batteries.

Exhibit 320.1 illustrates a method of performing maintenance that decreases the hazard to the worker. The degree of arc flash hazard increases as the capacity of the battery increases. The hazard/risk analysis must consider the number of cells associated with the work task.

#### (A) Method of Connection.

FPN No. 1: Batteries usually consist of a number of identical cells connected in series. The voltage of a series connection of cells is the voltage of a single cell multiplied by the number

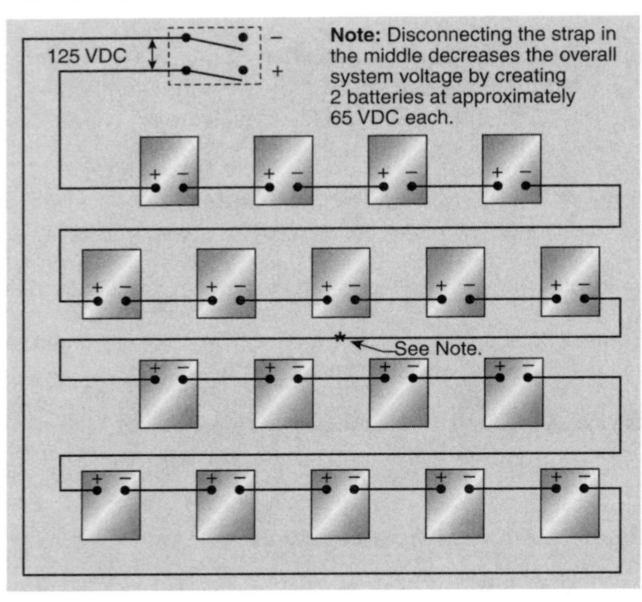


***EXHIBIT 320.1*** *A large battery installation.*

of cells. If cells of sufficiently large capacity are available, then two or more series-connected strings of equal numbers of cells could be connected in parallel to achieve the desired rated capacity. The rated capacity of such a battery is the sum of the capacities of a group of cells comprising a single cell from each of the parallel branches.

FPN No. 2: Cells of unequal capacity should not be connected in series.

FPN No. 3: Parallel connections of batteries are not recommended for constant current-charging applications.

FPN No. 4: Cells connected in series have high voltages that could produce a shock hazard.

**(B) Battery Short-Circuit Current.** The battery manufacturer shall be consulted regarding the sizing of the battery short-circuit protection.

*Exception: If information regarding the short-circuit protection of a battery is not available from the manufacturer, the prospective fault level at the battery terminals shall be considered to be twenty times the nominal battery capacity at the 3-hour rate.*

FPN: Battery short-circuit current = (battery voltage)/(internal resistance).

**(C) Connection Between Battery and DC Switching Equipment.**

**(1) General.** Any cable, busbar, or busway forming the connection between the battery terminal and the dc switching equipment shall be rated to withstand the prospective short-circuit current.

FPN: The available short-circuit current should be assumed for a time period of at least 1 second.

Outside busbars and cables should be both of the following:

(1) Insulated from the battery terminals to a height of 3.75 m (12 ft 4 in.), or to the battery room ceiling, whichever is lower
(2) Clearly identified and segregated from any other supply circuits

**(2) Cable.** Cables shall be effectively clamped and sufficient support shall be provided throughout the length of cables to minimize sag and prevent undue strain from being imposed on the cable.

**(3) Busbars.**

FPN: Busbars should be insulated throughout their length by an insulating material not affected by the acid fumes that are present in a battery room. The steelwork supporting the busbar system should be installed so as not to restrict access to the battery for the purpose of maintenance.

**(4) Busways.**

FPN: Busways should be fully enclosed and able to withstand high levels of fault current without danger.

**(D) DC Switching Equipment.** Switching equipment shall comply with applicable installation requirements.

FPN: For further information concerning electrical installation requirements, refer to *NFPA 70, National Electrical Code.*

**(E) Terminals and Connectors.** Intercell and battery terminal connections shall be constructed of materials, either intrinsically resistant to corrosion or suitably protected by surface finish against corrosion. The joining of materials that are incompatible in a corrosive atmosphere shall be avoided.

FPN No. 1: To prevent mechanical stress on the battery terminal posts, the connection between the battery and any busbar system or large cable should be by insulated flexible cable of suitable rating.

FPN No. 2: The takeoff battery terminals and busbar connections should be shrouded or protected by physical barriers to prevent accidental contact.

**(F) DC Systems Grounding and Ground-Fault Detection.** One of the four types of available dc grounding systems, described as Type 1 through Type 4, shall be used.

(1) Type 1. The ungrounded dc system in which neither pole of the battery is connected to ground

FPN: Work on such a system should be carried out with the battery isolated from the battery charger. If an intentional ground is placed at one end of the battery, an increased shock hazard would exist between the opposite end of the battery and ground. Also, if another ground develops within the system (e.g., dirt and acid touching the battery rack), it creates a short circuit that could cause a fire. An ungrounded dc system should be equipped with an alarm to indicate the presence of a ground fault.

(2) Type 2. The solidly grounded dc system where either the positive or negative pole of the battery is connected directly to ground

(3) Type 3. The resistance grounded dc system, where the battery is connected to ground through a resistance

FPN: The resistance is used to permit operation of a current relay, which in turn initiates an alarm.

(4) Type 4. A tapped solid ground, either at the center point or at another point to suit the load system

**(G) Protection of DC Circuits.** DC circuits shall be protected in accordance with the *NEC.*

**(H) Alarms.**

**(1) Abnormal Battery Conditions.** Alarms shall be provided for early warning of the following abnormal conditions of battery operation:

(1) For vented batteries:
  a. Overvoltage
  b. Undervoltage
  c. Overcurrent
  d. Ground fault
(2) For VRLA batteries, items (1)(a) through (1)(d) plus overtemperature, as measured at the pilot cell

**(2) Warning Signal.** The alarm system shall provide an audible alarm and visual indication at the battery location, and where applicable, at a remote manned control point.

### 320.4 Installations of Batteries

Installations using secondary batteries vary considerably in size, from large uninterruptible power supply systems, telecommunication systems, and demand load–leveling installations to small emergency lighting installations. Secondary batteries permanently installed in or on buildings, structures, or premises, having a nominal voltage exceeding 24 volts and a capacity exceeding 10 ampere-hours at the 1-hour rate, shall be installed in a battery room or battery enclosure.

**(A) Location.** Batteries shall be installed in one of the following:

(1) Dedicated battery rooms
(2) An area accessible only to authorized personnel
(3) An enclosure with lockable doors or a suitable housing that shall require a key or tool to gain access to the batteries and shall provide protection against electrical contact and damage to the battery

NFPA 111, *Standard on Stored Electrical Energy Emergency and Standby Power Systems,* provides valuable information for the installation, operation, and maintenance of battery rooms and systems. Compliance with these requirements can enhance worker safety in these environments. See also Article 480 of *NFPA 70®*, *National Electrical Code® (NEC®).*

**(B) Arrangement of Cells.** The arrangement of cells in a battery system shall meet the following requirements:

(1) All cells shall be readily accessible for such inspection and maintenance as is required by the manufacturer.
(2) The space between adjacent containers shall be no less than that recommended by the battery manufacturer or, where manufacturer guidance is not available, shall be at least 12.5 mm ($^1/_2$ in.).
(3) Each cell shall be readily accessible without having to reach over another cell, or alternatively, all exposed energized surfaces shall be shrouded.

**(C) Ventilation for Batteries of the Vented Type.**

**(1) Installation.** Batteries shall be located in rooms or enclosures with outside vents or in well-ventilated rooms, so arranged to prevent the escape of fumes, gases, or electrolyte spray into other areas.

**(2) Ventilation.** Ventilation shall be provided so as to prevent liberated hydrogen gas from exceeding 1 percent concentration.

(a) Adequacy. Room ventilation shall be adequate to assure that pockets of trapped hydrogen gas do not occur, particularly at the ceiling, to prevent the accumulation of an explosive mixture.

(b) Equipment Considerations. Exhaust air shall not pass over electrical equipment unless the equipment is listed for the use.

(c) Location of Inlets. Inlets shall be no higher than the tops of the battery cells and outlets at the highest level in the room.

FPN: The maximum hydrogen evolution rate for batteries should be obtained for the condition when the maximum charging current available from a constant current battery charger is applied into a fully charged battery or the current that would be expected from a constant voltage charger in boost/equalize mode. If possible, contact manufacturer for hydrogen evolution rates.

**(3) Mechanical Ventilation.** Where mechanical ventilation is installed, the following shall be required:

(1) Airflow sensors shall be installed to initiate an alarm if the ventilation fan becomes inoperative.
(2) Control equipment for the exhaust fan shall be located more than 1.8 m (6 ft) from the battery and a minimum of 100 mm (4 in.) below the lowest point of the highest ventilation opening.
(3) Where mechanical ventilation is used in a dedicated battery room, all exhaust air shall be discharged outside the building.
(4) Fans used to remove air from a battery room shall not be located in the duct unless the fan is listed for the use.

**(D) Ventilation for VRLA Type.**

**(1) Ventilation Requirements.** Ventilation shall be provided so as to prevent liberated hydrogen gas from exceeding a 1 percent concentration.

(a) Adequacy. Ventilation shall be adequate to ensure that pockets of trapped hydrogen gas do not occur, particularly at the ceiling of a room or at the top of a cabinet, to prevent the accumulation of an explosive mixture.

(b) Exhaust. Exhaust air shall not pass over electrical equipment unless the equipment is listed for the use.

(c) Inlets. Inlets shall be no higher than the tops of the battery cells and outlets at the highest level in the room.

**(2) Mechanical Ventilation.** Where mechanical ventilation is installed, the following shall be required:

(1) Airflow sensors shall be installed to initiate an alarm if the ventilation fan becomes inoperative.

(2) Control equipment for the exhaust fan in dedicated battery rooms shall be located more than 1.8 m (6 ft) from the battery and a minimum of 100 mm (4 in.) below the lowest point of the highest ventilation opening.

(3) Where mechanical ventilation is used in a dedicated battery room, all exhaust air shall be discharged outside the building.

(4) Fans used to remove air from a battery room shall not be located in the duct unless the fan is listed for the use.

**(3) Temperature Requirements.** Thermal management shall be provided to maintain battery design temperature to prevent thermal runaway that can cause cell meltdown, leading to a fire or explosion.

**320.5 Battery Room Requirements**

**(A) General.** The battery room shall be accessible only to authorized personnel and shall be locked when unoccupied.

**(1) Battery Rooms or Areas Restricted to Authorized Personnel.**

(a) Doors. The battery room and enclosure doors shall open outward. The doors shall be equipped with quick-release, quick-opening hardware.

(b) Foreign Piping. Foreign piping that is not protected against corrosion shall not pass through the battery room.

(c) Passageways. Passageways shall be of sufficient width to allow the replacement of all battery room equipment.

(d) Emergency Exits. Emergency exits shall be provided as required.

(e) Access. Access and entrance to working space about the battery shall be provided as required by 110.26 of *NFPA 70, National Electrical Code*.

FPN: Provision to include emergency services personnel and their equipment should be made.

**(2) Battery Enclosures.** All cells shall be readily accessible for inspection, cleaning, maintenance, and removal.

**(3) Battery Room Floor Loading.** Floor loading shall take into account the seismic activity.

**(4) Battery Room Floor Construction and Finish.** Battery systems containing free-flowing liquid electrolyte shall be provided with spill containment systems in accordance with the fire code.

FPN No. 1: The battery room floor should be of concrete construction. The floor should be graded so any spillage of electrolyte will drain to an area where the electrolyte could be neutralized before disposal. (The battery manufacturer should be consulted on the appropriate floor grading so as to reduce connection alignment problems.)

FPN No. 2: The floor should be covered with an electrolyte-resistant, durable, antistatic, and slip-resistant surface overall, to a height 100 mm (4 in.) on each wall. Where batteries are mounted against a wall, the wall behind and at each end of the battery should be coated to a distance of 500 mm (20 in.) around the battery with an electrolyte-resistant paint.

**(B) Battery Layout and Floor Area.** The battery layout and floor area shall meet the following requirements:

**(1) Battery Layout.** The installation shall be so designed that, unless there is a physical barrier, potential differences exceeding 120 volts shall be separated by a distance of not less than 900 mm (36 in.) measured in a straight line in any direction.

**(2) Floor Area.** The floor area shall allow for the following clearances:

(a) Aisle Width. The minimum aisle width shall be 900 mm (36 in.).

(b) Single-Row Batteries. In addition to the minimum aisle width, there shall be a minimum clearance of 25 mm (1 in.) between a cell and any wall or structure on the side not requiring access for maintenance. This required clearance does not preclude battery stands touching adjacent walls or structures, provided that the battery shelf has a free air space for no less than 90 percent of its length.

(c) Double-Row Batteries. The minimum aisle width shall be maintained on one end and both sides of the battery. The remaining end shall have a minimum clearance of 100 mm (4 in.) between any wall or structure and a cell.

(d) Tiered Batteries. Tiered batteries shall meet the requirements of 320.5(B)(2)(a), 320.5(B)(2)(b), and 320.5(B)(2)(c). In addition, there shall be a minimum clearance of 300 mm (12 in.) between the highest point of the battery located on the bottom tier and the lowest point of the underside of the upper runner bearers.

(e) Where a charger, or other associated electrical equipment, is located in a battery room, the aisle width between any battery and any part of the battery-charging equipment (including the doors when fully open) shall be at least 900 mm (36 in.).

**(C) Takeoff Battery Terminals and Outgoing Busbars and Cables.**

**(1) Takeoff Battery Terminals.** Outgoing busbars and cables shall meet the following requirements:

(1) Be insulated from the battery terminals to a height of 3.75 m (12 ft 4 in.) or the battery room ceiling, whichever is lower
(2) Be clearly identified and segregated from any other supply circuits
(3) Prevent mechanical stress on the battery posts

**(2) Outgoing Busbars and Cables.** The takeoff battery terminals and busbar connections shall comply with either of the following:

(1) Be shrouded
(2) Be protected by physical barriers to prevent accidental contact

**(D) Intertier and Interrow Connections.** The battery terminals and busbar and cable interconnections between rows shall comply with either of the following:

(1) Be shrouded
(2) Be protected by insulating barriers to prevent accidental contact

**(E) Barriers.** To avoid accidental contact with intercell connections, the following insulating barriers shall be installed:

**(1) Double-Row Batteries.** Insulating barriers between double-row batteries shall be installed for the entire length of the battery extending 100 mm (4 in.) past the end terminal unless those terminals are shrouded. The barrier shall extend vertically a minimum of 400 mm (16 in.) above the exposed portion of the intercell connections and a minimum of 25 mm (1 in.) below the top of the battery container.

**(2) Batteries Above 120 Volts.** Where the nominal voltage of the battery exceeds 120 volts, interblock barriers shall be installed to sectionalize the battery into voltage blocks not exceeding 120 volts. Barriers shall extend a minimum of 50 mm (2 in.) out from the exposed side of the battery and a minimum of 400 mm (16 in.) above the top of the container.

**(F) Illumination.**

**(1) Battery Room Lighting.** Battery room lighting shall be installed to provide a minimum level of illumination of 300 lux (30 ft-candles).

**(2) Emergency Lighting.** Emergency illumination shall be provided for safe egress from the battery room.

**(G) Location of Luminaires and Switches.** Luminaires shall not be installed directly over cells or exposed energized conductors and circuit parts. Switches for the controls of the luminaires shall be readily accessible.

**(H) Power.** General-purpose outlets shall be installed for the maintenance of the battery.

**(I) Location of General-Purpose Outlets.** General-purpose outlets shall be installed at least 1.8 m (6 ft) from the battery and a minimum of 100 mm (4 in.) below the lowest point of the highest ventilation opening.

**320.6 Battery Enclosure Requirements**

**(A) Enclosure Construction.**

**(1) General.** Where enclosures are designed to accommodate the battery, the battery charger, and other equipment, separate compartments shall be provided for each.

**(2) Ventilation.** The ventilation openings for the battery compartment shall:

- (a) Prevent the exchange of air within compartments containing electrical equipment
- (b) Prevent accumulation of flammable gas in pockets exceeding 1 percent concentration

**(B) Battery Takeoff Terminals and Outgoing Busbars and Cables.** Outgoing busbars and cables shall be fully insulated, and the battery takeoff terminals shall comply with the following:

(1) Takeoff terminals shall prevent excessive mechanical stress on the battery posts.
(2) Takeoff terminals shall comply with either of the following:
   (1) Be fully shrouded
   (2) Have physical barriers installed between them

**(C) Battery Compartment Circuits.** Only circuits associated with the battery shall be installed within a battery compartment of the enclosure.

### 320.7 Protection

**(A) General.**

**(1) Marking.** When the battery capacity exceeds 100 ampere-hours or where the nominal battery voltage is in excess of 50 volts, suitable warning notices indicating the battery voltage and the prospective short-circuit current of the installation shall be displayed.

**(2) Overcurrent Protection.** Each output conductor shall be individually protected by a fuse or circuit breaker positioned as close as practicable to the battery terminals.

**(3) Protective Equipment.** Protective equipment shall not be located in the battery compartment of the enclosure unless provided as part of a listed assembly.

**(B) Switching and Control Equipment.** Switching and control equipment shall comply with *NFPA 70, National Electrical Code*, and shall be listed for the application.

**(C) Ground-Fault Protection.** For an ungrounded battery of nominal voltage in excess of 120 volts, a ground-fault detector shall be provided to initiate a ground-fault alarm.

**(D) Main Isolating Switch.** The battery installation shall have an isolating switch installed as close as practicable to the main terminals of the battery. Where a busway system is installed, the isolating switch may be incorporated into the end of the busway.

**(E) Section Isolating Equipment.** Where the battery section exceeds 250 volts, the installation shall include an isolating switch, plugs, or links, as required, to isolate sections of the battery, or part of the battery, for maintenance.

**(F) Warning Signs.** The following signs shall be posted in appropriate locations:

(1) Electrical hazard warning signs indicating the shock hazard due to the battery voltage and the arc hazard due to the prospective short-circuit current
(2) Chemical hazard warning signs indicating the danger of hydrogen explosion from open flame and smoking and the danger of chemical burns from the electrolyte
(3) Notice for personnel to use and wear protective equipment and apparel
(4) Notice prohibiting access to unauthorized personnel

### 320.8 Personnel Protective Equipment

The following protective equipment shall be available to employees performing battery maintenance:

(1) Goggle and face shields
(2) Chemical-resistant gloves
(3) Protective aprons
(4) Protective overshoes
(5) Portable or stationary water facilities for rinsing eyes and skin in case of electrolyte spillage

### 320.9 Tools and Equipment

Tools and equipment for work on batteries shall comply with the following:

(1) Be of the nonsparking type
(2) Be equipped with handles listed as insulated for the maximum working voltage

# ARTICLE 330
# Safety-Related Work Practices for Use of Lasers

Article 330 is limited to tasks performed in the laboratory or in the shop using a laser. It does not cover the general application and use of lasers in the workplace. However, employees who work with lasers might be exposed to hazards associated with the laser output in addition to the electrical hazards associated with the equipment.

## 330.1 Scope

The requirements of this article shall apply to the use of lasers in the laboratory and the workshop.

## 330.2 Definitions

For the purposes of this article, the following definitions shall apply.

**Fail Safe.** The design consideration in which failure of a component does not increase the hazard. In the failure mode, the system is rendered inoperative or nonhazardous.

**Fail Safe Safety Interlock.** An interlock that in the failure mode does not defeat the purpose of the interlock, for example, an interlock that is positively driven into the off position as soon as a hinged cover begins to open, or before a detachable cover is removed, and that is positively held in the off position until the hinged cover is closed or the detachable cover is locked in the closed position.

**Laser.** Any device that can be made to produce or amplify electromagnetic radiation in the wavelength range from 100 nm to 1 mm primarily by the process of controlled stimulated emission.

**Laser Energy Source.** Any device intended for use in conjunction with a laser to supply energy for the excitation of electrons, ions, or molecules. General energy sources, such as electrical supply services or batteries, shall not be considered to constitute laser energy sources.

**Laser Product.** Any product or assembly of components that constitutes, incorporates, or is intended to incorporate a laser or laser system.

**Laser Radiation.** All electromagnetic radiation emitted by a laser product between 100 nm and 1 mm that is produced as a result of a controlled stimulated emission.

**Laser System.** A laser in combination with an appropriate laser energy source with or without additional incorporated components.

## 330.3 Safety Training

**(A) Personnel to Be Trained.** Employers shall provide training for all operator and maintenance personnel.

**(B) Scope of Training.** The training shall include, but is not limited to, the following:

(1) Familiarization with laser principles of operation, laser types, and laser emissions
(2) Laser safety, including the following:
    (1) System operating procedures
    (2) Hazard control procedures
    (3) The need for personnel protection

(4) Accident reporting procedures
(5) Biological effects of the laser upon the eye and the skin
(6) Electrical and other hazards associated with the laser equipment, including the following:
   a. High voltages (> 1 kV) and stored energy in the capacitor banks
   b. Circuit components, such as electron tubes, with anode voltages greater than 5 kV emitting X-rays
   c. Capacitor bank explosions
   d. Production of ionizing radiation
   e. Poisoning from the solvent or dye switching liquids or laser media
   f. High sound intensity levels from pulsed lasers

**(C) Proof of Qualification.** Proof of qualification of the laser equipment operator shall be available and in possession of the operator at all times.

Employees who work on lasers in the laboratory or in the shop must demonstrate an understanding of both electrical hazards and hazards associated with the laser output. Employees who have demonstrated this understanding should be issued some type of proof of qualification, such as a certificate indicating successful completion of safety training for work on or with the specific lasers available on site.

### 330.4 Safeguarding of Employees in the Laser Operating Area

**(A) Eye Protection.** Employees shall be provided with eye protection as required by federal regulation.

**(B) Warning Signs.** Warning signs shall be posted at the entrances to areas or protective enclosures containing laser products.

**(C) Master Control.** High power laser equipment shall include a key-operated master control.

**(D)** High-power laser equipment shall include a failsafe laser radiation emission audible and visible warning when it is switched on or if the capacitor banks are charged.

**(E)** Beam shutters or caps shall be utilized, or the laser switched off, when laser transmission is not required. The laser shall be switched off when unattended for 30 minutes or more.

**(F)** Laser beams shall not be aimed at employees.

**(G)** Laser equipment shall bear a label indicating its maximum output.

**(H)** Personnel protective equipment shall be provided for users and operators of high-power laser equipment.

### 330.5 Employee Responsibility

Employees shall be responsible for the following:

(1) Obtaining authorization for laser use
(2) Obtaining authorization for being in a laser operating area
(3) Observing safety rules
(4) Reporting laser equipment failures and accidents to the employer

# ARTICLE 340
# Safety-Related Work Practices: Power Electronic Equipment

Chapters 1 and 2 apply to electrical equipment that operates at frequencies normally supplied for consumer use. The reaction of the human body to current flow changes as the frequency increases or decreases. When an increase in frequency reaches the microwave band, joule heating can result in internal burns. Employees who work on or with equipment within the scope of Article 340 must be qualified to perform tasks on specific electronic equipment.

Employees who are qualified to work on or with this type of equipment should be trained to understand the unique hazards associated with the specific equipment on which they will perform work tasks. Workers should demonstrate understanding of the specific hazards and how to avoid exposure to them. The worker should be issued a certificate indicating successful completion of a safe work-practice training program.

### 340.1 Scope

This article shall apply to safety-related work practices around power electronic equipment, including the following:

(1) Electric arc welding equipment
(2) High-power radio, radar, and television transmitting towers and antenna
(3) Industrial dielectric and radio frequency (RF) induction heaters
(4) Shortwave or RF diathermy devices
(5) Process equipment that includes rectifiers and inverters such as the following:
 a. Motor drives
 b. Uninterruptible power supply systems
 c. Lighting controllers

### 340.2 Definition

For the purposes of this article, the following definition shall apply.

**Radiation Worker.** A person who is required to work in electromagnetic fields, the radiation levels of which exceed those specified for nonoccupational exposure.

### 340.3 Application

The purpose of this article is to provide guidance for safety personnel in preparing specific safety-related work practices within their industry.

### 340.4 Reference Standards

The following are reference standards for use in the preparation of specific guidance to employees:

(1) International Electrotechnical Commission IEC 60479, *Effects of Current Passing Through the Body*:
 a. 60479-1 Part 1: General aspects
 b. 60479-1-1 Chapter 1: Electrical impedance of the human body

c. 60479-1-2 Chapter 2: Effects of ac in the range of 15 Hz to 100 Hz

d. 60479-2 Part 2: Special aspects

e. 60479-2-4 Chapter 4: Effects of ac with frequencies above 100 Hz

f. 60479-2-5 Chapter 5: Effects of special waveforms of current

g. 60479-2-6 Chapter 6: Effects of unidirectional single impulse currents of short duration

(2) International Commission on Radiological Protection (ICRP) Publication 15, *Protection Against Ionizing Radiation from External Sources*

### 340.5 Hazardous Effects of Electricity on the Human Body

Employer and employees shall be aware of the following hazards associated with power electronic equipment.

(1) Results of Power Frequency Current.

a. At 5 mA, shock is perceptible.

b. At 10 mA, a person may not be able to voluntarily let go of the hazard.

c. At about 40 mA, the shock, if lasting for 1 second or longer, may be fatal due to ventricular fibrillation.

d. Further increasing current leads to burns and cardiac arrest.

(2) Results of Direct Current.

a. A dc current of 2 mA is perceptible.

b. A dc current of 10 mA is considered the threshold of the let-go current.

(3) Results of Voltage. A voltage of 30 V rms, or 60 V dc, is considered safe except when the skin is broken. The internal body resistance can be as low as 500 ohms, so fatalities can occur.

(4) Results of Short Contact.

a. For contact less than 0.1 second and with currents just greater than 0.5 mA, ventricular fibrillation may occur only if the shock is in a vulnerable part of the cardiac cycle.

b. For contact of less than 0.1 second and with currents of several amperes, ventricular fibrillation may occur if the shock is in a vulnerable part of the cardiac cycle.

c. For contact of greater than 0.8 second and with currents just greater than 0.5 A, cardiac arrest (reversible) may occur.

d. For contact greater than 0.8 second and with currents of several amperes, burns and death are probable.

(5) Results of ac at Frequencies Above 100 Hz. When the threshold of perception increases from 10 kHz to 100 kHz, the threshold of let-go current increases from 10 mA to 100 mA.

(6) Effects of Waveshape. Contact with voltages from phase controls usually causes effects between those of ac and dc sources.

(7) Effects of Capacitive Discharge.

a. A circuit of capacitance of 1 microfarad having a 10 kV capacitor charge may cause ventricular fibrillation.

b. A circuit of capacitance of 20 microfarad having a 10 kV capacitor charge may be dangerous and probably cause ventricular fibrillation.

### *CASE STUDY*

On an October Thursday, Ryan went to work anticipating the sixth World Series game.* Houston was down three games to two. Ryan had four tickets to the game and would be leaving Beaumont at 2:00 P.M. to drive to Houston with his family. He had planned to work in the morning before leaving for the game. He was as excited as his sons were.

Ryan was a senior electrician at a large chemical plant near Beaumont. He had "paid his dues" at the plant and felt that he had earned an easy morning. He was assigned to the electronics crew in the maintenance organization.

The solid-state drive room was air-conditioned in an effort to remove heat generated within the power electronics in the drives. Each section of the drive line-up had an exhaust duct at the top of the unit and a louvered door with an air filter on the inside of the door. This solid-state drive was old equipment. The equipment was physically very large. Electrically, the equipment was very large as well; the transformer supplying the rectifier was rated at 2000 kVA. The rectifier was close-coupled to the transformer, so there was no secondary overcurrent protection on the 480-volt bus.

The supervisor knew that Ryan was going to the game and wanted him to enjoy the outing. Ryan was spared all the heavy work on this Thursday. Instead, his work task was to change all of the air filters in the drive room. Ryan knew that the filters were all the same size: 12 by 18 inches. The filter material was man-made fiber, constrained in shape by aluminum screen, and the filters were light in weight. Ryan knew that the drive equipment would be running, but he also knew that would cause no problem. The filters were always changed while the equipment was running. All he had to do was open the door, change the filter, then close the door. There were no interlocks to defeat.

Ryan picked up the filters from the storeroom and put the box on his tool cart. Yessir, he would have an easy morning. He proceeded to the drive room and opened the first door. He changed the filter and discarded the dirty filter in the trash container nearby.

Ryan moved on to the second unit. He sure hoped that Houston would win the game. He hoped someone in his family would catch a foul ball. He knew the bat boy and would be able to get the ball autographed. As he reached for the door handle to open the door, he heard a noise. It seemed to be coming from the large rectifier at the other end of the room.

Ryan moved closer to the rectifier. Yes, the loud humming was coming from inside the rectifier. Instinctively, Ryan opened the door of the unit to look inside. He had done that a thousand times before. This time, however, there was a problem. He never learned the source of the humming. The explosion was immediate, as soon as the door was moved. A short circuit occurred somewhere on the bus from the transformer. Ryan's cotton clothing caught fire. No one else was in the room, but other workers came quickly when the noise from the explosion reverberated around the facility.

The pressure forces from the explosion blew Ryan against the adjacent wall. His lifeless body was transported to the hospital.

After an in-depth analysis of the incident, the chemical plant relocated the filters to the outside of the door. In the future, changing filters would become a job for two people. Arc-flash-resistant clothing would be provided for all employees who were required to enter the room. No rectifier or transformer door would be opened with the equipment running.

That action was too late to help Ryan. He never knew that Houston won the ball game.

* This account is based on an actual incident. The names, including the name of the facility, have all been changed to protect those involved. Any similarity to actual names or facilities is strictly coincidental.

Reprinted from Ray A. Jones and Jane G. Jones, *Electrical Safety in the Workplace,* 2000, with permission of Jones and Bartlett Publishers, Sudbury, MA.

### 340.6 Hazards Associated with Power Electronic Equipment

Employer and employees shall be aware of the hazards associated with the following:

(1) High voltages within the power supplies
(2) Radio frequency energy–induced high voltages
(3) Effects of radio frequency (RF) fields in the vicinity of antennas and antenna transmission lines, which can introduce electrical shock and burns
(4) Ionizing (X-radiation) hazards from magnetrons, klystrons, thyratrons, cathode-ray tubes, and similar devices
(5) Non-ionizing RF radiation hazards from the following:
    a. Radar equipment
    b. Radio communication equipment, including broadcast transmitters
    c. Satellite–earth-transmitters
    d. Industrial scientific and medical equipment
    e. RF induction heaters and dielectric heaters
    f. Industrial microwave heaters and diathermy radiators

### 340.7 Specific Measures for Personnel Safety

**(A) Employer Responsibility.** The employer shall be responsible for the following:

(1) Proper training and supervision by properly qualified personnel including the following:
    a. The nature of the associated hazard
    b. Strategies to minimize the hazard
    c. Methods of avoiding or protecting against the hazard
    d. The necessity of reporting any hazardous incident
(2) Properly installed equipment
(3) Proper access to the equipment
(4) Availability of the correct tools for operation and maintenance
(5) Proper identification and guarding of dangerous equipment
(6) Provision of complete and accurate circuit diagrams and other published information to the employee prior to the employee starting work (The circuit diagrams should be marked to indicate the hazardous components.)
(7) Maintenance of clear and clean work areas around the equipment to be worked
(8) Provision of adequate and proper illumination of the work area

**(B) Employee Responsibility.** The employee is responsible for the following:

(1) Being continuously alert and aware of the possible hazards
(2) Using the proper tools and procedures for the work

(3) Informing the employer of malfunctioning protective measures, such as faulty or inoperable enclosures and locking schemes
(4) Examining all documents provided by the employer relevant to the work, especially those documents indicating the hazardous components location
(5) Maintaining good housekeeping around the equipment and work space
(6) Reporting any hazardous incident

# ARTICLE 350 Safety-Related Work Requirements: Research and Development Laboratories

Article 350 has been added to this 2009 edition of *NFPA 70E*. It addresses unique conditions that might exist in laboratory and research conditions.

### 350.1 Scope

The requirements of this article shall apply to the electrical installations in those areas, with custom or special electrical equipment, designated by the facility management for research and development (R&D) or as laboratories.

The scope of this article covers all laboratory facilities, including those that exist in educational facilities. It also covers research and development facilities associated with universities and other institutions of higher learning.

### 350.2 Definitions

For the purposes of this article, the following definitions shall apply.

**Competent Person.** A person meeting all of the requirements of a qualified person, as defined in Article 100 in Chapter 1 of this document and, in addition, is responsible for all work activities or safety procedures related to custom or special equipment, and has detailed knowledge regarding the electrical hazard exposure, the appropriate controls for mitigating those hazards, and the implementation of those controls.

**Field Evaluated.** A thorough evaluation of nonlisted or modified equipment in the field that is performed by persons or parties acceptable to the authority having jurisdiction. The evaluation approval ensures that the equipment meets appropriate codes and standards, or is similarly found suitable for a specified purpose.

**Laboratory.** A building, space, room, or group of rooms intended to serve activities involving procedures for investigation, diagnostics, product testing, or use of custom or special electrical components, systems, or equipment.

**Research and Development (R&D).** An activity in an installation specifically designated for research or development conducted with custom or special electrical equipment.

### 350.3 Applications of Other Articles

The electrical system for R&D and laboratory applications shall meet the requirements of the remainder of this document, except as amended by Article 350.

All requirements of Chapters 1 and 2 apply to facilities within the scope of this article unless modified herein.

FPN: Examples of these applications include low voltage–high current power systems; high voltage–low current power systems; dc power supplies; capacitors; cable trays for signal cables and other systems, such as steam, water, air, gas, or drainage; and custom-made electronic equipment.

## 350.5 Specific Measures and Controls for Personnel Safety

Each laboratory or R&D system application shall be assigned a competent person as defined in this article to ensure the use of appropriate electrical safety-related work practices and controls.

As required in Chapter 1, the employer, through the competent person, should provide written procedures or other instructions as necessary to provide complete documentation of the electrical safety program.

## 350.6 Listing Requirements

The equipment or systems used in the R&D area or in the laboratory shall be listed or field evaluated prior to use.

FPN: Laboratory and R&D equipment or systems can pose unique electrical hazards that might require mitigation. Such hazards include ac and dc, low voltage and high amperage, high voltage and low current, large electromagnetic fields, induced voltages, pulsed power, multiple frequencies, and similar exposures.

ANNEX A

# Referenced Publications

Annex A identifies publications that serve as source documents for the content of the 2009 edition of *NFPA 70E.*

### A.1 General

The documents or portions thereof listed in this annex are referenced within this standard and shall be considered part of the requirements of this document.

### A.2 NFPA Publications

National Fire Protection Association, 1 Batterymarch Park, Quincy, MA 02169-7471.

*NFPA 70®, National Electrical Code®*, 2008.

### A.3 Other Publications

**A.3.1 ANSI Publications.** American National Standards Institute, Inc., 25 West 43rd Street, 4th Floor, New York, NY 10036.

ANSI A14.1, *Safety Requirements for Portable Wood Ladders,* 2000.
ANSI A14.3, *Safety Requirements for Fixed Ladders,* 2002.
ANSI A14.4, *Safety Requirements for Job-Made Ladders,* 2002.
ANSI A14.5, *Safety Requirement for Portable Reinforced Plastic Ladders,* 2000.
ANSI Z87.1, *Practice for Occupational and Educational Eye and Face Protection,* 2003.
ANSI Z89.1, *Requirements for Protective Headwear for Industrial Workers,* 2003.
ANSI Z535, *Series of Standards for Safety Signs and Tags,* 2006.

**A.3.2 ASTM Publications.** ASTM International, 100 Barr Harbor Drive, P.O Box C700, West Conshohocken, PA 19428-2959.

ASTM D 120, *Standard Specification for Rubber Insulating Gloves,* 2002a (R 2006).
ASTM D 1048, *Standard Specification for Rubber Insulating Blankets,* 2005.
ASTM D 1049, *Standard Specification for Rubber Covers,* 1998 (R 2002).
ASTM D 1050, *Standard Specification for Rubber Insulating Line Hoses,* 2005.
ASTM D 1051, *Standard Specification for Rubber Insulating Sleeves,* 2007.
ASTM F 478, *Standard Specification for In-Service Care of Insulating Line Hose and Covers,* 1999 (R 2007).
ASTM F 479, *Standard Specification for In-Service Care of Insulating Blankets,* 2006.
ASTM F 496, *Standard Specification for In-Service Care of Insulating Gloves and Sleeves,* 2006.
ASTM F 696, *Standard Specification for Leather Protectors for Rubber Insulating Gloves and Mittens,* 2006.
ASTM F 711, *Standard Specification for Fiberglass-Reinforced Plastic (FRP) Rod and Tube Used; in Line Tools,* 2002 (R 2007).

ASTM F 712, *Standard Test Methods and Specifications for Electrically Insulating Plastic Guard Equipment for Protection of Workers,* 2006.
ASTM F 855, *Standard Specification for Temporary Protective Grounds to Be Used on De-energized Electric Power Lines and Equipment,* 2004.
ASTM F 887, *Standard Specification for Personal Climbing Equipment*, 2005.
ASTM F 1117, *Standard Specification for Dielectric Overshoe Footwear,* 2003.
ASTM F 1236, *Standard Guide for Visual Inspection of Electrical Protective Rubber Products,* 2007.
ASTM F 1505, *Standard Specification for Insulated and Insulating Hand Tools,* 2007.
ASTM F 1506, *Standard Performance Specification for Flame Resistant Textile Materials for Wearing Apparel for Use by Electrical Workers Exposed to Momentary Electric Arc and Related Thermal Hazards,* 2002a.
ASTM F 1891, *Standard Specification for Arc and Flame Resistant Rainwear,* 2006.
ASTM F 1959, *Standard Test Method for Determining the Arc Thermal Performance Value of Materials for Clothing,* 2006.
ASTM F 2178, *Standard Test Method for Determining the Arc Rating and Standard Specification for Face Protective Products,* 2006.
ASTM F 2249, *Standard Specification for In-Service Test Methods for Temporary Grounding Jumper Assemblies Used on De-Energized Electric Power Lines and Equipment*, 2003.
ASTM F 2412, *Standard Test Methods for Foot Protections*, 2005.
ASTM F 2413, *Standard Specification for Performance Requirements for Foot Protection*, 2005.

**A.3.3 ICRP Publications.** International Commission on Radiological Protection, SE-171 16 Stockholm, Sweden.

ICRP 15, *Protection Against Ionizing Radiation from External Sources.*

**A.3.4 IEC Publications.** International Electrotechnical Commission, 3, rue de Varembé, P.O. Box 131, CH-1211 Geneva 20, Switzerland.

IEC 60479, *Effects of Current Passing Through the Body*, 1987.
60479-1 Part 1: General aspects
60479-1-1 Chapter 1: Electrical impedance of the human body
60479-1-2 Chapter 2: Effects of ac in the range of 15 Hz to 100 Hz
60479-2 Part 2: Special aspects
60479-2-4: Chapter 4: Effects of ac with frequencies above 100 Hz
60479-2-5 Chapter 5: Effects of special waveforms of current
60479-2-6 Chapter 6: Effects of unidirectional single impulse currents of short duration

**A.3.5 IEEE Publications.** Institute of Electrical and Electronics Engineers, IEEE Operations Center, 445 Hoes Lane, P. O. Box 1331, Piscataway, NJ 08855-1331.

IEEE C37.20.7, *Guide for Testing Metal-Enclosed Switchgear Rated up to 38 kV for Internal Arcing Faults*, 2007.

### A.4 References for Extracts in Mandatory Sections

*NFPA 70®*, *National Electrical Code®*, 2008.

ANNEX B

# Informational References

Annex B identifies important publications to provide additional information that might assist a user in developing understanding of the requirements.

## B.1 Referenced Publications

The following documents or portions thereof are referenced within this standard for informational purposes only and are thus not part of the requirements of this document unless also listed in Annex A.

**B.1.1 NFPA Publications.** National Fire Protection Association, 1 Batterymarch Park, Quincy, MA 02169-7471.

NFPA 1, *Fire Code*, 2009 edition.
*NFPA 70®*, *National Electrical Code®*, 2008 edition.
NFPA 70B, *Recommended Practice for Electrical Equipment Maintenance,* 2006 edition.

**B.1.2 ANSI Publications.** American National Standards Institute, Inc., 25 West 43rd Street, 4th Floor, New York, NY 10036.

ANSI/AIHA Z10, *American National Standard for Occupational Safety and Health Management Systems*, 2005.
ANSI/ASSE Z244.1, *Control of Hazardous Energy — Lockout/Tagout and Alternative Methods*, 2003.
ANSI/NETA MTS, *Standard for Maintenance Testing Specification,* 2007.

**B.1.3 ASTM Publications.** ASTM International, 100 Barr Harbor Drive, P.O. Box C 700, West Conshohocken, PA 19428-2959.

ASTM F 496, *Standard Specification for In-Service Care of Insulating Gloves and Sleeves*, 2006.
ASTM F 711, *Standard Specification for Fiberglass-Reinforced Plastic (FRP) Rod and Tube Used; in Line Tools*, 2002 (R 2007).
ASTM F 2249, *Standard Specification for In-Service Test Methods for Temporary Grounding Jumper Assemblies Used on De-Energized Electric Power Lines and Equipment*, 2003.

**B.1.4 IEEE Publications.** Institute of Electrical and Electronic Engineers, IEEE Operations Center, 445 Hoes Lane, P. O. Box 1331, Piscataway, NJ 08855-1331.

ANSI/IEEE C2, *National Electrical Safety Code,* 2007.
ANSI/IEEE C 37.20.6, *Standard for 4.76 kV to 38 kV-Rated Ground and Test Devices Used in Enclosures*, 2007.
ANSI/IEEE C84.1, *Electric Power Systems and Equipment — Voltage Ratings (60 Hz),* 1995.
IEEE 4, *Standard Techniques for High Voltage Testing,* 1978.

IEEE 4A, Amendment to IEEE 4, 2001.

IEEE 450, *IEEE Recommended Practice for Maintenance, Testing, and Replacement of Vented Lead-Acid Batteries for Stationary Applications*, 2002.

IEEE 484, *Recommended Practice for Installation Design and Installation of Vented Lead-Acid Batteries for Stationary Applications,* 2002.

IEEE 485, *IEEE Recommended Practice for Sizing Lead-Acid Storage Batteries for Stationary Applications*, 1997.

IEEE 516, *Guide for Maintenance Methods on Energized Power Lines,* 2003.

IEEE 937, *Recommended Practice for Installation and Maintenance of Lead-Acid Batteries for Photovoltaic Systems,* 2007.

IEEE 1106, *IEEE Recommended Practice for Installation, Maintenance, Testing, and Replacement of Vented Nickel-Cadmium Batteries for Stationary Applications*, 2005.

IEEE 1184, *IEEE Guide for Batteries for Uninterruptible Power Supply Systems*, 2006.

IEEE 1187, *Recommended Practice for Installation Design and Installation of Valve-Regulated Lead-Acid Storage Batteries for Stationary Applications,* 2002.

IEEE 1188, *IEEE Recommended Practice for Maintenance, Testing, and Replacement of Valve-Regulated Lead-Acid (VRLA) Batteries for Stationary Applications*, 2005.

IEEE 1189, *IEEE Guide for Selection of Valve-Regulated Lead-Acid (VRLA) Batteries for Stationary Applications*, 2007.

IEEE 1375, *IEEE Guide for Protection of Stationary Battery Systems*, 1998 (R 2003).

IEEE 1584, *Guide for Performing Arc Flash Calculations,* 2002.

IEEE 1584a, *Guide for Performing Arc Flash Hazard Calculations, Amendment 1*, 2004.

Anderson, W. E., "Risk Analysis Methodology Applied to Industrial Machine Development," *IEEE Trans. on Industrial Applications,* Vol. 41, No. 1, January/February 2005, pp. 180–187.

Doughty, R. L., T. E. Neal, and H. L. Floyd II, "Predicting Incident Energy to Better Manage the Electric Arc Hazard on 600 V Power Distribution Systems," *Record of Conference Papers IEEE IAS 45th* Annual Petroleum and Chemical Industry Conference, September 28–30, 1998.

Lee, Ralph, "The Other Electrical Hazard: Electrical Arc Flash Burns," IEEE *Trans. Industrial Applications*, Vol. 1A-18, No. 3, May/June 1982.

**B.1.5 ISA Publications.** Instrumentation, Systems, and Automation Society, 67 Alexander Drive, Research Triangle Park, NC 27709.

ANSI/ISA 61010-1, *Safety Requirements for Electrical Equipment for Measurement, Control, and Laboratory Use*, "Part 1: General Requirements," 2007.

**B.1.6 UL Publications.** Underwriters Laboratories Inc., 333 Pfingsten Road, Northbrook, IL 60062-2096.

UL 943, *Standard for Ground-Fault Circuit Interrupters*, 2006.

ANSI/UL 1203, *Explosion-Proof and Dust-Ignition-Proof Electrical Equipment for Use in Hazardous (Classified) Locations*, 2006.

**B.1.7 U.S. Government Publications.** U.S. Government Printing Office, Washington, DC 20402.

Title 29, Code of Federal Regulations, Part 1926, "Safety and Health Regulations for Construction," and Part 1910, "Occupational Safety and Health Standards."

OSHA 1910.137, *Personal Protective Equipment.*

OSHA 1910.305(j)(7), *Storage Batteries.*

OSHA 1926.441, *Batteries and Battery Charging.*

ANNEX C

# Limits of Approach

*This annex is not a part of the requirements of this NFPA document but is included for informational purposes only.*

Annex C provides information to illustrate the approach boundaries. The information is intended to provide users of the standard with suggestions regarding workers' safe approach to each limit. The approach limits are intended to trigger the need for greater control of work performed closer than the approach limit.

## C.1 Preparation for Approach

Observing a safe approach distance from exposed energized electrical conductors or circuit parts is an effective means of maintaining electrical safety. As the distance between a person and the exposed energized conductors or circuit parts decreases, the potential for electrical accident increases.

**C.1.1 Unqualified Persons, Safe Approach Distance.** Unqualified persons are safe when they maintain a distance from the exposed energized conductors or circuit parts, including the longest conductive object being handled, so that they cannot contact or enter a specified air insulation distance to the exposed energized electrical conductors or circuit parts. This safe approach distance is the Limited Approach Boundary. Further, persons must not cross the Arc Flash Protection Boundary unless they are wearing appropriate personal protective clothing and are under the close supervision of a qualified person. Only when continuously escorted by a qualified person should an unqualified person cross the Limited Approach Boundary. Under no circumstance should an unqualified person cross the Restricted Approach Boundary, where special shock protection techniques and equipment are required.

**C.1.2 Qualified Persons, Safe Approach Distance.**

**C.1.2.1** Determine the Arc Flash Protection Boundary and, if the boundary is to be crossed, appropriate flash-flame protection equipment must be utilized.

**C.1.2.2** For a person to cross the Limited Approach Boundary and enter the limited space, he or she must be qualified to perform the job/task.

**C.1.2.3** To cross the Restricted Approach Boundary and enter the restricted space, qualified persons must do the following:

(1) Have a plan that is documented and approved by authorized management
(2) Use personal protective equipment that is appropriate for working near exposed energized conductors or circuit parts and is rated for the voltage and energy level involved
(3) Be certain that no part of the body enters the prohibited space
(4) Minimize the risk from inadvertent movement by keeping as much of the body out of the restricted space as possible, using only protected body parts in the space as necessary to accomplish the work

**C.1.2.4** Crossing the Prohibited Approach Boundary and entering the prohibited space is considered the same as making contact with exposed energized conductors or circuit parts. *(See Figure C.1.2.4.)*

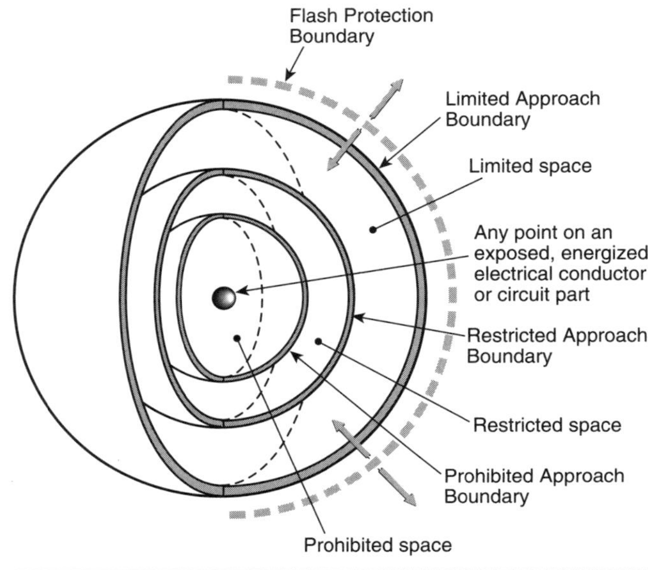


***FIGURE C.1.2.4*** *Limits of Approach.*

Therefore, qualified persons must do the following:

(1) Have specified training to work on energized conductors or circuit parts
(2) Have a documented plan justifying the need to work that close
(3) Perform a risk analysis
(4) Have the plan and the risk analysis approved by authorized management
(5) Use personal protective equipment that is appropriate for working on exposed energized conductors or circuit parts and is rated for the voltage and energy level involved

## C.2 Basis for Distance Values in Table 130.2(C)

Section C.2 illustrates how the information contained in Table 130.2(C) was derived.

**C.2.1 General Statement.** Columns 2 through 5 of Table 130.2(C) show various distances from the exposed energized electrical conductors or circuit parts. They include dimensions that are added to a basic minimum air insulation distance. Those basic minimum air insulation distances for voltages 72.5 kV and under are based on IEEE 4, *Standard Techniques for High Voltage Testing,* Appendix 2B; and voltages over 72.5 kV are based on IEEE 516, *Guide for Maintenance Methods on Energized Power Lines*. The minimum air insulation distances that are required to avoid flashover are as follows:

(1) ≤300 V: 1 mm (0 ft 0.03 in.)
(2) >300 V ≤750 V: 2 mm (0 ft 0.07 in.)
(3) >750 V ≤2 kV: 5 mm (0 ft 0.19 in.)
(4) >2 kV ≤15 kV: 39 mm (0 ft 1.5 in.)
(5) >15 kV ≤36 kV: 161 mm (0 ft 6.3 in.)

(6) >36 kV ≤48.3 kV: 254 mm (0 ft 10.0 in.)
(7) >48.3 kV ≤72.5 kV: 381 mm (1 ft 3.0 in.)
(8) >72.5 kV ≤121 kV: 640 mm (2 ft 1.2 in.)
(9) >138 kV ≤145 kV: 778 mm (2 ft 6.6 in.)
(10) >161 kV ≤169 kV: 915 mm (3 ft 0.0 in.)
(11) >230 kV ≤242 kV: 1.281 m (4 ft 2.4 in.)
(12) >345 kV ≤362 kV: 2.282 m (7 ft 5.8 in.)
(13) >500 kV ≤550 kV: 3.112 m (10 ft 2.5 in.)
(14) >765 kV ≤800 kV: 4.225 m (13 ft 10.3 in.)

**C.2.1.1 Column 1.** The voltage ranges have been selected to group voltages that require similar approach distances based on the sum of the electrical withstand distance and an inadvertent movement factor. The value of the upper limit for a range is the maximum voltage for highest nominal voltage in the range, based on ANSI/IEEE C84.1, *Electric Power Systems and Equipment — Voltage Ratings (60 Hz).* For single-phase systems, select the range that is equal to the system's maximum phase-to-ground voltage multiplied by 1.732.

**C.2.1.2 Column 2.** The distances in this column are based on OSHA's rule for unqualified persons to maintain a 3.05 m (10 ft) clearance for all voltages up to 50 kV (voltage-to-ground), plus 102 mm (4.0 in.) for each 1 kV over 50 kV.

**C.2.1.3 Column 3.** The distances are based on the following:

(1) ≤750 V: Use *NEC* Table 110.26(A)(1), Working Spaces, Condition 2 for 151 V–600 V range.
(2) >750 V ≤145 kV: Use *NEC* Table 110.34(A), Working Space, Condition 2.
(3) >145 kV: Use OSHA's 3.05 m (10 ft) rules as used in Column 2.

**C.2.1.4 Column 4.** The distances are based on adding to the flashover dimensions shown above the following inadvertent movement distance:

≤300 V: Avoid contact.

Based on experience and precautions for household 120/240 V systems:

>300 V ≤750 V: Add 304.8 mm (1 ft 0 in.) for inadvertent movement.

These values have been found to be adequate over years of use in ANSI/IEEE C2, *National Electrical Safety Code,* in the approach distances for communication workers.

>72.5 kV: Add 304.8 mm (1 ft 0 in.) for inadvertent movement.

These values have been found to be adequate over years of use in the *National Electrical Safety Code* in the approach distances for supply workers.

**C.2.1.5 Column 5.** The distances are based on the following:

(1) ≤300 V: Avoid contact.
(2) >300 ≤750 V: Use *NEC* Table 230.51(C), Clearances.

Between open conductors and surfaces, 600 V not exposed to weather.

(1) >750 V ≤2.0 kV: Select value that fits in with adjacent values.
(2) >2 kV ≤72.5 kV: Use *NEC* Table 490.24, Minimum Clearance of Live Parts, outdoor phase-to-ground values.
(3) >72.5 kV: Add 152.4 mm (0 ft 6 in.) for inadvertent movement.

These values have been found to be adequate over years of use where there has been a hazard/risk analysis, either formal or informal, of a special work procedure that allows a closer approach than that permitted by the Restricted Approach Boundary distance.

# Incident Energy and Flash Protection Boundary Calculation Methods

## ANNEX D

*This annex is not a part of the requirements of this NFPA document but is included for informational purposes only.*

Annex D illustrates how the Arc Flash Protection Boundary and incident energy might be calculated. These illustrations are not intended to limit the choice of calculation methods. All publicly known methods of calculating the Arc Flash Protection Boundary produce results that may not be completely accurate. The thermal hazard associated with an arcing fault is very complex, with many variable attributes having an impact on the calculation. The National Fire Protection Association (NFPA) and the Institute of Electrical and Electronic Engineers (IEEE) are sponsoring a joint research project on arc flash phenomena that is intended to better define the hazard and to refine the hazard assessment methods.

Workers who might be exposed to an arcing fault must wear flame-resistant clothing or other equipment to avoid a thermal injury. Protecting a worker from the thermal effects of an arcing fault does not necessarily protect the worker from injury. An arcing fault exhibits characteristics of other hazards. For instance, the arc generates a significant pressure wave. A worker could be injured by the pressure differential that could develop between the outside and inside of a person's body. The calculations illustrated in this annex do not offer protection from the effects of any pressure wave.

If the calculations illustrated in this annex (or from any other method) indicate that incident energy is 40 calories per square centimeter or greater, the pressure wave might be hazardous. That work task should not be performed unless an electrically safe work condition is established.

### D.1 Introduction

Annex D summarizes calculation methods available for calculating arc flash boundary and incident energy. It is important to investigate the limitations of any methods to be used. The limitations of methods summarized in Annex D are described in Table D.1.

Table D.1 identifies the source for various methods of performing the necessary calculations to determine the Arc Flash Protection Boundary and incident energy. The table identifies which section of Annex D covers each calculation method.

### D.2 Basic Equations for Calculating Arc Flash Protection Boundary Distances

The short-circuit symmetrical ampacity from a bolted 3-phase fault at the transformer terminals is calculated with the following formula:

$$I_{sc} = \{[MVA \text{ Base} \times 10^6] \div [1.732 \times V]\} \times \{100 \div \%Z\} \quad \textbf{[D.2(a)]}$$

where $I_{sc}$ is in amperes, $V$ is in volts, and $\%Z$ is based on the transformer $MVA$.

***TABLE D.1*** *Limitation of Calculation Methods*

| *Section* | *Source* | *Limitations/Parameters* |
|---|---|---|
| D.2, D.3, D.4 | Ralph Lee paper | Calculates Arc Flash Protection Boundary for arc in open air; conservative over 600 V and becomes more conservative as voltage increases |
| D.5 | Doughty/Neal paper | Calculates incident energy for 3-phase arc on systems rated 600 V and below; applies to short-circuit currents between 16 kA and 50 kA |
| D.6 | Ralph Lee paper | Calculates incident energy for 3-phase arc in open air on systems rated above 600 V; becomes more conservative as voltage increases |
| D.7 | IEEE Std. 1584 | Calculates incident energy and Arc Flash Protection Boundary for:<br>208 V to 15 kV; 3-phase; 50 Hz to 60 Hz; 700 A to 106,000 A short-circuit current; and 13 mm to 152 mm conductor gaps |
| D.8 | ANSI/IEEE C2 NESC-Section 410 Tables 410-1 and Table 410-2 | Calculates incident energy for open air phase-to-ground arcs 1 kV to 500 kV for live-line work |

A typical value for the maximum power (in MW) in a 3-phase arc can be calculated using the following formula:

$$P = [\text{maximum bolted fault in } MVA_{bf}] \times 0.707^2 \qquad \textbf{[D.2(b)]}$$

$$P = 1.732 \times V \times I_{sc} \times 10^{-6} \times 0.707^2 \qquad \textbf{[D.2(c)]}$$

The Flash Protection Boundary distance is calculated in accordance with the following formulae:

$$D_c = [2.65 \times MVA_{bf} \times t]^{1/2} \qquad \textbf{[D.2(d)]}$$

$$D_c = [53 \times MVA \times t]^{1/2} \qquad \textbf{[D.2(e)]}$$

where:

$D_c$ = distance in feet of person from arc source for a just curable burn (i.e., skin temperature remains less than 80°C)

$MVA_{bf}$ = bolted fault *MVA* at point involved

*MVA* = *MVA* rating of transformer. For transformers with *MVA* ratings below 0.75 *MVA*, multiply the transformer *MVA* rating by 1.25.

$t$ = time of arc exposure in seconds

The clearing time for a current limiting fuse is approximately $^1/_4$ cycle or 0.004 second if the arcing fault current is in the fuse's current limiting range. The clearing time of a 5 kV and 15 kV circuit breaker is approximately 0.1 second or 6 cycles if the instantaneous function is installed and operating. This can be broken down as follows: actual breaker time (approximately 2 cycles), plus relay operating time of approximately 1.74 cycles, plus an additional safety margin of 2 cycles, giving a total time of approximately 6 cycles. Additional time must be added if a time delay function is installed and operating.

The formulas used in this explanation are from Ralph Lee, "The Other Electrical Hazard: Electrical Arc Blast Burns," in *IEEE Trans. Industrial Applications*. Vol. 1A-18. No. 3, Page 246, May/June 1982. The calculations are based on the worst-case arc impedance. See Table D.2.

**D.3 Single Line Diagram of a Typical Petrochemical Complex**

***TABLE D.2*** *Flash Burn Hazard at Various Levels in a Large Petrochemical Plant*

| *(1)* | *(2)* | *(3)* | *(4)* | *(5)* | *(6)* | *(7)* | |
|---|---|---|---|---|---|---|---|
| ***Bus Nominal Voltage Levels*** | ***System (MVA)*** | ***Transformer (MVA)*** | ***System or Transformer (% Z)*** | ***Short Circuit Symmetrical (A)*** | ***Clearing Time of Fault (cycles)*** | ***Flash Protection Boundary Typical Distance**** | |
| | | | | | | ***SI*** | ***U.S.*** |
| 230 kV | 9000 | | 1.11 | 23,000 | 6.0 | 15 m | 49.2 ft |
| 13.8 kV | 750 | | 9.4 | 31,300 | 6.0 | 1.16 m | 3.8 ft |
| Load side of all 13.8 kV fuses | 750 | | 9.4 | 31,300 | 1.0 | 184 mm | 0.61 ft |
| 4.16 kV | | 10.0 | 5.5 | 25,000 | 6.0 | 2.96 m | 9.7 ft |
| 4.16 kV | | 5.0 | 5.5 | 12,600 | 6.0 | 1.4 m | 4.6 ft |
| Line side of incoming 600 V fuse | | 2.5 | 5.5 | 44,000 | 60.0–120.0 | 7 m–11 m | 23 ft–36 ft |
| 600 V bus | | 2.5 | 5.5 | 44,000 | 0.25 | 268 mm | 0.9 ft |
| 600 V bus | | 1.5 | 5.5 | 26,000 | 6.0 | 1.6 m | 5.4 ft |
| 600 V bus | | 1.0 | 5.57 | 17,000 | 6.0 | 1.2 m | 4 ft |

*Distance from an open arc to limit skin damage to a curable second-degree skin burn [less than 80°C (176°F) on skin] in free air.

The single line diagram *(see Figure D.3)* illustrates the complexity of a distribution system in a typical petrochemical plant.

## D.4 Sample Calculation

Many of the electrical characteristics of the systems and equipment are provided in Table D.2. The sample calculation is made on the 4160-volt bus 4A or 4B. Table D.2 tabulates the results of calculating the Flash Protection Boundary for each part of the system. For this calculation, based on Table D.2, the following results are obtained:

(1) Calculation is made on a 4160-volt bus.
(2) Transformer *MVA* (and base *MVA*) = 10 *MVA*.
(3) Transformer impedance on 10 *MVA* base = 5.5 percent.
(4) Circuit breaker clearing time = 6 cycles.

Using Equation D.2(a), calculate the short-circuit current:

$$\begin{aligned} I_{sc} &= \left\{\left[MVA \text{ Base} \times 10^6\right] \div \left[1.732 \times V\right]\right\} \times \left\{100 \div \%Z\right\} \\ &= \left\{\left[10 \times 10^6\right] \div \left[1.732 \times 4160\right]\right\} \times \left\{100 \div 5.5\right\} \\ &= 25{,}000 \text{ amperes} \end{aligned}$$

Using Equation D.2(b), calculate the power in the arc:

$$\begin{aligned} P &= 1.732 \times 4160 \times 25{,}000 \times 10^{-6} \times 0.707^2 \\ &= 91 \text{ MW} \end{aligned}$$

Using the Equation D.2(d), calculate the second-degree burn distance:

$$\begin{aligned} D_c &= \left\{2.65 \times \left[1.732 \times 25{,}000 \times 4160 \times 10^{-6}\right] \times 0.1\right\}^{1/2} \\ &= 6.9 \text{ or } 7.00 \text{ ft} \end{aligned}$$

Or, using Equation D.2(e), calculate the second-degree burn distance using an alternative

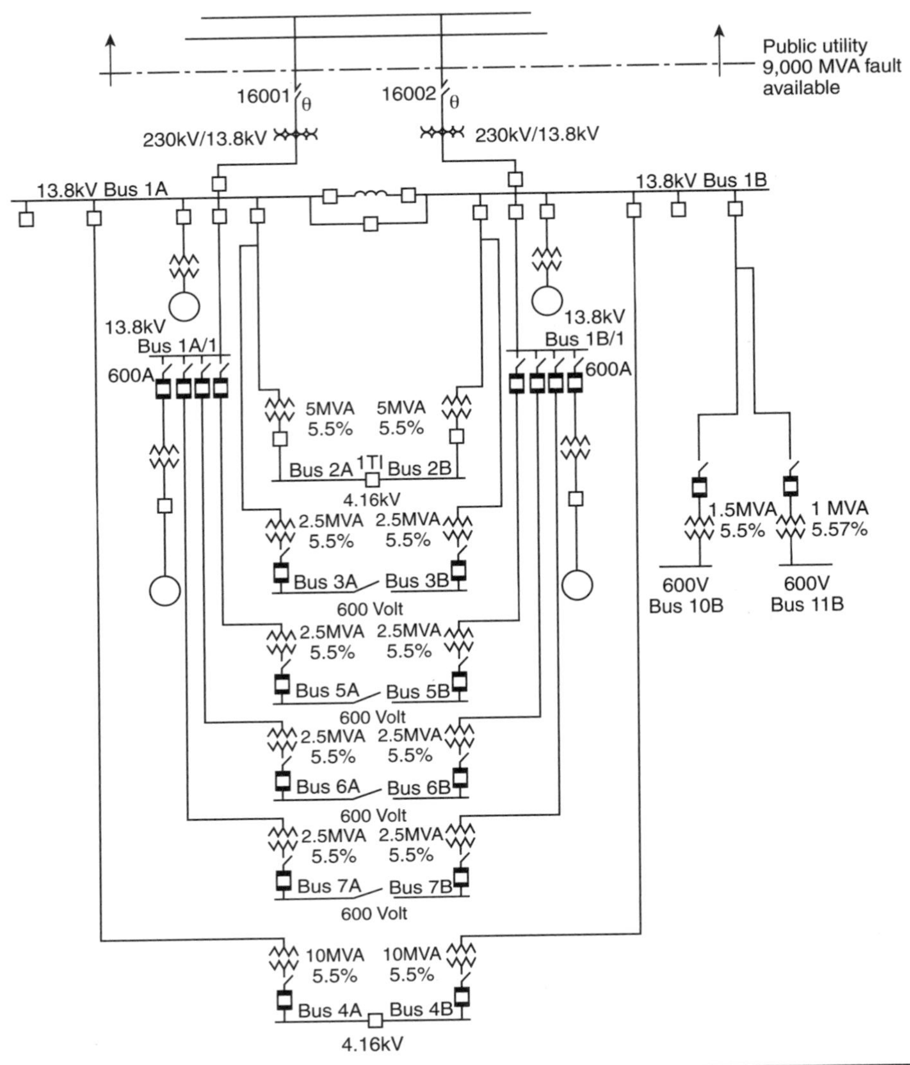


***FIGURE D.3*** *Single Line Diagram of a Typical Petrochemical Complex.*

method:

$$D_c = [53 \times 10 \times 0.1]^{1/2}$$
$$= 7.28 \text{ ft}$$

## D.5 Calculation of Incident Energy Exposure for an Arc Flash Hazard Analysis

The following equations can be used to predict the incident energy produced by a three-phase arc on systems rated 600 volts and below. The results of these equations might not represent the worst case in all situations. It is essential that the equations be used only within the limitations indicated in the definitions of the variables shown under the equations. The equations must be used only under qualified engineering supervision. (Note: Experimental testing continues to be performed to validate existing incident energy calculations and to determine new formulas.)

The parameters required to make the calculations follow:

(1) The maximum "bolted fault" three-phase short-circuit current available at the equipment and the minimum fault level at which the arc will self-sustain (Calculations should be made using the maximum value, and then at lowest fault level at which the arc is self-sustaining. For 480-volt systems, the industry accepted minimum level for a sustaining arcing fault is 38 percent of the available "bolted fault" three-phase short-circuit current. The highest incident energy exposure could occur at these lower levels where the overcurrent device could take seconds or minutes to open.)
(2) The total protective device clearing time (upstream of the prospective arc location) at the maximum short-circuit current, and at the minimum fault level at which the arc will sustain itself
(3) The distance of the worker from the prospective arc for the task to be performed

Typical working distances used for incident energy calculations are as follows:

(1) Low voltage (600 V and below) MCC and panelboards — 455 mm (18 in.)
(2) Low voltage (600 V and below) switchgear — 610 mm (24 in.)
(3) Medium voltage (above 600 V) switchgear — 910 mm (36 in.)

**D.5.1 Arc in Open Air.** The estimated incident energy for an arc in open air is

$$E_{MA} = 5271 D_A^{-1.9593} t_A \left[0.0016F^2 - 0.0076F + 0.8938\right] \quad \textbf{[D.5.1(a)]}$$

where:

$E_{MA}$ = maximum open arc incident energy, cal/cm$^2$
$D_A$ = distance from arc electrodes, in. (for distances 18 in. and greater)
$t_A$ = arc duration, seconds
$F$ = short-circuit current, kA (for the range of 16 kA to 50 kA)

Using Equation D.5.1(a), calculate the maximum open arc incident energy, cal/cm$^2$, where $D_A$ = 18 in., $t_A$ = 0.2 second, and $F$ = 20 kA.

$$\begin{aligned} E_{MA} &= 5271\ D_A^{-1.9593} t_A \left[0.0016F^2 - 0.0076F + 0.8938\right] \\ &= 5271 \times .0035 \times 0.2\left[0.0016 \times 400 - 0.0076 \times 20 + 0.8938\right] \quad \textbf{[D.5.1(b)]} \\ &= 3.69 \times \left[1.381\right] \\ &= 21.33\ \text{J/cm}^2\ \left(5.098\ \text{cal/cm}^2\right) \end{aligned}$$

**D.5.2 Arc in a Cubic Box.** The estimated incident energy for an arc in a cubic box (20 in. on each side, open on one end) is given in the following equation. This equation is applicable to arc flashes emanating from within switchgear, motor control centers, or other electrical equipment enclosures.

$$E_{MB} = 1038.7 D_B^{-1.4738} t_A\left[0.0093F^2 - 0.3453F + 5.9675\right] \quad \textbf{[D.5.2(a)]}$$

where:

$E_{MB}$ = maximum 20 in. cubic box incident energy, cal/cm$^2$
$D_B$ = distance from arc electrodes, inches (for distances 18 in. and greater)
$t_A$ = arc duration, seconds
$F$ = short circuit current, kA (for the range of 16 kA to 50 kA)

*Sample Calculation:* Using Equation D.5.2(a), calculate the maximum 20 in. cubic box incident energy, cal/cm$^2$, using the following:

(1) $D_B = 18$ in.
(2) $t_A = 0.2$ sec
(3) $F = 20$ kA

$$\begin{aligned} E_{MB} &= 1038.7 D_B^{-1.4738} t_A \left[0.0093F^2 - 0.3453F + 5.9675\right] \\ &= 1038 \times 0.0141 \times 0.2\left[0.0093 \times 400 - 0.3453 \times 20 + 5.9675\right] \quad \textbf{[D.5.2(b)]} \\ &= 2.928 \times \left[2.7815\right] \\ &= 34.1 \text{ J/cm}^2 \left(8.144 \text{ cal/cm}^2\right) \end{aligned}$$

**D.5.3 Reference.** The equations for this section were derived in the IEEE paper by R. L. Doughty, T. E. Neal, and H. L. Floyd, II, "Predicting Incident Energy to Better Manage the Electric Arc Hazard on 600 V Power Distribution Systems," *Record of Conference Papers IEEE IAS 45th Annual Petroleum and Chemical Industry Conference,* September 28–30, 1998.

Typical working distances used for incident energy calculations are given below:

(1) Low voltage (600 V and below) MCC and panelboards — 455 mm (18 in.)
(2) Low voltage (600 V and below) switchgear — 610 mm (24 in.)
(3) Medium voltage (above 600 V) switchgear — 910 mm (36 in.)

## D.6 Calculation of Incident Energy Exposure Greater Than 600 V for an Arc Flash Hazard Analysis

The following equation can be used to predict the incident energy produced by a three-phase arc in open air on systems rated above 600 V. The parameters required to make the calculations are as follows:

(1) The maximum "bolted fault" three-phase short-circuit current available at the equipment
(2) The total protective device clearing time (upstream of the prospective arc location) at the maximum short-circuit current
(3) The distance from the arc source
(4) Rated phase-to-phase voltage of the system:

$$E = \frac{793 \times F \times V \times t_A}{D^2}$$

where:

$E$ = incident energy, cal/cm$^2$
$F$ = bolted fault short-circuit current, kA
$V$ = system phase-to-phase voltage, kV
$t_A$ = arc duration, seconds
$D$ = distance from the arc source, inches

## D.7 Basic Equations for Calculating Incident Energy and Arc Flash Protection Boundary

This section offers equations for estimating incident energy and Flash Protection Boundaries based on statistical analysis and curve fitting of available test data. An IEEE working group produced the data from tests it performed to produce models of incident energy. Based on the selection of standard personal protective equipment (PPE) levels (1.2, 8, 25, and 40 cal/cm$^2$), it is estimated that the PPE is adequate or more than adequate to protect employees from second-degree burns in 95 percent of the cases.

FPN: When incident energy exceeds 40 cal/cm$^2$ at the working distance, greater emphasis than normal should be placed on de-energizing before working on or near the exposed electrical conductors or circuit parts.

The complete data, including a spreadsheet calculator to solve the equations, can be found in the IEEE 1584, *Guide for Performing Arc Flash Hazard Calculations*. It can be ordered from the Institute of Electrical and Electronics Engineers, Inc., 445 Hoes Lane, P.O. Box 1331, Piscataway, NJ 08855-1331.

**D.7.1 System Limits.** An equation for calculating incident energy can be empirically derived using statistical analysis of raw data along with a curve-fitting algorithm. It can be used for systems with the following limits:

(1) 0.208 kV to 15 kV, three-phase
(2) 50 Hz to 60 Hz
(3) 700 A to 106,000 A available short-circuit current
(4) 13 mm to 152 mm conductor gaps

For three-phase systems in open-air substations, open-air transmission systems, and distribution systems, a theoretically derived model is available. This theoretically derived model is intended for use with applications where faults escalate to three-phase faults. Where such an escalation is not possible or likely or where single-phase systems are encountered, this equation will likely provide conservative results.

**D.7.2 Arcing Current.** To determine the operating time for protective devices, find the predicted three-phase arcing current.

For applications with a system voltage under 1 kV, solve Equation D.7.2(a):

$$\lg I_a = K + 0.662 \lg I_{bf} + 0.0966V + 0.000526G + 0.5588V(\lg I_{bf}) - 0.00304G(\lg I_{bf}) \quad \textbf{[D.7.2(a)]}$$

where:

$lg$ = the $\log_{10}$
$I_a$ = arcing current in kA
$K$ = $-0.153$ for open air arcs; $-0.097$ for arcs-in-a-box
$I_{bf}$ = bolted three-phase available short-circuit current (symmetrical rms) (kA)
$V$ = system voltage in kV
$G$ = conductor gap (mm) *(See Table D.7.2.)*

For systems greater than or equal to 1 kV, use Equation D.7.2(b):

$$\lg I_a = 0.00402 + 0.983 \lg I_{bf} \quad \textbf{[D.7.2(b)]}$$

This higher voltage formula is utilized for both open-air arcs and for arcs-in-a-box.

Convert from lg:

$$I_a = 10^{\lg Ia} \quad \textbf{[D.7.2(c)]}$$

Use $0.85I_a$ to find a second arc duration. This second arc duration accounts for variations in the arcing current and the time for the overcurrent device to open. Calculate the incident energy using both arc durations ($I_a$ and $0.85\ I_a$), and use the higher incident energy.

***TABLE D.7.2*** *Factors for Equipment and Voltage Classes*

| *System Voltage (kV)* | *Type of Equipment* | *Typical Conductor Gap (mm)* | *Distance Exponent Factor X* |
|---|---|---|---|
| 0.208–1 | Open-air | 10–40 | 2.000 |
| | Switchgear | 32 | 1.473 |
| | MCCs and panels | 25 | 1.641 |
| | Cables | 13 | 2.000 |
| >1–5 | Open-air | 102 | 2.000 |
| | Switchgear | 13–102 | 0.973 |
| | Cables | 13 | 2.000 |
| >5–15 | Open-air | 13–153 | 2.000 |
| | Switchgear | 153 | 0.973 |
| | Cables | 13 | 2.000 |

**D.7.3 Incident Energy at Working Distance — Empirically Derived Equation.** To determine the incident energy using the empirically derived equation, determine the $\log^{10}$ of the normalized incident energy. This equation is based on data normalized for an arc time of 0.2 second and a distance from the possible arc point to the person of 610 mm:

$$\lg E_n = k_1 + k_2 + 1.081 \lg I_a + 0.0011G \qquad \textbf{[D.7.3(a)]}$$

where:

$E_n$ = incident energy (J/cm$^2$) normalized for time and distance

$k_1$ = −0.792 for open air arcs; −0.555 for arcs-in-a-box

$k_2$ = 0 for ungrounded and high-resistance grounded systems

= −0.113 for grounded systems

$G$ = the conductor gap (mm) *(See Table D.7.2.)*

Then,

$$E_n = 10^{\lg E_n} \qquad \textbf{[D.7.3(b)]}$$

Converting from normalized:

$$E = 4.184 C_f \mathrm{E}_n \left(\frac{t}{0.2}\right)\left(\frac{610^x}{D^x}\right) \qquad \textbf{[D.7.3(c)]}$$

where:

$E$ = incident energy in J/cm$^2$

$C_f$ = calculation factor

= 1.0 for voltages above 1 kV

= 1.5 for voltages at or below 1 kV

$E_n$ = incident energy normalized

$t$ = arcing time (seconds)

$D$ = distance (mm) from the arc to the person (working distance) *(See Table D7.3.)*

$X$ = the distance exponent from Table D.7.2

***TABLE D.7.3*** *Typical Working Distances*

| *Classes of Equipment* | *Typical Working Distance* (mm)* |
|---|---|
| 15kV switchgear | 910 |
| 5kV switchgear | 910 |
| Low-voltage switchgear | 610 |
| Low-voltage MCCs and panelboards | 455 |
| Cable | 455 |
| Other | To be determined in field |

* Typical working distance is the sum of the distance between the worker and the front of the equipment and the distance from the front of the equipment to the potential arc source inside the equipment.

**D.7.4 Incident Energy at Working Distance — Theoretical Equation.** The theoretically derived equation can be applied in cases where the voltage is over 15 kV or the gap is outside the range:

$$E = 2.142 \times 10^6 V I_{bf}\left(\frac{t}{D^2}\right) \qquad \textbf{[D.7.4]}$$

where:

$E$ = incident energy (J/cm²)

$V$ = system voltage (kV)

$t$ = arcing time (seconds)

$D$ = distance (mm) from the arc to the person (working distance)

$I_{bf}$ = available three-phase bolted-fault current

For voltages over 15 kV, arcing-fault current and bolted-fault current are considered equal.

**D.7.5 Arc Flash Protection Boundary.** The Arc Flash Protection Boundary is the distance at which a person is likely to receive a second-degree burn. The onset of a second-degree burn is assumed to be when the skin receives 5.0 J/cm² of incident energy.

For the empirically derived equation,

$$D_B = \left[4.184 C_f E_n \left(\frac{t}{0.2}\right)\left(\frac{610^x}{E_B}\right)\right]^{1/x} \qquad \textbf{[D.7.5(a)]}$$

For the theoretically derived equation,

$$D_B = \sqrt{2.142 \times 10^6 V I_{bf}\left(\frac{t}{E_B}\right)} \qquad \textbf{[D.7.5(b)]}$$

where:

$D_B$ = the distance (mm) of the Arc Flash Protection Boundary from the arcing point

$C_f$ = a calculation factor

= 1.0 for voltages above 1 kV

= 1.5 for voltages at or below 1 kV

$E_n$ = incident energy normalized
$E_B$ = incident energy in J/cm$^2$ at the distance of the Arc Flash Protection Boundary
$t$ = time (seconds)
$X$ = the distance exponent from Table D.7.2
$I_{bf}$ = bolted 3-phase available short-circuit current
$V$ = system voltage in kV

FPN: These equations could be used to determine whether selected PPE is adequate to prevent thermal injury at a specified distance in event of an arc flash.

**D.7.6 Current-Limiting Fuses.** The formulas in this section were developed for calculating arc-flash energies for use with current-limiting Class L and Class RK1 fuses. The testing was done at 600 volts and at a distance of 455 mm, using commercially available fuses from one manufacturer. The following variables are noted:

$I_{bf}$ = available three-phase bolted-fault current (symmetrical rms) (kA)
$E$ = incident energy (J/cm$^2$)

**(A) Class L Fuses 1,601 A–2,000 A.** Where $I_{bf}$ <22.6 kA, calculate the arcing current using Equation D.7.2(a), and use time-current curves to determine the incident energy using Equations D.7.3(a), D.7.3(b), and D.7.3(c).

Where 22.6 kA $\leq I_{bf} \leq$ 65.9 kA,

$$E = 4.184(-0.1284 I_{bf} + 32.262) \quad \textbf{[D.7.6(a)]}$$

Where 65.9 kA $< I_{bf} \leq$ 106 kA,

$$E = 4.184(-0.5177 I_{bf} + 57.917) \quad \textbf{[D.7.6(b)]}$$

Where $I_{bf}$ >106 kA, contact manufacturer.

**(B) Class L Fuses 1,201 A–1,600 A.** Where $I_{bf}$ <15.7 kA, calculate the arcing current using Equation D.7.2(a), and use time-current curves to determine the incident energy using Equations D.7.3(a), D.7.3(b), and D.7.3(c).

Where 15.7 kA $\leq I_{bf} \leq$ 31.8 kA,

$$E = 4.184(-0.1863 I_{bf} + 27.926) \quad \textbf{[D.7.6(c)]}$$

Where 31.8 kA $< I_{bf} <$ 44.1 kA,

$$E = 4.184(-1.5504 I_{bf} + 71.303) \quad \textbf{[D.7.6(d)]}$$

Where 44.1 kA $\leq I_{bf} \leq$ 65.9 kA,

$$E = 12.3\text{J/cm}^2(2.94\text{ cal/cm}^2) \quad \textbf{[D.7.6(e)]}$$

Where 65.9 kA $< I_{bf} \leq$ 106 kA,

$$E = 4.184(-0.0631 I_{bf} + 7.0878) \quad \textbf{[D.7.6(f)]}$$

Where $I_{bf}$ >106 kA, contact manufacturer.

**(C) Class L Fuses 801 A–1,200 A.** Where $I_{bf}$ <15.7 kA, calculate the arcing current per Equation D.7.2(a), and use time-current curves to determine the incident energy per Equations D.7.3(a), D.7.3(b), and D.7.3(c).

Where 15.7 kA $\leq I_{bf} \leq$22.6 kA,

$$E = 4.184(-0.1928I_{bf} + 14.226) \qquad \textbf{[D.7.6(g)]}$$

Where 22.6 kA $< I_{bf} \leq$44.1 kA,

$$E = 4.184(0.0143I_{bf}^{2} - 1.3919I_{bf} + 34.045) \qquad \textbf{[D.7.6(h)]}$$

Where 44.1 kA $< I_{bf} \leq$106 kA,

$$E = 1.63 \qquad \textbf{[D.7.6(i)]}$$

Where $I_{bf}$ >106 kA, contact manufacturer.

**(D) Class L Fuses 601 A–800 A.** Where $I_{bf}$ <15.7 kA, calculate the arcing current per Equation D.7.2(a), and use time-current curves to determine the incident energy using Equations D.7.3(a), D.7.3(b), and D.7.3(c).

Where 15.7 kA $\leq I_{bf} \leq$44.1 kA,

$$E = 4.184(-0.0601I_{bf} + 2.8992) \qquad \textbf{[D.7.6(j)]}$$

Where 44.1 kA $< I_{bf} \leq$106 kA,

$$E = 1.046 \qquad \textbf{[D.7.6(k)]}$$

Where $I_{bf}$ > 106 kA, contact manufacturer.

**(E) Class RK1 Fuses 401 A–600 A.** Where $I_{bf}$ <8.5 kA, calculate the arcing current using Equation D.7.2(a), and use time-current curves to determine the incident energy using Equations D.7.3(a), D.7.3(b), and D.7.3(c).

Where 8.5 kA $\leq I_{bf} \leq$14 kA,

$$E = 4.184(-3.0545I_{bf} + 43.364) \qquad \textbf{[D.7.6(l)]}$$

Where 14 kA $< I_{bf} \leq$15.7 kA,

$$E = 2.510 \qquad \textbf{[D.7.6(m)]}$$

Where 15.7 kA $< I_{bf} \leq$22.6 kA,

$$E = 4.184(-0.0507I_{bf} + 1.3964) \qquad \textbf{[D.7.6(n)]}$$

Where 22.6 kA $< I_{bf} \leq$106 kA,

$$E = 1.046 \qquad \textbf{[D.7.6(o)]}$$

Where $I_{bf}$ >106 kA, contact manufacturer.

**(F) Class RK1 Fuses 201 A–400 A.** Where $I_{bf}$ <3.16 kA, calculate the arcing current using Equation D.7.2(a), and use time-current curves to determine the incident energy using Equations D.7.3(a), D.7.3(b), and D.7.3(c).

Where 3.16 kA $\leq I_{bf} \leq$5.04 kA,

$$E = 4.184(-19.053I_{bf} + 96.808) \quad \textbf{[D.7.6(p)]}$$

Where 5.04 kA < $I_{bf} \leq$ 22.6 kA,

$$E = 4.184(-0.0302I_{bf} + 0.9321) \quad \textbf{[D.7.6(q)]}$$

Where 22.6 kA <$I_{bf} \leq$ 106 kA,

$$E = 1.046 \quad \textbf{[D.7.6(r)]}$$

Where $I_{bf}$ >106 kA, contact manufacturer.

**(G) Class RK1 Fuses 101 A–200 A.** Where $I_{bf}$ <1.16 kA, calculate the arcing current using Equation D.7.2(a), and use time-current curves to determine the incident energy using Equations D.7.3(a), D.7.3(b), and D.7.3(c).

Where 1.16 kA $\leq I_{bf} \leq$1.6 kA,

$$E = 4.184(-18.409I_{bf} + 36.355) \quad \textbf{[D.7.6(s)]}$$

Where 1.6 kA < $I_{bf} \leq$3.16 kA,

$$E = 4.184(-4.2628I_{bf} + 13.721) \quad \textbf{[D.7.6(t)]}$$

Where 3.16 kA <$I_{bf} \leq$106 kA,

$$E = 1.046 \quad \textbf{[D.7.6(u)]}$$

Where $I_{bf}$ > 106 kA, contact manufacturer.

**(H) Class RK1 Fuses 1 A–100 A.** Where $I_{bf}$ <0.65 kA, calculate the arcing current per Equation D.7.2(a), and use time-current curves to determine the incident energy using Equations D.7.3(a), D.7.3(b), and D.7.3(c).

Where 0.65 kA $\leq I_{bf} \leq$1.16 kA,

$$E = 4.184(-11.176I_{bf} + 13.565) \quad \textbf{[D.7.6(v)]}$$

Where 1.16 kA < $I_{bf} \leq$1.4 kA,

$$E = 4.184(-1.4583I_{bf} + 2.2917) \quad \textbf{[D.7.6(w)]}$$

Where 1.4 kA < $I_{bf} \leq$106 kA,

$$E = 1.046 \quad \textbf{[D.7.6(x)]}$$

Where $I_{bf}$ > 106 kA, contact manufacturer.

**D.7.7 Low-Voltage Circuit Breakers.** The equations in Table D.7.7 can be used for systems with low-voltage circuit breakers. The results of the equations will determine the incident energy and Arc Flash Protection Boundary when $I_{bf}$ is within the range as described. Time-current curves for the circuit breaker are not necessary within the appropriate range.

***TABLE D.7.7*** *Incident Energy and Arc Flash Protection Boundary by Circuit Breaker Type and Rating*

| Rating (A) | Breaker Type | Trip-Unit Type | 480 V and Lower | | 575–600 V | |
|---|---|---|---|---|---|---|
| | | | Incident Energy ($J/cm^2$)[a] | Flash Boundary (mm)[a] | Incident Energy ($J/cm^2$)[a] | Flash Boundary (mm)[a] |
| 100–400 | MCCB | TM or M | $0.189\ I_{bf} + 0.548$ | $9.16\ I_{bf} + 194$ | $0.271\ I_{bf} + 0.180$ | $11.8\ I_{bf} + 196$ |
| 600–1,200 | MCCB | TM or M | $0.223\ I_{bf} + 1.590$ | $8.45\ I_{bf} + 364$ | $0.335\ I_{bf} + 0.380$ | $11.4\ I_{bf} + 369$ |
| 600–1,200 | MCCB | E, LI | $0.377\ I_{bf} + 1.360$ | $12.50\ I_{bf} + 428$ | $0.468\ I_{bf} + 4.600$ | $14.3\ I_{bf} + 568$ |
| 1,600–6,000 | MCCB or ICCB | TM or E, LI | $0.448\ I_{bf} + 3.000$ | $11.10\ I_{bf} + 696$ | $0.686\ I_{bf} + 0.165$ | $16.7\ I_{bf} + 606$ |
| 800–6,300 | LVPCB | E, LI | $0.636\ I_{bf} + 3.670$ | $14.50\ I_{bf} + 786$ | $0.958\ I_{bf} + 0.292$ | $19.1\ I_{bf} + 864$ |
| 800–6,300 | LVPCB | E, LS[b] | $4.560\ I_{bf} + 27.230$ | $47.20\ I_{bf} + 2660$ | $6.860\ I_{bf} + 2.170$ | $62.4\ I_{bf} + 2930$ |

MCCB: Molded-case circuit breaker.
ICCB: Insulated-case circuit breaker.
LVPC: Low-voltage power circuit breaker.
TM: Thermal-magnetic trip units.
M: Magnetic (instantaneous only) trip units.
E: Electronic trip units have three characteristics that may be used separately or in combination: L: Long-time, S: Short-time, I: Instantaneous.
[a] $I_{bf}$ is in kA; working distance is 455 mm (18 in.).
[b] Short-time delay is assumed to be set at maximum.

When the bolted-fault current is below the range indicated, calculate the arcing current per Equation D.7.2(a), and use time-current curves to determine the incident energy using Equations D.7.3(a), D.7.3(b), and D.7.3(c).

The range of available three-phase bolted-fault currents is from 700 A to 106,000 A. Each equation is applicable for the range

$$I_1 < I_{bf} < I_2$$

where:

$I_2$ = the interrupting rating of the CB at the voltage of interest.

$I_1$ = the minimum available three-phase, bolted, short-circuit current at which this method can be applied. $I_1$ is the lowest available three-phase, bolted, short-circuit current level that causes enough arcing current for instantaneous tripping to occur or for circuit breakers with no instantaneous trip, that causes short-time tripping to occur.

To find $I_1$, the instantaneous trip ($I_t$) of the circuit breaker must be found. This can be determined from the time-current curve, or it can be assumed to be 10 times the rating of the circuit breaker for circuit breakers rated above 100 amperes. For circuit breakers rated 100 amperes and below, a value of $I_t$ = 1,300 A can be used. When short-time delay is utilized, $I_t$ is the short-time pick-up current.

The corresponding bolted-fault current, $I_{bf}$, is found by solving the equation for arc current for box configurations by substituting $I_t$ for arcing current. The 1.3 factor in Equation D.7.7(b) adjusts current to the top of the tripping band.

$$\lg(1.3I_t = 0.084 + 0.096V + 0.586(\lg I_b) + 0.559V\left(\lg I_{bf}\right) \quad \textbf{[D.7.7(a)]}$$

At 600 V,

$$\lg I_1 = 0.0281 + 1.09\lg(1.3I_t) \quad \textbf{[D.7.7(b)]}$$

At 480 V and lower,

$$\lg I_1 = 0.0407 + 1.17\ \lg(1.3I_t) \quad \textbf{[D.7.7(c)]}$$

$$I_{bf} = I_1 = 10^{\lg I1} \quad \textbf{[D.7.7(d)]}$$

**D.7.8 References.** The complete data, including a spreadsheet calculator to solve the equations, may be found in IEEE 1584, *Guide for Performing Arc Flash Hazard Calculations*. IEEE publications are available from the Institute of Electrical and Electronics Engineers, 445 Hoes Lane, P.O. Box 1331, Piscataway, NJ 08855-1331, USA (http://standards.ieee.org/).

## D.8 Estimated Incident Energy Exposures for Live Line Work on Overhead Open Air Systems 1 kV to 800 kV

Table D.8(1) and Table D.8(2) list the heat flux rate in cal/cm²/sec derived from the ANSI/IEEE C2 Tables 410-1 and 410-2. To estimate the incident energy, multiply the heat flux rate in the tables by the maximum clearing time (in seconds).

*TABLE D.8(1)*

| *Max Fault Current (kA)* | *Phase-to-Phase Voltage (kV)* | | | |
|---|---|---|---|---|
| | *1 to 15* | *15.1 to 25* | *25.1 to 36* | *36.1 to 46* |
| | *Heat Flux Rate (cal/cm²/sec)* | | | |
| 5 | 4.9 | 8.7 | 11.6 | 14.8 |
| 10 | 12.5 | 20.8 | 27.1 | 34.5 |
| 15 | 22.2 | 35.6 | 45.4 | 56.2 |
| 20 | 34 | 52.8 | 66.4 | 78.7 |

Notes:

(1) These calculations are based on open air phase-to-ground arcs. This table is not intended for phase-to-phase arcs or enclosed arc (arc in a box).

(2) These calculations are based on a 15-in. separation distance from the arc to the employee and arc gaps as follows: 1 to 15 kV = 2 in., 15.1 to 25 kV = 4 in., 25.1–36 kV = 6 in., 36.1 to 46 kV = 9 in. *(See IEEE 4.)*

(3) These calculations were derived using a commercially available computer software program. Other methods are available to estimate arc exposure values and may yield slightly different but equally acceptable results.

*TABLE D.8(2)*

| *Max Fault Current (kA)* | *Phase-to-Phase Voltage (kV)* | | | | | | | |
|---|---|---|---|---|---|---|---|---|
| | *46.1 to 72.5* | *72.6 to 121* | *138 to 145* | *161 to 169* | *230 to 242* | *345 to 362* | *500 to 550* | *765 to 800* |
| | *Heat Flux Rate (cal/cm²/sec)* | | | | | | | |
| 20 | 12.4 | 24.2 | 19.4 | 21.1 | 17.7 | 8.3 | 9.8 | 8.2 |
| 30 | 22.3 | 42.1 | 33.5 | 34.2 | 28.7 | 13.5 | 15.8 | 13.3 |
| 40 | 34.7 | 63.6 | 50.4 | 49 | 41.1 | 19.3 | 22.7 | 19 |
| 50 | 49.5 | 88.7 | 70 | 65.2 | 54.7 | 25.6 | 30.2 | 25.3 |

Notes:

(1) These calculations are based on open-air phase-to-ground arcs. This table is not intended for phase-to-phase arcs or enclosed arc (arc in a box).

(2) Arc gap is calculated by using the phase-to-ground voltage of the circuit and dividing by 10. The dielectric strength of air is taken at 10 kV per inch. *(See IEEE 4.)*

(3) Distance from the arc to the employee is calculated by using the minimum approach distance from ANSI/IEEE C2, Table 441-2, and subtracting two times the assumed arc gap length.

(4) These calculations were derived using a commercially available computer software program. Other methods are available to estimate arc exposure values and may yield slightly different but equally acceptable results.

## D.9 Guideline for the use of Hazard/Risk Category (HRC) 2 and HRC 4 Personal Protective Equipment

**D.9.1 Tables for Guidance on the Use of Personal Protective Equipment.** The following tables can be used to determine the suitability of Hazard/Risk Category (HRC) 2 and HRC 4 personal protective equipment on systems rated up to 15 kV, line-to-line. See Table D.9.1 and Table D.9.2 for the recommended limitations based on bolted 3-phase short-circuit currents at the listed fault-clearing times. The limitations listed below are based on IEEE 1584 calculation methods.

***TABLE D.9.1*** *Low-Voltage Systems – Maximum Three-Phase Bolted-Fault Current Limits at Various System Voltages and Fault-Clearing Times of Circuit Breakers, for the Recommended Use of Hazard/Risk Category (HRC) 2 and HRC 4 Personal Protective Equipment in an "Arc in a Box" Situation*

| *System Voltage (volts, phase-to-phase)* | *Upstream Protection Fault-Clearing Time (sec)* | *Maximum 3-Phase Bolted-Fault Current for Use of HRC 2 PPE (8 cal/cm²)* | *Maximum 3-Phase Bolted-Fault Current for Use of HRC 4 PPE (40 cal/cm²)* |
|---|---|---|---|
| 690 | 0.05 | 39 kA | 180 kA |
| | 0.10 | 20 kA | 93 kA |
| | 0.20 | 10 kA | 48 kA |
| | 0.33 | Not Recommended | 29 kA |
| | 0.50 | Not Recommended | 20 kA |
| 600 | 0.05 | 48 kA | 200 kA* |
| | 0.10 | 24 kA | 122 kA |
| | 0.20 | 12 kA | 60 kA |
| | 0.33 | Not Recommended | 36 kA |
| | 0.50 | Not Recommended | 24 kA |
| 480 | 0.05 | 68 kA | 200 kA* |
| | 0.10 | 32 kA | 183 kA |
| | 0.20 | 15 kA | 86 kA |
| | 0.33 | 8 kA | 50 kA |
| | 0.50 | Not Recommended | 32 kA |
| 400 | 0.05 | 87 kA | 200 kA* |
| | 0.10 | 39 kA | 200 kA* |
| | 0.20 | 18 kA | 113 kA |
| | 0.33 | 10 kA | 64 kA |
| | 0.50 | Not Recommended | 39 kA |
| 208 | 0.05 | 200 kA* | Not applicable |
| | 0.10 | 104 kA | 200 kA* |

Notes:

(1) Three-phase "bolted fault" value is at the terminals of the equipment on which work is to be done.

(2) "Upstream Protection Fault-Clearing Time" is normally the "short-time delay" setting on the trip unit of the low-voltage power circuit breaker upstream of the equipment on which work is to be done.

(3) For application of this table, the recommended maximum setting (pick-up) of either the instantaneous or short-delay protection of the circuit breaker's trip unit is 30% of the actual available 3-phase bolted fault current at the specific work location.

(4) Working distance for the arc-flash exposures is assumed to be 455 mm (18 in.).

(5) Flash Protection Boundary (threshold distance for a second-degree skin burn) is 1.7 m (6 ft) for HRC 2 and 4.9 m (16 ft) for HRC 4. PPE is required for all personnel working within the Flash Protection Boundary.

(6) Instantaneous circuit breaker trip unit(s) have no intentional time delay, and the circuit breaker will clear the fault within 0.050 sec of initiation. Application of circuit breakers with faster clearing times or the use of current-limiting circuit breakers or fuses should permit the use of HRC 2 and HRC 4 PPE at greater fault currents than listed.

(7) Systems are assumed to be resistance grounded, except for 208 V (solidly grounded system). This assumption results in conservative application if the table is used on a solidly grounded system, since the incident energy on a solidly grounded system is lower for the same bolted fault current availability.

*Maximum equipment short-circuit current rating available.

***TABLE D.9.2*** *High-Voltage Systems – Maximum Three-Phase Bolted-Fault Current Limits at Various System Voltages and Fault-Clearing Times of Circuit Breakers, for the Recommended Use of Hazard / Risk Category (HRC) 2 and HRC 4 Personal Protective Equipment in an "Arc in a Box" Situation*

| ***System Voltage (volts, phase-to-phase)*** | ***Upstream Protection Fault-Clearing Time (sec)*** | ***Maximum 3-Phase Bolted-Fault Current for Use of HRC 2 PPE (8 cal/cm²)*** | ***Maximum 3-Phase Bolted-Fault Current for Use of HRC 4 PPE (40 cal/cm²)*** |
|---|---|---|---|
| 15 kV Class and 12 kV Class | 0.10 | 45 kA | 63 kA*(11.4 cal/cm²) |
| | 0.35 | 13 kA | 63 kA |
| | 0.70 | 7 kA | 32 kA 23 kA |
| | 1.0 | 5 kA | 32 kA 23 kA |
| 5 kV Class | 0.10 | 50 kA | 63 kA* (10 cal/cm²) |
| | 0.35 | 15 kA | 63 kA*(35 cal/cm²) |
| | 0.70 | 8 kA | 37 kA |
| | 1.0 | 5 kA | 26 kA |

Notes:

(1) "Upstream Protection Fault-Clearing Time" is the protective relaying operating time at 90% of the actual available 3-phase bolted fault current at the specific work location (the time for the output contact operating the trip coil of the circuit breaker to be closed), plus the circuit breaker operating time (upstream of the equipment on which work is to be done).

(2) Working distance for the above arc-flash exposures is assumed to be 0.92 m (3 ft).

(3) Systems are assumed to be resistance grounded. This assumption results in conservative application if the table is used on a solidly grounded system, since the incident energy on a solidly grounded system is lower.

(4) The cal/cm² in parentheses in the last column are calculated at the maximum equipment short-circuit current ratings available.

* Maximum equipment short-circuit current rating available.

ANNEX E

# Electrical Safety Program

*This annex is not a part of the requirements of this NFPA document but is included for informational purposes only.*

Annex E provides information that could be used as the foundation for an electrical safety program.

See 110.7, Electrical Safety Program.

## E.1 Typical Electrical Safety Program Principles

Electrical safety program principles include, but are not limited to, the following:

(1) Inspect/evaluate the electrical equipment
(2) Maintain the electrical equipment's insulation and enclosure integrity
(3) Plan every job and document first-time procedures
(4) Deenergize, if possible *(see* 120.1*)*
(5) Anticipate unexpected events
(6) Identify and minimize the hazard
(7) Protect the employee from shock, burn, blast, and other hazards due to the working environment
(8) Use the right tools for the job
(9) Assess people's abilities
(10) Audit these principles

## E.2 Typical Electrical Safety Program Controls

Electrical safety program controls can include, but are not limited to, the following:

(1) Every electrical conductor or circuit part is considered energized until proven otherwise.
(2) No bare-hand contact is to be made with exposed energized electrical conductors or circuit parts above 50 volts to ground, unless the "bare-hand method" is properly used.
(3) Deenergizing an electrical conductor or circuit part and making it safe to work on is in itself a potentially hazardous task.
(4) Employer develops programs, including training, and employees apply them.
(5) Use procedures as "tools" to identify the hazards and develop plans to eliminate/control the hazards.
(6) Train employees to qualify them for working in an environment influenced by the presence of electrical energy.

(7) Identify/categorize tasks to be performed on or near exposed energized electrical conductors and circuit parts.
(8) Use a logical approach to determine potential hazard of task.
(9) Identify and use precautions appropriate to the working environment.

**E.3 Typical Electrical Safety Program Procedures.**

Electrical safety program procedures can include, but are not limited to, the following:

(1) Purpose of task
(2) Qualifications and number of employees to be involved
(3) Hazardous nature and extent of task
(4) Limits of approach
(5) Safe work practices to be utilized
(6) Personal protective equipment involved
(7) Insulating materials and tools involved
(8) Special precautionary techniques
(9) Electrical diagrams
(10) Equipment details
(11) Sketches/pictures of unique features
(12) Reference data

# Hazard/Risk Evaluation Procedure

# ANNEX F

*This annex is not a part of the requirements of this NFPA document but is included for informational purposes only.*

Before any task is performed on or near exposed energized conductors or circuit parts, the worker should determine the degree to which he or she might be exposed to a safety hazard. If the worker might be exposed to a hazard, the worker should determine whether the risk of injury is significant.

Section F.1 illustrates a series of questions in the form of a flow chart that is intended to identify electrical hazards and the degree of risk as a worker performs a hazard/risk analysis. Once the hazard and degree of risk are identified, personal protective equipment (PPE) must be selected. The worker and his or her supervisor then can use the degree of risk to evaluate whether the risk of injury is sufficiently low to accept the risk.

**F.1** See 110.7(F), Hazard/Risk Evaluation Procedure. Figure F.1 illustrates the steps of a hazard/risk analysis evaluation procedure flow chart.

**F.2** See 110.7(F), Hazard/Risk Assessment Procedure. Figure F.2.1 illustrates a hazard/risk analysis procedure.

Section F.2, which is new in the 2009 edition of *NFPA 70E*, provides one method that can be used to evaluate risk. When an employer determines the degree of risk associated with a type of work, he or she should perform an overall assessment of the amount of risk associated with the type of work. The employer first should determine if the risk of injury to the worker is sufficient to avoid performing the work in an energized condition. If the risk can be reduced to an acceptable level, the employer must determine the necessary actions to mitigate the risk, such as additional authorization(s), or control(s), procedural requirements, PPE, special training, and similar measures. If the risk cannot be sufficiently reduced to ensure lack of injury, the work must not be performed.

A Hazard/Risk Evaluation is an analytical tool consisting of a number of discrete steps intended to ensure that hazards are properly identified and evaluated, and that appropriate measures are taken to reduce those hazards to a tolerable level (adapted from ANSI/ASSE Z244.1).

This procedure is a comprehensive review of the task and associated foreseeable hazards that use event severity, frequency, probability, and avoidance to determine the level of safe practices employed. This procedure includes:

(1) Gathering task information and determining task limits
(2) Documenting hazards associated with each task
(3) Estimating the risk factors for each hazard/task pair

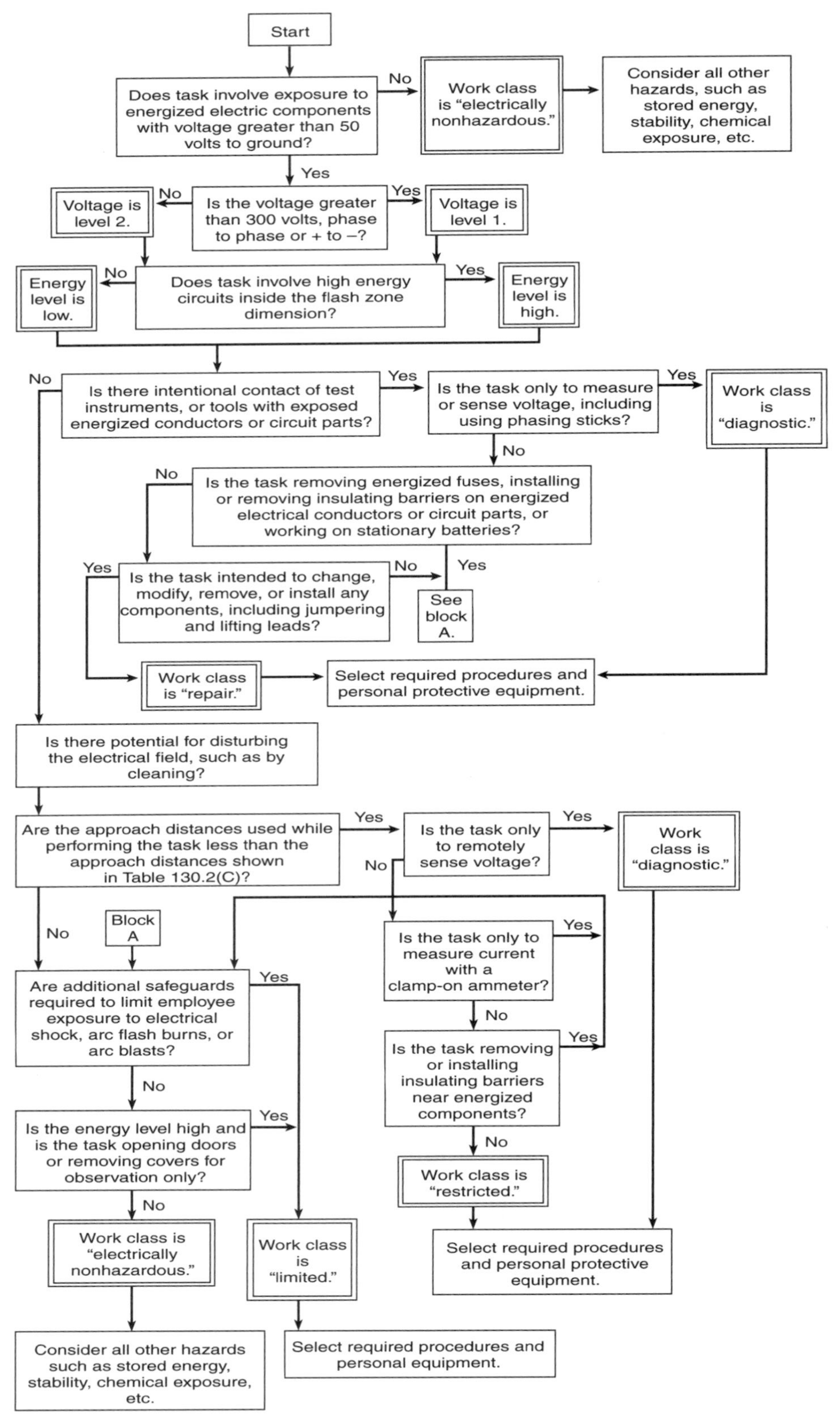


***FIGURE F.1*** *Hazard/Risk Analysis Evaluation Procedure Flow Chart.*

## HAZARD / RISK EVALUATION PROCEDURE

Task: Voltage testing

Equipment:

Issued by:

Date:

Document no.:

Part of:

Pre–risk assessment
Intermediate risk assessment
Follow-up risk assessment

Black area = Safety measures required Grey area = Safety measures recommended

| Consequences | Severity Se | Class Cl 3–4 | Class Cl 5–7 | Class Cl 8–10 | Class Cl 11–13 | Class Cl 14–15 | Frequency Fr | Probability Pr | Avoidance Av |
|---|---|---|---|---|---|---|---|---|---|
| Irreversible trauma, death | 4 | | | | | | Daily 5 | Common 5 | |
| Permanent, third-degree burn | 3 | | | | | | Weekly 4 | Likely 4 | |
| Reversible, second-degree burn | 2 | | | | | | Monthly 3 | Possible 3 | Impossible 5 |
| Reversible, first aid | 1 | | | | | | Yearly 2 | Rarely 2 | Possible 3 |
| | | | | | | | Less 1 | Negligible 1 | Likely 1 |

| Hzd. No. | Hazard | Se | Fr + | Pr + | Av + | = Cl | Severity Mitigators | Safe |
|---|---|---|---|---|---|---|---|---|
| 1 | Human factors | 4 | 5 | 3 | 5 | 13 | Use appropriate PPE and follow established safety procedures. | Y |
| 2 | Shortened test leads | 3 | 5 | 2 | 5 | 12 | Inspect leads before each use. | Y |
| 3 | Meter misapplication | 4 | 5 | 3 | 5 | 13 | Ensure that the meter is rated for the level of voltage being tested. | Y |
| 4 | Meter malfunctions | 3 | 5 | 2 | 5 | 12 | Ensure that the meter is CAT rated to the appropriate hazard level. | Y |
| | | | | | | | | |
| | | | | | | | | |
| | | | | | | | | |
| | | | | | | | | |

Comments:

PPE required: Voltage rated gloves and leather protectors, face and head protection, clothing rated for the incident energy exposure.



NFPA 70E

***FIGURE F.2.1*** *Sample Hazard/Risk Evaluation Procedure Form. (Source: Figure 5 in Anderson, W. E. "Risk Analysis Methodology Applied to Industrial Machine Development," IEEE Transactions on Industrial Applications.)*

(4) Assigning a safety measure for each hazard to attain an acceptable or tolerable level of risk

While this procedure might not result in a reduction of PPE for a task, it can help in understanding of the specific hazards associated with a task to a greater degree and thus allow for a more comprehensive assessment to occur.

While severity, frequency, and avoidance factors are straightforward, consideration of probability includes the following estimators:

(1) Hazard exposures
(2) Human factors
(3) Task history
(4) Workplace culture
(5) Safeguard reliability
(6) Ability to maintain or defeat protective measures
(7) Preventive maintenance history

Reduction strategies to be employed if an unacceptable risk cannot be achieved include the following hierarchy of controls:

(1) Eliminate the hazard
(2) Reduce the risk by design
(3) Apply safeguards
(4) Implement administrative controls
(5) Use personal protective equipment

Figure F.2.2 illustrates the steps of a hazard/risk evaluation assessment procedure. Figure F.2.3 is included as a blank version of Figure F.2.1.

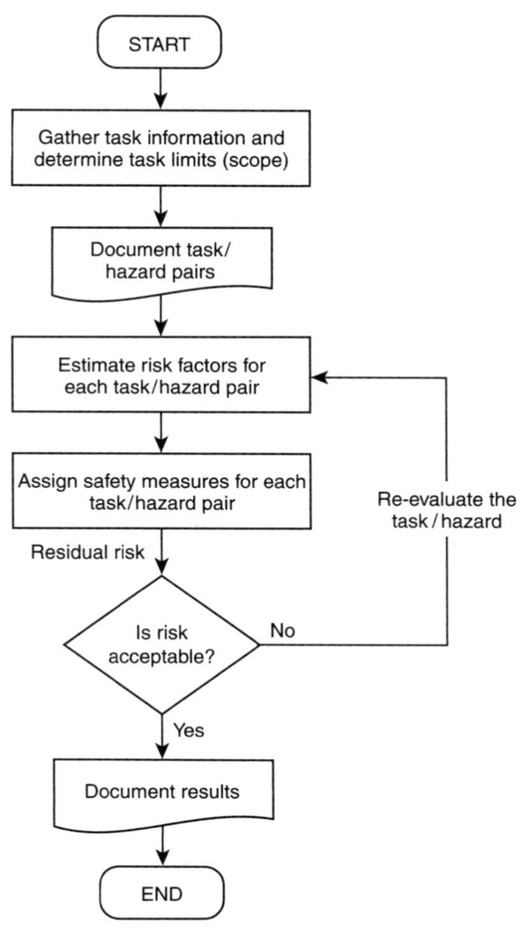


***FIGURE F.2.2*** *The Steps of a Hazard/Risk Evaluation Assessment Procedure.*

## HAZARD/RISK EVALUATION PROCEDURE

Task: ______________________________

Equipment: ______________________________

Issued by: ______________________________

Date: ______________________________

Document No.: ______________________________

Part of: ______________________________

❑ Pre–risk assessment
❑ Intermediate risk assessment
❑ Follow-up risk assessment

Black area = Safety measures required
Grey area = Safety measures recommended

| Consequences | Severity Se | Class Cl | | | | | Frequency Fr | | Probability Pr | | Avoidance Av | |
|---|---|---|---|---|---|---|---|---|---|---|---|---|
| | | 3–4 | 5–7 | 8–10 | 11–13 | 14–15 | | | | | | |
| Irreversible trauma, death | 4 | Black | Black | Black | Black | Black | Daily | 5 | Common | 5 | | |
| Permanent, third-degree burn | 3 | | Grey | Black | Black | Black | Weekly | 4 | Likely | 4 | | |
| Reversible, second-degree burn | 2 | | | Grey | Black | Black | Monthly | 3 | Possible | 3 | Impossible | 5 |
| Reversible, first aid | 1 | | | | Grey | Black | Yearly | 2 | Rarely | 2 | Possible | 3 |
| | | | | | | | Less | 1 | Negligible | 1 | Likely | 1 |

| Hzd. No. | Hazard | Se | Fr + | Pr + | Av + | = Cl | Severity Mitigators | Safe |
|---|---|---|---|---|---|---|---|---|
| | | | | | | | | |
| | | | | | | | | |
| | | | | | | | | |
| | | | | | | | | |
| | | | | | | | | |
| | | | | | | | | |
| | | | | | | | | |
| | | | | | | | | |
| | | | | | | | | |
| | | | | | | | | |
| | | | | | | | | |
| | | | | | | | | |

**Comments**

______________________________

______________________________

______________________________

______________________________

______________________________

 NFPA 70E

***FIGURE F.2.3*** *Hazard/Risk Evaluation Procedure (Blank Form for Use). (Source: Figure 5 in Anderson, W. E. "Risk Analysis Methodology Applied to Industrial Machine Development," IEEE Transactions on Industrial Applications.)*

# Sample Lockout/Tagout Procedure

# ANNEX G

*This annex is not a part of the requirements of this NFPA document but is included for informational purposes only.*

Annex G is provided to illustrate how a lockout/tagout procedure might be published. An employer could "fill in the blanks" and publish the resulting procedure for his or her organization.

Lockout is the preferred method of controlling personnel exposure to electrical energy hazards. Tagout is an alternative method that is available to employers. To assist employers in developing a procedure that meets the requirement of 120.2 of *NFPA 70E*, the following sample procedure is provided for use in lockout or tagout programs. This procedure can be used for an individual employee control, a simple lockout/tagout, or as part of a complex lockout/tagout. Where a job or task is under the control of one person, the individual employee control procedure can be used in lieu of a lockout/tagout procedure. A more comprehensive plan will need to be developed, documented, and utilized for the complex lockout/tagout.

LOCKOUT (TAGOUT) PROCEDURE FOR ABC COMPANY
OR
TAGOUT PROCEDURE FOR _____ COMPANY

**1.0 Purpose.** This procedure establishes the minimum requirements for lockout (tagout) of electrical energy sources. It is to be used to ensure that conductors and circuit parts are disconnected from sources of electrical energy, locked (tagged), and tested before work begins where employees could be exposed to dangerous conditions. Sources of stored energy, such as capacitors or springs, shall be relieved of their energy, and a mechanism shall be engaged to prevent the re-accumulation of energy.

**2.0 Responsibility.** All employees shall be instructed in the safety significance of the lockout (tagout) procedure. All new or transferred employees and all other persons whose work operations are or might be in the area shall be instructed in the purpose and use of this procedure. [Include the name(s) of person(s) or job title(s) of employees with responsibility] shall ensure that appropriate personnel receive instructions on their roles and responsibilities. All persons installing a lockout (tagout) device shall sign their names and the date on the tag (or state how the name of the individual or person in charge will be available).

**3.0 Preparation for Lockout (Tagout)**

**3.1** Review current diagrammatic drawings (or other equally effective means), tags, labels, and signs to identify and locate all disconnecting means to determine that power is interrupted by a physical break and not deenergized by a circuit interlock. Make a list of disconnecting means to be locked (tagged).

**3.2** Review disconnecting means to determine adequacy of their interrupting ability. Determine if it will be possible to verify a visible open point, or if other precautions will be necessary.

**3.3** Review other work activity to identify where and how other personnel might be exposed to sources of electrical energy hazards. Review other energy sources in the physical area to determine employee exposure to sources of other types of energy. Establish energy control methods for control of other hazardous energy sources in the area.

**3.4** Provide an adequately rated voltage detector to test each phase conductor or circuit part to verify that they are deenergized. *(See 12.3.)* Provide a method to determine that the voltage detector is operating satisfactorily.

**3.5** Where the possibility of induced voltages or stored electrical energy exists, call for grounding the phase conductors or circuit parts before touching them. Where it could be reasonably anticipated that contact with other exposed energized conductors or circuit parts is possible, call for applying ground connecting devices.

**4.0 Individual Employee Control Procedure.** The individual employee control procedure can be used when equipment with exposed conductors and circuit parts is deenergized for minor maintenance, servicing, adjusting, cleaning, inspection, operating corrections, and the like, and the work shall be permitted to be performed without the placement of lockout/tagout devices on the disconnecting means, provided the disconnecting means is adjacent to the conductor, circuit parts, and equipment on which the work is performed, the disconnecting means is clearly visible to all employees involved in the work, and the work does not extend beyond the work shift.

**5.0 Simple Lockout/Tagout.** The simple lockout/tagout procedure will involve paragraphs 1.0 through 3.0, 5.0 through 9.0, and 11.0 through 13.0.

**6.0 Sequence of Lockout (Tagout) System Procedures.**

**6.1** The employees shall be notified that a lockout (tagout) system is going to be implemented and the reason therefor. The qualified employee implementing the lockout (tagout) shall know the disconnecting means location for all sources of electrical energy and the location of all sources of stored energy. The qualified person shall be knowledgeable of hazards associated with electrical energy.

**6.2** If the electrical supply is energized, the qualified person shall deenergize and disconnect the electric supply and relieve all stored energy.

**6.3** Lockout (tagout) all disconnecting means with lockout (tagout) devices.

FPN: For tagout, one additional safety measure must be employed, such as opening, blocking, or removing an additional circuit element.

**6.4** Attempt to operate the disconnecting means to determine that operation is prohibited.

**6.5** A voltage-detecting instrument shall be used. *(See 12.3.)* Inspect the instrument for visible damage. Do not proceed if there is an indication of damage to the instrument until an undamaged device is available.

**6.6** Verify proper instrument operation and then test for absence of voltage.

**6.7** Verify proper instrument operation after testing for absence of voltage.

**6.8** Where required, install grounding equipment/conductor device on the phase conductors or circuit parts, to eliminate induced voltage or stored energy, before touching them. Where it has been determined that contact with other exposed energized conductors or circuit parts is possible, apply ground connecting devices rated for the available fault duty.

**6.9** The equipment and/or electrical source is now locked out (tagged out).

**7.0 Restoring the Equipment and/or Electrical Supply to Normal Condition**

**7.1** After the job or task is complete, visually verify that the job or task is complete.

**7.2** Remove all tools, equipment, and unused materials and perform appropriate housekeeping.

**7.3** Remove all grounding equipment/conductor/devices.

**7.4** Notify all personnel involved with the job or task that the lockout (tagout) is complete, that the electrical supply is being restored, and to remain clear of the equipment and electrical supply.

**7.5** Perform any quality control tests or checks on the repaired or replaced equipment and/or electrical supply.

**7.6** Remove lockout (tagout) devices by the person who installed them.

**7.7** Notify the equipment and/or electrical supply owner that the equipment and/or electrical supply is ready to be returned to normal operation.

**7.8** Return the disconnecting means to their normal condition.

**8.0 Procedure Involving More Than One Person**

For a simple lockout/tagout and where more than one person is involved in the job or task, each person shall install his or her own personal lockout (tagout) device.

**9.0 Procedure Involving More Than One Shift**

When the lockout (tagout) extends for more than one day, the lockout (tagout) shall be verified to be still in place at the beginning of the next day. Where the lockout (tagout) is continued on successive shifts, the lockout (tagout) is considered to be a complex lockout (tagout).

For complex lockout (tagout), the person in charge shall identify the method for transfer of the lockout (tagout) and of communication with all employees.

**10.0 Complex Lockout (Tagout)**

A complex lockout/tagout plan is required where one or more of the following exist:

(1) Multiple energy sources (more than one)
(2) Multiple crews
(3) Multiple crafts
(4) Multiple locations
(5) Multiple employers
(6) Unique disconnecting means
(7) Complex or particular switching sequences
(8) Continues for more than one shift, that is, new workers

**10.1** All complex lockout/tagout procedures shall require a written plan of execution. The plan will include the requirements in 1.0 through 3.0, 6.0, 7.0, and 9.0 through 13.0.

**10.2** A person in charge shall be involved with a complex lockout/tagout procedure. The person in charge shall be at this location.

**10.3** The person in charge shall develop a written plan of execution and communicate that plan to all persons engaged in the job or task. The person in charge shall be held accountable for safe execution of the complex lockout/tagout plan. The complex lockout/tagout plan must address all the concerns of employees who might be exposed, and they must understand how electrical energy is controlled. The person in charge shall ensure that each person understands the hazards to which they are exposed and the safety-related work practices they are to use.

**10.4** All complex lockout/tagout plans identify the method to account for all persons who might be exposed to electrical hazards in the course of the lockout/tagout.

Select which of the following methods is to be used:

(1) Each individual will install his or her own personal lockout or tagout device.
(2) The person in charge shall lock his/her key in a "lock box."
(3) The person in charge shall maintain a sign in/out log for all personnel entering the area.
(4) Another equally effective methodology.

**10.5** The person in charge can install locks/tags, or direct their installation on behalf of other employees.

**10.6** The person in charge can remove locks/tags or direct their removal on behalf of other employees, only after all personnel are accounted for and ensured to be clear of potential electrical hazards.

**10.7** Where the complex lockout (tagout) is continued on successive shifts, the person in charge shall identify the method for transfer of the lockout and of communication with all employees.

**11.0 Discipline**

**11.1** Knowingly violating this procedure will result in _____ (state disciplinary actions that will be taken).

**11.2** Knowingly operating a disconnecting means with an installed lockout device (tagout device) will result in _____ (state disciplinary actions to be taken).

**12.0 Equipment**

**12.1** Locks shall be _____ (state type and model of selected locks).

**12.2** Tags shall be _____ (state type and model to be used).

**12.3** Voltage detecting device(s) to be used shall be _____ (state type and model).

**13.0 Review.** This procedure was last reviewed on ______ and is scheduled to be reviewed again on _____ (not more than one year from the last review).

**14.0 Lockout/Tagout Training.** Recommended training can include, but is not limited to, the following:

(1) Recognizing lockout/tagout devices
(2) Installing lockout/tagout devices
(3) Duty of employer in writing procedures
(4) Duty of employee in executing procedures
(5) Duty of person in charge
(6) Authorized and unauthorized removal of locks/tags
(7) Enforcing execution of lockout/tagout procedures

(8) Individual employee control of energy
(9) Simple lockout/tagout
(10) Complex lockout/tagout
(11) Using single line and diagrammatic drawings to identify sources of energy
(12) Use of tags and warning signs
(13) Release of stored energy
(14) Personnel accounting methods
(15) Grounding needs and requirements
(16) Safe use of voltage detecting instruments

# Simplified, Two-Category, Flame-Resistant (FR) Clothing System

## ANNEX H

*This annex is not a part of the requirements of this NFPA document but is included for informational purposes only.*

After an employer develops and publishes a procedure that describes how thermal protection is determined, the employer must develop a system that enables the procedure requirements to be administered. Annex H is intended to illustrate one method that enables a personal protective equipment (PPE) program to be administered.

### H.1 Use of Simplified Approach

The use of Table H.1 is suggested as a simplified approach to provide minimum personal protective equipment for electrical workers within facilities with large and diverse electrical systems. The clothing listed in Table H.1 fulfills the minimum FR clothing requirements of Table 130.7(C)(9) and Table 130.7(C)(10). The clothing systems listed in this table should be used with the other PPE appropriate for the Hazard/Risk Category. *[See Table 130.7(C)(10).]* The notes at the bottom of Table 130.7(C)(9) must apply as shown in that table.

***TABLE H.1*** *Simplified, Two-Category, Flame-Resistant Clothing System*

| *Clothing*[a] | *Applicable Tasks* |
|---|---|
| **Everyday Work Clothing**<br>FR long-sleeve shirt with FR pants (minimum arc rating of 8)<br>*or*<br>FR coveralls (minimum arc rating of 8) | All Hazard/Risk Category 1, 2 and 2* tasks listed in Table 130.7(C)(9)[b] |
| **Arc Flash Suit**<br>A total clothing system consisting of FR shirt and pants and/or FR coveralls and/or arc flash coat and pants (clothing system minimum arc rating of 40) | All Hazard/Risk Category 3 and 4 tasks listed in Table 130.7(C)(9)[b] |

[a]Note other PPE required for the specific tasks listed in Tables 130.7(C)(9) and 130.7(C)(10), which includes arc-rated face shields or arc flash suit hoods, FR hardhat liners, safety glasses or safety goggles, hard hat, hearing protection, leather gloves, voltage-rated gloves, and voltage-rated tools. Arc rating for a garment is expressed in cal/cm$^2$.

[b]The assumed short-circuit current capacities and fault clearing times for various tasks are listed in the text and notes to Table 130.7(C)(9). For tasks not listed, or for power systems with greater than the assumed short-circuit capacity or with longer than the assumed fault clearing times, an arc flash hazard analysis is required in accordance with 130.3.

# Job Briefing and Planning Checklist

## ANNEX I

*This annex is not a part of the requirements of this NFPA document but is included for informational purposes only.*

Annex I is intended to illustrate the various subjects that should be discussed when a job briefing is held. Other subjects might need to be discussed. The purpose of this checklist is to help facilitate the conversation.

**I.1** Figure I.1 illustrates considerations for a Job Briefing and Planning Checklist.

**Identify**

- ❑ The hazards
- ❑ The voltage levels involved
- ❑ Skills required
- ❑ Any "foreign" (secondary source) voltage source
- ❑ Any unusual work conditions
- ❑ Number of people needed to do the job
- ❑ The shock protection boundaries
- ❑ The available incident energy
- ❑ Potential for arc flash (Conduct an arc flash-hazard analysis.)
- ❑ Arc flash protection boundary

**Ask**

- ❑ Can the equipment be de-energized?
- ❑ Are backfeeds of the circuits to be worked on possible?
- ❑ Is a "standby person" required?

**Check**

- ❑ Job plans
- ❑ Single-line diagrams and vendor prints
- ❑ Status board
- ❑ Information on plant and vendor resources is up to date
- ❑ Safety procedures
- ❑ Vendor information
- ❑ Individuals are familiar with the facility

**Know**

- ❑ What the job is
- ❑ Who else needs to know—Communicate!
- ❑ Who is in charge

**Think**

- ❑ About the unexpected event . . . What if?
- ❑ Lock — Tag — Test — Try
- ❑ Test for voltage — FIRST
- ❑ Use the right tools and equipment, including PPE
- ❑ Install and remove grounds
- ❑ Install barriers and barricades
- ❑ What else . . . ?

**Prepare for an emergency**

- ❑ Is the standby person CPR trained?
- ❑ Is the required emergency equipment available? Where is it?
- ❑ Where is the nearest telephone?
- ❑ Where is the fire alarm?
- ❑ Is confined space rescue available?
- ❑ What is the exact work location?
- ❑ How is the equipment shut off in an emergency?
- ❑ Are the emergency telephone numbers known?
- ❑ Where is the fire extinguisher?
- ❑ Are radio communications available?

***FIGURE I.1*** *Sample Job Briefing and Planning Checklist.*

# Energized Electrical Work Permit

## ANNEX J

*This annex is not a part of the requirements of this NFPA document but is included for informational purposes only.*

Annex J illustrates an energized electrical work permit. The content of the work permit is not fixed by requirement, although many employers have used this permit successfully. The basic purpose of the permit is to ensure that people in responsible positions are involved in the decision of whether or not to accept the increased risk associated with working on or near exposed live parts. A side benefit of the work permit is that its review might produce a decision to perform the work deenergized.

**J.1** Figure J.1 illustrates considerations for an Energized Electrical Work Permit.

**J.2** Figure J.2 illustrates items to consider when determining the need for an Energized Electrical Work Permit.

## ENERGIZED ELECTRICAL WORK PERMIT

**PART I: TO BE COMPLETED BY THE REQUESTER:**

Job/Work Order Number ________________

(1) Description of circuit/equipment/job location: ________________

(2) Description of work to be done: ________________

(3) Justification of why the circuit/equipment cannot be de-energized or the work deferred until the next scheduled outage: ________________

Requester/Title ________________ Date ________________

**PART II: TO BE COMPLETED BY THE ELECTRICALLY QUALIFIED PERSONS *DOING* THE WORK:**

| | Check when complete |
|---|---|
| (1) Detailed job description procedure to be used in performing the above detailed work: | ☐ |
| (2) Description of the Safe Work Practices to be employed: | ☐ |
| (3) Results of the Shock Hazard Analysis: | ☐ |
| (4) Determination of Shock Protection Boundaries: | ☐ |
| (5) Results of the Arc Flash Hazard Analysis: | ☐ |
| (6) Determination of the Arc Flash Protection Boundary: | ☐ |
| (7) Necessary personal protective equipment to safely perform the assigned task: | ☐ |
| (8) Means employed to restrict the access of unqualified persons from the work area: | ☐ |
| (9) Evidence of completion of a Job Briefing including discussion of any job-related hazards: | ☐ |

(10) Do you agree the above described work can be done safely? ❑ Yes ❑ No (If *no,* return to requester)

Electrically Qualified Person(s) ________________ Date ________________

Electrically Qualified Person(s) ________________ Date ________________

**PART III: APPROVAL(S) TO PERFORM THE WORK WHILE ELECTRICALLY ENERGIZED:**

Manufacturing Manager ________________ Maintenance/Engineering Manager ________________

Safety Manager ________________ Electrically Knowledgeable Person ________________

General Manager ________________ Date ________________

Note: Once the work is complete, forward this form to the site Safety Department for review and retention.

 NFPA 70E

***FIGURE J.1*** *Sample Permit for Energized Electrical Work.*

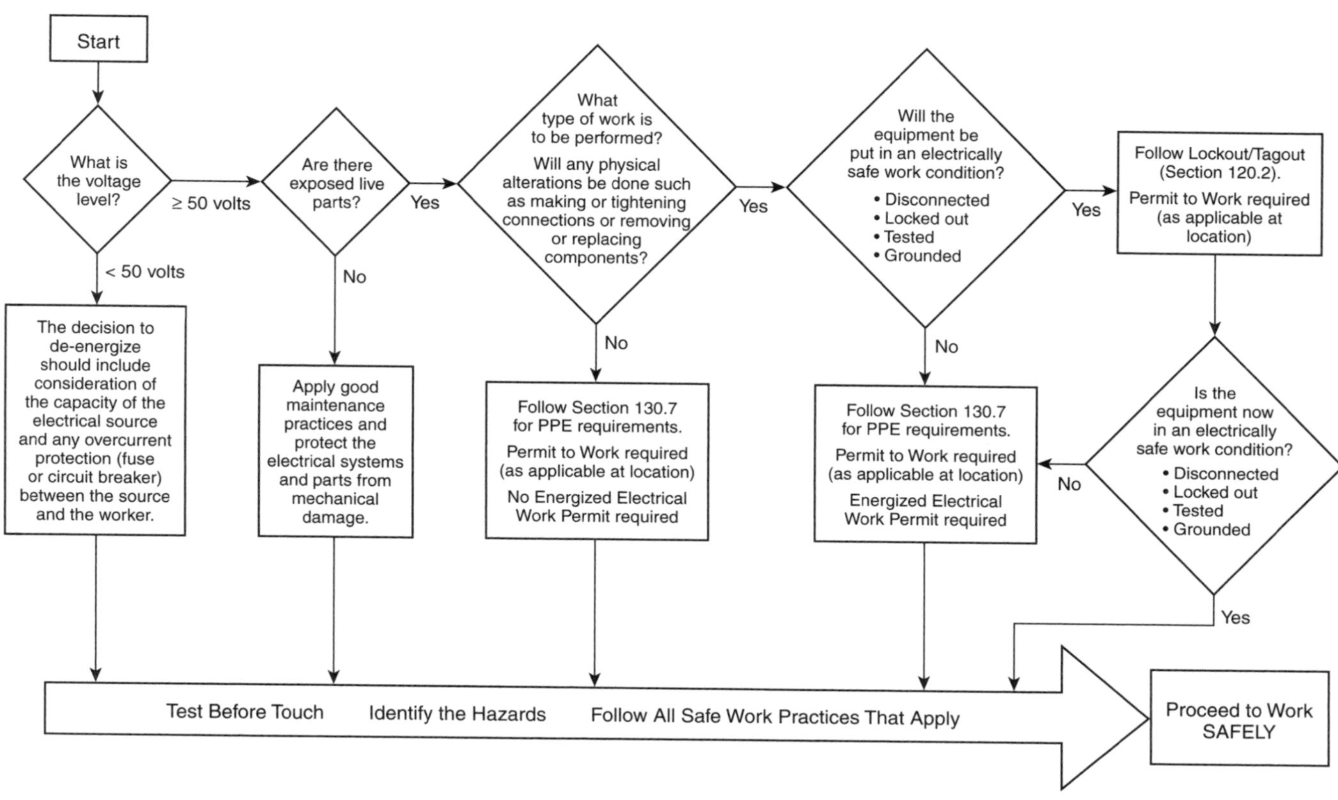


*FIGURE J.2 Energized Electrical Work Permit Flow Chart.*

# General Categories of Electrical Hazards

# ANNEX K

*This annex is not a part of the requirements of this NFPA document but is included for informational purposes only.*

Annex K provides an abbreviated discussion of known electrical hazards. The discussion is intended to provide critical information for a worker to use in his or her argument for an improved electrical safety program.

## K.1 General Categories

There are three general categories of electrical hazards: electrical shock, arc-flash, and arc-blast.

## K.2 Electric Shock

Approximately 30,000 nonfatal electrical shock accidents occur each year. The National Safety Council estimates that about 1000 fatalities each year are due to electrocution, more than half of them while servicing energized systems of less than 600 volts.

Electrocution is the fourth leading cause of industrial fatalities, after traffic, homicide, and construction accidents. The current required to light a $7\frac{1}{2}$ watt, 120 volt lamp, if passed across the chest, is enough to cause a fatality. The most damaging paths through the body are through the lungs, heart, and brain.

## K.3 Arc-Flash

When an electric current passes through air between ungrounded conductors or between ungrounded conductors and grounded conductors, the temperatures can reach 35,000°F. Exposure to these extreme temperatures both burns the skin directly and causes ignition of clothing, which adds to the burn injury. The majority of hospital admissions due to electrical accidents are from arc-flash burns, not from shocks. Each year more than 2000 people are admitted to burn centers with severe arc-flash burns. Arc-flashes can and do kill at distances of 3 m (10 ft).

## K.4 Arc-Blast

The tremendous temperatures of the arc cause the explosive expansion of both the surrounding air and the metal in the arc path. For example, copper expands by a factor of 67,000 times when it turns from a solid to a vapor. The danger associated with this expansion is one of high pressures, sound, and shrapnel. The high pressures can easily exceed hundreds or even thousands of pounds per square foot, knocking workers off ladders, rupturing eardrums, and collapsing lungs. The sounds associated with these pressures can exceed 160 dB. Finally, material and molten metal is expelled away from the arc at speeds exceeding 1600 km/hr (700 mph), fast enough for shrapnel to completely penetrate the human body.

# Typical Application of Safeguards in the Cell Line Working Zone

## ANNEX L

*This annex is not a part of the requirements of this NFPA document but is included for informational purposes only.*

Annex L discusses and illustrates the application of typical safeguards in a process area associated with a dc electrical system that cannot be deenergized.

### L.1 Application of Safeguards.

This annex permits a typical application of safeguards in electrolytic areas where hazardous electrical conditions exist. Take, for example, an employee working on an energized cell. The employee uses manual contact to make adjustments and repairs. Consequently, the exposed energized cell and grounded metal floor could present a hazardous electrical condition. Safeguards for this employee can be provided in several ways:

(1) Protective boots can be worn that isolate the employee's feet from the floor and that provide a safeguard from the hazardous electrical condition.
(2) Protective gloves can be worn that isolate the employee's hands from the energized cell and that provide a safeguard.
(3) If the work task causes severe deterioration, wear, or damage to personal protective equipment, the employee might have to wear both protective gloves and boots.
(4) A permanent or temporary insulating surface can be provided for the employee to stand on to provide a safeguard.
(5) The design of the installation can be modified to provide a conductive surface for the employee to stand on. If the conductive surface is bonded to the cell, the hazardous electrical condition will be removed and a safeguard will be provided by voltage equalization.
(6) Safe work practices can provide safeguards. If protective boots are worn, the employee should not make long reaches over energized (or grounded) surfaces such that his or her elbow bypasses the safeguard. If such movements are required, protective sleeves, protective mats, or special tools should be utilized. Training on the nature of hazardous electrical conditions and proper use and condition of safeguards is in itself a safeguard.
(7) The energized cell can be temporarily bonded to ground to remove the hazardous electrical condition.

### L.2 Electrical Power Receptacles.

Power supply circuits and receptacles in the cell line area for portable electric equipment should meet the requirements of *NFPA 70, National Electrical Code*, Section 668.21. However, it is recommended that receptacles for portable electric equipment not be installed in electrolytic cell areas and that only pneumatic powered portable tools and equipment be used.

# Layering of Protective Clothing and Total System Arc Rating

## ANNEX M

*This annex is not a part of the requirements of this NFPA document but is included for informational purposes only.*

Annex M discusses how layering of protective clothing can impact the overall rating of the layered protection. When layers of clothing are worn, some air space is captured between the layers. Air is a good thermal insulator and modifies the protective nature of the sum of the protective clothing. The resulting protection offered by the layers of clothing can be greater or less than the sum of the protection afforded by the individual layers. This annex is intended to discuss the protection; however, consultation with the clothing manufacturer is recommended.

### M.1 Layering of Protective Clothing

Layering of flame-resistant (FR) clothing is an effective approach to achieving the required arc rating. The use of all FR clothing layers will result in achieving the required arc rating with the lowest number of layers and lowest clothing system weight.

**M.1.1** The total system of protective clothing can be selected to take credit for the protection provided of all the layers of clothing that are worn. For example, to achieve an arc rating of 40 cal/cm$^2$, an arc flash suit with an arc rating of 40 cal/cm$^2$ could be worn over a cotton shirt and cotton pants. Alternatively, an arc flash suit with a 25 cal/cm$^2$ arc rating could be worn over an FR shirt and FR pants with an arc rating of 8 cal/cm$^2$ to achieve a total system arc rating of 40 cal/cm$^2$. This latter approach provides the required arc rating at a lower weight and with fewer total layers of fabric, and consequently would provide the required protection with a higher level of worker comfort.

### M.2 Layering Using FR Clothing over Natural Fiber Clothing Under Layers

Under some exposure conditions, natural fiber under layers can ignite even when they are worn under FR clothing.

**M.2.1** If the arc flash exposure is sufficient to break open all the FR clothing outer layer or layers, the natural fiber under layer can ignite and cause more severe burn injuries to an expanded area of the body. This is due to the natural fiber under layers burning onto areas of the worker's body that were not exposed by the arc flash event. This can occur when the natural fiber underlayer continues to burn underneath FR clothing layers even in areas in which the FR clothing layer or layers are not broken open due to a "chimney effect."

### M.3 Total System Arc Rating

The total system arc rating is the arc rating obtained when all clothing layers worn by a worker are tested as a multilayer test sample. An example of a clothing system is an FR coverall worn over an FR shirt and FR pants in which all of the garments are constructed from the same FR fabric. For this two-layer FR clothing system the arc rating would typically be more than three times higher than the arc rating of the individual layers, that is, if the arc rating of the FR coverall, shirt, and pants were all in the range of 5 to 6 cal/cm$^2$, the total two-layer system arc rating would be over 20 cal/cm$^2$.

**M.3.1** It is important to understand that the total system arc rating cannot be determined by adding the arc ratings of the individual layers. In a few cases, it has been observed that the total system arc rating actually decreased when another FR layer of a specific type was added to the system as the outermost layer. The only way to determine the total system arc rating is to conduct a multilayer arc test on the combination of all of the layers assembled as they would be worn.

# Example Industrial Procedures and Policies for Working Near Overhead Electrical Lines and Equipment

## ANNEX N

*This annex is not a part of the requirements of this NFPA document but is included for informational purposes only.*

Annex N illustrates electrical safety measures that might be appropriate for working near overhead lines. Although the content of this annex is not enforceable, the measures contained in the annex are effective.

### N.1 Introduction

This annex is an example of an industrial procedure for working near overhead electrical systems. Areas covered include operations that could expose employees or equipment to contact with overhead electrical systems.

When working near electrical lines or equipment, avoid direct or indirect contact. Direct contact is contact with any part of the body. Indirect contact is when part of the body touches or is in dangerous proximity to any object in contact with energized electrical equipment. Two assumptions should always be made: (1) Lines are "live" (energized), and (2) lines are operating at high voltage (over 1000 volts).

As the voltage increases, the minimum working clearances increase. Through arc-over, injuries or fatalities may occur even if actual contact with high voltage lines or equipment is not made. Potential for arc-over increases as the voltage increases.

### N.2 Overhead Powerline Policy (OPP)

This annex applies to all overhead conductors, regardless of voltage and requires the following:

(1) That employees not place themselves in close proximity to overhead powerlines. "Close proximity" is within a distance of 10 ft for systems up to 50 kilovolts, and 4 in. for every 10 kilovolts above 50 kilovolts.
(2) That employees be informed of the hazards and precautions when working near overhead lines.
(3) That warning decals be posted on cranes and similar equipment regarding minimum clearance of 10 ft.
(4) That a "spotter" be designated when equipment is working near the proximity of overhead lines. This person's responsibility is to observe safe working clearances around all overhead lines and to direct the operator accordingly.
(5) That warning cones be used as visible indicators of the 10 ft safety zone when working near overhead powerlines.

*Note: "Working near," for the purpose of this annex, is defined as working within a distance from any overhead powerline that is less than the combined height or length of the lifting device, the associated load length, and the required minimum clearance distance [as stated in N.2(1)].*

***Required Clearance = Lift Equipment Height or Length + Load Length + At Least 10 ft***

(6) Notify the local responsible person at least 24 hours before any work begins to allow time to identify voltages and clearances, or to place the line in an electrically safe work condition.

## N.3 Policy

All employees and contractors shall conform to the OPP. The first line of defense in preventing electrical contact accidents is to remain outside the Limited Approach Boundary. Because most company and contractor employees are not qualified to determine the system voltage level, a qualified person shall be called to establish voltages and minimum clearances, and take appropriate action to make the work zone safe.

## N.4 Procedures

**N.4.1 General.** Prior to the start of all operations where potential contact with overhead electrical systems is possible, the person in charge shall identify overhead lines or equipment, reference their location with respect to prominent physical features, or physically mark the area directly in front of the overhead lines with safety cones, survey tape, or other means. Electrical line location shall be discussed at a pre-work safety meeting of all employees on the job (through a job briefing). All company employees and contractors shall attend this meeting and require their employees to conform to electrical safety standards. New or transferred employees shall be informed of electrical hazards and proper procedures during orientations.

On construction projects, the contractor shall identify and reference all potential electrical hazards and document such actions with the on-site employers. The location of overhead electrical lines and equipment shall be conspicuously marked by the person in charge. New employees shall be informed of electrical hazards and of proper precautions and procedures.

Where there is potential for contact with overhead electrical systems, local area management shall be called to decide whether to place the line in an electrically safe work condition, or to otherwise protect the line against accidental contact. Where there is a suspicion of lines with low clearance (height under 20 ft), the local on-site electrical supervisor shall be notified to verify and take appropriate action.

All electrical contact incidents, including "near misses," shall be reported to the local area health and safety specialist.

**N.4.2 LOOK UP AND LIVE Flags.** In order to prevent accidental contacts of overhead lines, all aerial lifts, cranes, boom trucks, service rigs, and similar equipment shall use "LOOK UP AND LIVE" flags. The flags are visual indicators that the equipment is currently being used or has been returned to its "stowed or cradled" position. The flags shall be yellow with black lettering and shall state in bold lettering "LOOK UP AND LIVE."

The procedure for the use of the flag shall be:

(1) When the boom or lift is in its stowed or cradled position, the flag shall be located on the load hook or boom end.

(2) Prior to operation of the boom or lift, the operator of the equipment shall assess the work area to determine the location of all overhead lines and communicate this information to all crews on site. Once completed, the operator shall remove the flag from the load hook or boom and transfer the flag to the steering wheel of the vehicle. Once the flag is placed on the steering wheel, the operator may begin to operate the equipment.

(3) After successfully completing the work activity and returning the equipment to its stowed or cradled position, the operator shall return the flag to the load hook.
(4) The operator of the equipment is responsible for the placement of the LOOK UP AND LIVE flag.

### N.4.3 High Risk Tasks.

**N.4.3.1. Heavy Mobile Equipment.** Prior to the start of each workday, a high visibility marker (orange safety cones or other devices) shall be temporarily placed on the ground to mark the location of overhead wires. The supervisors shall discuss electrical safety with appropriate crew members at on-site tailgate safety talks. When working in the proximity of overhead lines, a spotter shall be positioned in a conspicuous location to direct movement and observe for contact with the overhead wires. The spotter, equipment operator, and all other employees working on the job location shall be alert for overhead wires and remain at least 3 m (10 ft) from the mobile equipment.

All mobile equipment shall display a warning decal regarding electrical contact. Independent truck drivers delivering materials to field locations shall be cautioned about overhead electrical line before beginning work, and a properly trained on-site or contractor employee shall assist in the loading or off-loading operation. Trucks that have emptied their material shall not leave the work location until the boom, lift, or box is down and is safely secured.

**N.4.3.2 Aerial Lifts, Cranes, and Boom Devices.** Where there is potential for near operation or contact with overhead lines or equipment, work shall not begin until a safety meeting is conducted and appropriate steps taken to identify, mark, and warn against accidental contact. The supervisor will review operations daily to ensure compliance.

Where the operator's visibility is impaired, a spotter shall guide the operator. Hand signals shall be used and clearly understood between operator and spotter. When visual contact is impaired, the spotter and operator shall be in radio contact. Aerial lifts, cranes, and boom devices shall have appropriate warning decals and shall use warning cones or similar devices to indicate the location of overhead lines and identify the 3 m (10 ft) minimum safe working boundary.

**N.4.3.3 Tree Work.** Wires shall be treated as live and operating at high voltage until verified as otherwise by the local area on-site employer. The local maintenance organization or an approved electrical contractor shall remove branches touching wires before work begins. Limbs and branches shall not be dropped onto overhead wires. If limbs or branches fall across electrical wires, all work shall stop immediately and the local area maintenance organization called. When climbing or working in trees, pruners shall try to position themselves so that the trunk or limbs are between their bodies and electrical wires. If possible, pruners shall not work with their backs toward electrical wires. An insulated bucket truck is the preferential method of pruning when climbing poses a greater threat of electrical contact. Personal protective equipment shall be used while working on or near lines.

**N.4.4 Underground Electrical Lines and Equipment.** Before excavation starts and where there exists reasonable possibility of contacting electrical or utility lines or equipment, the local area supervision (or USA DIG organization, when appropriate), shall be called and a request made for identifying/marking the line location(s).

When USA DIG is called, telephone operators will need the following:

(1) Minimum of 2 working days' notice prior to start of work, name of county, name of city, name and number of street or highway marker, and nearest intersection
(2) Type of work
(3) Date and time work is to begin

(4) Caller's name, contractor/department name and address
(5) Telephone number for contact
(6) Special instructions

Utilities that do not belong to USA DIG must be contacted separately. USA DIG may not have a complete list of utility owners. Utilities discovered shall be marked before work begins. Supervisors shall periodically refer their location to all workers, including new employees, subject to exposure.

**N.4.5 Vehicles with Loads in Excess of a Height of 4.25 m (14 ft).** This policy requires that all vehicles with loads in excess of 4.25 m (14 ft) use specific procedures to maintain safe working clearances when in transit below overhead lines.

The specific procedures for moving loads in excess of 4.25 m (14 ft) or via routes with lower clearance heights are listed below:

(1) Prior to movement of any load in excess of 4.25 m (14 ft), the local health and safety department, along with the local person in charge, shall be notified of the equipment move.
(2) An on-site electrician, electrical construction representative, or qualified electrical contractor should check the intended route to the next location before relocation.
(3) Check the new site for overhead lines and clearances.
(4) Powerlines and communication lines shall be noted and extreme care used when traveling beneath the lines.
(5) The company moving the load or equipment will provide a driver responsible for measuring each load and ensuring each load is secured and transported in a safe manner.
(6) An on-site electrician, electrical construction representative, or qualified electrical contractor shall escort the first load to the new location, ensuring safe clearances. A service company representative shall be responsible for subsequent loads to follow the same safe route.

If proper working clearances cannot be maintained, the job must be shut down until a safe route can be established or the necessary repairs or relocations have been completed to ensure that a safe working clearance has been achieved.

**All work requiring movement of loads in excess of 4.25 m (14 ft) are required to begin only after a General Work Permit has been completed, detailing all pertinent information about the move.**

**N.4.6 Emergency Response.** If an overhead line falls or is contacted, the following precautions should be taken:

(1) Keep everyone at least 3 m (10 ft) away.
(2) Use flagging to protect motorists, spectators, and other individuals from fallen or low wires.
(3) Call the local area electrical department or electric utility immediately.
(4) Place barriers around the area.
(5) Do not attempt to move the wire(s).
(6) Do not touch anything that is touching the wire(s).
(7) Be alert to water or other conductors present.
(8) Crews shall have emergency numbers readily available. These numbers shall include local area electrical department, utility, police/fire, and medical assistance.
(9) If an individual becomes energized, DO NOT TOUCH the individual or anything in contact with the person. Call for emergency medical assistance and call the local utility im-

mediately. If the individual is no longer in contact with the energized conductors, CPR, rescue breathing, or first aid should be administered immediately, but only by a trained person. It is safe to touch the victim once contact is broken or the source is known to be de-energized.

(10) Wires that contact vehicles or equipment will cause arcing, smoke, and possibly fire. Occupants should remain in the cab and wait for the local area electrical department or utility. If it becomes necessary to exit the vehicle, leap with both feet as far away from the vehicle as possible, without touching the equipment. Jumping free of the vehicle is the last resort.

(11) If operating the equipment and an overhead wire is contacted, stop the equipment immediately and if safe to do so jump free and clear of the equipment. Maintain your balance, keep your feet together and either shuffle or bunny hop away from the vehicle another 3 m (10 ft) or more. Do not return to the vehicle or allow anyone else for any reason to return to the vehicle until the local utility has removed the powerline from the vehicle and has confirmed that the vehicle is no longer in contact with the overhead lines.

# Safety-Related Design Requirements

## ANNEX O

*This annex is not a part of the requirements of this NFPA document but is included for informational purposes only.*

Annex O indicates that the design of a facility, equipment, or circuit could determine if a work task can be performed safely. In a large measure, the facility and circuit design determines whether or how a worker is or might be exposed to an electrical hazard when performing tasks necessary to troubleshoot, repair, or maintain a facility.

The circuit and equipment design determines the amount of incident energy that might be available at various points in the system and whether an electrically safe work condition can be created for sectors of the circuit. The location of components and isolating or insulating barriers determine whether or how a worker might be exposed to shock or electrocution.

### O.1 Introduction

This annex addresses the responsibilities of the facility owner or manager, or the employer having responsibility for facility ownership or operations management, to apply electrical hazard analysis during the design of electrical systems and installations.

**O.1.1** This annex covers employee safety–related design concepts for electrical equipment and installations in workplaces covered by the scope of this standard. This annex discusses design considerations that have impact on the application of the safety-related work practices only.

**O.1.2** This annex does not discuss specific design requirements. The facility owner or manager, or the employer should choose design options that eliminate or reduce exposure risks and enhance the effectiveness of safety-related work practices.

### O.2 General Design Considerations

Employers, facility owners, and managers who have responsibility for facilities and installation having electrical energy as a potential hazard to employees and other personnel should ensure the application of 110.8(B)(1), Electrical Hazard Analysis, during the design of electrical systems and installations.

**O.2.2** The application of 110.8(B)(1) should be used to compare design options and choices to facilitate design decisions that serve to eliminate risk, reduce frequency of exposure, reduce magnitude or severity of exposure, enable the ability to achieve an electrically safe work condition, and otherwise serve to enhance the effectiveness of the safety-related work practices contained in this standard.

## PART TWO

# Supplements

In addition to the 2009 edition of *NFPA 70E®*, *Standard for Electrical Safety in the Workplace®*, and commentary presented in Part One, the *Handbook for Electrical Safety in the Workplace* includes two supplements.

Supplement 1 is an extract from NFPA® 70B, *Recommended Practice for Electrical Equipment Maintenance*. Supplement 2 is an extract from the 2008 edition of the *National Electrical Code® Handbook*. These supplements are not part of the standard. They are included as additional information for handbook users.

# SUPPLEMENT 1

# Electrical Preventive Maintenance Programs

***Editor's Note:*** *This supplement is an extract of Chapter 5 and Chapter 6 from the 2006 edition of NFPA® 70B, Recommended Practice for Electrical Equipment Maintenance. NFPA 70B provides detailed information on the maintenance of electrical equipment, and Chapter 5 and Chapter 6 provide specific information on how to set up a maintenance program. Businesses can no longer tolerate equipment failures as the signal that maintenance of equipment is due. Just-in-time (JIT) manufacturing and six sigma programs dictate a need for an effective preventive maintenance program that ensures continuity of operations and prevents equipment breakdowns. Breakdowns not only affect production schedules, they often affect production quality. A company's failure to meet its production schedule also disrupts their customers' supply chain. JIT falls apart when suppliers cannot perform. Scheduled maintenance tends to be more systematic and orderly. Breakdown maintenance is often performed under stressful conditions that may tempt workers to take dangerous safety shortcuts.*

## CHAPTER 5 WHAT IS AN EFFECTIVE ELECTRICAL PREVENTIVE MAINTENANCE (EPM) PROGRAM?

### 5.1 Introduction.

An effective electrical preventive maintenance (EPM) program should enhance safety and also reduce equipment failure to a minimum consistent with good economic judgment.

**5.1.1** An effective electrical preventive maintenance program should include the following basic ingredients:

(1) Personnel qualified to carry out the program
(2) Regularly scheduled inspection, testing, and servicing of equipment

**5.1.2** A successful program should also include the following:

(1) Application of sound judgment in evaluating and interpreting results of inspections and tests
(2) Keeping of concise but complete records

### 5.2 Planning an EPM Program.

The basic factors listed in 5.2.1 through 5.2.3 should be considered in the planning of an EPM program.

**5.2.1 Personnel Safety.** Will an equipment failure endanger or threaten the safety of any personnel? What can be done to ensure personnel safety?

**5.2.2 Equipment Loss.** Is installed equipment — both electrical and mechanical — complex or so unique that required repairs would be unusually expensive?

**5.2.3 Production Economics.** Will breakdown repairs or replacement of failed equipment require extensive downtime? How many production dollars will be lost in the event of an equipment failure? Which equipment is most vital to production?

### 5.3 Main Parts of an EPM Program.

An EPM program should consist of the following essential ingredients:

(1) Responsible and qualified personnel
(2) Survey and analysis of electrical equipment and systems to determine maintenance requirements and priorities
(3) Programmed routine inspections and suitable tests
(4) Accurate analysis of inspection and test reports so that proper corrective measures can be prescribed
(5) Performance of necessary work
(6) Concise but complete records

**5.3.1 Personnel.**

**5.3.1.1** A well-qualified individual should be in charge of the program.

**5.3.1.2** Personnel assigned to inspection and testing duties should be selected from the best maintenance personnel in the plant.

**5.3.1.3** Where in-plant personnel are not qualified, a maintenance contractor should be employed.

**5.3.2 Survey and Analysis.**

**5.3.2.1** Survey and analysis should cover equipment and systems that have been determined to be essential in accordance with a priority plan.

**5.3.2.2** Regardless of the size of the program being contemplated, the EPM supervisor should determine the extent of the work to be done and where to begin.

**5.3.2.3** All electrical equipment — motors, transformers, circuit breakers, controls, and the like — should receive a thorough inspection and evaluation to permit the EPM supervisor to make a qualified judgment as to how, where, and when each piece of equipment should fit into the program.

**5.3.2.4** In addition to determining the equipment's physical condition, the survey should determine if the equipment is operating within its rating.

**5.3.2.5** In the course of the survey, the condition of electrical protective devices such as fuses, circuit breakers, protective relays, and motor overload relays should be checked. These devices are the safety valves of an electrical system, and their proper operating condition ensures the safety of personnel, protection of equipment, and reduction of economic loss.

**5.3.2.6** After the survey has been completed, data should be evaluated to determine equipment condition. Equipment condition will reveal repair work to be done, as well as the nature and frequency of required inspections and tests.

**5.3.3 Programmed Inspections.**

Inspection and testing procedures should be carefully tailored to requirements. In some plants, regularly scheduled tests will call for scheduled outages of production or process equipment. In such cases, close coordination between maintenance and production personnel is necessary.

**5.3.3.1 Analysis of Inspection and Test Reports.** Analysis of inspection and test reports should be followed by implementation of appropriate corrective measures. Follow-through with necessary repairs, replacement, and adjustment is the end purpose of an effective EPM program.

**5.3.3.2 Records.**

**5.3.3.2.1** Records should be accurate and contain all vital information.

**5.3.3.2.2** Care should be taken to ensure that extraneous information does not become part of the record, because excessive record keeping can hamper the program.

**5.3.4 EPM Support Procedures.**

**5.3.4.1 Design for Ease of Maintenance.** Effective electrical preventive maintenance begins with good design. In the design of new facilities, a conscious effort to ensure optimum maintainability is recommended. Dual circuits, tie circuits, auxiliary power sources, and drawout protective devices make it easier to schedule maintenance and to perform maintenance work with minimum interruption of production. Other effective design techniques include equipment rooms to provide environmental protection, grouping of equipment for more convenience and accessibility, and standardization of equipment and components.

**5.3.5 Training for Safety and Technical Skills.**

**5.3.5.1 Training Requirements.**

**5.3.5.1.1** All employees who face a risk of electrical hazard should be trained to understand the specific hazards associated with electrical energy.

**5.3.5.1.2** All employees should be trained in safety-related work practices and required procedures as necessary to provide protection from electrical hazards associated with their jobs or task assignments.

**5.3.5.1.3** Employees should be trained to identify and understand the relationship between electrical hazards and possible injury.

**5.3.5.1.4** Refresher training should be provided as required.

**5.3.5.2 Type of Training.** The training can be in the classroom, on the job, or both. The type of training should be determined by the needs of the employee.

**5.3.5.3 Emergency Procedures.**

**5.3.5.3.1** Employees working on or near exposed energized electrical conductors or circuit parts should be trained in methods of release of victims from contact with exposed energized conductors or circuit parts.

**5.3.5.3.2** Employees working on or near exposed energized electrical conductors or circuit parts should be instructed regularly in methods of first aid and emergency procedures, such as approved methods of resuscitation.

**5.3.5.4 Training Scope.** Employees should be trained and knowledgeable in the following:

(1) Construction and operation of equipment
(2) Specific work method
(3) Electrical hazards that can be present with respect to specific equipment or work method
(4) Proper use of special precautionary techniques, personal protective equipment, insulating and shielding materials, and insulated tools and test equipment
(5) Skills and techniques necessary to distinguish exposed, energized parts from other parts of electrical equipment
(6) Skills and techniques necessary to determine the nominal voltage of exposed energized parts
(7) Decision-making process necessary to determine the degree and extent of hazard
(8) Job planning necessary to perform the task safely

**5.3.5.5 Record Keeping.** Records of training should be maintained for each employee.

**5.3.6 Outside Service Agencies.** Some maintenance and testing operations, such as relay and circuit-breaker inspection and testing, require specialized skills and special equipment. In small organizations, it might be impractical to develop the skills and acquire the equipment needed for this type of work. In such cases, it might be advisable to contract the work to firms that specialize in providing such services.

**5.3.7 Tools and Instruments.** Proper tools and instruments are an important part of an EPM program, and safety protective gear is an essential part of the necessary equipment. Proper tools, instruments, and other equipment should be used to ensure maximum safety and productivity from the maintenance crew. Where specialized instruments and test equipment are needed only occasionally, they can be rented from a variety of sources.

## CHAPTER 6 PLANNING AND DEVELOPING AN ELECTRICAL PREVENTIVE MAINTENANCE (EPM) PROGRAM

**6.1 Introduction.**

**6.1.1** The purpose of an EPM program is to reduce hazard to life and property that can result from the failure or malfunction of electrical systems and equipment. The first part of these recommendations for an effective EPM program has been prepared with the intent of providing a better understanding of benefits — both direct and intangible — that can be derived from a well-administered EPM program. This chapter explains the function, requirements, and economic considerations that can be used to establish such a program.

**6.1.2** The following four basic steps should be taken in the planning and development of an EPM program. In their simplest form, they are as follows:

(1) Compile a listing of all equipment and systems.
(2) Determine which equipment and systems are most critical and most important.
(3) Develop a system for keeping up with what needs to be done.
(4) Train people for the work that needs to be done or contract the special services that are needed.

**6.1.3** The success of an EPM program depends on the caliber of personnel responsible for its implementation.

**6.1.3.1** The primary responsibility for EPM program implementation and its success should lie with a single individual.

**6.1.3.2** This individual responsible for the EPM program should be given the authority to do the job and should have the cooperation of management, production, and other departments whose operations might affect the EPM program.

**6.1.3.3** Ideally, the person designated to head the EPM program should have the following qualifications:

(1) *Technical competence.* The person should, by education, training, and experience, be well-rounded in all aspects of electrical maintenance.
(2) *Administrative and supervisory skills.* The person should be skilled in the planning and development of long-range objectives to achieve specific results and should be able to command respect and solicit the cooperation of all persons involved in the program.

**6.1.4** The maintenance supervisor should have open lines of communication with design supervision. Frequently, an unsafe installation or one that requires excessive maintenance can be traced to improper design or construction methods or misapplication of hardware.

**6.1.5** The work center of each maintenance work group should be conveniently located. This work center should contain the following:

(1) Copies of all the inspection and testing procedures for that zone
(2) Copies of previous reports
(3) Single-line diagrams
(4) Schematic diagrams
(5) Records of complete nameplate data
(6) Vendors' catalogs
(7) Facility stores' catalogs
(8) Supplies of report forms

**6.1.5.1** There should be adequate storage facilities for tools and test equipment that are common to the group.

**6.1.6** In a continuously operating facility, running inspections (inspections made with equipment operating) play a vital role in the continuity of service. The development of running inspection procedures varies with the type of operation. Running inspection procedures should be as thorough as practicable within the limits of safety and the skill of the craftsman. These procedures should be reviewed regularly in order to keep them current. Each failure of electrical equipment, be it an electrical or a mechanical failure, should be reviewed against the running inspection procedure to determine if some other inspection technique would have indicated the impending failure. If so, the procedure should be modified to reflect the findings.

**6.1.7** Supervisors find their best motivational opportunities through handling the results of running inspections. When the electrical maintenance supervisor initiates corrective action, the craftsperson should be so informed. The craftsperson who found the condition will then feel that his or her job was worthwhile and will be motivated to try even harder. However, if nothing is done, individual motivation might be affected adversely.

**6.1.8** Trends in failure rates are hard to change and take a long time to reverse. For this reason, the inspection should continue and resulting work orders should be written, even though the work force might have been reduced. Using the backlog of work orders as an indicator, the electrical maintenance supervisor can predict trends before they develop. With the accumulation of a sizable backlog of work orders, an increase in electrical failures and production downtime can be expected.

**6.2 Survey of Electrical Installation.**

**6.2.1 Data Collection.**

**6.2.1.1** The first step in organizing a survey should be to take a look at the total package. Will the available manpower permit the survey of an entire system, process, or building, or should it be divided into segments?

**6.2.1.2** Next, a priority should be assigned to each segment. Segments found to be sequential should be identified before the actual work commences.

**6.2.1.3** The third step should be the assembling of all documentation. This might necessitate a search of desks, cabinets, and such, and might also require that manufacturers be contacted, to replace lost documents. All of the documents should be brought to a central location and marked immediately with some form of effective identification.

**6.2.2 Diagrams and Data.** The availability of up-to-date, accurate, and complete diagrams is the foundation of a successful EPM program. No EPM program can operate without them, and their importance cannot be overemphasized. The diagrams discussed in 6.2.2.1 through 6.2.2.8.2 are some of those in common use.

**6.2.2.1** Single-line diagrams should show the electrical circuitry down to, and often including, the major items of utilization equipment. They should show all electrical equipment in the power system and give all pertinent ratings. In making this type of diagram, it is basic that voltage, frequency, phase, and normal operating position be included. No less important, but perhaps less obvious, are items such as transformer impedance, available short-circuit current, and equipment continuous and interrupting ratings. Other items include current and potential transformers and their ratios, surge capacitors, and protective relays. If one diagram cannot cover all the equipment involved, additional diagrams, appropriately noted on the main diagram, can be drawn.

**6.2.2.2** Short-circuit and coordination studies are important. Many managers have the misconception that these engineering studies are part of the initial facility design, after which the subject can be forgotten. However, a number of factors can affect the available short-circuit current in an electrical system. Among these factors are changes in the supply capacity of the utility company, changes in size or percent impedance of transformers, changes in conductor size, addition of motors, and changes in system operating conditions.

**(A)** In the course of periodic maintenance testing of protective equipment, such as relays and series or shunt-trip devices, equipment settings should be evaluated. Along with the proper sizing of fuses, this evaluation is part of the coordination study.

**(B)** In a small facility, one receiving electrical energy at utilization voltage or from a single step-down transformer, the short-circuit study is simple. The available incoming short-circuit current can be obtained from the utility company sales engineer.

**(C)** In a larger system, it might be desirable to develop a computerized short-circuit study to improve accuracy and reduce engineering time. Should resources not be available within the plant organization, the short-circuit study can be performed on a contract basis. The short-circuit data are used to determine the required momentary and interrupting ratings of circuit breakers, fuses, and other equipment.

**(D)** Fuses are rated on the basis of their current-carrying and interrupting capacities. These ratings should be determined and recorded. Other protective devices are usually adjustable as to pickup point and time-current characteristics. The settings of such protective devices should be determined, verified by electrical tests, and recorded for future reference.

**(E)** Personnel performing the tests should be trained and qualified in proper test procedures. Various organizations and manufacturers of power and test equipment periodically schedule seminars in which participants are taught the principles of maintenance and testing of electrical protective devices.

**(F)** Additional guidance on electrical systems can be found in Chapter 25.

**6.2.2.3** Circuit-routing diagrams, cable maps, or raceway layouts should show the physical location of conductor runs. In addition to voltage, such diagrams should also indicate the type of raceway, number and size of conductors, and type of insulation.

**(A)** Where control conductors or conductors of different systems are contained within the same raceway, the coding appropriate to each conductor should be noted.

**(B)** Vertical and horizontal runs with the location of taps, headers, and pull boxes should be shown.

**(C)** Access points should be noted where raceways pass through tunnels or shafts with limited access.

**6.2.2.4** Layout diagrams, plot plans, equipment location plans, or facility maps should show the physical layout (and in some cases, the elevations) of all equipment in place.

**(A)** Switching equipment, transformers, control panels, mains, and feeders should be identified.

**(B)** Voltage and current ratings should be shown for each piece of equipment.

**6.2.2.5** Schematic diagrams should be arranged for simplicity and ease of understanding circuits without regard for the actual physical location of any components. The schematic should always be drawn with switches and contacts shown in a deenergized position.

**6.2.2.6** Wiring diagrams, like schematics, should show all components in the circuit but arranged in their actual physical location.

**6.2.2.6.1** Electromechanical components and strictly mechanical components interacting with electrical components should be shown. Of particular value is the designation of terminals and terminal strips with their appropriate numbers, letters, or colors.

**6.2.2.6.2** Wiring diagrams should identify all equipment parts and devices by standard methods, symbols, and markings.

**6.2.2.7** An effective EPM program should have manufacturers' service manuals and instructions. These manuals should include recommended practices and procedures for the following:

(1) Installation
(2) Disassembly/assembly (interconnections)
(3) Wiring diagrams, schematics, bills of materials
(4) Operation (set-up and adjustment)
(5) Maintenance (including parts list and recommended spares)
(6) Software program (if applicable)
(7) Troubleshooting

**6.2.2.8 Electrical Equipment Installation Change.** The documentation of the changes that result from engineering decisions, planned revisions, and so on, should be the responsibility of the engineering group that initiates the revisions.

**6.2.2.8.1** Periodically, changes occur as a result of an EPM program. The EPM program might also uncover undocumented practices or installations.

**6.2.2.8.2** A responsibility of the EPM program is to highlight these changes, note them in an appropriate manner, and formally submit the revisions to the organization responsible for the maintenance of the documentation.

**6.2.3 System Diagrams.** System diagrams should be provided to complete the data being assembled. The importance of the system determines the extent of information shown. The information can be shown on the most appropriate type of diagram but should include the same basic information, source and type of power, conductor and raceway information, and switching and protective devices with their physical locations. It is vital to show where the system might interface with another system, such as with emergency power; hydraulic, pneumatic, or mechanical systems; security and fire-alarm systems; and monitoring and control systems. Some of the more common of these are described in 6.2.3.1 through 6.2.3.4.

**6.2.3.1 Lighting System Diagrams.** Lighting system diagrams (normal and emergency) can terminate at the branch circuit panelboard, listing the number of fixtures, type and lamp size for each area, and design lighting level. The diagram should show watchman lights and probably an automatic transfer switch to the emergency power system.

**6.2.3.2 Ventilation.** Ventilation systems normally comprise the heating, cooling, and air-filtering system. Exceptions include furnace, dryer, oven, casting, and similar areas where process heat is excessive and air conditioning is not practical. Numerous fans are used to exhaust the heated and possibly foul air. In some industries, such as chemical plants and those using large amounts of flammable solvents, large volumes of air are needed to remove hazardous vapors. Basic information, including motor and fan sizes, motor or pneumatically operated dampers, and so on, should be shown. Additionally, many safety features can be involved to ensure that fans start before the process — airflow switches to shut down an operation on loss of ventilation and other interlocks of similar nature. Each of these should be identified with respect to type, function, physical location, and operating limits.

**6.2.3.3 Heating and Air Conditioning.** Heating and air-conditioning systems are usually manufactured and installed as a unit, furnished with diagrams and operating and maintenance manuals. This information should be updated as the system is changed or modified. Because these systems are often critical to the facility operation, additional equipment might have been incorporated: for example, humidity, lint, and dust control for textile, electronic, and similar processes and corrosive and flammable vapor control for chemical and related industries. Invariably, these systems interface with other electrical or nonelectrical systems: pneumatic or electromechanical operation of dampers, valves, and so on; electric operation for normal and abnormal temperature control; and manual control stations for emergency smoke removal are just a few. There might be others, but all should be shown and complete information given for each.

**6.2.3.4 Control and Monitoring.** Control and monitoring system diagrams should be provided to describe how these complicated systems function. They usually are in the form of a schematic diagram and can refer to specific wiring diagrams. Maximum benefit can be obtained only when every switching device is shown, its function is indicated, and it is identified for ease in finding a replacement. These devices often involve interfaces with other systems, whether electromechanical (heating or cooling medium) pumps and valves, electro-pneumatic temperature and damper controls, or safety and emergency operations. A sequence-of-operation chart and a list of safety precautions should be included to promote the safety of personnel and equipment. Understanding these complex circuits is best accomplished by breaking down the circuits into their natural functions, such as heating, cooling, process, or humidity controls. The knowledge of how each function relates to another enables the craftsperson to have a better concept of the entire system and thus perform assignments more efficiently.

**6.2.4 Emergency Procedures.** Emergency procedures should list, step by step, the action to be taken in case of emergency or for the safe shutdown or start-up of equipment or systems. Optimum use of these procedures is made when they are bound for quick reference and posted in the area of the equipment or systems. Some possible items to consider for inclusion in the emergency procedures are interlock types and locations, interconnections with other systems, and tagging procedures of the equipment or systems. Accurate single-line diagrams posted in strategic places are particularly helpful in emergency situations. The production of such diagrams in anticipation of an emergency is essential to a complete EPM program.

Diagrams are a particularly important training tool in developing a state of preparedness. Complete and up-to-date diagrams provide a quick review of the emergency plan. During an actual emergency, when time is at a premium, they provide a simple, quick reference guide.

**6.2.5 Test and Maintenance Equipment.**

**6.2.5.1** All maintenance work requires the use of proper tools and equipment to properly perform the task to be done. In addition to their ordinary tools, craftspersons (such as carpenters, pipe fitters, and machinists) use special tools or equipment based on the nature of the work to be performed. The electrician is no exception, but for EPM, additional equipment not found in the toolbox should be readily available. The size of the plant, the nature of its operations, and the extent of its maintenance, repair, and test facilities are all factors that determine the use frequency of the equipment. Economics seldom justify purchasing an infrequently used, expensive tool when it can be rented. However, a corporation having a number of plants in the area might well justify common ownership of the same device for joint use, making it quickly available at any time to any plant. Typical examples might be high-current or dc high-potential test equipment or a ground-fault locator.

**6.2.5.2** Because a certain amount of mechanical maintenance is often a part of the EPM program being conducted on associated equipment, the electrical craftsperson should have ready access to such items as the following:

(1) Assorted lubrication tools and equipment
(2) Various types and sizes of wrenches
(3) Nonmetallic hammers and blocks to protect against injury to machined surfaces
(4) Wheel pullers
(5) Feeler gauges to function as inside-and outside-diameter measuring gauges
(6) Instruments for measuring torque, tension, compression, vibration, and speed
(7) Standard and special mirrors with light sources for visual inspection
(8) Industrial-type portable blowers and vacuums having insulated nozzles for removal of dust and foreign matter
(9) Nontoxic, nonflammable cleaning solvents
(10) Clean, lint-free wiping cloths

**6.2.5.3** The use of well-maintained safety equipment is essential and should be mandatory for work on or near live electrical equipment. Prior to performing maintenance on or near live electrical equipment, NFPA 70E, *Standard for Electrical Safety in the Workplace,* should be used to identify the degree of personal protective equipment (PPE) required. Some of the more important equipment that should be provided includes the following:

(1) Heavy leather gloves
(2) Insulating gloves, mats, blankets, baskets, boots, jackets, and coats
(3) Insulated hand tools such as screwdrivers and pliers
(4) Nonmetallic hard hats with clear insulating face shields for protection against arcs
(5) Poles with hooks and hot sticks to safely open isolating switches

**6.2.5.3.1** A statiscope is recommended to indicate the presence of high voltage on certain types of equipment.

**6.2.5.4** Portable electric lighting should be provided, particularly in emergencies involving the power supply. Portable electric lighting used for maintenance areas that are normally wet or where personnel will be working within grounded metal structures such as drums, tanks, and vessels should be operated at an appropriate low voltage from an isolating transformer or other isolated source. This voltage level is a function of the ambient condition in which the portable lighting is used. The aim is to limit the exposure of personnel to hazardous current levels by limiting the voltage. Ample supply of battery lanterns and extra batteries should be available. Suitable extension cords should be provided.

**6.2.5.5** Portable meters and instruments are necessary for testing and troubleshooting, especially on circuits of 600 volts or less. These include general-purpose volt meters, volt-ohmmeters, and clamp-on-type ammeters with multi-scale ranges. In addition to conventional instruments, recording meters are useful for measuring magnitudes and fluctuations of current, voltage, power factor, watts, and volt-amperes versus time values. These instruments are a definite aid in defining specific electrical problems and determining if equipment malfunction is due to abnormal electrical conditions. Other valuable test equipment includes devices to measure the insulation resistance of motors and similar equipment in the megohm range and similar instruments in the low range for determining ground resistance, lightning protection systems, and grounding systems. Continuity testers are particularly valuable for checking control circuits and for circuit identification.

**6.2.5.6** Special instruments can be used to test the impedance of the grounding circuit conductor or the grounding path of energized low-voltage distribution systems and equipment. These instruments can be used to test the equipment-grounding circuit path of electrical equipment.

**6.2.5.7** Insulation-resistance-measuring equipment should be used to indicate insulation values at the time equipment is put into service. Later measurements might indicate any deterioration trend of the insulation values of the equipment. High-potential ac and dc testers are used effectively to indicate dielectric strength and insulation resistance of the insulation, respectively. It should be recognized that the possibility of breakdown under test due to concealed weakness is always present. High-potential testing should be performed with caution and only by qualified operators.

**6.2.5.8** Portable ground-fault locators can be used to test ungrounded power systems. Such devices will indicate ground location while the power system is energized. They thus provide a valuable aid for safe operation by indicating where to take corrective steps before an insulation breakdown occurs on another phase.

**6.2.5.9** Receptacle circuit testers are devices that, by a pattern of lights, indicate some types of incorrect wiring of 15-and 20-ampere, 125-volt grounding-type receptacles.

**6.2.5.9.1** Although these test devices can provide useful and easily acquired information, some have limitations, and the test results should be used with caution. For example, a high-resistance ground can give a correct wiring display, as can some multiple wiring errors. An incorrect display can be considered a valid indication that there is an incorrect situation, but a correct wiring display should not be accepted without further investigation.

### 6.3 Identification of Critical Equipment.

**6.3.1** Equipment (electric or otherwise) should be considered critical if its failure to operate normally and under complete control will cause a serious threat to people, property, or the product. Electric power, like process steam, water, and so forth, might be essential to the operation of a machine, but unless loss of one or more of these supplies causes the machine to become hazardous to people, property, or production, that machine might not be critical. The combined knowledge and experience of several people might be needed to make this determination. In a small plant, the plant engineer or master mechanic working with the operating superintendent should be able to make this determination.

**6.3.1.1** A large operation should use a team comprising the following qualified people:

(1) The electrical foreman or superintendent
(2) Production personnel thoroughly familiar with the operation capabilities of the equipment and the effect its loss will have on final production
(3) The senior maintenance person who is generally familiar with the maintenance and repair history of the equipment or process
(4) A technical person knowledgeable in the theoretical fundamentals of the process and its hazards (in a chemical plant, a chemist; in a mine, a geologist; etc.)
(5) A safety engineer or the person responsible for the overall security of the plant and its personnel against fire and accidents of all kinds

**6.3.1.2** The team should go over the entire plant or each of its operating segments in detail, considering each unit of equipment as related to the entire operation and the effect of its loss on safety and production.

**6.3.2** There are entire systems that might be critical by their very nature. Depending on the size and complexity of the operation, a plant can contain any or all of the following examples: emergency power, emergency lighting, fire-alarm systems, fire pumps, and certain communications systems. There should be no problem in establishing whether a system is critical and in having the proper amount of emphasis placed on its maintenance.

**6.3.3** More difficult to identify are the parts of a system that are critical because of the function of the utilization equipment and its associated hardware. Some examples are as follows:

(1) The agitator drive motor for a kettle-type reactor can be extremely critical in that, if it fails to run for some period of time, when the charge materials are added to the reactor, the catalyst stratifies. If the motor is then started, a rapid reaction, rather than a slow, controlled reaction, could result that might run away, overpressurize, and destroy the reactor.
(2) The cooling water source of an exothermic reactor might have associated with it some electrical equipment such as a drive motor, solenoid valves, controls, or the like. Failure of the cooling water might allow the exothermic reaction to go beyond the stable point and overpressurize and destroy the vessel.
(3) A process furnace recirculating fan drive motor or fan might fail, nullifying the effects of temperature-sensing points and thus allowing hot spots to develop, with serious side reactions.
(4) The failure of gas analysis equipment and interlocks in a drying oven or annealing furnace might allow the atmosphere in the drying oven or furnace to become flammable, with the possibility of an explosion.
(5) The failure of any of the safety combustion controls on a large firebox, such as a boiler or an incinerator, can cause a serious explosion.
(6) Two paralleled pump motors might be needed to provide the total requirements of a continuous process. Failure of either motor can cause a complete shutdown, rather than simply reduce production.

**6.3.4** There are parts of the system that are critical because they reduce the widespread effect of a fault in electrical equipment. The determination of these parts should be primarily the responsibility of the electrical person on the team. Among the things that fall into this category are the following:

(1) Source overcurrent protective devices, such as circuit breakers or fuses, including the relays, control circuits, and coordination of trip characteristics of the devices
(2) Automatic bus transfer switches or other transfer switches that would supply critical loads with power from the emergency power source if the primary source failed; includes instrument power supplies as well as load power supplies

**6.3.5** Parts of the control system are critical because they monitor the process and automatically shut down equipment or take other action to prevent catastrophe. These items are the interlocks, cutout devices, or shutdown devices installed throughout the plant or operation. Each interlock or shutdown device should be considered carefully by the entire team to establish whether it is a critical shutdown or a "convenience" shutdown. The maintenance group should thoroughly understand which shutdowns are critical and which are convenience. Critical shutdown devices are normally characterized by a sensing device separate from the normal control device. They probably have separate, final, or end devices that cause action to take place. Once the critical shutdown systems have been determined, they should be distinctly identified on drawings, on records, and on the hardware itself. Some examples of critical shutdown devices are overspeed trips; high or low temperature, pressure, flow, or level trips; low-lube-oil pressure trips; pressure-relief valves; over-current trips; and low-voltage trips.

**6.3.6** There are parts of the system that are critical because they alert operating personnel to dangerous or out-of-control conditions. These are normally referred to as alarms. Like shutdown devices, alarms fall into at least three categories: (1) those that signify a true pending catastrophe, (2) those that indicate out-of-control conditions, and (3) those that indicate the end of an operation or similar condition. The entire team should consider each alarm in the system with the same thoroughness with which they have considered the shutdown circuits. A truly critical alarm should be characterized by its separate sensing device, a separate readout device, and, preferably, separate circuitry and power source. The maintenance department should thoroughly understand the critical level of each alarm. The critical alarms and their significance should be distinctly marked on drawings, in records, and on the operating unit. For an alarm to be critical does not necessarily mean that it is complex or related to complex action. A simple valve position indicator can be one of the most critical alarms in an operating unit.

**6.4 Establishment of a Systematic Program.**

The purpose of any inspection and testing program is to establish the condition of equipment to determine what work should be done and to verify that it will continue to function until the next scheduled servicing occurs. Inspection and testing are best done in conjunction with routine maintenance. In this way, many minor items that require no special tools, training, or equipment can be corrected as they are found. The inspection and testing program is probably the most important function of a maintenance department in that it establishes what should be done to keep the system in service to perform the function for which it is required.

**6.4.1 Atmosphere or Environment.**

**6.4.1.1** The atmosphere or environment in which electrical equipment is located has a definite effect on its operating capabilities and the degree of maintenance required. An ideal environment is one in which the air is (1) clean or filtered to remove dust, harmful vapor, excess moisture, and so on; (2) maintained in the temperature range of 15°C to 29°C (60°F to 85°F); and (3) in the range of 40 percent to 70 percent humidity. Under such conditions, the need for maintenance will be minimized. Where these conditions are not maintained, the performance of electrical equipment will be adversely affected. Good housekeeping contributes to a good environment and reduced maintenance.

**6.4.1.2** Dust can foul cooling passages and thus reduce the capabilities of motors, transformers, switchgear, and so on, by raising their operating temperatures above rated limits, decreasing operating efficiencies, and increasing fire hazard. Similarly, chemicals and vapors can coat and reduce the heat transfer capabilities of heating and cooling equipment. Chemicals, dusts, and vapors can be highly flammable, explosive, or conductive, increasing the hazard of fire, explosion, ground faults, and short circuits. Chemicals and corrosive vapors can cause high contact resistance that will decrease contact life and increase contact power losses with possible fire hazard or false overload conditions due to excess heat. Large temperature changes combined with high humidity can cause condensation problems, malfunction of operating and safety devices, and lubrication problems. High ambient temperatures in areas where thermally sensitive protective equipment is located can cause such protective equipment to operate below its intended operating point. Ideally, both the electrical apparatus and its protective equipment should be located within the same ambient temperature. Where the ambient-temperature difference between equipment and its protective device is extreme, compensation in the protective equipment should be made.

**6.4.1.3** Electrical equipment installed in hazardous (classified) locations as described in NFPA 70, *National Electrical Code,* requires special maintenance considerations. *(See Section 23.2.)*

**6.4.2 Load Conditions.**

**6.4.2.1** Equipment is designed and rated to perform satisfactorily when subjected to specific operating and load conditions. A motor designed for safe continuous operation at rated load might not be satisfactory for frequent intermittent operation, which can produce excessive winding temperatures or mechanical trouble. The resistance grid or transformer of a reduced-voltage starter will overheat if left in the starting position. So-called "jogging" or "inching" service imposes severe demands on equipment such as motors, starters, and controls. Each type of duty influences the type of equipment used and the extent of maintenance required. The five most common types of duty are defined in NFPA 70, *National Electrical Code,* and they are repeated in 6.4.2.2.

**6.4.2.2** The following definitions can be found in Chapter 3 and are unique to this chapter:

(1) Continuous duty *(See 3.3.13.1.)*
(2) Intermittent duty *(See 3.3.13.2.)*
(3) Periodic duty *(See 3.3.13.3.)*
(4) Short-time duty *(See 3.3.13.4.)*
(5) Varying duty *(See 3.3.13.5.)*

**6.4.2.3** Some devices used in establishing a proper maintenance period are running-time meters (to measure total "on" or "use" time); counters to measure number of starts, stops, or load-on, load-off, and rest periods; and recording ammeters to graphically record load and no-load conditions. These devices can be applied to any system or equipment and will help classify the duty. They will help establish a proper frequency of preventive maintenance.

**6.4.2.4** Safety and limit controls are devices whose sole function is to ensure that values remain within the safe design level of the system. Because these devices function only during an abnormal situation in which an undesirable or unsafe condition is reached, each device should be periodically and carefully inspected, checked, and tested to be certain that it is in reliable operating condition.

**6.4.3** Wherever practical, a history of each electrical system should be developed for all equipment or parts of a system vital to a plant's operation, production, or process. The record should include all pertinent information for proper operation and maintenance. This information is useful in developing repair cost trends, items replaced, design changes or modifications, significant trouble or failure patterns, and replacement parts or devices that should be stocked. System and equipment information should include the following:

(1) Types of electrical equipment, such as motors, starters, contactors, heaters, relays
(2) Types of mechanical equipment, such as valves, controls, and so on, and driven equipment, such as pumps, compressors, fans, and whether they are direct, geared, or belt driven
(3) Nameplate data
(4) Equipment use
(5) Installation date
(6) Available replacement parts
(7) Maintenance test and inspection dates: type and frequency of lubrication; electrical inspections, test, and repair; mechanical inspections, test, and repair; replacement parts list with manufacturer's identification; electrical and mechanical drawings for assembly, repair, and operation

**6.4.4 Inspection Frequency.** Those pieces of equipment found to be critical should require the most frequent inspections and tests. Depending on the degree of reliability required, other items can be inspected and tested much less frequently.

**6.4.4.1** Manufacturers' service manuals should have a recommended frequency of inspection. The frequency given is based on standard or usual operating conditions and environments. It would be impossible for a manufacturer to list all combinations of environmental and operating conditions. However, a manufacturer's service manual is a good basis from which to begin considering the frequency for inspection and testing.

**6.4.4.2** There are several points to consider in establishing the initial frequency of inspections and tests. Electrical equipment located in a separate air-conditioned control room or switch room certainly would not be considered normal, so the inspection interval might be extended 30 percent. However, if the equipment is located near another unit or operating plant that discharges dust or corrosive vapors, this time might be reduced by as much as 50 percent.

**6.4.4.3** Continuously operating units with steady loads or with less than the rated full load tend to operate much longer and more reliably than intermittently operated or standby units. For this reason, the interval between inspections might be extended 10 to 20 percent for continuously operating equipment and possibly reduced by 20 to 40 percent for standby or infrequently operated equipment.

**6.4.4.4** Once the initial frequency for inspection and tests has been established, this frequency should be adhered to for at least four maintenance cycles unless undue failures occur. For equipment that has unexpected failures, the interval between inspections should be reduced by 50 percent as soon as the trouble occurs. On the other hand, after four cycles of inspections have been completed, a pattern should have developed. If equipment consistently goes through more than two inspections without requiring service, the inspection period can be extended by 50 percent. Loss of production due to an emergency shutdown is almost always more expensive than loss of production due to a planned shutdown. Accordingly, the interval between inspections should be planned to avoid the diminishing returns of either too long or too short an interval.

**6.4.4.5** Adjustment in the interval between inspections should continue until the optimum interval is reached. This adjustment time can be minimized and the optimum interval approximated more closely initially by providing the person responsible for establishing the first interval with as much pertinent history and technology as possible.

**6.4.4.6** The frequency of inspection for similar equipment operating under differing conditions can differ widely. Typical examples are as follows:

(1) In a continuously operating plant having a good load factor and located in a favorable environment, the high-voltage oil circuit breakers might need an inspection only every two years. On the other hand, an electrolytic process plant using similar oil circuit breakers for controlling furnaces might find it necessary to inspect and service them as frequently as every 7 to 10 days.
(2) An emergency generator to provide power for noncritical loads can be tested on a monthly basis. Yet the same generator in another plant having processes sensitive to explosion on loss of power might need to be tested during each shift.

### 6.5 Methods and Procedures.

#### 6.5.1 General.

**6.5.1.1** If a system is to operate without failure, not only should the discrete components of the system be maintained, but the connections between these components also should be covered by a thorough set of methods and procedures. Overlooking this important link in the system causes many companies to suffer high losses every year.

**6.5.1.2** Other areas where the maintenance department should develop its own procedures are shutdown safeguards, interlocks, and alarms. Although the individual pieces of equipment can have testing and calibrating procedures furnished by the manufacturer, the application is probably unique, so the system per se should have an inspection and testing procedure developed for it.

#### 6.5.2 Forms.

**6.5.2.1** A variety of forms can go along with the inspection, testing, and repair (IT&R) procedure; these forms should be detailed and direct, yet simple and durable enough to be used in the field. Field notes should be legibly transcribed. One copy of reports should go in the working file of the piece of equipment and one in the master file maintained by first line supervision. These forms should be used by the electrical maintenance personnel; they are not for general distribution. If reports to production or engineering are needed, they should be separate, and inspection reports should not be used.

**6.5.2.2** The IT&R procedure folder for a piece of equipment should list the following items:

(1) All the special tools, materials, and equipment necessary to do the job
(2) The estimated or actual average time to do the job
(3) Appropriate references to technical manuals
(4) Previous work done on the equipment
(5) Points for special attention indicated by previous IT&R
(6) If major work was predicted at the last IT&R, a copy of the purchase order and receiving reports for the parts to do the work and references to unusual incidents reported by production that might be associated with the equipment

**6.5.2.3** Special precautions relative to operation, such as the following, should be part of the IT&R document:

(1) What other equipment is affected and in what way?
(2) Who has to be informed that the IT&R is going to be done?
(3) How long will the equipment be out of service if all goes well? How long if major problems are uncovered?

#### 6.5.3 Planning.

**6.5.3.1** After the IT&R procedures have been developed and the frequency has been established (even though preliminary), the task of scheduling should be handled. Scheduling in a continuous-process plant (as opposed to a batch-process plant) is most critically affected by availability of equipment in blocks consistent with maintenance manpower capabilities. In general, facilities should be shut down on some regular basis for overall maintenance and repair. Some of the electrical maintenance items should be done at this time. IT&R that could be done while equipment is in service should be done prior to shutdown. Only work that needs to be done during shutdown should be scheduled at that time, to level out manpower requirements and limit downtime.

**6.5.3.2** The very exercise of scheduling IT&R will point out design weaknesses that require excessive manpower during critical shutdown periods or that require excessive downtime to do the job with the personnel available. Once these weaknesses have been uncovered, consideration can be given to rectifying them. For example, the addition of one circuit breaker and a little cable can change a shutdown from three days to one day.

**6.5.3.3** Availability of spare equipment affects scheduling in many ways. Older facilities might have installed spares for a major part of the equipment, or the facility might be made up of many parallel lines so that they can be shut down, one at a time, without seriously curtailing operations. This concept is particularly adaptable to electrical distribution. The use of a circuit breaker and a transfer bus can extend the interval between total shutdown on a main transformer station from once a year to once in 5 years or more.

**6.5.3.4** In many continuous-process plants, particularly newer ones, the trend is toward a large single-process line with no installed spares. This method of operation requires inspections and tests, since there will be a natural desire to extend the time between maintenance shutdowns. Downtime in such plants is particularly costly, so it is desirable to build as much monitoring into the electrical systems as possible.

**6.5.3.5** Planning running inspections can vary from a simple desk calendar to a computer program. Any program for scheduling should have the following four facets:

(1) A reminder to order parts and equipment with sufficient lead time to have them on the job when needed
(2) The date and man-hours to do the job
(3) A check to see that the job has been completed
(4) Noticing if parts will be needed for the next IT&R and when they should be ordered

**6.5.3.6** Planning shutdown IT&R is governed by the time between shutdowns, established by the limitations of the process or production units involved. Reliability of electrical equipment can and should be built in to correspond to almost any length of time.

**6.5.3.7** Small plants should use, in an abbreviated form, the following shutdown recommendations of a large plant IT&R.

(1) Know how many personnel-shifts the work will take.
(2) Know how many persons will be available.
(3) Inform production of how many shifts the electrical maintenance will require.
(4) Have all the necessary tools, materials, and spare parts assembled on the job site. Overage is better than shortage.
(5) Plan the work so that each person is used to best suit his or her skills.
(6) Plan what each person will be doing during each hour the shutdown. Allow sufficient off time so that if a job not finished as scheduled, the person working on that job can be held over without becoming overtired for the next shift. This procedure will allow the schedule to be kept.
(7) Additional clerical people during shutdown IT&R will make the job go more smoothly, help prevent omission some important function, and allow an easier transition back to normal.
(8) Supply copies of the electrical group plan to the overall shutdown coordinator so it can be incorporated into the overall plan. The overall plan should be presented in form that is easy to use by all levels of supervision. In large, complex operation, a critical path program or some similar program should be used.

**6.5.3.8** Automatic shutdown systems and alarm systems that have been determined as critical should be designed and maintained so that nuisance tripping does not destroy operator confidence. Loss of operator confidence can and will cause these systems to be bypassed and the intended safety lost. Maintenance should prove that each operation was valid and caused by an unsafe condition.

**6.5.3.9** A good electrical preventive maintenance program should identify the less critical jobs, so it is clear to first-line supervision which EPM can be delayed to make personnel available for emergency breakdown repair.

**6.5.4 Analysis of Safety Procedures.**

**6.5.4.1** It is beyond the scope of this recommended practice to cover the details of safety procedures for each IT&R activity. Manufacturers' instructions contain safety procedures required in using their test equipment.

**6.5.4.2** The test equipment (high voltage, high current, or other uses) should be inspected in accordance with vendor recommendations before the job is started. Any unsafe condition should be corrected before proceeding.

**6.5.4.3** The people doing the IT&R should be briefed to be sure that all facets of safety before, during, and after the IT&R are understood. It is important that all protective equipment is in good condition and is on the job.

**6.5.4.4** Screens, ropes, guards, and signs needed to protect people other than the IT&R team should be provided and used.

**6.5.4.5** A procedure should be developed, understood, and used for leaving the test site in a safe condition when unattended at times such as a smoke break, a lunch break, or overnight.

**6.5.4.6** A procedure should be developed, understood, and used to ensure safety to and from the process before, during, and after the IT&R. The process or other operation should be put in a safe condition for the IT&R by the operating people before the work is started. The procedure should include such checks as are necessary to ensure that the unit is ready for operation after the IT&R is completed and before the operation is restarted.

**6.5.5 Records.**

**6.5.5.1 General.** Sufficient records should be kept by maintenance management to evaluate results. Analysis of the records should guide the spending level for EPM and breakdown repair.

**6.5.5.2 Records of Cost.** Figures should be kept to show the total cost of each breakdown. This should be the actual cost plus an estimated cost of the business interruption. This figure is a powerful indicator for the guidance of expenditures for EPM.

**6.5.5.3 Records Kept by First-Line Supervisor of EPM.** Of the many approaches to this phase of the program, the following approach is a typical one that fulfills the minimum requirements.

**(A) Inspection Schedule.** The first-line supervisor should maintain, in some easy-to-use form, a schedule of inspections so that he or she can plan manpower requirements.

**(B) Work Order Log.** An active log should be kept of unfinished work orders. A greater susceptibility to imminent breakdown is indicated by a large number of outstanding work orders resulting from the inspection function.

**(C) Unusual Event Log.** As the name implies, this log lists unusual events that affect the electrical system in any way. This record is derived from reports of operating and other personnel and is a good tool for finding likely problems after the supervisor has learned to interpret and evaluate the reports. Near misses can be recorded and credit given for averting trouble.

**6.5.6 Emergency Procedures.** It should be recognized that properly trained electrical maintenance personnel have the potential to make an important contribution in the emergency situations that are most likely to occur. However, most such situations will also involve other crafts and disciplines, such as operating personnel, pipe fitters, and mechanics. An overall emergency procedure for each anticipated emergency situation should be developed cooperatively by the qualified personnel of each discipline involved, detailing steps to be followed, sequence of steps, and assignment of responsibility. The total procedure should then be run periodically as an emergency drill to ensure that all involved personnel are kept thoroughly familiar with the tasks they are to perform.

**6.6 Maintenance of Imported Electrical Equipment.**

Imported equipment poses some additional maintenance considerations.

**6.6.1** Quick delivery of replacement parts cannot be taken for granted. Suppliers should be identified, and the replacement parts problem should be reflected in the in-plant spare parts inventory. In addition to considering possible slow delivery of replacement parts, knowledgeable outside sources of engineering services for the imported equipment should be established.

**6.6.2** Parts catalogs, maintenance manuals, and drawings should be available in the language of the user. Documents created in a different language and then translated should not be presumed to be understandable. Problems in translation should be identified as soon as literature is received to ensure that material will be fully understood later, when actual maintenance must be performed.

**6.7 Maintenance of Electrical Equipment for Use in Hazardous (Classified) Locations.**

See Section 23.2.

# SUPPLEMENT 2

# Safe Electrical Installations

***Editor's Note:*** *This supplement is an extract that provides some requirements and commentary from the 2008 edition of the National Electrical Code® Handbook. Worker safety begins with the use of safe products and safe installations. Product standards provide construction criteria that products must meet to be listed by a qualified testing laboratory. They also provide the test criteria for conformity assessment. Since 1897, NFPA 70®, National Electrical Code® (NEC®), has provided rules for safe electrical installations. The code rules provide safety from electric shock and fire hazards.*

*All of the prior editions of NFPA 70E®, Standard for Electrical Safety in the Workplace®, contained some of the installation rules from the NEC. Although the entire code applies to a safe installation, some have a more direct bearing on worker safety. Examples include space around electrical equipment, disconnecting means, warning signs, grounding, and temporary wiring. Requirements for space around electrical equipment are intended to provide sufficient space for a worker to work on equipment, while minimizing exposure to hazard. Requirements for access and exits provide a means to escape the area if there is an arcing fault in equipment. Requirements for accessible disconnecting means are located in several articles in the code. These requirements are intended to protect workers by ensuring that the disconnect is available and accessible. Requirements also exist for lockable disconnects, which a worker can apply a lock to as part of a lockout/tagout program. These worker safety requirements are an essential part of the installation code to make sure that they are part of the initial installation.*

*The NEC and NFPA 70E are on different adoption cycles. The different cycles could result in conflicts in rules between the two documents because NFPA 70E could exist for a time with a requirement that was based on a different edition of the code. In addition, the entire NEC applies, not just those sections that apply to worker safety. The NEC rules were provided as a convenience because, as of the 2009 edition, the NEC-based requirements no longer form a part of the text of NFPA 70E. For the purpose of this handbook, it is useful to provide some of those requirements that affect persons working on existing electrical installations.*

*While the entire code does apply, the following code sections may require special attention for workers who may be exposed to an electrical hazard. Many of these requirements also have explanatory commentary, which is shown in blue. The reader should also note that any references to other sections or to tables and figures are in the 2008 edition of the National Electrical Code Handbook and not this handbook.*

## Article 110 Requirements for Electrical Installations

**110.2 Approval.** The conductors and equipment required or permitted by this *Code* shall be acceptable only if approved.

FPN: See 90.7, Examination of Equipment for Safety, and 110.3, Examination, Identification, Installation, and Use of Equipment. See definitions of *Approved, Identified, Labeled,* and *Listed.*

All electrical equipment is required to be approved as defined in Article 100 and, as such, to be acceptable to the authority having jurisdiction (also defined in Article 100). Section 110.3 provides guidance for the evaluation of equipment and recognizes listing or labeling as a means of establishing suitability.

Approval of equipment is the responsibility of the electrical inspection authority, and many such approvals are based on tests and listings of testing laboratories. Unique equipment is often approved following a field evaluation by a qualified third-party laboratory or qualified individual.

**110.3 Examination, Identification, Installation, and Use of Equipment.**

For wire-bending and connection space in cabinets and cutout boxes, see 312.6, Table 312.6(A), Table 312.6(B), 312.7, 312.9, and 312.11. For wire-bending and connection space in other equipment, see the appropriate *NEC* article and section. For example, see 314.16 and 314.28 for outlet, device, pull, and junction boxes, as well as conduit bodies; 404.3 and 404.18 for switches; 408.3(F) for switchboards and panelboards; and 430.10 for motors and motor controllers.

**(A) Examination.** In judging equipment, considerations such as the following shall be evaluated:

(1) Suitability for installation and use in conformity with the provisions of this *Code*

FPN: Suitability of equipment use may be identified by a description marked on or provided with a product to identify the suitability of the product for a specific purpose, environment, or application. Suitability of equipment may be evidenced by listing or labeling.

(2) Mechanical strength and durability, including, for parts designed to enclose and protect other equipment, the adequacy of the protection thus provided
(3) Wire-bending and connection space
(4) Electrical insulation
(5) Heating effects under normal conditions of use and also under abnormal conditions likely to arise in service
(6) Arcing effects
(7) Classification by type, size, voltage, current capacity, and specific use
(8) Other factors that contribute to the practical safeguarding of persons using or likely to come in contact with the equipment

**(B) Installation and Use.** Listed or labeled equipment shall be installed and used in accordance with any instructions included in the listing or labeling.

Manufacturers usually supply installation instructions with equipment for use by general contractors, erectors, electrical contractors, electrical inspectors, and others concerned with an installation. It is important to follow the listing or labeling installation instructions. For example, 210.52, second paragraph, permits permanently installed electric baseboard heaters to be equipped with receptacle outlets that meet the requirements for the wall space utilized by such heaters. The installation instructions for such permanent baseboard heaters indicate that the heaters should not be mounted beneath a receptacle. In dwelling units, it is common to use low-density heating units that measure in excess of 12 ft in length. Therefore, to meet the provisions of 210.52(A) and also the installation instructions, a receptacle must either be part of the heating unit or be installed in the floor close to the wall but not above the heating unit. (See 210.52, FPN, and Exhibit 210.24 for more specific details.)

In itself, 110.3 does not require listing or labeling of equipment. It does, however, require considerable evaluation of equipment. Section 110.2 requires that equipment be acceptable only if approved. The term *approved* is defined in Article 100 as acceptable to the authority having jurisdiction (AHJ). Before issuing approval, the authority having jurisdiction may require evidence of compliance with 110.3(A). The most common form of evidence considered acceptable by authorities having jurisdiction is a listing or labeling by a third party.

Some sections in the *Code* require listed or labeled equipment. For example, 250.8 specifies "listed pressure connectors, pressure connectors listed as grounding and bonding equipment, or other listed means" as connection methods for grounding and bonding conductors.

**110.7 Wiring Integrity.** Completed wiring installations shall be free from short circuits, ground faults, or any connections to ground other than as required or permitted elsewhere in this *Code*.

Insulation is the material that prevents the flow of electricity between points of different potential in an electrical system. Failure of the insulation system is one of the most common causes of problems in electrical installations, in both high-voltage and low-voltage systems.

Insulation tests are performed on new or existing installations to determine the quality or condition of the insulation of conductors and equipment. The principal causes of insulation failures are heat, moisture, dirt, and physical damage (abrasion or nicks) occurring during and after installation. Insulation can also fail due to chemical attack, sunlight, and excessive voltage stresses.

Insulation integrity must be maintained during overcurrent conditions. Overcurrent protective devices must be selected and coordinated using tables of insulation thermal-withstand ability to ensure that the damage point of an insulated conductor is never reached. These tables, entitled "Allowable Short-Circuit Currents for Insulated Copper (or Aluminum) Conductors," are contained in the Insulated Cable Engineers Association's publication ICEA P-32-382. See 110.10 for other circuit components.

In an insulation resistance test, a voltage ranging from 100 to 5000 (usually 500 to 1000 volts for systems of 600 volts or less), supplied from a source of constant potential, is applied across the insulation. A megohmmeter is usually the potential source, and it indicates the insulation resistance directly on a scale calibrated in megohms (MΩ). The quality of the insulation is evaluated based on the level of the insulation resistance.

The insulation resistance of many types of insulation varies with temperature, so the field data obtained should be corrected to the standard temperature for the class of equipment being tested. The megohm value of insulation resistance obtained is inversely proportional to the volume of insulation tested. For example, a cable 1000 ft long would be expected to have one-tenth the insulation resistance of a cable 100 ft long if all other conditions are identical.

The insulation resistance test is relatively easy to perform and is useful on all types and classes of electrical equipment. Its main value lies in the charting of data from periodic tests, corrected for temperature, over a long period so that deteriorative trends can be detected.

Manuals on this subject are available from instrument manufacturers. Thorough knowledge in the use of insulation testers is essential if the test results are to be meaningful. Exhibit 110.1 shows a typical megohmmeter insulation tester.

**110.9 Interrupting Rating.** Equipment intended to interrupt current at fault levels shall have an interrupting rating sufficient for the nominal circuit voltage and the current that is available at the line terminals of the equipment.

Equipment intended to interrupt current at other than fault levels shall have an interrupting rating at nominal circuit voltage sufficient for the current that must be interrupted.

The interrupting rating of overcurrent protective devices is determined under standard test conditions. It is important that the test conditions match the actual installation needs. Section 110.9 requires that all fuses and circuit breakers intended to interrupt the circuit at fault levels have an adequate interrupting rating wherever they are used in the electrical system. Fuses or circuit breakers that do not have adequate interrupting ratings could rupture while attempting to clear a short circuit.

Interrupting ratings should not be confused with short-circuit current ratings. Short-circuit current ratings are further explained in the commentary following 110.10.

**110.10 Circuit Impedance and Other Characteristics.** The overcurrent protective devices, the total impedance, the component short-circuit current ratings, and other characteristics of the circuit to be protected shall be selected and coordinated to permit the circuit-protective devices used to clear a fault to do so without extensive damage to the electrical components of the circuit. This fault shall be assumed to be either between two or more of the circuit conductors or between any circuit conductor and the grounding conductor or enclosing metal raceway. Listed products applied in accordance with their listing shall be considered to meet the requirements of this section.

Short-circuit current ratings are marked on equipment such as panelboards, switchboards, busways, contactors, and starters. The last sentence of 110.10 is meant to address concerns of what exactly constitutes "extensive damage." Because, under product safety requirements, electrical equipment is evaluated for indications of extensive damage, listed products used within their ratings are considered to have met the requirements of 110.10.

The basic purpose of overcurrent protection is to open the circuit before conductors or conductor insulation is damaged when an overcurrent condition occurs. An overcurrent condition can be the result of an overload, a ground fault, or a short circuit and must be eliminated before the conductor insulation damage point is reached.

Overcurrent protective devices (such as fuses and circuit breakers) should be selected to ensure that the short-circuit current rating of the system components is not exceeded should a short circuit or high-level ground fault occur.

System components include wire, bus structures, switching, protection and disconnect devices, and distribution equipment, all of which have limited short-circuit ratings and would be damaged or destroyed if those short-circuit ratings were exceeded. Merely providing overcurrent protective devices with sufficient interrupting ratings would not ensure adequate short-circuit protection for the system components. When the available short-circuit current exceeds the short-circuit current rating of an electrical component, the overcurrent protective device must limit the let-through energy to within the rating of that electrical component.

Utility companies usually determine and provide information on available short-circuit current levels at the service equipment. Literature on how to calculate short-circuit currents at each point in any distribution generally can be obtained by contacting the manufacturers of overcurrent protective devices or by referring to IEEE 141-1993, *IEEE Recommended Practice for Electric Power Distribution for Industrial Plants* (Red Book).

For a typical one-family dwelling with a 100-ampere service using 2 AWG aluminum supplied by a 37½ kVA transformer with 1.72 percent impedance located at a distance of 25 ft, the available short-circuit current would be approximately 6000 amperes.

Available short-circuit current to multifamily structures, where pad-mounted transformers are located close to the multimetering location, can be relatively high. For example, the line-to-line fault current values close to a low-impedance transformer could exceed 22,000 amperes. At the secondary of a single-phase, center-tapped transformer, the line-to-neutral fault current is approximately one and one-half times that of the line-to-line fault current. The short-circuit current rating of utilization equipment located and connected near the service equipment should be known. For example, HVAC equipment is tested at 3500 amperes through a 40-ampere load rating and at 5000 amperes for loads rated more than 40 amperes.

Adequate short-circuit protection can be provided by fuses, molded-case circuit breakers, and low-voltage power circuit breakers, depending on specific circuit and installation requirements.

**110.11 Deteriorating Agents.** Unless identified for use in the operating environment, no conductors or equipment shall be located in damp or wet locations; where exposed to gases, fumes, vapors, liquids, or other agents that have a deteriorating effect on the conductors or equipment; or where exposed to excessive temperatures.

FPN No. 1: See 300.6 for protection against corrosion.

FPN No. 2: Some cleaning and lubricating compounds can cause severe deterioration of many plastic materials used for insulating and structural applications in equipment.

Equipment not identified for outdoor use and equipment identified only for indoor use, such as "dry locations," "indoor use only," "damp locations," or enclosure Types 1, 2, 5, 12, 12K, and/or 13, shall be protected against permanent damage from the weather during building construction.

FPN No. 3: See Table 110.20 for appropriate enclosure-type designations.

This section was expanded for the 2008 *Code* to make it clear that several enclosure types must be protected from the weather during construction.

**110.12 Mechanical Execution of Work.** Electrical equipment shall be installed in a neat and workmanlike manner.

FPN: Accepted industry practices are described in ANSI/NECA 1-2006, *Standard Practices for Good Workmanship in Electrical Contracting,* and other ANSI-approved installation standards.

The requirement in 110.12 calling for "neat and workmanlike" installations has appeared in the *NEC* as currently worded for more than a half-century. It stands as a basis for pride in one's work and has been emphasized by persons involved in the training of apprentice electricians for many years.

Many *Code* conflicts or violations have been cited by the authority having jurisdiction based on the authority's interpretation of "neat and workmanlike manner." Many electrical inspection authorities use their own experience or precedents in their local areas as the basis for their judgments.

### Application Example

Installations that do not qualify as "neat and workmanlike" include exposed runs of cables or raceways that are improperly supported (e.g., sagging between supports or use of improper support methods); field-bent and kinked, flattened, or poorly measured raceways; or cabinets, cutout boxes, and enclosures that are not plumb or not properly secured.

The FPN directs the user to an industry-accepted ANSI standard that clearly describes and illustrates "neat and workmanlike" electrical installations. See Exhibit 110.2.

**(A) Unused Openings.** Unused openings, other than those intended for the operation of equipment, those intended for mounting purposes, or those permitted as part of the design for listed equipment, shall be closed to afford protection substantially equivalent to the wall of the equipment. Where metallic plugs or plates are used with nonmetallic enclosures, they shall be recessed at least 6 mm (¼ in.) from the outer surface of the enclosure.

This section was revised for the 2008 *Code*. It now requires all *unused openings* other than those openings used for mounting, cooling, or drainage to be closed up.

See 408.7 for requirements on unused openings in switchboard and panelboard enclosures.

**(B) Integrity of Electrical Equipment and Connections.** Internal parts of electrical equipment, including busbars, wiring terminals, insulators, and other surfaces, shall not be damaged or contaminated by foreign materials such as paint, plaster, cleaners, abrasives, or corrosive residues. There shall be no damaged parts that may adversely affect safe operation or mechanical strength of the equipment such as parts that are broken; bent; cut; or deteriorated by corrosion, chemical action, or overheating.

**110.13 Mounting and Cooling of Equipment.**

**(A) Mounting.** Electrical equipment shall be firmly secured to the surface on which it is mounted. Wooden plugs driven into holes in masonry, concrete, plaster, or similar materials shall not be used.

**(B) Cooling.** Electrical equipment that depends on the natural circulation of air and convection principles for cooling of exposed surfaces shall be installed so that room airflow over such surfaces is not prevented by walls or by adjacent installed equipment. For equipment designed for floor mounting, clearance between top surfaces and adjacent surfaces shall be provided to dissipate rising warm air.

Electrical equipment provided with ventilating openings shall be installed so that walls or other obstructions do not prevent the free circulation of air through the equipment.

*Ventilated* is defined in Article 100. Panelboards, transformers, and other types of equipment are adversely affected if enclosure surfaces normally exposed to room air are covered or tightly enclosed. Ventilating openings in equipment are provided to allow the circulation of room air around internal components of the equipment; the blocking of such openings can cause dangerous overheating. For example, a ventilated busway must be located where there are no walls or other objects that might interfere with the natural circulation of air and convection principles for cooling. Ventilation for motor locations is covered in 430.14(A) and 430.16. Ventilation for transformer locations is covered in 450.9 and 450.45. In addition to 110.13, proper placement of equipment requiring ventilation becomes enforceable using the requirements of 110.3(B).

**110.14 Electrical Connections.** Because of different characteristics of dissimilar metals, devices such as pressure terminal or pressure splicing connectors and soldering lugs shall be identified for the material of the conductor and shall be properly installed and used. Conductors of dissimilar metals shall not be intermixed in a terminal or splicing connector where physical contact occurs between dissimilar conductors (such as copper and aluminum, copper and copper-clad aluminum, or aluminum and copper-clad aluminum), unless the device is identified for the purpose and conditions of use. Materials such as solder, fluxes, inhibitors, and compounds, where employed, shall be suitable for the use and shall be of a type that will not adversely affect the conductors, installation, or equipment.

FPN: Many terminations and equipment are marked with a tightening torque.

Section 110.3(B) applies where terminations and equipment are marked with tightening torques.

For the testing of wire connectors for which the manufacturer has not assigned another value appropriate for the design, Commentary Tables 110.1 through 110.4 provide data on the tightening torques that Underwriters Laboratories uses. These tables should be used for guidance only if no tightening information on a specific wire connector is available. They should not be used to replace the manufacturer's instructions, which should always be followed.

The information in the tables was taken from UL 486A-486B, *Wire Connectors.*

**(A) Terminals.** Connection of conductors to terminal parts shall ensure a thoroughly good connection without damaging the conductors and shall be made by means of pressure connectors (including set-screw type), solder lugs, or splices to flexible leads. Connection by means of wire-binding screws or studs and nuts that have upturned lugs or the equivalent shall be permitted for 10 AWG or smaller conductors.

Terminals for more than one conductor and terminals used to connect aluminum shall be so identified.

**(B) Splices.** Conductors shall be spliced or joined with splicing devices identified for the use or by brazing, welding, or soldering with a fusible metal or alloy. Soldered splices shall first be spliced or joined so as to be mechanically and electrically secure without solder and then be soldered. All splices and joints and the free ends of conductors shall be covered with an insulation equivalent to that of the conductors or with an insulating device identified for the purpose.

Wire connectors or splicing means installed on conductors for direct burial shall be listed for such use.

Field observations and trade magazine articles indicate that electrical connection failures have been determined to be the cause of many equipment burnouts and fires. Many of these failures are attributable to improper terminations, poor workmanship, the differing characteristics of dissimilar metals, and improper binding screws or splicing devices.

**(C) Temperature Limitations.** The temperature rating associated with the ampacity of a conductor shall be selected and coordinated so as not to exceed the lowest temperature rating of any connected termination, conductor, or device. Conductors with temperature ratings higher than specified for terminations shall be permitted to be used for ampacity adjustment, correction, or both.

**(1) Equipment Provisions.** The determination of termination provisions of equipment shall be based on 110.14(C)(1)(a) or (C)(1)(b). Unless the equipment is listed and marked otherwise, conductor ampacities used in determining equipment termination provisions shall be based on Table 310.16 as appropriately modified by 310.15(B)(6).

(a) Termination provisions of equipment for circuits rated 100 amperes or less, or marked for 14 AWG through 1 AWG conductors, shall be used only for one of the following:

(1) Conductors rated 60°C (140°F).
(2) Conductors with higher temperature ratings, provided the ampacity of such conductors is determined based on the 60°C (140°F) ampacity of the conductor size used.
(3) Conductors with higher temperature ratings if the equipment is listed and identified for use with such conductors.
(4) For motors marked with design letters B, C, or D, conductors having an insulation rating of 75°C (167°F) or higher shall be permitted to be used, provided the ampacity of such conductors does not exceed the 75°C (167°F) ampacity.

(b) Termination provisions of equipment for circuits rated over 100 amperes, or marked for conductors larger than 1 AWG, shall be used only for one of the following:

(1) Conductors rated 75°C (167°F)
(2) Conductors with higher temperature ratings, provided the ampacity of such conductors does not exceed the 75°C (167°F) ampacity of the conductor size used, or up to their ampacity if the equipment is listed and identified for use with such conductors

**(2) Separate Connector Provisions.** Separately installed pressure connectors shall be used with conductors at the ampacities not exceeding the ampacity at the listed and identified temperature rating of the connector.

FPN: With respect to 110.14(C)(1) and (C)(2), equipment markings or listing information may additionally restrict the sizing and temperature ratings of connected conductors.

Section 110.14(C)(1) states that where conductors are terminated in equipment, the selected conductor ampacities must be based on Table 310.16, unless the equipment is specifically listed and marked otherwise. The intent of this requirement is to clarify which ampacities are used to determine the proper conductor size at equipment terminations.

When equipment of 600 volts or less is evaluated relative to the appropriate temperature characteristics of the terminations, conductors sized according to Table 310.16 are required to be used. The UL *General Information Directory* (White Book) clearly indicates that the 60°C and 75°C provisions for equipment have been determined using conductors from Table 310.16. However, installers or designers unaware of the UL guide card information might attempt to select conductors based on a table other than Table 310.16, especially if a wiring method that allows the use of ampacities such as those in Table 310.17 is used. That use can result in overheated terminations at the equipment. Clearly, the ampacities shown in other tables (such as Table 310.17) could be used for various conditions to which the wiring method is subject (ambient, ampacity correction, etc.), but the conductor size at the termination must be based on ampacities from Table 310.16. This change does not introduce any new impact on the equipment or the wiring methods; it simply adds a rule from the listing information into the *Code* because it is an installation and equipment selection issue.

Section 110.14(C)(1)(a) requires that conductor terminations, as well as conductors, be rated for the operating temperature of the circuit. For example, the load on an 8 AWG THHN, 90°C copper wire is limited to 40 amperes where connected to a disconnect switch with terminals rated at 60°C. The same 8 AWG THHN, 90°C wire is limited to 50 amperes where connected to a fusible switch with terminals rated at 75°C. The conductor ampacities were selected from Table 310.16. Not only does this requirement apply to conductor terminations of breakers and fusible switches, but the equipment enclosure must also permit terminations above 60°C. Exhibit 110.6 shows an example of termination temperature markings.

**110.16 Flash Protection.** Electrical equipment, such as switchboards, panelboards, industrial control panels, meter socket enclosures, and motor control centers, that are in other than dwelling occupancies, and are likely to require examination, adjustment, servicing, or maintenance while energized shall be field marked to warn qualified persons of potential electric arc flash hazards. The marking shall be located so as to be clearly visible to qualified persons before examination, adjustment, servicing, or maintenance of the equipment.

This requirement was revised in the 2008 *Code* by adding the words "Electrical equipment, such as" to make it clear that the requirement was not limited to the equipment on the list. Arc flash hazards could exist in other equipment. Field marking that warns electrical workers of potential electrical arc flash hazards is required because significant numbers of electricians have been seriously burned or killed by accidental electrical arc flash while working on

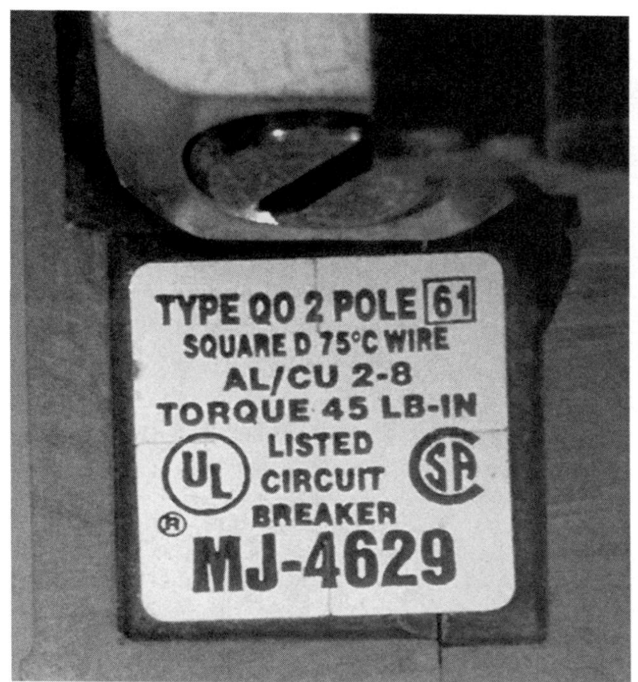


***EXHIBIT 110.6*** *An example of termination temperature markings on a main circuit breaker. (Courtesy of Square D/Schneider Electric)*

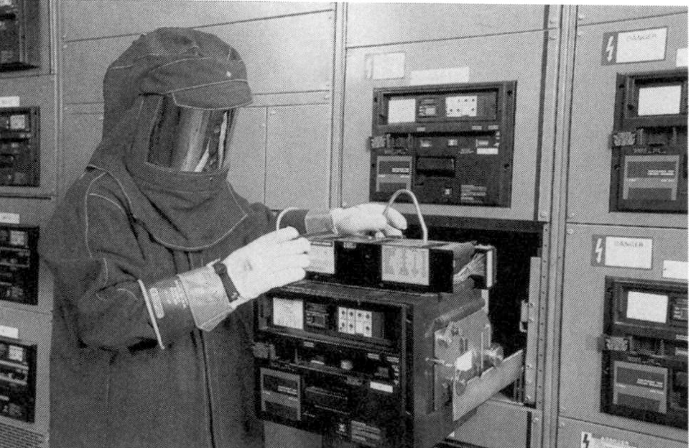

***EXHIBIT 110.7*** *Electrical worker clothed in personal protective equipment (PPE) appropriate for the hazard involved.*

⚠WARNING

**Arc Flash Hazard**

**15.5 cal/cm² incident energy**

**at 18" working distance**

©Clarion Safety Systems, LLC clarionsafety.com 800-748-0241 xxxxx Reorder No. NFPA2

***EXHIBIT 110.8*** *One example of an arc flash warning sign required by 110.16. (Courtesy of the International Association of Electrical Inspectors)*

"hot" (energized) equipment. Most of those accidents could have been prevented or their severity significantly reduced if electricians had been wearing the proper type of protective clothing. Requiring switchboards, panelboards, and motor control centers to be individually field marked with proper warning labels will raise the level of awareness of electrical arc flash hazards and thereby decrease the number of accidents.

Exhibit 110.7 shows an electrical employee working inside the flash protection boundary and in front of a large-capacity service-type switchboard that has not been de-energized and that is not under the lockout/tagout procedure. The worker is wearing personal protective equipment (PPE) considered appropriate flash protection clothing for the flash hazard involved. Suitable PPE appropriate to a particular hazard is described in NFPA 70E, *Standard for Electrical Safety in the Workplace*.

Exhibit 110.8 displays one example of a warning sign required by 110.16.

Accident reports continue to confirm the fact that workers responsible for the installation or maintenance of electrical equipment often do not turn off the power source before working on the equipment. Working electrical equipment energized is a major safety concern in the electrical industry. The real purpose of this additional *Code* requirement is to alert electrical contractors, electricians, facility owners and managers, and other interested parties to some of the hazards of working on or near energized equipment and to emphasize the importance of turning off the power before working on electrical circuits.

The information in fine print notes is not mandatory. Employers can be assured that they are providing a safe workplace for their employees if safety-related work practices required by NFPA 70E have been implemented and are being followed. (See also the commentary following the definition of *qualified person* in Article 100.)

In addition to the standards referenced in the fine print notes and their individual bibliographies, additional information on this subject can be found in the 1997 report "Hazards of Working Electrical Equipment Hot," published by the National Electrical Manufacturers Association.

FPN No. 1: NFPA 70E-2004, *Standard for Electrical Safety in the Workplace*, provides assistance in determining severity of potential exposure, planning safe work practices, and selecting personal protective equipment.

FPN No. 2: ANSI Z535.4-1998, *Product Safety Signs and Labels*, provides guidelines for the design of safety signs and labels for application to products.

**110.18 Arcing Parts.** Parts of electrical equipment that in ordinary operation produce arcs, sparks, flames, or molten metal shall be enclosed or separated and isolated from all combustible material.

Examples of electrical equipment that may produce sparks during ordinary operation include open motors having a centrifugal starting switch, open motors with commutators, and collector rings. Adequate separation from combustible material is essential if open motors with those features are used.

FPN: For hazardous (classified) locations, see Articles 500 through 517. For motors, see 430.14.

**110.21 Marking.** The manufacturer's name, trademark, or other descriptive marking by which the organization responsible for the product can be identified shall be placed on all electrical equipment. Other markings that indicate voltage, current, wattage, or other ratings shall be provided as specified elsewhere in this *Code*. The marking shall be of sufficient durability to withstand the environment involved.

The *Code* requires that equipment ratings be marked on the equipment and that such markings be located so as to be visible or easily accessible during or after installation.

**110.22 Identification of Disconnecting Means.**

**(A) General.** Each disconnecting means shall be legibly marked to indicate its purpose unless located and arranged so the purpose is evident. The marking shall be of sufficient durability to withstand the environment involved.

Proper identification needs to be specific. For example, the marking should indicate not simply "motor" but rather "motor, water pump"; not simply "lights" but rather "lights, front lobby." Consideration also should be given to the form of identification. Marking often fades or is covered by paint after installation. See 408.4 and its associated commentary for further information on circuit directories for switchboards and panelboards.

**(B) Engineered Series Combination Systems.** Where circuit breakers or fuses are applied in compliance with series combination ratings selected under engineering supervision and marked on the equipment as directed by the engineer, the equipment enclosure(s) shall be legibly marked in the field to indicate the equipment has been applied with a series combination rating. The marking shall be readily visible and state the following:

CAUTION — ENGINEERED SERIES COMBINATION SYSTEM RATED ________ AMPERES. IDENTIFIED REPLACEMENT COMPONENTS REQUIRED.

FPN: See 240.86(A) for engineered series combination systems.

Section 110.22(B) requires the enclosures of engineered series-rated overcurrent devices to be legibly marked. If the ratings are determined under engineering supervision, the equipment must have a label, as specified in 110.22(B), to indicate that the series combination rating has been used. It is important that the warnings on replacement components be heeded in order to maintain the level of protection provided by the design.

**(C) Tested Series Combination Systems.** Where circuit breakers or fuses are applied in compliance with the series combination ratings marked on the equipment by the manufacturer, the equipment enclosure(s) shall be legibly marked in the field to indicate the equipment has been applied with a series combination rating. The marking shall be readily visible and state the following:

CAUTION — SERIES COMBINATION SYSTEM RATED ________ AMPERES. IDENTIFIED REPLACEMENT COMPONENTS REQUIRED.

FPN: See 240.86(B) for tested series combination systems.

Section 110.22(C) requires the enclosures of tested series combination systems to be legibly marked. The equipment manufacturer can mark the equipment to be used with series combination ratings. If the equipment is installed in the field at its marked series combination rating, the equipment must have an additional label, as specified in 110.22, to indicate that the series combination rating has been used.

**110.26 Spaces About Electrical Equipment.** Sufficient access and working space shall be provided and maintained about all electrical equipment to permit ready and safe operation and maintenance of such equipment.

Key to understanding 110.26 is the division of requirements for spaces about electrical equipment in two separate and distinct categories: working space and dedicated equipment space. The term *working space* generally applies to the protection of the worker, and *dedicated equipment space* applies to the space reserved for future access to electrical equipment and to protection of the equipment from intrusion by nonelectrical equipment. The performance requirements for all spaces about electrical equipment are set forth in this section. Storage of material that blocks access or prevents safe work practices must be avoided at all times.

**(A) Working Space.** Working space for equipment operating at 600 volts, nominal, or less to ground and likely to require examination, adjustment, servicing, or maintenance while energized shall comply with the dimensions of 110.26(A)(1), (A)(2), and (A)(3) or as required or permitted elsewhere in this *Code*.

The intent of 110.26(A) is to provide enough space for personnel to perform any of the operations listed without jeopardizing worker safety. These operations include examination, adjustment, servicing, and maintenance of equipment. Examples of such equipment include panelboards, switches, circuit breakers, controllers, and controls on heating and air-conditioning equipment. It is important to understand that the word *examination*, as used in 110.26(A), includes such tasks as checking for the presence of voltage using a portable voltmeter.

Minimum working clearances are not required if the equipment is such that it is not likely to require examination, adjustment, servicing, or maintenance while energized. However, "sufficient" access and working space are still required by the opening paragraph of 110.26.

**(1) Depth of Working Space.** The depth of the working space in the direction of live parts shall not be less than that specified in Table 110.26(A)(1) unless the requirements of 110.26(A)(1)(a), (A)(1)(b), or (A)(1)(c) are met. Distances shall be measured from the exposed live parts or from the enclosure or opening if the live parts are enclosed.

For the 2008 *Code*, the minimum clear distances were revised to reflect an accurate metric conversion (also known as a soft conversion).

Included in these clearance requirements is the step-back distance from the face of the equipment. Table 110.26(A)(1) provides requirements for clearances away from the equipment, based on the circuit voltage to ground and whether there are grounded or ungrounded objects in the step-back space or exposed live parts across from each other. The voltages to ground consist of two groups: 0 to 150, inclusive, and 151 to 600, inclusive. Examples of common electrical supply systems covered in the 0 to 150 volts to ground group include 120/240-volt, single-phase, 3-wire and 208Y/120-volt, 3-phase, 4-wire. Examples of common electrical supply systems covered in the 151 to 600 volts to ground group include 240-volt, 3-phase, 3-wire; 480Y/277-volt, 3-phase, 4-wire; and 480-volt, 3-phase, 3-wire (ungrounded and corner grounded). Remember, where an ungrounded system is utilized, the voltage to ground (by definition) is the greatest voltage between the given conductor and any other conductor of the circuit. For example, the voltage to ground for a 480-volt ungrounded delta system is 480 volts. See Exhibit 110.9 for the general working clearance requirements for each of the three conditions listed in Table 110.26(A)(1).

(a) *Dead-Front Assemblies.* Working space shall not be required in the back or sides of assemblies, such as dead-front switchboards or motor control centers, where all connections and all renewable or adjustable parts, such as fuses or switches, are accessible from locations other than the back or sides. Where rear access is required to work on nonelectrical parts on the back of enclosed equipment, a minimum horizontal working space of 762 mm (30 in.) shall be provided.

The intent of this section is to point out that work space is required only from the side(s) of the enclosure that requires access. The general rule still applies: Equipment that requires front, rear, or side access for the electrical activities described in 110.26(A) must meet the requirements of Table 110.26(A)(1). In many cases, equipment of "dead-front" assemblies requires only front access. For equipment that requires rear access for nonelectrical activity, however, a reduced working space of at least 30 in. must be provided. Exhibit 110.10 shows a reduced working space of 30 in. at the rear of equipment to allow work on nonelectrical parts.

(b) *Low Voltage.* By special permission, smaller working spaces shall be permitted where all exposed live parts operate at not greater than 30 volts rms, 42 volts peak, or 60 volts dc.

(c) *Existing Buildings.* In existing buildings where electrical equipment is being replaced, Condition 2 working clearance shall be permitted between dead-front switchboards, panelboards, or motor control centers located across the aisle from each other where conditions of maintenance and supervision ensure that written procedures have been adopted to prohibit equipment on both sides of the aisle from being open at the same time and qualified persons who are authorized will service the installation.

This section permits some relief for installations that are being upgraded. When assemblies such as dead-front switchboards, panelboards, or motor-control centers are replaced in an existing building, the working clearance allowed is that required by Table 110.26(A)(1), Condition 2. The reduction from a Condition 3 to a Condition 2 clearance is allowed only where a written procedure prohibits facing doors of equipment from being open at the same time and where only authorized and qualified persons service the installation. Exhibit 110.11 illustrates this relief for existing buildings.

**(2) Width of Working Space.** The width of the working space in front of the electrical equipment shall be the width of the equipment or 762 mm (30 in.), whichever is greater. In all cases, the work space shall permit at least a 90 degree opening of equipment doors or hinged panels.

Regardless of the width of the electrical equipment, the working space cannot be less than 30 in. wide. This space allows an individual to have at least shoulder-width space in front of the equipment. The 30-in. measurement can be made from either the left or the right edge of the equipment and can overlap other electrical equipment, provided the other equipment does not extend beyond the clearance required by Table 110.26(A)(1). If the equipment is wider than 30 in., the left-to-right space must be equal to the width of the equipment. See Exhibit 110.12 for an explanation of the 30-in. width requirement.

**(3) Height of Working Space.** The work space shall be clear and extend from the grade, floor, or platform to the height required by 110.26(E). Within the height requirements of this section, other equipment that is associated with the electrical installation and is located above or below the electrical equipment shall be permitted to extend not more than 150 mm (6 in.) beyond the front of the electrical equipment.

In addition to requiring a working space to be clear from the floor to a height of 6½ ft or to the height of the equipment, whichever is greater, 110.26(A)(3) permits electrical equipment located above or below other electrical equipment to extend into the working space not more than 6 in. This requirement allows the placement of a 12 in. × 12 in. wireway on the wall directly above or below a 6 in.-deep panelboard without impinging on the working space or compromising practical working clearances. The requirement continues to prohibit large differences in depth of equipment below or above other equipment that specifically requires working space. To minimize the amount of space required for electrical equipment, it was not uncommon to find installations of large free-standing, dry-type transformers within the required work space for a wall-mounted panelboard. Clear access to the panelboard is compromised by the location of the transformer with its grounded enclosure and this type of installation and is clearly not permitted by this section. Electrical equipment that produces heat or that otherwise requires ventilation also must comply with 110.3(B) and 110.13.

**(B) Clear Spaces.** Working space required by this section shall not be used for storage. When normally enclosed live parts are exposed for inspection or servicing, the working space, if in a passageway or general open space, shall be suitably guarded.

Section 110.26(B), as well as the rest of 110.26, does not prohibit the placement of panelboards in corridors or passageways. For that reason, when the covers of corridor-mounted panelboards are removed for servicing or other work, access to the area around the panelboard should be guarded or limited to protect unqualified persons using the corridor.

Equipment that requires servicing while energized must be located in an area that is not used for storage, as shown in Exhibit 110.14.

*EXHIBIT 110.14 Equipment location that is free of storage to allow the equipment to be worked on safely. (Courtesy of the International Association of Electrical Inspectors)*

**(C) Entrance to and Egress from Working Space.**

**(1) Minimum Required.** At least one entrance of sufficient area shall be provided to give access to and egress from working space about electrical equipment.

This section was revised for the 2008 *Code*. The requirements are intended to provide access to electrical equipment. However, the primary intent is to provide egress from the area so that workers can escape if there is an arc flash incident.

**(2) Large Equipment.** For equipment rated 1200 amperes or more and over 1.8 m (6 ft) wide that contains overcurrent devices, switching devices, or control devices, there shall be one entrance to and egress from the required working space not less than 610 mm (24 in.) wide and 2.0 m (6½ ft) high at each end of the working space.

A single entrance to and egress from the required working space shall be permitted where either of the conditions in 110.26(C)(2)(a) or (C)(2)(b) is met.

The stipulation that large equipment is that which is over 6 ft wide was added back into the 2008 *Code*. Now, for the purposes of this section, large equipment is equipment that is rated 1200 amperes or more and is over 6 ft wide.

Where the entrance(s) to the working space is through a door, each door must comply with the requirements for swinging open in the direction of egress and have door opening hardware that does not require turning of a door knob or similar action that may preclude quick exit from the area in the event of an emergency.

This requirement affords safety for workers exposed to energized conductors by allowing an injured worker to safely and quickly exit an electrical room without having to turn knobs or pull doors open.

For a graphical explanation of access and entrance requirements to a working space, see Exhibits 110.15 and 110.16. Notice the unacceptable and hazardous situation shown in Exhibit 110.17.

(a) *Unobstructed Egress.* Where the location permits a continuous and unobstructed way of egress travel, a single entrance to the working space shall be permitted.

(b) *Extra Working Space.* Where the depth of the working space is twice that required by 110.26(A)(1), a single entrance shall be permitted. It shall be located such that the distance from the equipment to the nearest edge of the entrance is not less than the minimum clear distance specified in Table 110.26(A)(1) for equipment operating at that voltage and in that condition.

For an explanation of paragraphs 110.26(C)(2)(a) and 110.26(C)(2)(b), see Exhibits 110.18 and 110.19.

**(3) Personnel Doors.** Where equipment rated 1200 A or more that contains overcurrent devices, switching devices, or control devices is installed and there is a personnel door(s) intended for entrance to and egress from the working space less than 7.6 m (25 ft) from the nearest edge of the working space, the door(s) shall open in the direction of egress and be equipped with panic bars, pressure plates, or other devices that are normally latched but open under simple pressure.

This section is new for the 2008 *Code*. The requirement is based only on equipment rated 1200 A or more, not on its width. The need to have the panic hardware on the door is independent of the need for two exits from the working space.

The measurement for the personnel door is measured from the nearest edge of the working space. Not every electrical installation is in an equipment room. This section requires personnel doors that are up to 25 ft from the working space to have panic hardware and to open in the direction of egress from the area. Exhibit 110.20 shows one of the required exits.

***EXHIBIT 110.20*** *An installation of large equipment showing one of the required exits. (Courtesy of the International Association of Electrical Inspectors)*

**(D) Illumination.** Illumination shall be provided for all working spaces about service equipment, switchboards, panelboards, or motor control centers installed indoors. Additional lighting outlets shall not be required where the work space is illuminated by an adjacent light source or as permitted by 210.70(A)(1), Exception No. 1, for switched receptacles. In electrical equipment rooms, the illumination shall not be controlled by automatic means only.

**(E) Headroom.** The minimum headroom of working spaces about service equipment, switchboards, panelboards, or motor control centers shall be 2.0 m (6½ ft). Where the electrical equipment exceeds 2.0 m (6½ ft) in height, the minimum headroom shall not be less than the height of the equipment.

*Exception: In existing dwelling units, service equipment or panelboards that do not exceed 200 amperes shall be permitted in spaces where the headroom is less than 2.0 m (6½ ft).*

**(F) Dedicated Equipment Space.** All switchboards, panelboards, distribution boards, and motor control centers shall be located in dedicated spaces and protected from damage.

*Exception: Control equipment that by its very nature or because of other rules of the Code must be adjacent to or within sight of its operating machinery shall be permitted in those locations.*

**(1) Indoor.** Indoor installations shall comply with 110.26(F)(1)(a) through (F)(1)(d).

(a) *Dedicated Electrical Space.* The space equal to the width and depth of the equipment and extending from the floor to a height of 1.8 m (6 ft) above the equipment or to the structural ceiling, whichever is lower, shall be dedicated to the electrical installation. No piping, ducts, leak protection apparatus, or other equipment foreign to the electrical installation shall be located in this zone.

*Exception: Suspended ceilings with removable panels shall be permitted within the 1.8-m (6-ft) zone.*

(b) *Foreign Systems.* The area above the dedicated space required by 110.26(F)(1)(a) shall be permitted to contain foreign systems, provided protection is installed to avoid damage to the electrical equipment from condensation, leaks, or breaks in such foreign systems.

(c) *Sprinkler Protection.* Sprinkler protection shall be permitted for the dedicated space where the piping complies with this section.

(d) *Suspended Ceilings.* A dropped, suspended, or similar ceiling that does not add strength to the building structure shall not be considered a structural ceiling.

The dedicated electrical space includes the space defined by extending the footprint of the switchboard or panelboard from the floor to a height of 6 ft above the height of the equipment or to the structural ceiling, whichever is lower. This reserved space permits busways, conduits, raceways, and cables to enter the equipment. The dedicated electrical space must be clear of piping, ducts, leak protection apparatus, or equipment foreign to the electrical installation. Plumbing, heating, ventilation, and air-conditioning piping, ducts, and equipment must be installed outside the width and depth zone.

Foreign systems installed directly above the dedicated space reserved for electrical equipment must include protective equipment that ensures that occurrences such as leaks, condensation, and even breaks do not damage the electrical equipment located below.

Sprinkler protection is permitted for the dedicated spaces as long as the sprinkler or other suppression system piping complies with 110.26(F)(1)(c). A dropped, suspended, or similar ceiling is permitted to be located directly in the dedicated space, as are building structural members.

The electrical equipment also must be protected from physical damage. Damage can be caused by activities performed near the equipment, such as material handling by personnel or the operation of a forklift or other mobile equipment. See 110.27(B) for other provisions relating to the protection of electrical equipment.

Exhibits 110.21, 110.22, and 110.23 illustrate the two distinct indoor installation spaces required by 110.26(A) and 110.26(F), that is, the working space and the dedicated electrical space.

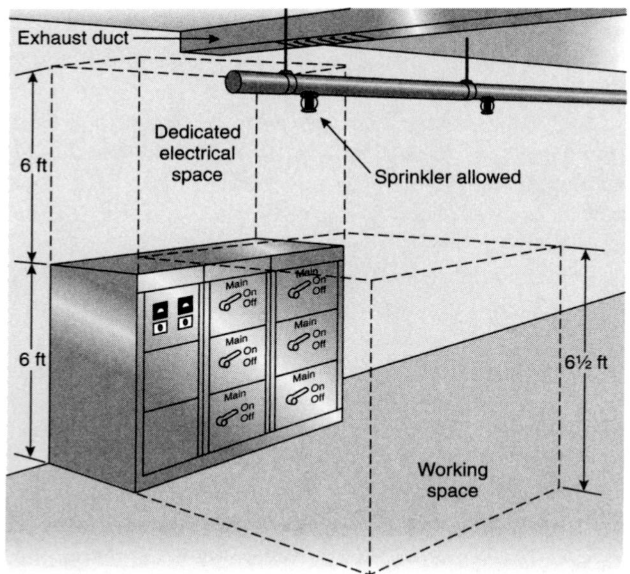


***EXHIBIT 110.21** The two distinct indoor installation spaces required by 110.26(A) and 110.26(F): the working space and the dedicated electrical space.*

In Exhibit 110.21, the dedicated electrical space required by 110.26(F) is the space outlined by the width and the depth of the equipment (the footprint) and extending from the floor to 6 ft above the equipment or to the structural ceiling (whichever is lower). The dedicated electrical space is reserved for the installation of electrical equipment and for the installation of conduits, cable trays, and so on, entering or exiting that equipment. The outlined area in front of the electrical equipment in Exhibit 110.21 is the working space required by 110.26(A). Note that sprinkler protection is afforded the entire dedicated electrical space and working space without actually entering either space. Also note that the exhaust duct is not located in or directly above the dedicated electrical space. Although not specifi-

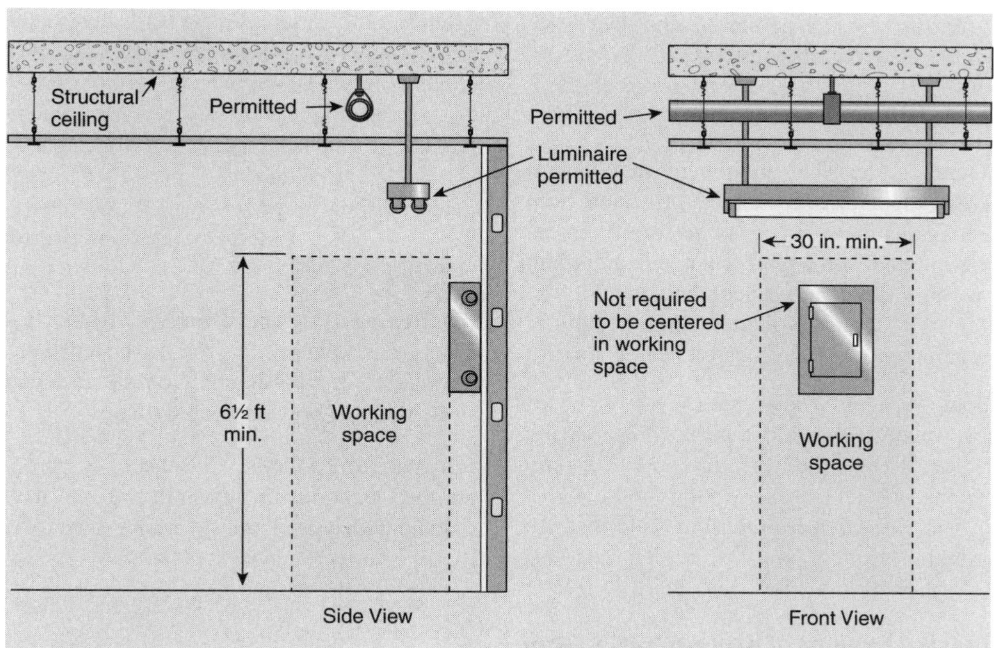


***EXHIBIT 110.22*** *The working space required by 110.26(A) in front of a panelboard. This illustration supplements the dedicated electrical space shown in Exhibit 110.21.*

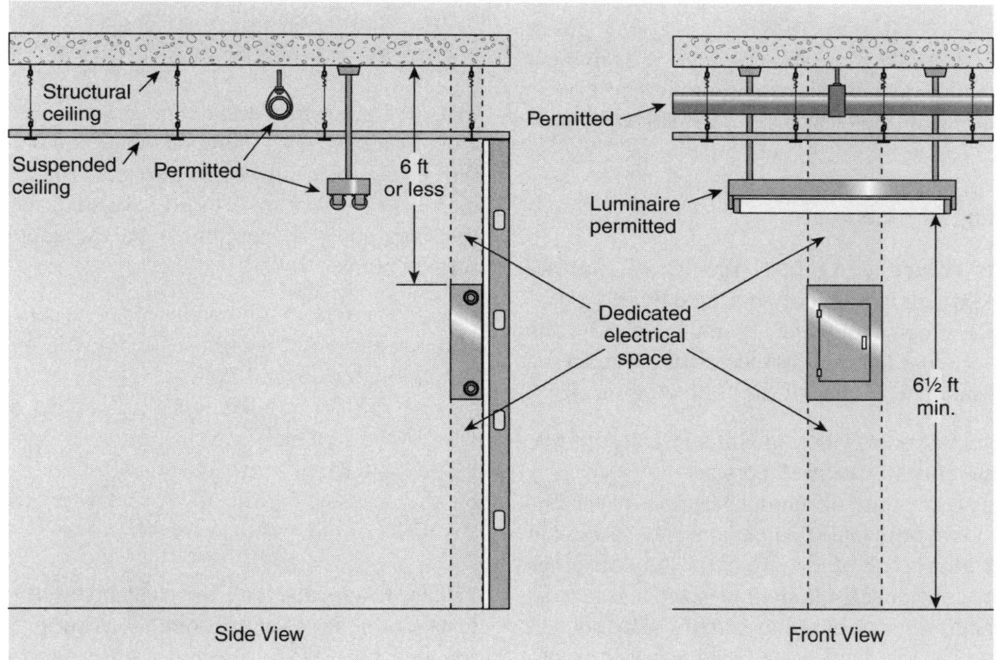


***EXHIBIT 110.23*** *The dedicated electrical space required by 110.26(F)(1) above and below a panelboard.*

cally required to be located here, this duct location may be a cost-effective solution that avoids the substantial physical protection requirements of 110.26(F)(1)(b).

Exhibit 110.22 illustrates the working space required in front of the panelboard by 110.26(A). No equipment, electrical or otherwise, is allowed in the working space.

Exhibit 110.23 illustrates the dedicated electrical space above and below the panelboard required by 110.26(F)(1). This space is for the cables, raceways, and so on, that run to and from the panelboard.

**(2) Outdoor.** Outdoor electrical equipment shall be installed in suitable enclosures and shall be protected from accidental contact by unauthorized personnel, or by vehicular traffic, or by accidental spillage or leakage from piping systems. The working clearance space shall include the zone described in 110.26(A). No architectural appurtenance or other equipment shall be located in this zone.

Extreme care should be taken where protection from unauthorized personnel or vehicular traffic is added to existing installations in order to comply with 110.26(F)(2). Any excavation or driving of steel into the ground for the placement of fencing, vehicle stops, or bollards should be done only after a thorough investigation of the belowgrade wiring.

**(G) Locked Electrical Equipment Rooms or Enclosures.** Electrical equipment rooms or enclosures housing electrical apparatus that are controlled by a lock(s) shall be considered accessible to qualified persons.

This requirement was relocated from the first paragraph of 110.26 for the 2008 *Code*. The intent is to allow equipment enclosures and rooms to be locked to prevent unauthorized access. Such rooms or enclosures are nonetheless considered accessible if qualified personnel have access.

**110.27 Guarding of Live Parts.**

**(A) Live Parts Guarded Against Accidental Contact.** Except as elsewhere required or permitted by this *Code*, live parts of electrical equipment operating at 50 volts or more shall be guarded against accidental contact by approved enclosures or by any of the following means:

(1) By location in a room, vault, or similar enclosure that is accessible only to qualified persons.
(2) By suitable permanent, substantial partitions or screens arranged so that only qualified persons have access to the space within reach of the live parts. Any openings in such partitions or screens shall be sized and located so that persons are not likely to come into accidental contact with the live parts or to bring conducting objects into contact with them.
(3) By location on a suitable balcony, gallery, or platform elevated and arranged so as to exclude unqualified persons.
(4) By elevation of 2.5 m (8 ft) or more above the floor or other working surface.

Contact conductors used for traveling cranes are permitted to be bare by 610.13(B) and 610.21(A). Although contact conductors obviously have to be bare for contact shoes on the moving member to make contact with the conductor, it is possible to place guards near the conductor to prevent its accidental contact with persons and still have slots or spaces through which the moving contacts can operate. The *Code* also recognizes the guarding of live parts by elevation.

**(B) Prevent Physical Damage.** In locations where electrical equipment is likely to be exposed to physical damage, enclosures or guards shall be so arranged and of such strength as to prevent such damage.

**(C) Warning Signs.** Entrances to rooms and other guarded locations that contain exposed live parts shall be marked with conspicuous warning signs forbidding unqualified persons to enter.

FPN: For motors, see 430.232 and 430.233. For over 600 volts, see 110.34.

Live parts of electrical equipment should be covered, shielded, enclosed, or otherwise protected by covers, barriers, mats, or platforms to prevent the likelihood of contact by persons or objects. See the definitions of *dead front* and *isolated (as applied to location)* in Article 100.

**110.30 General..** Conductors and equipment used on circuits over 600 volts, nominal, shall comply with Part I of this article and with 110.30 through 110.40, which supplement or modify Part I. In no case shall the provisions of this part apply to equipment on the supply side of the service point.

See "Over 600 volts" in the index to this *Handbook* for articles, parts, and sections that include requirements for installations over 600 volts.

Equipment on the supply side of the service point is outside the scope of the *NEC*. Such equipment is covered by ANSI C2, *National Electrical Safety Code*, published by the Institute of Electrical and Electronics Engineers (IEEE).

**110.31 Enclosure for Electrical Installations.** Electrical installations in a vault, room, or closet or in an area surrounded by a wall, screen, or fence, access to which is controlled by a lock(s) or other approved means, shall be considered to be accessible to qualified persons only. The type of enclosure used in a given case shall be designed and constructed according to the nature and degree of the hazard(s) associated with the installation.
For installations other than equipment as described in

110.31(D), a wall, screen, or fence shall be used to enclose an outdoor electrical installation to deter access by persons who are not qualified. A fence shall not be less than 2.1 m (7 ft) in height or a combination of 1.8 m (6 ft) or more of fence fabric and a 300-mm (1-ft) or more extension utilizing three or more strands of barbed wire or equivalent. The distance from the fence to live parts shall be not less than given in Table 110.31.

FPN: See Article 450 for construction requirements for transformer vaults.

**(A) Fire Resistance of Electrical Vaults.** The walls, roof, floors, and doorways of vaults containing conductors and equipment over 600 volts, nominal, shall be constructed of materials that have adequate structural strength for the conditions, with a minimum fire rating of 3 hours. The floors of vaults in contact with the earth shall be of concrete that is not less than 4 in. (102 mm) thick, but where the vault is constructed with a vacant space or other stories below it, the floor shall have adequate structural strength for the load imposed on it and a minimum fire resistance of 3 hours. For the purpose of this section, studs and wallboards shall not be considered acceptable.

**(B) Indoor Installations.**

**(1) In Places Accessible to Unqualified Persons.** Indoor electrical installations that are accessible to unqualified persons shall be made with metal-enclosed equipment. Metal-enclosed switchgear, unit substations, transformers, pull boxes, connection boxes, and other similar associated equipment shall be marked with appropriate caution signs. Openings in ventilated dry-type transformers or similar openings in other equipment shall be designed so that foreign objects inserted through these openings are deflected from energized parts.

**(2) In Places Accessible to Qualified Persons Only.** Indoor electrical installations considered accessible only to qualified persons in accordance with this section shall comply with 110.34, 110.36, and 490.24.

**(C) Outdoor Installations.**

**(1) In Places Accessible to Unqualified Persons.** Outdoor electrical installations that are open to unqualified persons shall comply with Parts I, II, and III of Article 225.

**(2) In Places Accessible to Qualified Persons Only.** Outdoor electrical installations that have exposed live parts shall be accessible to qualified persons only in accordance with the first paragraph of this section and shall comply with 110.34, 110.36, and 490.24.

**(D) Enclosed Equipment Accessible to Unqualified Persons.** Ventilating or similar openings in equipment shall be designed such that foreign objects inserted through these openings are deflected from energized parts. Where exposed to physical damage from vehicular traffic, suitable guards shall be provided. Nonmetallic or metal-enclosed equipment located outdoors and accessible to the general public shall be designed such that exposed nuts or bolts cannot be readily removed, permitting access to live parts. Where nonmetallic or metal-enclosed equipment is accessible to the general public and the bottom of the enclosure is less than 2.5 m (8 ft) above the floor or grade level, the enclosure door or hinged cover shall be kept locked. Doors and covers of enclosures used solely as pull boxes, splice boxes, or junction boxes shall be locked, bolted, or screwed on. Underground box covers that weigh over 45.4 kg (100 lb) shall be considered as meeting this requirement.

**110.32 Work Space About Equipment.** Sufficient space shall be provided and maintained about electrical equipment to permit ready and safe operation and maintenance of such equipment. Where energized parts are exposed, the minimum clear work space shall be not less than 2.0 m (6½ ft) high (measured vertically from the floor or platform) or not less than 914 mm (3 ft) wide (measured parallel to the equipment). The depth shall be as required in 110.34(A). In all cases, the work space shall permit at least a 90 degree opening of doors or hinged panels.

**110.33 Entrance to Enclosures and Access to Working Space.**

**(A) Entrance.** At least one entrance to enclosures for electrical installations as described in 110.31 not less than 610 mm (24 in.) wide and 2.0 m (6½ ft) high shall be provided to give access to the working space about electrical equipment.

**(1) Large Equipment.** On switchboard and control panels exceeding 1.8 m (6 ft) in width, there shall be one entrance at each end of the equipment. A single entrance to the required working space shall be permitted where either of the conditions in 110.33(A)(1)(a) or (A)(1)(b) is met.

(a) *Unobstructed Exit.* Where the location permits a continuous and unobstructed way of exit travel, a single entrance to the working space shall be permitted.

(b) *Extra Working Space.* Where the depth of the working space is twice that required by 110.34(A), a single entrance shall be permitted. It shall be located so that the distance from the equipment to the nearest edge of the entrance is not less than the minimum clear distance specified in Table 110.34(A) for equipment operating at that voltage and in that condition.

**(2) Guarding.** Where bare energized parts at any voltage or insulated energized parts above 600 volts, nominal, to ground are located adjacent to such entrance, they shall be suitably guarded.

Section 110.33(A) contains requirements very similar to those of 110.26(C). For further information, see the commentary following 110.26(C)(2), most of which also is valid for installations over 600 volts.

**(3) Personnel Doors.** Where there is a personnel door(s) intended for entrance to and egress from the working space less than 7.6 m (25 ft) from the nearest edge of the working space, the door(s) shall open in the direction of egress and be equipped with panic bars, pressure plates, or other devices that are normally latched but open under simple pressure.

This section is new for the 2008 *Code*. The dimension for the personnel door is measured from the nearest edge of the working space. Not every electrical installation is in an equipment room. This section will require personnel doors that are up to 25 ft from the working space to have panic hardware and to open in the direction of egress from the area.

**(B) Access.** Permanent ladders or stairways shall be provided to give safe access to the working space around electrical equipment installed on platforms, balconies, or mezzanine floors or in attic or roof rooms or spaces.

**110.34 Work Space and Guarding.**

**(A) Working Space.** Except as elsewhere required or permitted in this *Code,* equipment likely to require examination, adjustment, servicing, or maintenance while energized shall have clear working space in the direction of access to live parts of the electrical equipment and shall be not less than specified in Table 110.34(A). Distances shall be measured from the live parts, if such are exposed, or from the enclosure front or opening if such are enclosed.

*Exception: Working space shall not be required in back of equipment such as dead-front switchboards or control assemblies where there are no renewable or adjustable parts (such as fuses or switches) on the back and where all connections are accessible from locations other than the back. Where rear access is required to work on de-energized parts on the back of enclosed equipment, a minimum working space of 762 mm (30 in.) horizontally shall be provided.*

The provisions of 110.34 are conditional, just like the requirements in 110.26; that is, some of the requirements are applicable only where the equipment "is likely to require examination, adjustment, servicing, or maintenance while energized."

**(B) Separation from Low-Voltage Equipment.** Where switches, cutouts, or other equipment operating at 600 volts, nominal, or less are installed in a vault, room, or enclosure where there are exposed live parts or exposed wiring operating at over 600 volts, nominal, the high-voltage equipment shall be effectively separated from the space occupied by the low-voltage equipment by a suitable partition, fence, or screen.

*Exception: Switches or other equipment operating at 600 volts, nominal, or less and serving only equipment within the high-voltage vault, room, or enclosure shall be permitted to be installed in the high-voltage vault, room, or enclosure without a partition, fence, or screen if accessible to qualified persons only.*

**(C) Locked Rooms or Enclosures.** The entrance to all buildings, vaults, rooms, or enclosures containing exposed live parts or exposed conductors operating at over 600 volts, nominal, shall be kept locked unless such entrances are under the observation of a qualified person at all times.

Where the voltage exceeds 600 volts, nominal, permanent and conspicuous warning signs shall be provided, reading as follows:

DANGER — HIGH VOLTAGE — KEEP OUT

Equipment used on circuits over 600 volts, nominal, and containing exposed live parts or exposed conductors is required to be located in a locked room or in an enclosure. The provisions for locking are not required if the room or enclosure is under observation at all times, as is the case with some engine rooms. Where the room or enclosure is accessible to other than qualified persons, the entry to the room and equipment is required to be provided with warning labels. See Exhibit 110.24.

**(D) Illumination.** Illumination shall be provided for all working spaces about electrical equipment. The lighting outlets shall be arranged so that persons changing lamps or making repairs on the lighting system are not endangered by live parts or other equipment.

The points of control shall be located so that persons are not likely to come in contact with any live part or moving part of the equipment while turning on the lights.

**(E) Elevation of Unguarded Live Parts.** Unguarded live parts above working space shall be maintained at elevations not less than required by Table 110.34(E).

*EXHIBIT 110.24 Enclosures of equipment operations at over 600 volts required to have the warning label specified in 110.31(B)(1). (Courtesy of the International Association of Electrical Inspectors)*

**(F) Protection of Service Equipment, Metal-Enclosed Power Switchgear, and Industrial Control Assemblies.** Pipes or ducts foreign to the electrical installation and requiring periodic maintenance or whose malfunction would endanger the operation of the electrical system shall not be located in the vicinity of the service equipment, metal-enclosed power switchgear, or industrial control assemblies. Protection shall be provided where necessary to avoid damage from condensation leaks and breaks in such foreign systems. Piping and other facilities shall not be considered foreign if provided for fire protection of the electrical installation.

**110.51 General.**

**(A) Covered.** The provisions of this part shall apply to the installation and use of high-voltage power distribution and utilization equipment that is portable, mobile, or both, such as substations, trailers, cars, mobile shovels, draglines, hoists, drills, dredges, compressors, pumps, conveyors, underground excavators, and the like.

**(B) Other Articles.** The requirements of this part shall be additional to, or amendatory of, those prescribed in Articles 100 through 490 of this *Code*.

**(C) Protection Against Physical Damage.** Conductors and cables in tunnels shall be located above the tunnel floor and so placed or guarded to protect them from physical damage.

**110.54 Bonding and Equipment Grounding Conductors.**

**(A) Grounded and Bonded.** All non–current-carrying metal parts of electrical equipment and all metal raceways and cable sheaths shall be solidly grounded and bonded to all metal pipes and rails at the portal and at intervals not exceeding 300 m (1000 ft) throughout the tunnel.

**(B) Equipment Grounding Conductors.** An equipment grounding conductor shall be run with circuit conductors inside the metal raceway or inside the multiconductor cable jacket. The equipment grounding conductor shall be permitted to be insulated or bare.

**110.56 Energized Parts.** Bare terminals of transformers, switches, motor controllers, and other equipment shall be enclosed to prevent accidental contact with energized parts.

**110.57 Ventilation System Controls.** Electrical controls for the ventilation system shall be arranged so that the airflow can be reversed.

**110.58 Disconnecting Means.** A switch or circuit breaker that simultaneously opens all ungrounded conductors of the circuit shall be installed within sight of each transformer or motor location for disconnecting the transformer or motor. The switch or circuit breaker for a transformer shall have an ampere rating not less than the ampacity of the transformer supply conductors. The switch or circuit breaker for a motor shall comply with the applicable requirements of Article 430.

## Article 200 Use and Identification of Grounded Conductors

**200.2 General.** All premises wiring systems, other than circuits and systems exempted or prohibited by 210.10, 215.7, 250.21, 250.22, 250.162, 503.155, 517.63, 668.11, 668.21, and 690.41, Exception, shall have a grounded conductor that is identified in accordance with 200.6. The grounded conductor shall comply with 200.2(A) and (B).

**(A) Insulation.** The grounded conductor, where insulated, shall have insulation that is (1) suitable, other than color, for any ungrounded conductor of the same circuit on circuits of less than 1000 volts or impedance grounded neutral systems of 1 kV and over, or (2) rated not less than 600 volts for solidly grounded neutral systems of 1 kV and over as described in 250.184(A).

**(B) Continuity.** The continuity of a grounded conductor shall not depend on a connection to a metallic enclosure, raceway, or cable armor.

Section 200.2(B) requires the grounded conductor to connect to a terminal or busbar that is specifically intended for connection of grounded or neutral conductors. Some electrical distribution equipment used as service equipment has equipment grounding busbars that are separate from the grounded conductor busbar. The two are connected through the *main bonding jumper.* This section prohibits using the main bonding jumper and the metal enclosure of service equipment (or system bonding jumper for a separately derived system) as a means to connect a load side grounded conductor to the supply side grounded conductor as would be the case where a grounded conductor is terminated on an equipment grounding busbar. Such a connection results in the entire metal enclosure becoming a current-carrying conductor.

### 200.6 Means of Identifying Grounded Conductors.

**(A) Sizes 6 AWG or Smaller.** An insulated grounded conductor of 6 AWG or smaller shall be identified by a continuous white or gray outer finish or by three continuous white stripes on other than green insulation along its entire length. Wires that have their outer covering finished to show a white or gray color but have colored tracer threads in the braid identifying the source of manufacture shall be considered as meeting the provisions of this section. Insulated grounded conductors shall also be permitted to be identified as follows:

(1) The grounded conductor of a mineral-insulated, metal-sheathed cable shall be identified at the time of installation by distinctive marking at its terminations.
(2) A single-conductor, sunlight-resistant, outdoor-rated cable used as a grounded conductor in photovoltaic power systems as permitted by 690.31 shall be identified at the time of installation by distinctive white marking at all terminations.
(3) Fixture wire shall comply with the requirements for grounded conductor identification as specified in 402.8.
(4) For aerial cable, the identification shall be as above, or by means of a ridge located on the exterior of the cable so as to identify it.

The use of white insulation or white marking is the most common method of identifying the grounded conductor. However, Article 200 provides a number of alternative identification means, including the use of gray insulation or markings, the use of three continuous white stripes along the conductor insulation, surface markings, and colored braids or separators. The required identification of the grounded conductor is performed either by the wire or cable manufacturer or by the installer at the time of installation.

The general rule of 200.6(A) requires insulated conductors 6 AWG or smaller to be white or gray for their entire length where they are used as grounded conductors. The *Code* also permits three continuous white stripes along the entire length of conductor insulation that is colored other than green as a means to identify a conductor as the grounded conductor. Using three white stripes for identification is permitted for all conductor sizes and is the method most typically employed by a wire or cable manufacturer.

Other methods of identification are also permitted in 200.6(A). For example, the grounded conductor of mineral-insulated (MI) cable, due to its unique construction, is permitted to be identified at the time of installation. Aerial cable may have its grounded conductor identified by a ridge along its insulated surface, and fixture wires are permitted to have the grounded conductor identified by various methods, including colored insulation, stripes on the insulation, colored braid, colored separator, and tinned conductors. These identification methods are found in 402.8 and explained in detail in 400.22(A) through (E).

For 6 AWG or smaller, identification of the grounded conductor solely by distinctive white or gray marking at the time of installation is not permitted except as described for flexible cords and multiconductor cables in 200.6(C) and (E) and for single conductors in outdoor photovoltaic power installations in accordance with 200.6(A)(2).

**(B) Sizes Larger Than 6 AWG.** An insulated grounded conductor larger than 6 AWG shall be identified by one of the following means:

The general rule of 200.6(B) requires that insulated grounded conductors larger than 6 AWG be identified by one of three acceptable methods. As allowed by 200.6(A) for 6 AWG and smaller insulated conductors, 200.6(B) permits the use of a continuous white or gray color along the entire length of the conductor insulation or the use of three continuous white stripes on the entire length of the insulated (other than green colored insulation) conductor. The most common method used by installers to identify a single conductor as a grounded conductor is application of a white or gray marking to the insulation at all termination points at the time of installation. To be clearly visible, this field-applied white or gray marking must completely encircle the conductor insulation. This coloring can be applied by using marking tape or by painting the insulation. This method of identification is shown in Exhibit 200.1.

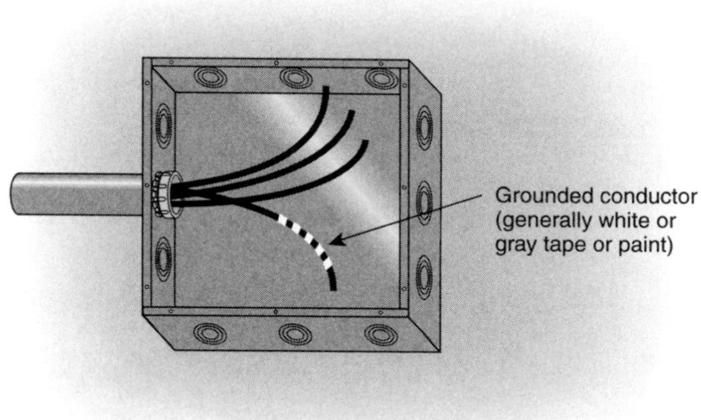


***EXHIBIT 200.1*** *Field-applied identification, as permitted by 200.6(B), of a 4 AWG conductor to identify it as the grounded conductor.*

(1) By a continuous white or gray outer finish.
(2) By three continuous white stripes along its entire length on other than green insulation.
(3) At the time of installation, by a distinctive white or gray marking at its terminations. This marking shall encircle the conductor or insulation.

**(C) Flexible Cords.** An insulated conductor that is intended for use as a grounded conductor, where contained within a flexible cord, shall be identified by a white or gray outer finish or by methods permitted by 400.22.

**(D) Grounded Conductors of Different Systems.** Where grounded conductors of different systems are installed in the same raceway, cable, box, auxiliary gutter, or other type of enclosure, each grounded conductor shall be identified by system. Identification that distinguishes each system grounded conductor shall be permitted by one of the following means:

(1) One system grounded conductor shall have an outer covering conforming to 200.6(A) or (B).
(2) The grounded conductor(s) of other systems shall have a different outer covering conforming to 200.6(A) or 200.6(B) or by an outer covering of white or gray with a readily distinguishable colored stripe other than green running along the insulation.
(3) Other and different means of identification as allowed by 200.6(A) or (B) that will distinguish each system grounded conductor.

This means of identification shall be permanently posted at each branch-circuit panelboard.

Exhibit 200.2 is an example of an enclosure containing grounded conductors of two different systems that are distinguished from each other by white and gray colored insulation. Gray and white colored insulation on conductors is considered to be two separate means of identifying grounded conductors.

The use of colored stripes (other than green) on white insulation for the entire conductor insulation length is also an acceptable method to distinguish one system grounded conductor from one with white insulation or white marking. It is important to note that this requirement applies only where grounded conductors of different systems are installed in a common enclosure, such as a junction or pull box or a wireway.

The *Code* requires the identification method or scheme used to distinguish the grounded conductors of different systems to be posted at all panelboards that supply branch circuits. In addition, industry practice of using white for lower-voltage systems and gray for higher-voltage systems is permitted but not mandated by the *Code*. Exhibit 200.3 is an example of a permanent label posted at electrical distribution equipment, indicating the identification scheme for the grounded conductors of each nominal voltage system.

**(E) Grounded Conductors of Multiconductor Cables.** The insulated grounded conductors in a multiconductor

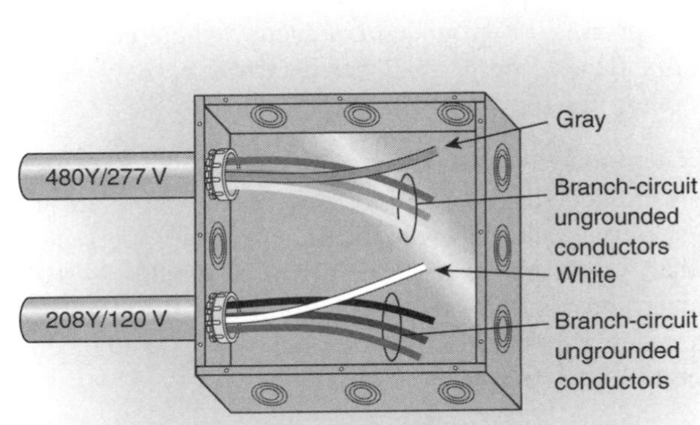


*EXHIBIT 200.2 Grounded conductors of different systems in the same enclosure. The grounded conductors of the different systems are identified by color through the use of white and gray colored insulation, one of the methods specified by 200.6(D).*

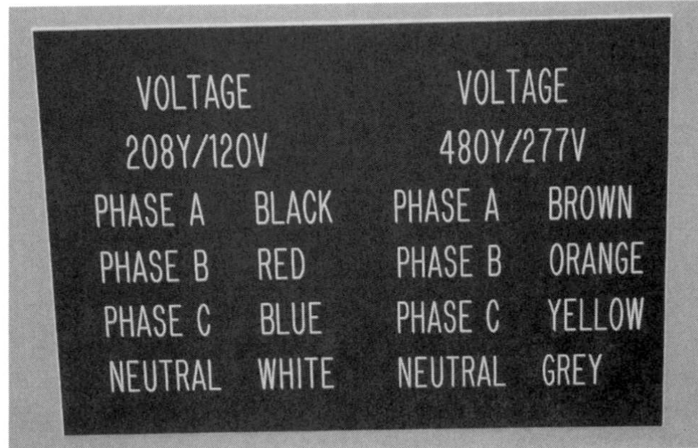


*EXHIBIT 200.3 A label providing the required information on how grounded and ungrounded conductors of different nominal voltage systems are identified at this particular site.*

cable shall be identified by a continuous white or gray outer finish or by three continuous white stripes on other than green insulation along its entire length. Multiconductor flat cable 4 AWG or larger shall be permitted to employ an external ridge on the grounded conductor.

*Exception No. 1: Where the conditions of maintenance and supervision ensure that only qualified persons service the installation, grounded conductors in multiconductor cables shall be permitted to be permanently identified at their terminations at the time of installation by a distinctive white marking or other equally effective means.*

Exception No. 1 to 200.6(E) introduces the concept of identifying grounded conductors of multiconductor cables at termination locations. This exception allows identification of a conductor that is part of a multiconductor cable as the grounded conductor at the time of installation by use of a distinctive white marking or other equally effective means, such as numbering, lettering, or tagging, as shown in Exhibit 200.4. Exception No. 1 to 200.6(E) is intended to apply to installations in facilities that have a regulated system of maintenance and supervision to ensure that only qualified persons service the installation. Permission to reidentify a conductor within a cable assembly is not predicated on the conductor size.

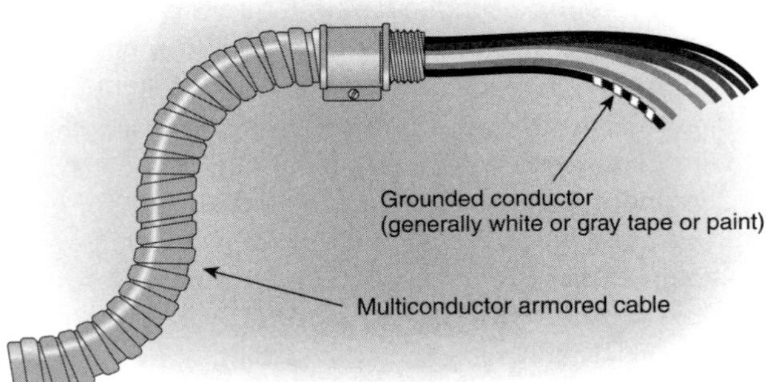


*EXHIBIT 200.4 Field-applied identification to the conductor of a multiconductor armored cable to be used as the grounded conductor as permitted by 200.6(E), Exception No. 1.*

*Exception No. 2: The grounded conductor of a multiconductor varnished-cloth-insulated cable shall be permitted to be identified at its terminations at the time of installation by a distinctive white marking or other equally effective means.*

FPN: The color gray may have been used in the past as an ungrounded conductor. Care should be taken when working on existing systems.

The term *natural gray* was changed to *gray* in the 2002 *Code* because the phrase "natural gray outer finish" was deemed obsolete. This change reserves all shades of gray insulation and marking for grounded conductors. The FPN following 200.6 warns the user to exercise caution when working on existing systems because gray may have been used on those existing systems.

### 200.7 Use of Insulation of a White or Gray Color or with Three Continuous White Stripes.

**(A) General.** The following shall be used only for the grounded circuit conductor, unless otherwise permitted in 200.7(B) and (C):

(1) A conductor with continuous white or gray covering
(2) A conductor with three continuous white stripes on other than green insulation
(3) A marking of white or gray color at the termination

**(B) Circuits of Less Than 50 Volts.** A conductor with white or gray color insulation or three continuous white stripes or having a marking of white or gray at the termination for circuits of less than 50 volts shall be required to be grounded only as required by 250.20(A).

**(C) Circuits of 50 Volts or More.** The use of insulation that is white or gray or that has three continuous white stripes for other than a grounded conductor for circuits of 50 volts or more shall be permitted only as in (1) through (3).

(1) If part of a cable assembly and where the insulation is permanently reidentified to indicate its use as an ungrounded conductor, by painting or other effective means at its termination, and at each location where the conductor is visible and accessible. Identification shall encircle the insulation and shall be a color other than white, gray, or green.
(2) Where a cable assembly contains an insulated conductor for single-pole, 3-way or 4-way switch loops and the conductor with white or gray insulation or a marking of three continuous white stripes is used for the supply to the switch but not as a return conductor from the switch to the switched outlet. In these applications, the conductor with white or gray insulation or with three continuous white stripes shall be permanently reidentified to indicate its use by painting or other effective means at its terminations and at each location where the conductor is visible and accessible.

Previous editions of the *Code* permitted switch loops using a white insulated conductor to serve as an ungrounded conductor supplying the switch but not as the return ungrounded conductor to supply the lighting outlet. Prior to the 1999 *NEC*, re-identification of a white conductor used for this purpose was not required. However, electronic switching devices with small power supplies are available that can be installed at switch locations. These devices require a grounded conductor in order to power the internal components. To avoid confusion and improper wiring where such devices are installed, re-identification of any conductor, used as an ungrounded conductor, with a white, gray, or white with an identifiable stripe colored insulation is required at every termination point to avoid confusion and improper wiring at the time a switching device is installed or replaced. The required re-identification must be effective, permanent, and suitable for the environment, to clearly identify the insulated conductor as an ungrounded conductor.

(3) Where a flexible cord, having one conductor identified by a white or gray outer finish or three continuous white stripes or by any other means permitted by 400.22, is used for connecting an appliance or equipment permitted by 400.7. This shall apply to flexible cords connected to outlets whether or not the outlet is supplied by a circuit that has a grounded conductor.

FPN: The color gray may have been used in the past as an ungrounded conductor. Care should be taken when working on existing systems.

**200.9 Means of Identification of Terminals.** The identification of terminals to which a grounded conductor is to be connected shall be substantially white in color. The identification of other terminals shall be of a readily distinguishable different color.

*Exception: Where the conditions of maintenance and supervision ensure that only qualified persons service the installations, terminals for grounded conductors shall be permitted to be permanently identified at the time of installation by a distinctive white marking or other equally effective means.*

**200.11 Polarity of Connections.** No grounded conductor shall be attached to any terminal or lead so as to reverse the designated polarity.

## Article 210 Branch Circuits

### 210.4 Multiwire Branch Circuits.

**(A) General.** Branch circuits recognized by this article shall be permitted as multiwire circuits. A multiwire circuit shall be permitted to be considered as multiple circuits. All conductors of a multiwire branch circuit shall originate from the same panelboard or similar distribution equipment.

FPN: A 3-phase, 4-wire, wye-connected power system used to supply power to nonlinear loads may necessitate that the power system design allow for the possibility of high harmonic currents on the neutral conductor.

The power supplies for equipment such as computers, printers, and adjustable-speed motor drives can introduce harmonic currents in the system neutral conductor. The resulting total harmonic distortion current could exceed the load current of the device itself. See the commentary following 310.15(B)(4)(c) for a discussion of neutral conductor ampacity.

**(B) Disconnecting Means.** Each multiwire branch circuit shall be provided with a means that will simultaneously disconnect all ungrounded conductors at the point where the branch circuit originates.

Multiwire branch circuits can be dangerous when not all the ungrounded circuit conductors are de-energized and equipment supplied from a multiwire circuit is being serviced. For this reason, all ungrounded conductors of a multiwire branch circuit must be simultaneously disconnected to reduce the risk of shock to personnel working on equipment supplied by a multiwire branch circuit. The simultaneous disconnecting means requirement takes the guesswork out of ensuring safe conditions for maintenance. In former editions of the *NEC*, this requirement applied only where the multiwire branch circuit supplied equipment mounted to a common yoke or strap.

For a single-phase installation, the simultaneous disconnection can be achieved by two single-pole circuit breakers with an identified handle tie, as shown in Exhibit 210.1 (top), or by a 2-pole switch or circuit breaker, as shown in Exhibit 210.1 (bottom). For a 3-phase installation, a 3-pole circuit breaker or three single-pole circuit breakers with an identified handle tie provides the required simultaneous disconnection of the ungrounded conductors. Where fuses are used for the branch-circuit overcurrent protection, a 2-pole or 3-pole switch is required.

The simultaneous opening of both "hot" conductors at the panelboard effectively protects personnel from inadvertent contact with an energized conductor or device terminal during servicing. The simultaneous disconnection can be achieved by a 2-pole switch or circuit breaker, as shown in Exhibit 210.1 (bottom), or by two single-pole circuit breakers with an identified handle tie, as shown in Exhibit 210.1 (top). Where fuses are used for the branch circuit overcurrent protection, a 2-pole disconnect switch is required.

**(C) Line-to-Neutral Loads.** Multiwire branch circuits shall supply only line-to-neutral loads.

*Exception No. 1: A multiwire branch circuit that supplies only one utilization equipment.*

*Exception No. 2: Where all ungrounded conductors of the multiwire branch circuit are opened simultaneously by the branch-circuit overcurrent device.*

FPN: See 300.13(B) for continuity of grounded conductor on multiwire circuits.

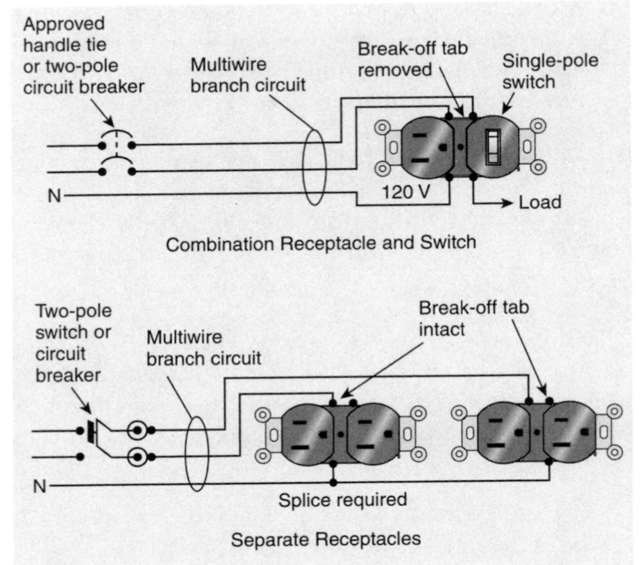


***EXHIBIT 210.1** Examples where 210.4(B) requires the simultaneous disconnection of all ungrounded conductors to multiwire branch circuits supplying more than one device or equipment.*

The term *multiwire branch circuit* is defined in Article 100 as "a branch circuit that consists of two or more ungrounded conductors that have a voltage between them, and a grounded conductor that has equal voltage between it and each ungrounded conductor of the circuit and that is connected to the neutral or grounded conductor of the system." Although defined as "a" branch circuit, 210.4(A) permits a multiwire branch circuit to be considered as multiple circuits and could be used, for instance, to satisfy the requirement for providing two small-appliance branch circuits for countertop receptacle outlets in a dwelling-unit kitchen.

The circuit most commonly used as a multiwire branch circuit consists of two ungrounded conductors and one grounded conductor supplied from a 120/240-volt, single-phase, 3-wire system. Such multiwire circuits supply appliances that have both line-to-line and line-to-neutral connected loads, such as electric ranges and clothes dryers, and also supply loads that are line-to-neutral connected only, such as the split-wired combination device shown in Exhibit 210.1 (bottom). A multiwire branch circuit is also permitted to supply a device with a 250-volt receptacle and a 125-volt receptacle, as shown in Exhibit 210.2, provided the branch-circuit overcurrent device simultaneously opens both of the ungrounded conductors.

Multiwire branch circuits have many advantages, including using three wires to do the work of four (in place

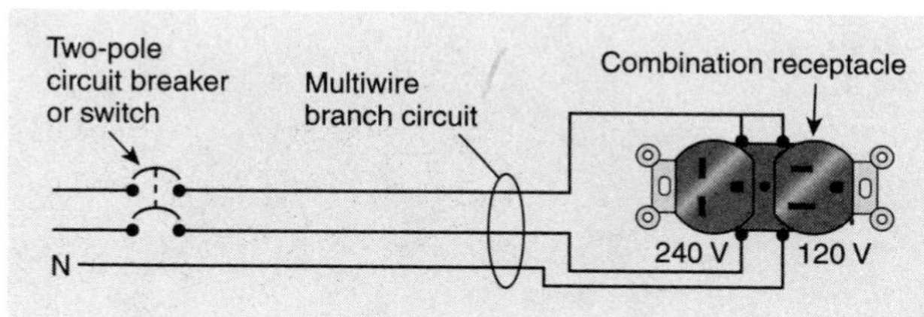


*EXHIBIT 210.2 An example of 210.4(C), Exception No. 2, which permits a multiwire branch circuit to supply line-to-neutral and line-to-line connected loads, provided the ungrounded conductors are opened simultaneously by the branch-circuit overcurrent device.*

of two 2-wire circuits), less raceway fill, easier balancing and phasing of a system, and less voltage drop. See the commentary following 215.2(A)(3), FPN No. 3, for further information on voltage drop for branch circuits.

Multiwire branch circuits may be derived from a 120/240-volt, single-phase; a 208Y/120-volt and 480Y/277-volt, 3-phase, 4-wire; or a 240/120-volt, 3-phase, 4-wire delta system. Section 210.11(B) requires multiwire branch circuits to be properly balanced. If two ungrounded conductors and a common neutral are used as a multiwire branch circuit supplied from a 208Y/120-volt, 3-phase, 4-wire system, the neutral carries the same current as the phase conductor with the highest current and, therefore, should be the same size. The neutral for a 2-phase, 3-wire or a 2-phase, 5-wire circuit must be sized to carry 140 percent of the ampere rating of the circuit, as required by 220.61(A) Exception. See the commentary following 210.4(A), FPN, for further information on 3-phase, 4-wire system neutral conductors.

If loads are connected line-to-line (i.e., utilization equipment connected between 2 or 3 phases), 2-pole or 3-pole circuit breakers are required to disconnect all ungrounded conductors simultaneously. In testing 240-volt equipment, it is quite possible not to realize that the circuit is still energized with 120 volts if one pole of the overcurrent device is open. See 210.10 and 240.15(B) for further information on circuit breaker overcurrent protection of ungrounded conductors. Other precautions concerning device removal on multiwire branch circuits are found in the commentary following 300.13(B).

**(D) Grouping.** The ungrounded and grounded conductors of each multiwire branch circuit shall be grouped by wire ties or similar means in at least one location within the panelboard or other point of origination.

*Exception: The requirement for grouping shall not apply if the circuit enters from a cable or raceway unique to the circuit that makes the grouping obvious.*

**210.8 Ground-Fault Circuit-Interrupter Protection for Personnel.**

Section 210.8 is the main rule for application of ground-fault circuit interrupters (GFCIs). Since the introduction of the GFCI in the 1971 *Code*, these devices have proved to their users and to the electrical community that they are worth the added cost during construction or remodeling. Published data from the U.S. Consumer Product Safety Commission show a decreasing trend in the number of electrocutions in the United States since the introduction of GFCI devices. Unfortunately, no statistics are available for the actual number of lives saved or injuries prevented by GFCI devices. However, most safety experts agree that GFCIs are directly responsible for a substantial number of saved lives and prevented injuries.

Exhibit 210.6 shows a typical circuit arrangement of a GFCI. The line conductors are passed through a sensor and are connected to a shunt-trip device. As long as the current in the conductors is equal, the device remains in a closed position. If one of the conductors comes in contact with a grounded object, either directly or through a person's body, some of the current returns by an alternative path, resulting in an unbalanced current. The toroidal coil senses the unbalanced current, and a circuit is established to the shunt-trip mechanism that reacts and opens the circuit. Note that the circuit design does not require the presence of an equipment grounding conductor, which is the reason 406.3(D)(3)(b) permits the use of GFCIs as replacements for receptacles where a grounding means does not exist.

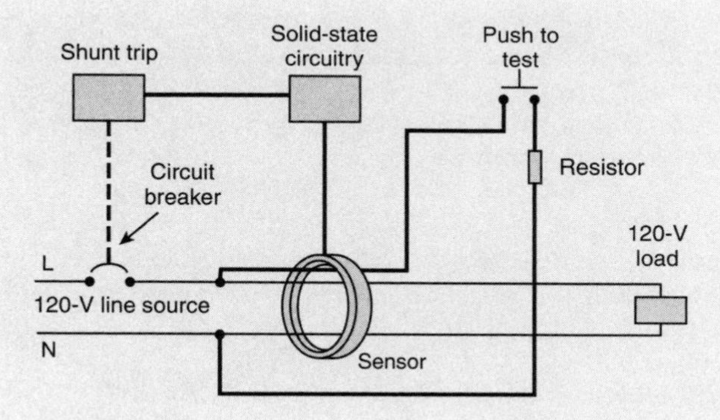


*EXHIBIT 210.6 The circuitry and components of a typical GFCI.*

GFCIs operate on currents of 5 mA. The listing standard permits a differential of 4 to 6 mA. At trip levels of 5 mA (the instantaneous current could be much higher), a shock can be felt during the time of the fault. The shock can lead to involuntary reactions that may cause secondary accidents such as falls. GFCIs do not protect persons from shock hazards where contact is between phase and neutral or between phase-to-phase conductors.

A variety of GFCIs are available, including portable and plug-in types and circuit-breaker types, types built into attachment plug caps, and receptacle types. Each type has a test switch so that units can be checked periodically to ensure proper operation. See Exhibits 210.7 and 210.8.

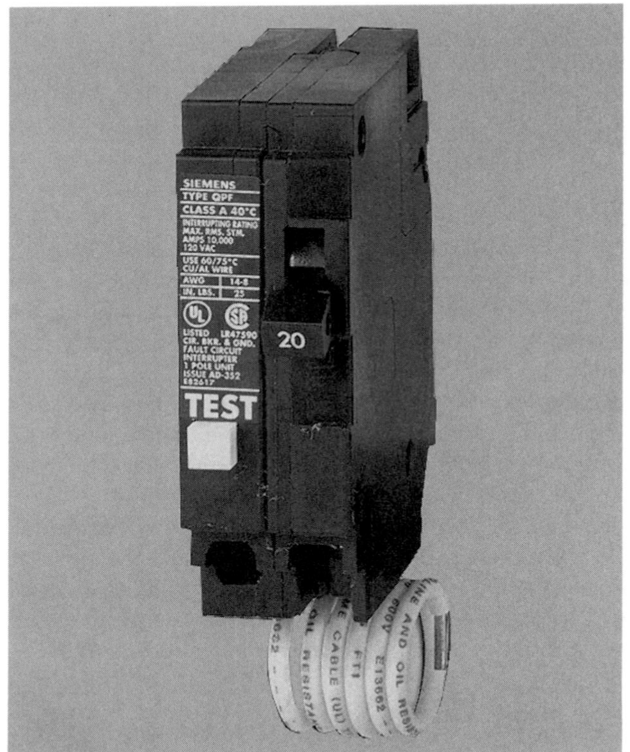


***EXHIBIT 210.7*** *A GFCI circuit breaker, which provides protection at all outlets supplied by the branch circuit. (Courtesy of Siemens)*

Although 210.8 is the main rule for GFCIs, other specific applications require the use of GFCIs. These additional specific applications are listed in Commentary Table 210.1.

**(B) Other Than Dwelling Units.** All 125-volt, single-phase, 15- and 20-ampere receptacles installed in the locations specified in (1) through (5) shall have ground-fault circuit-interrupter protection for personnel:

(1) Bathrooms

If receptacles are provided in bathroom areas of hotels and motels, GFCI-protected receptacles are required. Lavatories in airports, commercial buildings, industrial facilities, and other nondwelling occupancies are required to have *all* their receptacles GFCI protected. The only exception

***EXHIBIT 210.8*** *A 15-ampere duplex receptacle with integral GFCI that also protects downstream loads. (Courtesy of Pass & Seymour/Legrand®)*

to this requirement is found in 517.21, which permits receptacles in hospital critical care areas to be non-GFCI if the toilet and basin are installed in the patient room rather than in a separate bathroom. Some motel and hotel bathrooms, like the one shown in Exhibit 210.15, place the basin outside the door to the room containing the tub, toilet, or

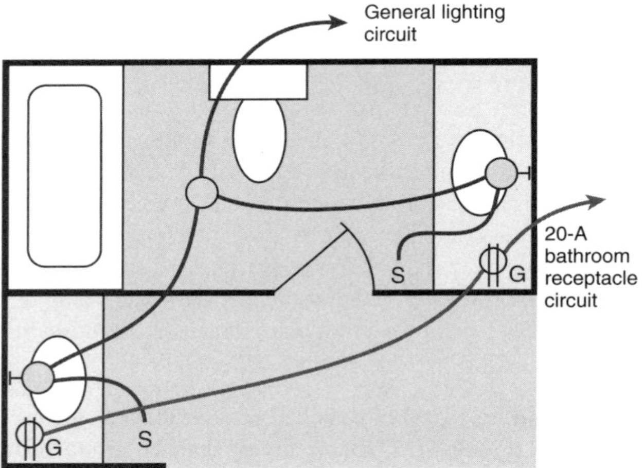


G = GFCI protection required

***EXHIBIT 210.15*** *GFCI protection of receptacles in a motel/hotel bathroom where one basin is located outside the door to the rest of the bathroom area, in accordance with 210.8(B)(1).*

another basin. The definition of *bathroom* as found in Article 100 applies to motel and hotel bathrooms, as does the GFCI requirement of 210.8(B)(1).

(2) Kitchens

Section 210.8(B)(2) requires all 15- and 20-ampere, 125-volt receptacles in nondwelling-type kitchens to be GFCI protected. This requirement applies to all 15- and 20-ampere, 125-volt kitchen receptacles, whether or not the receptacle serves countertop areas.

Accident data related to electrical incidents in non-dwelling kitchens reveal the presence of many hazards, including poorly maintained electrical apparatus, damaged electrical cords, wet floors, and employees without proper electrical safety training. Mandating some limited form of GFCI protection for high-hazard areas such as nondwelling kitchens should help prevent electrical accidents. This requirement provides specific information on what is considered to be a commercial or institutional kitchen. A location with a sink and a portable cooking appliance (e.g., cord-and-plug-connected microwave oven) is not considered a commercial or institutional kitchen for the purposes of applying this requirement. Kitchens in restaurants, hotels, schools, churches, dining halls, and similar facilities are examples of the types of kitchens covered by this requirement.

(3) Rooftops

Section 210.8(B)(3) requires all rooftop 15- and 20-ampere receptacles in nondwelling occupancies to be GFCI protected. For rooftops that also have heating, air-conditioning, and refrigeration equipment, see 210.63.

(4) Outdoors

Electrocution and electrical shock accident data provided by the U.S. Consumer Product Safety Commission indicate that such accidents are occurring at locations other than dwelling units and construction sites. In addition, 210.63, which requires the installation of a 125-volt receptacle within 25 ft of heating, air-conditioning, and refrigeration (HACR) equipment for use by service personnel, has been expanded since its first appearance in the *Code*, from applying to only equipment installed on rooftop to now applying to any location where HACR equipment is installed, including all outdoor locations. This GFCI requirement correlates with the expanded coverage of 210.63 and affords service personnel a permanently installed, GFCI-protected receptacle for servicing outdoor HACR equipment for all occupancies not covered by the dwelling unit requirements in 210.8(A)(3).

*Exception No. 1 to (3) and (4): Receptacles that are not readily accessible and are supplied from a dedicated branch circuit for electric snow-melting or deicing equipment shall be permitted to be installed without GFCI protection.*

*Exception No. 2 to (4): In industrial establishments only, where the conditions of maintenance and supervision ensure that only qualified personnel are involved, an assured equipment grounding conductor program as specified in 590.6(B)(2) shall be permitted for only those receptacle outlets used to supply equipment that would create a greater hazard if power is interrupted or having a design that is not compatible with GFCI protection.*

(5) Sinks — where receptacles are installed within 1.8 m (6 ft) of the outside edge of the sink.

As is the case for dwelling units, GFCI protection is now required for all 125-volt, 15- and 20-ampere receptacles installed within 6 ft of the outside edge of all sinks that are located in other occupancies. This requirement covers receptacles installed near sinks in lunchrooms, janitors' closets, classrooms, and all other areas that are not covered by the bathroom and kitchen provisions of 210.8(B)(1) and (B)(2).

*Exception No. 1 to (5): In industrial laboratories, receptacles used to supply equipment where removal of power would introduce a greater hazard shall be permitted to be installed without GFCI protection.*

*Exception No. 2 to (5): For receptacles located in patient care areas of health care facilities other than those covered under 210.8(B)(1), GFCI protection shall not be required.*

**210.21 Outlet Devices.** Outlet devices shall have an ampere rating that is not less than the load to be served and shall comply with 210.21(A) and (B).

**(A) Lampholders.** Where connected to a branch circuit having a rating in excess of 20 amperes, lampholders shall be of the heavy-duty type. A heavy-duty lampholder shall have a rating of not less than 660 watts if of the admedium type, or not less than 750 watts if of any other type.

The intent of 210.21(A) is to restrict a fluorescent lighting branch-circuit rating to not more than 20 amperes because most lampholders manufactured for use with fluorescent lights have a rating less than that required for heavy-duty lampholders (660 watts for admedium type or 750 watts for all other types).

Branch-circuit conductors for fluorescent electric-discharge lighting are usually connected to ballasts rather than to lampholders, and, by specifying a wattage rating for these lampholders, a limit of 20 amperes is applied to ballast circuits.

Only the admedium-base lampholder is recognized as heavy duty at the rating of 660 watts. Other lampholders are required to have a rating of not less than 750 watts to be recognized as heavy duty. The requirement of 210.21(A) prohibits the use of medium-base screw shell lampholders on branch circuits that are in excess of 20 amperes.

**(B) Receptacles.**

**(1) Single Receptacle on an Individual Branch Circuit.** A single receptacle installed on an individual branch circuit shall have an ampere rating not less than that of the branch circuit.

*Exception No. 1: A receptacle installed in accordance with 430.81(B).*

*Exception No. 2: A receptacle installed exclusively for the use of a cord-and-plug-connected arc welder shall be permitted to have an ampere rating not less than the minimum branch-circuit conductor ampacity determined by 630.11(A) for arc welders.*

FPN: See the definition of *receptacle* in Article 100.

**(2) Total Cord-and-Plug-Connected Load.** Where connected to a branch circuit supplying two or more receptacles or outlets, a receptacle shall not supply a total cord-and-plug-connected load in excess of the maximum specified in Table 210.21(B)(2).

**(3) Receptacle Ratings.** Where connected to a branch circuit supplying two or more receptacles or outlets, receptacle ratings shall conform to the values listed in Table 210.21(B)(3), or where larger than 50 amperes, the receptacle rating shall not be less than the branch-circuit rating.

*Exception No. 1: Receptacles for one or more cord-and-plug-connected arc welders shall be permitted to have ampere ratings not less than the minimum branch-circuit conductor ampacity permitted by 630.11(A) or (B) as applicable for arc welders.*

*Exception No. 2: The ampere rating of a receptacle installed for electric discharge lighting shall be permitted to be based on 410.62(C).*

A single receptacle installed on an individual branch circuit must have an ampere rating not less than that of the branch circuit. For example, a single receptacle on a 20-ampere individual branch circuit must be rated at 20 amperes in accordance with 210.21(B)(1); however, two or more 15-ampere single receptacles or a 15-ampere duplex receptacle are permitted on a 20-ampere branch circuit in accordance with 210.21(B)(3). This requirement does not apply to specific types of cord-and-plug-connected arc welders.

**(4) Range Receptacle Rating.** The ampere rating of a range receptacle shall be permitted to be based on a single range demand load as specified in Table 220.55.

**210.50 General.** Receptacle outlets shall be installed as specified in 210.52 through 210.63.

**(A) Cord Pendants.** A cord connector that is supplied by a permanently connected cord pendant shall be considered a receptacle outlet.

**(B) Cord Connections.** A receptacle outlet shall be installed wherever flexible cords with attachment plugs are used. Where flexible cords are permitted to be permanently connected, receptacles shall be permitted to be omitted for such cords.

Flexible cords are permitted to be permanently connected to boxes or fittings where specifically permitted by the *Code*. However, plugging a cord into a lampholder by inserting a screw-plug adapter is not permitted, because 410.90 requires lampholders of the screw shell type to be installed for use as lampholders only.

**(C) Appliance Receptacle Outlets.** Appliance receptacle outlets installed in a dwelling unit for specific appliances, such as laundry equipment, shall be installed within 1.8 m (6 ft) of the intended location of the appliance.

See 210.52(F) and 210.11(C)(2) for requirements regarding laundry receptacle outlets and branch circuits.

## Article 225 Outside Branch Circuits and Feeders

**225.18 Clearance for Overhead Conductors and Cables** Overhead spans of open conductors and open multiconductor cables of not over 600 volts, nominal, shall have a clearance of not less than the following:

(1) 3.0 m (10 ft) — above finished grade, sidewalks, or from any platform or projection from which they might be reached where the voltage does not exceed 150 volts to ground and accessible to pedestrians only
(2) 3.7 m (12 ft) — over residential property and driveways, and those commercial areas not subject to truck traffic where the voltage does not exceed 300 volts to ground
(3) 4.5 m (15 ft) — for those areas listed in the 3.7-m (12-ft) classification where the voltage exceeds 300 volts to ground
(4) 5.5 m (18 ft) — over public streets, alleys, roads, parking areas subject to truck traffic, driveways on

other than residential property, and other land traversed by vehicles, such as cultivated, grazing, forest, and orchard

**225.19 Clearances from Buildings for Conductors of Not over 600 Volts, Nominal.**

**(A) Above Roofs.** Overhead spans of open conductors and open multiconductor cables shall have a vertical clearance of not less than 2.5 m (8 ft) above the roof surface. The vertical clearance above the roof level shall be maintained for a distance not less than 900 mm (3 ft) in all directions from the edge of the roof.

*Exception No. 1: The area above a roof surface subject to pedestrian or vehicular traffic shall have a vertical clearance from the roof surface in accordance with the clearance requirements of 225.18.*

*Exception No. 2: Where the voltage between conductors does not exceed 300, and the roof has a slope of 100 mm in 300 mm (4 in. in 12 in.) or greater, a reduction in clearance to 900 mm (3 ft) shall be permitted.*

*Exception No. 3: Where the voltage between conductors does not exceed 300, a reduction in clearance above only the overhanging portion of the roof to not less than 450 mm (18 in.) shall be permitted if (1) not more than 1.8 m (6 ft) of the conductors, 1.2 m (4 ft) horizontally, pass above the roof overhang and (2) they are terminated at a through-the-roof raceway or approved support.*

*Exception No. 4: The requirement for maintaining the vertical clearance 900 mm (3 ft) from the edge of the roof shall not apply to the final conductor span where the conductors are attached to the side of a building.*

**(B) From Nonbuilding or Nonbridge Structures.** From signs, chimneys, radio and television antennas, tanks, and other nonbuilding or nonbridge structures, clearances — vertical, diagonal, and horizontal — shall not be less than 900 mm (3 ft).

**(C) Horizontal Clearances.** Clearances shall not be less than 900 mm (3 ft).

**(D) Final Spans.** Final spans of feeders or branch circuits shall comply with 225.19(D)(1), (D)(2), and (D)(3).

**(1) Clearance from Windows.** Final spans to the building they supply, or from which they are fed, shall be permitted to be attached to the building, but they shall be kept not less than 900 mm (3 ft) from windows that are designed to be opened, and from doors, porches, balconies, ladders, stairs, fire escapes, or similar locations.

*Exception: Conductors run above the top level of a window shall be permitted to be less than the 900-mm (3-ft) requirement.*

**(2) Vertical Clearance.** The vertical clearance of final spans above, or within 900 mm (3 ft) measured horizontally of, platforms, projections, or surfaces from which they might be reached shall be maintained in accordance with 225.18.

**(3) Building Openings.** The overhead branch-circuit and feeder conductors shall not be installed beneath openings through which materials may be moved, such as openings in farm and commercial buildings, and shall not be installed where they obstruct entrance to these buildings' openings.

**(E) Zone for Fire Ladders.** Where buildings exceed three stories or 15 m (50 ft) in height, overhead lines shall be arranged, where practicable, so that a clear space (or zone) at least 1.8 m (6 ft) wide will be left either adjacent to the buildings or beginning not over 2.5 m (8 ft) from them to facilitate the raising of ladders when necessary for fire fighting.

## Article 230 Services

**230.9 Clearances on Buildings.** Service conductors and final spans shall comply with 230.9(A), (B), and (C).

**(A) Clearances.** Service conductors installed as open conductors or multiconductor cable without an overall outer jacket shall have a clearance of not less than 900 mm (3 ft) from windows that are designed to be opened, doors, porches, balconies, ladders, stairs, fire escapes, or similar locations.

*Exception: Conductors run above the top level of a window shall be permitted to be less than the 900-mm (3-ft) requirement.*

As illustrated in Exhibit 230.16, the clearance of 3 ft applies to open conductors, not to a raceway or to a cable assembly that has an overall outer jacket such as Types SE, MC, and MI cables. The intent is to protect the conductors from physical damage and to protect personnel from accidental contact with the conductors. The exception permits service conductors, including drip loops and service-drop conductors, to be located just above window openings, because they are considered out of reach.

**(B) Vertical Clearance.** The vertical clearance of final spans above, or within 900 mm (3 ft) measured horizontally of, platforms, projections, or surfaces from which they might be reached shall be maintained in accordance with 230.24(B).

*NOTICE: For full text, see the 2008 edition of the NECH*

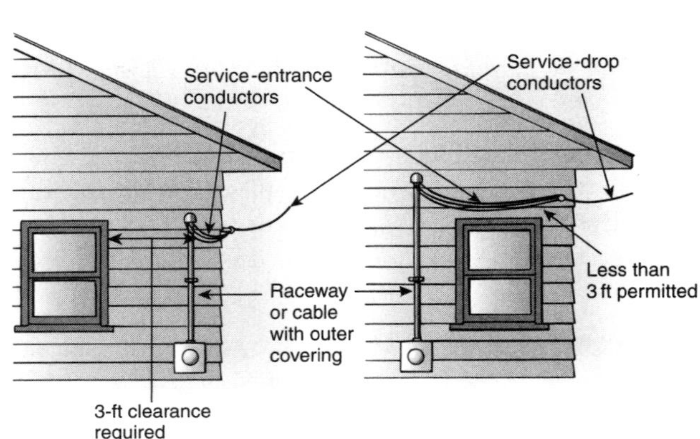


***EXHIBIT 230.16*** *Required dimensions for service conductors located alongside a window (left) and service conductors above the top level of a window designed to be opened (right).*

Where service conductors are located within 3 ft measured horizontally of a balcony, stair landing, or other platform, clearance to the platform must be at least 10 ft, as shown in Exhibit 230.17. See 230.24(B) for vertical clearances from ground.

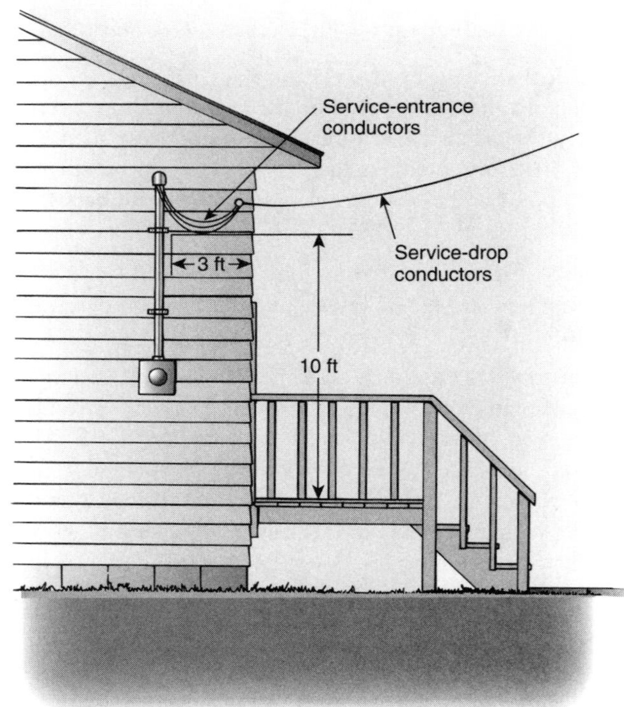


***EXHIBIT 230.17*** *Required dimensions for service conductors located above a stair landing, according to 230.9(B) and 230.24(B).*

**(C) Building Openings.** Overhead service conductors shall not be installed beneath openings through which materials may be moved, such as openings in farm and commercial buildings, and shall not be installed where they obstruct entrance to these building openings.

Elevated openings in buildings through which materials may be moved, such as barns and storage buildings, are often high enough for the service conductors to be installed below the opening. However, 230.9(C) prohibits such placement in order to reduce the likelihood of damage to the service conductors and the potential for electric shock to persons using the openings.

**230.10 Vegetation as Support.** Vegetation such as trees shall not be used for support of overhead service conductors.

**230.24 Clearances.** Service-drop conductors shall not be readily accessible and shall comply with 230.24(A) through (D) for services not over 600 volts, nominal.

**(A) Above Roofs.** Conductors shall have a vertical clearance of not less than 2.5 m (8 ft) above the roof surface. The vertical clearance above the roof level shall be maintained for a distance of not less than 900 mm (3 ft) in all directions from the edge of the roof.

Service-drop conductors are not permitted to be readily accessible. This main rule applies to services rated 600 volts and less, grounded or ungrounded. An 8-ft vertical clearance is required over the roof surface, extending 3 ft in all directions from the edge. Note that Exception No. 4 to 230.24(A) allows the final span of the service drop to enter this space in order to attach to the building or service mast.

*Exception No. 1: The area above a roof surface subject to pedestrian or vehicular traffic shall have a vertical clearance from the roof surface in accordance with the clearance requirements of 230.24(B).*

Exception No. 1 to 230.24(A) requires service-drop conductor clearance above a roof surface subject to vehicular or pedestrian traffic, such as the rooftop parking area shown in Exhibit 230.18, to meet the clearance requirements of 230.24(B).

*Exception No. 2: Where the voltage between conductors does not exceed 300 and the roof has a slope of 100 mm in 300 mm (4 in. in 12 in.) or greater, a reduction in clearance to 900 mm (3 ft) shall be permitted.*

Exception No. 2 to 230.24(A) permits a reduction in service-drop conductor clearance above the roof from 8 ft to 3 ft,

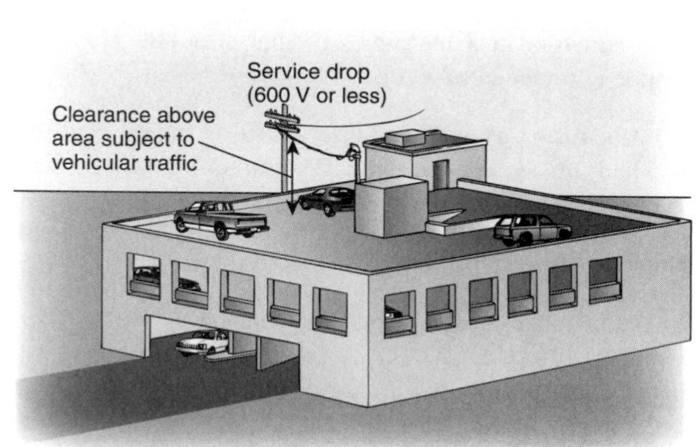


*EXHIBIT 230.18 Service-drop conductor clearance required by 230.24(A), Exception No. 1.*

as illustrated in Exhibit 230.19, where the voltage between conductors does not exceed 300 volts (e.g., 120/240, 208Y/120 services) and the roof is sloped not less than 4 in. vertically in 12 in. horizontally. Steeply sloped roofs are less likely to be walked on by other than those who have to work on the roof. There are no restrictions on the length of the conductors over the roof.

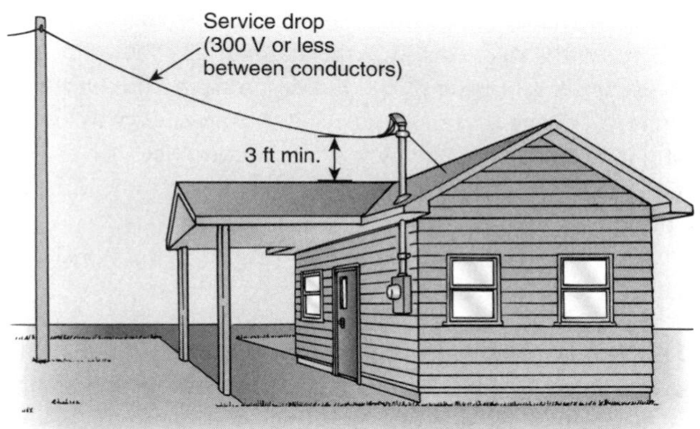


*EXHIBIT 230.19 Reduction in clearance above a roof as permitted by 230.24(A), Exception No. 2.*

*Exception No. 3: Where the voltage between conductors does not exceed 300, a reduction in clearance above only the overhanging portion of the roof to not less than 450 mm (18 in.) shall be permitted if (1) not more than 1.8 m (6 ft) of service-drop conductors, 1.2 m (4 ft) horizontally, pass above the roof overhang, and (2) they are terminated at a through-the-roof raceway or approved support.*

FPN: See 230.28 for mast supports.

Exception No. 3 to 230.24(A) permits a reduction of service-drop conductor clearances to 18 in. above the roof, as illustrated in Exhibit 230.20. This reduction is for service-mast (through-the-roof) installations where the voltage between conductors does not exceed 300 volts (e.g., 120/240, 208Y/120 services) and the mast is located within 4 ft of the edge of the roof, measured horizontally. Exception No. 3 applies to sloped or flat roofs that are easily walked on. Not more than 6 ft of conductors is permitted to pass over the roof.

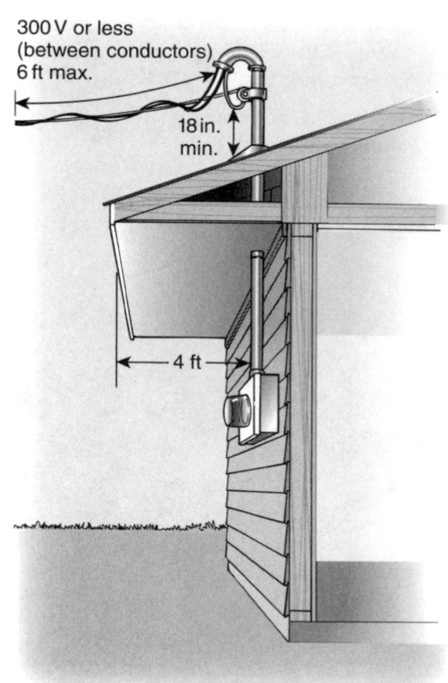


*EXHIBIT 230.20 Reduction in clearance above a roof as permitted by 230.24(A), Exception No. 3.*

*Exception No. 4: The requirement for maintaining the vertical clearance 900 mm (3 ft) from the edge of the roof shall not apply to the final conductor span where the service drop is attached to the side of a building.*

Section 230.24(A) applies to the vertical clearance above roofs for service-drop conductors up to 600 volts. This main rule requires a vertical clearance of 8 ft above the roof, including those areas 3 ft in all directions beyond the edge of the roof.

Exception No. 4 to 230.24(A) exempts the final span of a service drop attached to the side of a building from the 8-ft and 3-ft requirements, to allow the service conductors to be attached to the building, as illustrated in Exhibit 230.21. Exception No. 2 and Exception No. 3 permit lesser clearances for service drops of 300 volts or less, as illustrated in Exhibits 230.19 and 230.20.

If the roof is subject to pedestrian or vehicular traffic, the vertical clearance of the service drop must be the same as the vertical clearance from the ground, in accordance with 230.24(B).

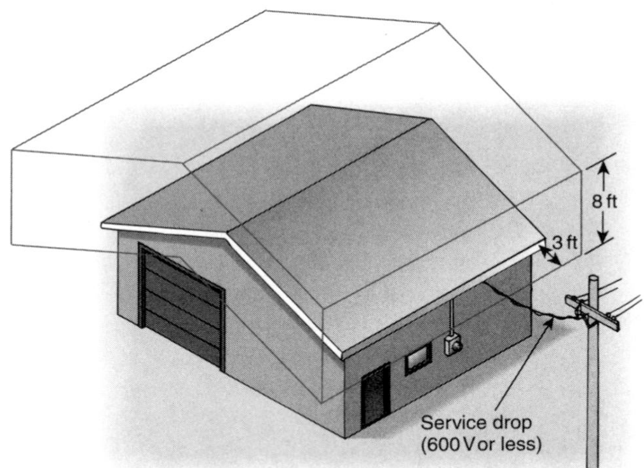


*EXHIBIT 230.21 Clearance of the final span of a service drop, as permitted by 230.24(A), Exception No. 4.*

**(B) Vertical Clearance for Service-Drop Conductors.** Service-drop conductors, where not in excess of 600 volts, nominal, shall have the following minimum clearance from final grade:

(1) 3.0 m (10 ft) — at the electrical service entrance to buildings, also at the lowest point of the drip loop of the building electrical entrance, and above areas or sidewalks accessible only to pedestrians, measured from final grade or other accessible surface only for service-drop cables supported on and cabled together with a grounded bare messenger where the voltage does not exceed 150 volts to ground
(2) 3.7 m (12 ft) — over residential property and driveways, and those commercial areas not subject to truck traffic where the voltage does not exceed 300 volts to ground
(3) 4.5 m (15 ft) — for those areas listed in the 3.7-m (12-ft) classification where the voltage exceeds 300 volts to ground
(4) 5.5 m (18 ft) — over public streets, alleys, roads, parking areas subject to truck traffic, driveways on other than residential property, and other land such as cultivated, grazing, forest, and orchard

**(C) Clearance from Building Openings.** See 230.9.

**(D) Clearance from Swimming Pools.** See 680.8.

**230.70 General.** Means shall be provided to disconnect all conductors in a building or other structure from the service-entrance conductors.

**(A) Location.** The service disconnecting means shall be installed in accordance with 230.70(A)(1), (A)(2), and (A)(3).

No maximum distance is specified from the point of entrance of service conductors to a readily accessible location for the installation of a service disconnecting means. The authority enforcing this *Code* has the responsibility for, and is charged with, making the decision on how far inside the building the service-entrance conductors are allowed to travel to the service disconnecting means. The length of service-entrance conductors should be kept to a minimum inside buildings, because power utilities provide limited overcurrent protection. In the event of a fault, the service conductors could ignite nearby combustible materials.

Some local jurisdictions have ordinances that allow service-entrance conductors to run within the building up to a specified length to terminate at the disconnecting means. The authority having jurisdiction may permit service conductors to bypass fuel storage tanks or gas meters and the like, permitting the service disconnecting means to be located in a readily accessible location.

However, if the authority judges the distance as being excessive, the disconnecting means may be required to be located on the outside of the building or near the building at a readily accessible location that is not necessarily nearest the point of entrance of the conductors. See also 230.6 and Exhibit 230.15 for conductors considered to be outside a building.

See 404.8(A) for mounting-height restrictions for switches and for circuit breakers used as switches.

**(1) Readily Accessible Location.** The service disconnecting means shall be installed at a readily accessible location either outside of a building or structure or inside nearest the point of entrance of the service conductors.

**(2) Bathrooms.** Service disconnecting means shall not be installed in bathrooms.

**(3) Remote Control.** Where a remote control device(s) is used to actuate the service disconnecting means, the service disconnecting means shall be located in accordance with 230.70(A)(1).

**(B) Marking.** Each service disconnect shall be permanently marked to identify it as a service disconnect.

**(C) Suitable for Use.** Each service disconnecting means shall be suitable for the prevailing conditions. Service equipment installed in hazardous (classified) locations shall comply with the requirements of Articles 500 through 517.

**230.71 Maximum Number of Disconnects.**

**(A) General.** The service disconnecting means for each service permitted by 230.2, or for each set of service-entrance conductors permitted by 230.40, Exception No. 1, 3, 4, or 5, shall consist of not more than six switches or sets of circuit breakers, or a combination of not more than six switches and sets of circuit breakers, mounted in a single enclosure, in a group of separate enclosures, or in or on a switchboard. There shall be not more than six sets of disconnects per service grouped in any one location.

Section 230.71(A) covers the maximum number of disconnects permitted as the disconnecting means for the service conductors that supply the building or structure. One set of service-entrance conductors, either overhead or underground, is permitted to supply two to six service disconnecting means in lieu of a single main disconnect. A single-occupancy building can have up to six disconnects for each set of service-entrance conductors. Multiple-occupancy buildings (residential or other than residential) can be provided with one main service disconnect or up to six main disconnects for each set of service-entrance conductors.

Multiple-occupancy buildings may have service-entrance conductors run to each occupancy, and each such set of service-entrance conductors may have from one to six disconnects (see 230.40, Exception No. 1).

Where service-entrance conductors are routed outside the building (see 230.6 and Exhibit 230.15), each set of service-entrance conductors is permitted to supply not more than six disconnecting means at each occupancy of a multiple-occupancy building. See Exhibit 230.2 through Exhibit 230.13 for examples of permitted service configurations.

Exhibit 230.26 shows a single enclosure for grouping service equipment that consists of six circuit breakers or six fused switches. This arrangement does not require a main switch. Six separate enclosures also would be permitted as the service equipment. Where factory-installed switches that disconnect power to surge protective devices and power monitoring equipment are included as part of listed equipment, the last sentence of 230.71(A) specifies that the disconnect switch for such equipment installed as part of the listed equipment does *not* count as one of the six service disconnecting means permitted by 230.71(A). The disconnecting means for the control circuit of ground-fault protection equipment or for a power-operable service disconnecting means are also not considered to be service disconnecting means where such disconnecting means are installed as a component of listed equipment.

***EXHIBIT 230.26** An enclosure for grouping service equipment consisting of six circuit breakers or six fused switches.*

For the purpose of this section, disconnecting means installed as part of listed equipment and used solely for the following shall not be considered a service disconnecting means:

(1) Power monitoring equipment
(2) Surge-protective device(s)
(3) Control circuit of the ground-fault protection system
(4) Power-operable service disconnecting means

**(B) Single-Pole Units.** Two or three single-pole switches or breakers, capable of individual operation, shall be permitted on multiwire circuits, one pole for each ungrounded conductor, as one multipole disconnect, provided they are equipped with identified handle ties or a master handle to disconnect all conductors of the service with no more than six operations of the hand.

FPN: See 408.36, Exception No. 1 and Exception No. 3, for service equipment in certain panelboards, and see 430.95 for service equipment in motor control centers.

## Article 240 Overcurrent Protection

**240.4 Protection of Conductors.** Conductors, other than flexible cords, flexible cables, and fixture wires, shall be protected against overcurrent in accordance with their ampacities specified in 310.15, unless otherwise permitted or required in 240.4(A) through (G).

**(A) Power Loss Hazard.** Conductor overload protection shall not be required where the interruption of the circuit would create a hazard, such as in a material-handling magnet circuit or fire pump circuit. Short-circuit protection shall be provided.

FPN: See NFPA 20-2007, *Standard for the Installation of Stationary Pumps for Fire Protection.*

**(B) Devices Rated 800 Amperes or Less.** The next higher standard overcurrent device rating (above the ampacity of the conductors being protected) shall be permitted to be used, provided all of the following conditions are met:

(1) The conductors being protected are not part of a multioutlet branch circuit supplying receptacles for cord-and-plug-connected portable loads.
(2) The ampacity of the conductors does not correspond with the standard ampere rating of a fuse or a circuit breaker without overload trip adjustments above its rating (but that shall be permitted to have other trip or rating adjustments).
(3) The next higher standard rating selected does not exceed 800 amperes.

Table 210.24 summarizes the requirements for the size of conductors and the size of the overcurrent protection for branch circuits where two or more outlets are required. The first footnote indicates that the wire sizes are for copper conductors. Section 210.3 indicates that branch-circuit conductors rated 15, 20, 30, 40, and 50 amperes must be protected at their ratings. Section 210.19(A) requires that branch-circuit conductors have an ampacity not less than the rating of the branch circuit and not less than the maximum load to be served. These specific requirements take precedence over 240.4(B), which applies generally.

Tables 310.16 through 310.86 list the ampacities of conductors. Section 240.6 lists the standard ratings of overcurrent devices. Where the ampacity of the conductor specified in these tables does not match the rating of the standard overcurrent device, 240.4 permits the use of the next larger standard overcurrent device. All three conditions in 240.4(B) must be met for this permission to apply. However, if the ampacity of a conductor matches the standard rating of 240.6, that conductor must be protected at the standard size device. For example, in Table 310.16, 3 AWG, 75°C copper, Type THWN, the ampacity is listed as 100 amperes. That conductor would be protected by a 100-ampere overcurrent device.

The provisions of 240.4(B) do not modify or change the allowable ampacity of the conductor — they only serve to provide a reasonable increase in the permitted overcurrent protective device rating where the allowable ampacity and the standard overcurrent protective device ratings do not correspond. For circuits rated 600 volts and under, the allowable ampacity of branch circuit, feeder, or service conductors always has to be capable of supplying the calculated load in accordance with the requirements of 210.19(A)(1), 215.2(A)(1), and 230.42(A). For example, a 500-kcmil THWN copper conductor has an allowable ampacity of 380 amperes from Table 310.16. This conductor can supply a load not exceeding 380 amperes and, in accordance with 240.4(B), can be protected by a 400-ampere overcurrent protective device.

In contrast to 240.4(B), 310.15(B)(6) does permit the conductor types and sizes specified in Table 310.15(B)(6) to supply calculated loads based on their ratings from that table that exceed their allowable ampacities specified in Table 310.16. The overcurrent protection for these residential supply conductors is also permitted to be based on the increased rating allowed by this Article 310 table. Application of 310.15(B)(6) and its table is permitted only for single-phase, 120/240-volt, residential services and main power feeders. The increased ratings given in Table 310.15(B)(6) are based on the significant diversity inherent to most dwelling unit loads and the fact that only the two ungrounded service or feeder conductors are considered to be current carrying.

**(C) Devices Rated over 800 Amperes.** Where the overcurrent device is rated over 800 amperes, the ampacity of the conductors it protects shall be equal to or greater than the rating of the overcurrent device defined in 240.6.

**(D) Small Conductors.** Unless specifically permitted in 240.4(E) or (G), the overcurrent protection shall not exceed that required by (D)(1) through (D)(7) after any correction factors for ambient temperature and number of conductors have been applied.

**(1) 18 AWG Copper.** 7 amperes, provided all the following conditions are met:

(1) Continuous loads do not exceed 5.6 amperes.
(2) Overcurrent protection is provided by one of the following:
   a. Branch-circuit-rated circuit breakers listed and marked for use with 18 AWG copper wire
   b. Branch-circuit-rated fuses listed and marked for use with 18 AWG copper wire
   c. Class CC, Class J, or Class T fuses

**(2) 16 AWG Copper.** 10 amperes, provided all the following conditions are met:

(1) Continuous loads do not exceed 8 amperes.
(2) Overcurrent protection is provided by one of the following:
   a. Branch-circuit-rated circuit breakers listed and marked for use with 16 AWG copper wire
   b. Branch-circuit-rated fuses listed and marked for use with 16 AWG copper wire
   c. Class CC, Class J, or Class T fuses

**(3) 14 AWG Copper.** 15 amperes

**(4) 12 AWG Aluminum and Copper-Clad Aluminum.** 15 amperes

**(5) 12 AWG Copper.** 20 amperes

**(6) 10 AWG Aluminum and Copper-Clad Aluminum.** 25 amperes

**(7) 10 AWG Copper.** 30 amperes

**(E) Tap Conductors.** Tap conductors shall be permitted to be protected against overcurrent in accordance with the following:

(1) 210.19(A)(3) and (A)(4), Household Ranges and Cooking Appliances and Other Loads
(2) 240.5(B)(2), Fixture Wire
(3) 240.21, Location in Circuit
(4) 368.17(B), Reduction in Ampacity Size of Busway
(5) 368.17(C), Feeder or Branch Circuits (busway taps)
(6) 430.53(D), Single Motor Taps

**(F) Transformer Secondary Conductors.** Single-phase (other than 2-wire) and multiphase (other than delta-delta, 3-wire) transformer secondary conductors shall not be considered to be protected by the primary overcurrent protective device. Conductors supplied by the secondary side of a single-phase transformer having a 2-wire (single-voltage) secondary, or a three-phase, delta-delta connected transformer having a 3-wire (single-voltage) secondary, shall be permitted to be protected by overcurrent protection provided on the primary (supply) side of the transformer, provided this protection is in accordance with 450.3 and does not exceed the value determined by multiplying the secondary conductor ampacity by the secondary-to-primary transformer voltage ratio.

The fundamental requirement of 240.4 specifies that conductors are to be protected against overcurrent in accordance with their ampacity, and 240.21 requires that the protection be provided at the point the conductor receives its supply. Section 240.4(F) permits the secondary circuit conductors from a transformer to be protected by overcurrent devices in the primary circuit conductors of the transformer only in the following two special cases:

1. A transformer with a 2-wire primary and a 2-wire secondary, provided the transformer primary is protected in accordance with 450.3
2. A 3-phase, delta-delta-connected transformer having a 3-wire, single-voltage secondary, provided its primary is protected in accordance with 450.3

Except for those two special cases, transformer secondary conductors must be protected by the use of overcurrent devices, because the primary overcurrent devices do not provide such protection. As an example, consider a single-phase transformer with a 2-wire secondary that is provided with primary overcurrent protection rated at 50 amperes. The transformer is rated 480/240 volts. Conductors supplied by the secondary have an ampacity of 100 amperes. Is the 50-ampere overcurrent protection allowed to protect the conductors that are connected to the secondary?

The secondary-to-primary voltage ratio in this example is 240 ÷ 480, a ratio of 0.5. Multiplying the secondary conductor ampacity of 100 amperes by 0.5 yields 50 amperes. Thus, the maximum rating of the overcurrent device allowed on the primary of the transformer that will also provide overcurrent protection for the secondary conductors is 50 amperes. These secondary conductors are not tap conductors, are not limited in length, and do not require overcurrent protection where they receive their supply, which is at the transformer secondary terminals.

However, if the secondary consisted of a 3-wire, 240/120-volt system, a 120-volt line-to-neutral load could draw up to 200 amperes before the overcurrent device in the primary actuated. That would be the result of the 1:4 secondary-to-primary voltage ratio of the 120-volt winding of the transformer secondary, which can cause dangerous overloading of the secondary conductors.

**(G) Overcurrent Protection for Specific Conductor Applications.** Overcurrent protection for the specific conductors shall be permitted to be provided as referenced in Table 240.4(G).

**240.5 Protection of Flexible Cords, Flexible Cables, and Fixture Wires.** Flexible cord and flexible cable, including tinsel cord and extension cords, and fixture wires shall be protected against overcurrent by either 240.5(A) or (B).

**(A) Ampacities.** Flexible cord and flexible cable shall be protected by an overcurrent device in accordance with their ampacity as specified in Table 400.5(A) and Table 400.5(B). Fixture wire shall be protected against overcurrent in accordance with its ampacity as specified in Table 402.5. Supplementary overcurrent protection, as covered in 240.10, shall be permitted to be an acceptable means for providing this protection.

**(B) Branch-Circuit Overcurrent Device.** Flexible cord shall be protected, where supplied by a branch circuit, in accordance with one of the methods described in 240.5(B)(1), (B)(3), or (B)(4). Fixture wire shall be protected, where supplied by a branch circuit, in accordance with 240.5(B)(2).

**(1) Supply Cord of Listed Appliance or Luminaire.** Where flexible cord or tinsel cord is approved for and used with a specific listed appliance or luminaire, it shall be considered to be protected when applied within the appliance or luminaire listing requirements. For the purposes of this section, a luminaire may be either portable or permanent.

**(2) Fixture Wire.** Fixture wire shall be permitted to be tapped to the branch-circuit conductor of a branch circuit in accordance with the following:

(1) 20-ampere circuits — 18 AWG, up to 15 m (50 ft) of run length
(2) 20-ampere circuits — 16 AWG, up to 30 m (100 ft) of run length
(3) 20-ampere circuits — 14 AWG and larger
(4) 30-ampere circuits — 14 AWG and larger

Section 240.5(A) references Tables 400.5(A) and (B) for flexible cords and flexible cables, and Table 402.5 for fixture wire ampacity. Supplementary protection, as described in 240.10, is also acceptable as an alternative for protection of either flexible cord or fixture wire.

Sections 240.5(B)(1) through (B)(4) permit smaller conductors to be connected to branch circuits of a greater rating. For flexible cords, 240.5(B)(1) and (B)(3) now specify that flexible cord connected to a listed appliance or portable lamp or used in a listed extension cord set is considered to be protected as long as the appliance, lamp, or extension cord is used in accordance with its listing requirements. These listing requirements are developed by the third-party testing and listing organizations with technical input from cord, appliance, and lamp manufacturers. For other than field-assembled extension cords, the *Code* no longer contains specific provisions for the overcurrent protection of flexible cord based on cord conductor size. For fixture wire, 240.5(B)(2) establishes a maximum protective device rating based on a minimum conductor size and a maximum conductor length.

(5) 40-ampere circuits — 12 AWG and larger
(6) 50-ampere circuits — 12 AWG and larger

**(3) Extension Cord Sets.** Flexible cord used in listed extension cord sets shall be considered to be protected when applied within the extension cord listing requirements.

**(4) Field Assembled Extension Cord Sets.** Flexible cord used in extension cords made with separately listed and installed components shall be permitted to be supplied by a branch circuit in accordance with the following:

20-ampere circuits — 16 AWG and larger

Field-assembled extension cords are permitted, provided the conductors are 16 AWG or larger and the overcurrent protection for the branch circuit to which the cord is connected does not exceed 20 amperes. The cord and the cord caps and connectors used for this type of assembly are required to be listed.

**240.40 Disconnecting Means for Fuses.** Cartridge fuses in circuits of any voltage where accessible to other than qualified persons, and all fuses in circuits over 150 volts to ground, shall be provided with a disconnecting means on their supply side so that each circuit containing fuses can be independently disconnected from the source of power. A current-limiting device without a disconnecting means shall be permitted on the supply side of the service disconnecting means as permitted by 230.82. A single disconnecting means shall be permitted on the supply side of more than one set of fuses as permitted by 430.112, Exception, for group operation of motors and 424.22(C) for fixed electric space-heating equipment.

A single disconnect switch is allowed to serve more than one set of fuses, such as in multimotor installations or for electric space-heating equipment where the heating element load is required to be subdivided, each element with its own set of fuses. The installation of cable limiters or similar current-limiting devices on the supply side of the service disconnecting means is permitted by 230.82(1). No disconnecting means is required on the supply side of such devices.

**240.41 Arcing or Suddenly Moving Parts.** Arcing or suddenly moving parts shall comply with 240.41(A) and (B).

**(A) Location.** Fuses and circuit breakers shall be located or shielded so that persons will not be burned or otherwise injured by their operation.

**(B) Suddenly Moving Parts.** Handles or levers of circuit breakers, and similar parts that may move suddenly in such a way that persons in the vicinity are likely to be injured by being struck by them, shall be guarded or isolated.

Arcing or sudden-moving parts are usually associated with switchboards or control boards that may be of the open type. Switchboards and control boards should be under competent supervision and accessible only to qualified persons. Fuses or circuit breakers must be located or shielded so that, under an abnormal condition, the subsequent arc across the opening device will not injure persons in the vicinity.

Guardrails may be provided in the vicinity of disconnecting means because sudden-moving handles may be capable of causing injury. Modern switchboards, for example, are equipped with removable handles. See Article 100 for the definition of *guarded*. See also 110.27 for the guarding of live parts (600 volts, nominal, or less).

**240.86 Series Ratings.** Where a circuit breaker is used on a circuit having an available fault current higher than the marked interrupting rating by being connected on the load side of an acceptable overcurrent protective device having a higher rating, the circuit breaker shall meet the requirements specified in (A) or (B), and (C).

A series rated system is a combination of circuit breakers or a combination of fuses and circuit breakers that can be applied at available short-circuit levels above the interrupting rating of the load-side circuit breakers but not above that of the main or line-side device. Series rated systems can consist of fuses that protect circuit breakers or of circuit breakers that protect circuit breakers. The arrangement of protective components in a series rated system can be as specified in 240.86(A) for engineered systems applied to existing installations or in 240.86(B) for tested combinations that can be applied in any new or existing installation.

**(A) Selected Under Engineering Supervision in Existing Installations.** The series rated combination devices shall be selected by a licensed professional engineer engaged primarily in the design or maintenance of electrical installations. The selection shall be documented and stamped by the professional engineer. This documentation shall be available to those authorized to design, install, inspect, maintain, and operate the system. This series combination rating, including identification of the upstream device, shall be field marked on the end use equipment.

For calculated applications, the engineer shall ensure that the downstream circuit breaker(s) that are part of the series combination remain passive during the interruption period of the line side fully rated, current-limiting device.

This provision allows for an engineering solution at existing facilities where an increase in the available fault current (due to factors such as increases in transformer size, lowering of transformer impedances, and changes in utility distribution systems) puts the existing circuit overcurrent protection equipment at peril in regard to interrupting fault currents as required by 110.9. The objective of this "engineered system" is to maintain compliance with 110.9 by redesigning the overcurrent protection scheme to accommodate the increase in available fault current and not having to undertake a wholesale replacement of electrical distribution equipment. Where the increase in fault current causes existing equipment to be "underrated," the engineering approach is to provide upstream protection that functions in concert with the existing protective devices to safely open the circuit under fault conditions. The requirement specifies that the design of such systems is to be performed only by licensed professional engineers whose credentials substantiate their ability to perform this type of engineering. Documentation in the form of stamped drawings and field marking of end-use equipment to indicate it is a component of a series rated system is required.

Designing a series rated system requires careful consideration of the fault-clearing characteristics of the existing protective devices and their ability to interact with the newly installed upstream protective device(s) when subjected to fault conditions. This new provision does not ensure that an engineered series rated system can be applied to all existing installations. The operating parameters of the existing overcurrent protection equipment dictate what can be done in a field-engineered protection scheme.

Compatibility with series rated systems will in all likelihood be limited to circuit breakers that (1) remain closed during the interruption period of the fully rated OCPD installed on their line side and (2) have an interrupting rating that is not less than the let-through current of an upstream protective device (such as a current-limiting fuse). In those cases where the opening of a circuit breaker, under any level of fault current, begins in less than ½ cycle, the use of a field engineered series rated system will in all likelihood be contrary to acceptable application practices specified by the circuit breaker manufacturer. The passivity of the downstream circuit breakers during interruption of a fault by an upstream device has to be ensured by the design engineer in order for this engineered approach to be employed. Where there is any doubt about the proper application of existing downstream circuit breakers with new upstream overcurrent protective devices, the manufacturers of the existing circuit breakers and the new upstream overcurrent protective devices must be consulted.

The safety objective of any overcurrent protection scheme is to ensure compliance with 110.9.

**(B) Tested Combinations.** The combination of line-side overcurrent device and load-side circuit breaker(s) is tested and marked on the end use equipment, such as switchboards and panelboards.

FPN to (A) and (B): See 110.22 for marking of series combination systems.

Section 240.86(B) requires that, when a series rating is used, the switchboards, panelboards, and load centers be marked for use with the series rated combinations that may be used. Therefore, the enclosures must have a label affixed by the equipment manufacturer that provides the series rating of the combination(s). Because there is often not enough room in the equipment to show all the legitimate series rated combinations, UL 67, *Standard for Panelboards*, allows a bulletin to be referenced and supplied with the panelboard. These bulletins typically provide all the acceptable combinations. Note that the installer of a series rated system also must provide the additional labeling on equipment enclosures required by 110.22, indicating that the equipment has been applied in a series rated system.

**(C) Motor Contribution.** Series ratings shall not be used where

(1) Motors are connected on the load side of the higher-rated overcurrent device and on the line side of the lower-rated overcurrent device, and
(2) The sum of the motor full-load currents exceeds 1 percent of the interrupting rating of the lower-rated circuit breaker.

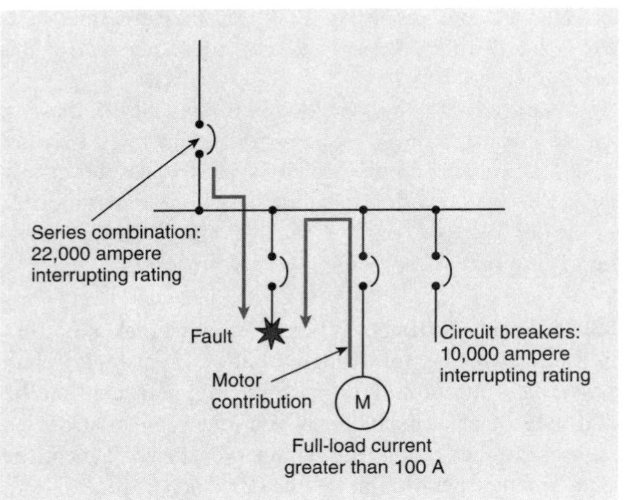


*EXHIBIT 240.14 Example of installation where level of motor contribution exceeds 1 percent of interrupting rating for the lowest-rated circuit breaker in this series rated system.*

One critical requirement limits the use of series rated systems in which motors are connected between the line-side (protecting) device and the load-side (protected) circuit breaker. Section 240.86(C) requires that series ratings developed under the parameters of either 240.86(A) or (B) are not to be used where the sum of motor full-load currents exceeds 1 percent of the interrupting rating of the load-side (protected) circuit breaker, as illustrated in Exhibit 240.14.

**240.101 Additional Requirements for Feeders.**

**(A) Rating or Setting of Overcurrent Protective Devices.** The continuous ampere rating of a fuse shall not exceed three times the ampacity of the conductors. The long-time trip element setting of a breaker or the minimum trip setting of an electronically actuated fuse shall not exceed six times the ampacity of the conductor. For fire pumps, conductors shall be permitted to be protected for overcurrent in accordance with 695.4(B).

**(B) Feeder Taps.** Conductors tapped to a feeder shall be permitted to be protected by the feeder overcurrent device where that overcurrent device also protects the tap conductor.

## Article 250 Grounding and Bonding

**250.20 Alternating-Current Systems to Be Grounded.** Alternating-current systems shall be grounded as provided for in 250.20(A), (B), (C), (D), or (E). Other systems shall be permitted to be grounded. If such systems are grounded, they shall comply with the applicable provisions of this article.

FPN: An example of a system permitted to be grounded is a corner-grounded delta transformer connection. See 250.26(4) for conductor to be grounded.

**(A) Alternating-Current Systems of Less Than 50 Volts.** Alternating-current systems of less than 50 volts shall be grounded under any of the following conditions:

(1) Where supplied by transformers, if the transformer supply system exceeds 150 volts to ground
(2) Where supplied by transformers, if the transformer supply system is ungrounded
(3) Where installed outside as overhead conductors

**(B) Alternating-Current Systems of 50 Volts to 1000 Volts.** Alternating-current systems of 50 volts to 1000 volts that supply premises wiring and premises wiring systems shall be grounded under any of the following conditions:

(1) Where the system can be grounded so that the maximum voltage to ground on the ungrounded conductors does not exceed 150 volts

Exhibit 250.4 illustrates the grounding requirements of 250.20(B)(1) as applied to a 120-volt, single-phase, 2-wire system and to a 120/240-volt, single-phase, 3-wire system. The selection of which conductor to be grounded is covered by 250.26.

(2) Where the system is 3-phase, 4-wire, wye connected in which the neutral conductor is used as a circuit conductor
(3) Where the system is 3-phase, 4-wire, delta connected in which the midpoint of one phase winding is used as a circuit conductor

Exhibit 250.5 illustrates which conductor is required to be grounded for all wye systems if the neutral is used as a circuit conductor. Where the midpoint of one phase of a 3-phase, 4-wire delta system is used as a circuit conductor, it must be grounded and the high-leg conductor must be identified. See 250.20(B)(2) and (B)(3), as well as 250.26.

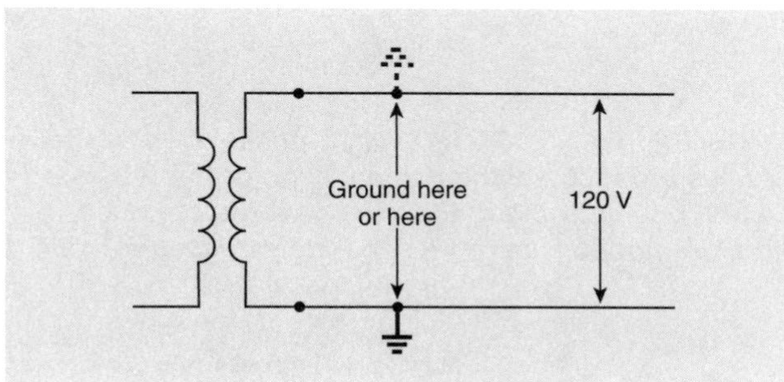


120-V, single-phase, 2-wire system

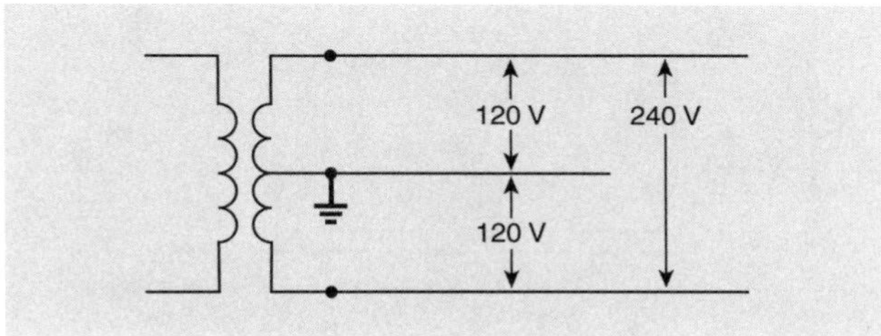


120/240-V, single-phase, 3-wire system

***EXHIBIT 250.4*** *Typical systems required to be grounded in accordance with 250.20(B)(1). The conductor to be grounded is in accordance with 250.26.*

**(C) Alternating-Current Systems of 1 kV and Over.** Alternating-current systems supplying mobile or portable equipment shall be grounded as specified in 250.188. Where supplying other than mobile or portable equipment, such systems shall be permitted to be grounded.

**(D) Separately Derived Systems.** Separately derived systems, as covered in 250.20(A) or (B), shall be grounded as specified in 250.30(A). Where an alternate source such as an on-site generator is provided with transfer equipment that includes a grounded conductor that is not solidly interconnected to the service-supplied grounded conductor, the alternate source (derived system) shall be grounded in accordance with 250.30(A).

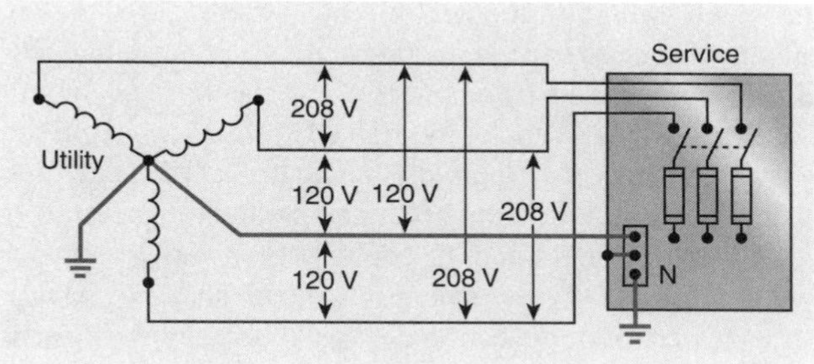


208Y/120-V, 3-phase, 4-wire wye system

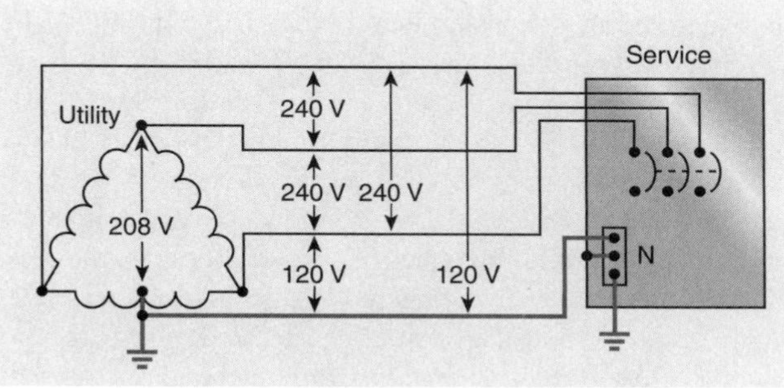


120/240-V, 3-phase, 4-wire delta system

***EXHIBIT 250.5*** *Typical systems required to be grounded by 250.20(B)(2) and 250.20(B)(3). The conductor to be grounded is in accordance with 250.26.*

Two of the most common sources of separately derived systems in premises wiring are transformers and generators. An autotransformer or step-down transformer that is part of electrical equipment and that does not supply premises wiring is not the source of a separately derived system. See the definition of *premises wiring* in Article 100.

FPN No. 1: An alternate ac power source such as an on-site generator is not a separately derived system if the grounded conductor is solidly interconnected to a service-supplied system grounded conductor. An example of such situations is where alternate source transfer equipment does not include a switching action in the grounded conductor and allows it to remain solidly connected to the service-supplied grounded conductor when the alternate source is operational and supplying the load served.

Exhibits 250.6 and 250.7 depict a 208Y/120-volt, 3-phase, 4-wire electrical service supplying a service disconnecting means to a building. The system is fed through a transfer switch connected to a generator intended to provide power for an emergency or standby system.

In Exhibit 250.6, the neutral conductor from the generator to the load is not disconnected by the transfer switch. The system has a direct electrical connection between the normal grounded system conductor (neutral) and the generator neutral through the neutral bus in the transfer switch, thereby grounding the generator neutral. Because the generator is grounded by connection to the normal system ground, it is not a separately derived system, and there are no requirements for grounding the neutral at the generator.

In Exhibit 250.7, the grounded conductor (neutral) is connected to the switching contacts of a 4-pole transfer switch. Therefore, the generator system does not have a direct electrical connection to the other supply system grounded conductor (neutral), and the system supplied by the generator is considered separately derived. This separately derived system (3-phase, 4-wire, wye-connected system that supplies line-to-neutral loads) is required to be grounded in accordance with 250.20(B) and (D). The methods for grounding the system are specified in 250.30(A).

Section 250.30(A)(1) requires separately derived systems to have a system bonding jumper connected between the generator frame and the grounded circuit conductor (neutral). The grounding electrode conductor from the generator is required to be connected to a grounding electrode. This conductor and the grounding electrode are to be located as close to the generator as practicable, according to 250.30(A)(7). If the generator is in a building, the grounding electrode is required to be one of the following, depending on which grounding electrode is closest to the generator location: (1) effectively grounded structural metal member or (2) the first 5 ft of water pipe into a building where the piping is effectively grounded. [The exception to 250.52(A)(1) permits the grounding connection to the water piping beyond the first 5 ft.] For buildings or structures in which the preferred electrodes are not available, the choice can be made from any of the grounding electrodes specified in 250.52(A)(3) through (A)(8). See 250.35(B) for more information on the grounding/bonding conductor installed between the generator and the transfer switch.

FPN No. 2: For systems that are not separately derived and are not required to be grounded as specified in 250.30, see 445.13 for minimum size of conductors that must carry fault current.

**(E) Impedance Grounded Neutral Systems.** Impedance grounded neutral systems shall be grounded in accordance with 250.36 or 250.186.

**250.21 Alternating-Current Systems of 50 Volts to 1000 Volts Not Required to Be Grounded.**

**(A) General.** The following ac systems of 50 volts to 1000 volts shall be permitted to be grounded but shall not be required to be grounded:

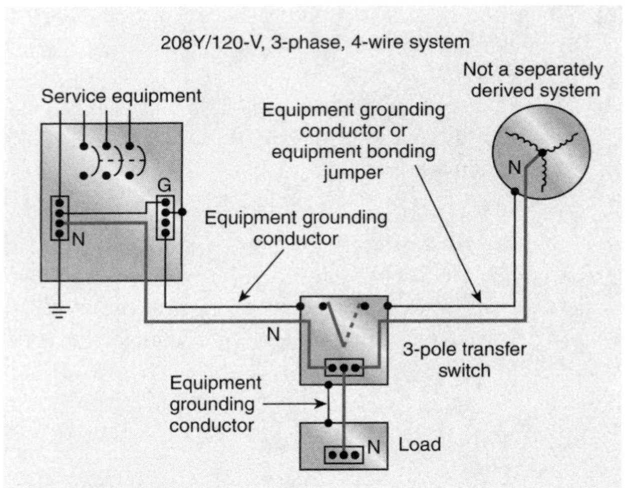


***EXHIBIT 250.6*** *A 208Y/120-volt, 3-phase, 4-wire system that has a direct electrical connection of the grounded circuit conductor (neutral) to the generator and is therefore not considered a separately derived system.*

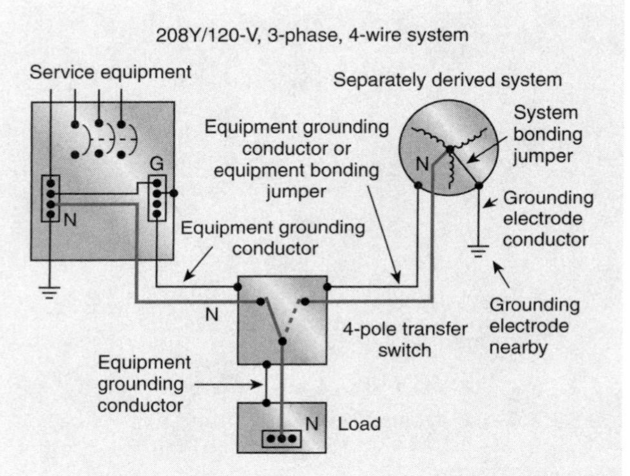


***EXHIBIT 250.7*** *A 208Y/120-volt, 3-phase, 4-wire system that does not have a direct electrical connection of the grounded circuit conductor (neutral) to the generator and is therefore considered a separately derived system.*

(1) Electrical systems used exclusively to supply industrial electric furnaces for melting, refining, tempering, and the like
(2) Separately derived systems used exclusively for rectifiers that supply only adjustable-speed industrial drives
(3) Separately derived systems supplied by transformers that have a primary voltage rating less than 1000 volts, provided that all the following conditions are met:

a. The system is used exclusively for control circuits.
b. The conditions of maintenance and supervision ensure that only qualified persons service the installation.
c. Continuity of control power is required.

(4) Other systems that are not required to be grounded in accordance with the requirements of 250.20(B)

**(B) Ground Detectors.** Ungrounded alternating current systems as permitted in 250.21(A)(1) through (A)(4) operating at not less than 120 volts and not exceeding 1000 volts shall have ground detectors installed on the system.

Ungrounded electrical systems as permitted in 250.21 are required to be provided with ground detectors. In the 2002 and previous editions of the *Code*, the installation of ground detectors was required only for some very specific applications of ungrounded systems (and in impedance grounded neutral systems), but only a recommendation was given that they be installed on all ungrounded electrical systems. For further information on what is considered to be the voltage to ground in an ungrounded system, see the definition of *voltage to ground* in Article 100.

Ungrounded electrical systems are permitted by the *NEC* for the specific functions described in 250.21(1), (2), and (3) and for general power distribution systems in accordance with 250.21(4). Delta-connected, 3-phase, 3-wire, 240-volt and 480-volt systems are examples of common electrical distribution systems that are permitted but are not required to have a circuit conductor that is intentionally grounded. The operational advantage in using an ungrounded electrical system is continuity of operation, which in some processes might create a safer condition than would be achieved by the automatic and unplanned opening of the supply circuit. The disadvantage of operating systems ungrounded is increased susceptibility to high transient voltages that can hasten insulation deterioration. As stated in 250.4(A)(1), limiting voltage impressed on the system due to lightning or line surges is a primary function of system grounding.

Unlike solidly grounded systems, in which the first line-to-ground fault causes the overcurrent protective device to automatically open the circuit, the same line-to-ground fault in an ungrounded system does not result in the operation of the overcurrent device — it simply results in the faulted circuit conductor becoming a grounded conductor until a repair of the damaged conductor insulation can be performed. However, this latent ground-fault condition will remain undetected unless ground detectors are installed in the ungrounded system or until another insulation failure on a different ungrounded conductor results in a line-to-line-to-ground fault, with the potential for more extensive damage to electrical equipment.

Ground detectors are used to provide a visual indication, an audible signal, or both, to alert system operators and maintainers of a ground-fault condition in the electrical system. With notification of the ground-fault condition, rather than automatic interruption of the circuit, the operators of the process supplied by the ungrounded system can then take the necessary steps to effect an orderly shutdown, determine where the ground fault is located in the system, and safely perform the necessary repair.

It should be noted that ungrounded systems are simply systems without an intentionally grounded circuit conductor that is part of normal circuit operation, as is the case in 120/240-volt, single-phase, 3-wire; 208Y/120-volt, 3-phase, 4-wire; and 480Y/277-volt, 3-phase, 4-wire systems in which there is a grounded conductor that is used as a circuit conductor. The fact that a system operates without a grounded conductor does not exempt that system from complying with all of the applicable requirements in Article 250 for establishing a grounding electrode system and for equipment grounding. These protective features are required for grounded and ungrounded electrical distribution systems.

**250.22 Circuits Not to Be Grounded.** The following circuits shall not be grounded:

(1) Circuits for electric cranes operating over combustible fibers in Class III locations, as provided in 503.155
(2) Circuits in health care facilities as provided in 517.61 and 517.160
(3) Circuits for equipment within electrolytic cell working zone as provided in Article 668
(4) Secondary circuits of lighting systems as provided in 411.5(A)
(5) Secondary circuits of lighting systems as provided in 680.23(A)(2).

**250.24 Grounding Service-Supplied Alternating-Current Systems.**

**(A) System Grounding Connections.** A premises wiring system supplied by a grounded ac service shall have a grounding electrode conductor connected to the grounded service conductor, at each service, in accordance with 250.24(A)(1) through (A)(5).

**(1) General.** The grounding electrode conductor connection shall be made at any accessible point from the load end of the service drop or service lateral to and including the terminal or bus to which the grounded service conductor is connected at the service disconnecting means.

FPN: See definitions of *Service Drop* and *Service Lateral* in Article 100.

The grounded conductor of an ac service is connected to a grounding electrode system to limit the voltage to ground imposed on the system by lightning, line surges, and (unintentional) high-voltage crossovers. Another reason for requiring this connection is to stabilize the voltage to ground during normal operation, including short circuits. These performance requirements are stated in 250.4(A) and 250.4(B).

The actual connection of the grounded service conductor to the grounded electrode conductor is permitted to be made at various locations, according to 250.24(A)(1). Allowing various locations for the connection to be made continues to meet the overall objectives for grounding while allowing the installer a variety of practical solutions. Exhibit 250.8 illustrates three possible connection point locations to where the grounded conductor of the service could be connected to the grounding electrode conductor.

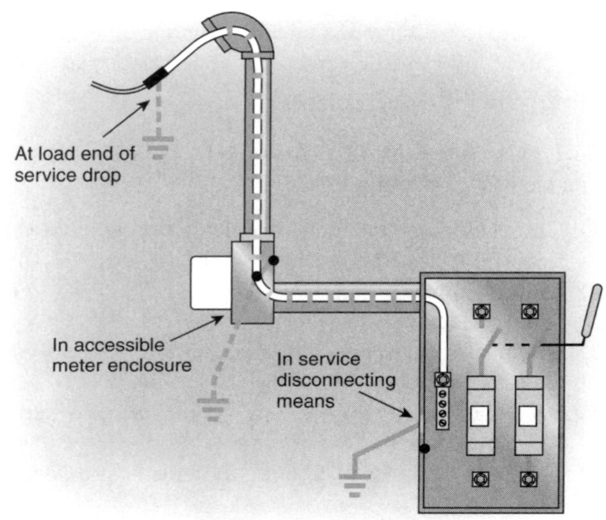


*EXHIBIT 250.8 An ac service supplied from an overhead distribution system illustrating three accessible connection points where the grounded service conductor is connected to the grounding electrode conductor according to 250.24(A)(1).*

**(2) Outdoor Transformer.** Where the transformer supplying the service is located outside the building, at least one additional grounding connection shall be made from the grounded service conductor to a grounding electrode, either at the transformer or elsewhere outside the building.

See Exhibit 250.9 for an illustration of an outdoor distribution system transformer connected to an additional grounding electrode.

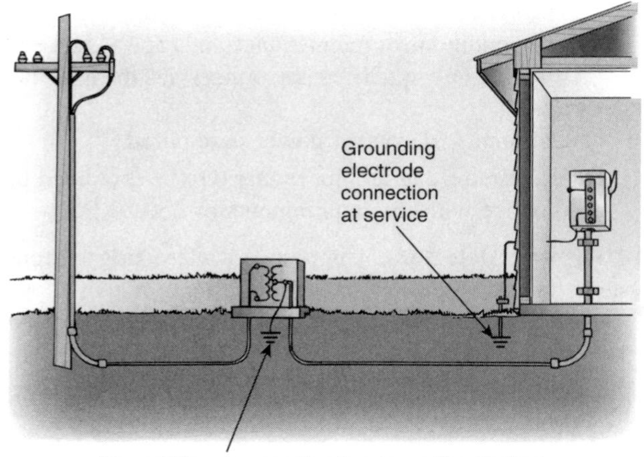


*EXHIBIT 250.9 A 3-wire, 120/240-volt ac, single-phase, secondary distribution system in which grounding connections are required on the secondary side of the transformer according to 250.24(A)(2) and on the supply side of the service disconnecting means according to 250.24(A)(1).*

*Exception: The additional grounding electrode conductor connection shall not be made on high-impedance grounded neutral systems. The system shall meet the requirements of 250.36.*

**(3) Dual-Fed Services.** For services that are dual fed (double ended) in a common enclosure or grouped together in separate enclosures and employing a secondary tie, a single grounding electrode conductor connection to the tie point of the grounded conductor(s) from each power source shall be permitted.

**(4) Main Bonding Jumper as Wire or Busbar.** Where the main bonding jumper specified in 250.28 is a wire or busbar and is installed from the grounded conductor terminal bar or bus to the equipment grounding terminal bar or bus in the service equipment, the grounding electrode conductor shall be permitted to be connected to the equipment grounding terminal, bar, or bus to which the main bonding jumper is connected.

**(5) Load-Side Grounding Connections.** A grounded conductor shall not be connected to normally non–current-carrying metal parts of equipment, to equipment grounding conductor(s), or be reconnected to ground on the load side of the service disconnecting means except as otherwise permitted in this article.

FPN: See 250.30(A) for separately derived systems, 250.32 for connections at separate buildings or structures, and 250.142 for use of the grounded circuit conductor for grounding equipment.

The power for ac premises wiring systems is either separately derived, in accordance with 250.20(D), or supplied by the service. See the definition of *service* in Article 100. Section 250.30 covers grounding requirements for separately derived ac systems, and 250.24(A) covers system grounding requirements for service-supplied ac systems.

According to 250.24, a premises wiring system supplied by an ac service that is required to be grounded must have a grounding electrode conductor at each service connected to the grounding electrodes that meets the requirements in Part III of Article 250. Note that the grounding electrode requirements for a grounded separately derived system are specified in 250.30(A)(3), (A)(4), (A)(5), and (A)(7).

The grounding electrode conductor connection to the grounded conductor is specific. The *Code* requires that the connection be made to the grounded service conductor and describes where this connection is permitted. Where the transformer supplying a service is located outside of a building or structure, a grounding connection must be made at the transformer secondary or at another outdoor location under the conditions specified in 250.24(A)(2). In addition, the conductor that is grounded at the transformer is required to be grounded again at the building or structure, according to 250.24(A)(1).

Section 250.24(A)(5) prohibits regrounding of the grounded conductor on the load side of the service disconnecting means. This requirement correlates with the requirement of 250.142(B), which is a general prohibition on the use of the grounded conductor for grounding equipment.

**(B) Main Bonding Jumper.** For a grounded system, an unspliced main bonding jumper shall be used to connect the equipment grounding conductor(s) and the service-disconnect enclosure to the grounded conductor within the enclosure for each service disconnect in accordance with 250.28.

Where the service equipment of a grounded system consists of multiple disconnecting means, a main bonding jumper for each separate service disconnecting means is required to connect the grounded service conductor, the equipment grounding conductor, and the service equipment enclosure. The size of the main bonding jumper in each enclosure is selected from Table 250.66 based on the size of the ungrounded service conductors, or it is calculated in accordance with 250.28(D)(1). See Exhibits 250.10 and 250.11, which accompany the commentary following 250.24(C), Exception.

*Exception No. 1: Where more than one service disconnecting means is located in an assembly listed for use as service equipment, an unspliced main bonding jumper shall bond the grounded conductor(s) to the assembly enclosure.*

Where multiple service disconnecting means are part of an assembly listed as service equipment, all grounded service conductors are required to be run to and bonded to the assembly. However, only one section of the assembly is required to have the main bonding jumper connection. See Exhibit 250.12, which accompanies the commentary following 250.28(D).

*Exception No. 2: Impedance grounded neutral systems shall be permitted to be connected as provided in 250.36 and 250.186.*

**(C) Grounded Conductor Brought to Service Equipment.** Where an ac system operating at less than 1000 volts is grounded at any point, the grounded conductor(s) shall be run to each service disconnecting means and shall be connected to each disconnecting means grounded conductor(s) terminal or bus. A main bonding jumper shall connect the grounded conductor(s) to each service disconnecting means enclosure. The grounded conductor(s) shall be installed in accordance with 250.24(C)(1) through (C)(3).

*Exception: Where more than one service disconnecting means are located in a single assembly listed for use as service equipment, it shall be permitted to run the grounded conductor(s) to the assembly common grounded conductor(s) terminal or bus. The assembly shall include a main bonding jumper for connecting the grounded conductor(s) to the assembly enclosure.*

If the utility service that supplies premises wiring is grounded, the grounded conductor, whether or not it is used to supply a load, must be run to the service equipment, be bonded to the equipment, and be connected to a grounding electrode system. Exhibit 250.10 shows an example of the main rule in 250.24(C), which requires the grounded service conductor to be brought in and bonded to each service disconnecting means enclosure. This requirement is based on the grounded conductor being used to complete the ground-fault current path between the service equipment and the utility source. The grounded service conductor's other function, as a circuit conductor for normal loads, is covered in 200.3 and 220.61.

The exception to 250.24(C) permits a single connection of the grounded service conductor to a listed service

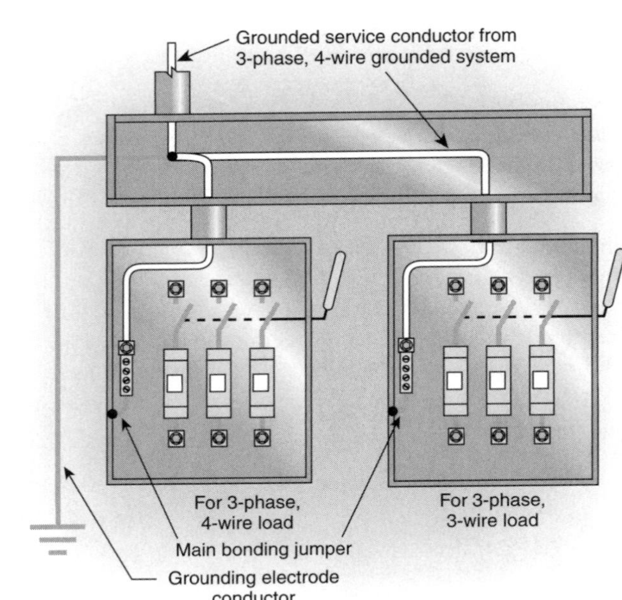


*EXHIBIT 250.10 A grounded system in which the grounded service conductor is brought into a 3-phase, 4-wire service equipment enclosure and to the 3-phase, 3-wire service equipment enclosure, where it is bonded to each service disconnecting means.*

assembly (such as a switchboard) that contains more than one service disconnecting means, as shown in Exhibit 250.11.

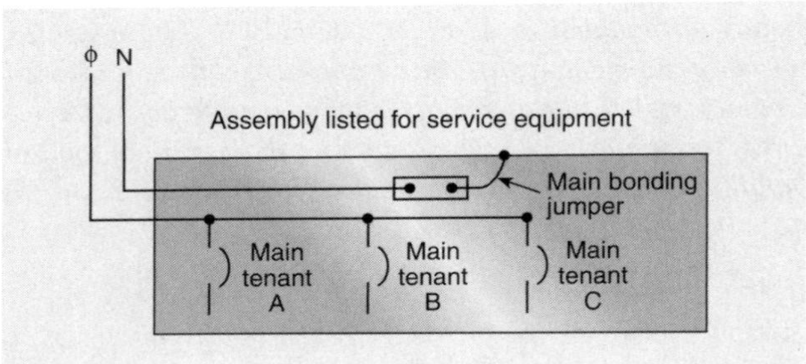


*EXHIBIT 250.11 One connection of the grounded service conductor to a listed service assembly containing multiple service disconnecting means, in accordance with 250.24(C), Exception.*

**(1) Routing and Sizing.** This conductor shall be routed with the phase conductors and shall not be smaller than the required grounding electrode conductor specified in Table 250.66 but shall not be required to be larger than the largest ungrounded service-entrance phase conductor. In addition, for service-entrance phase conductors larger than 1100 kcmil copper or 1750 kcmil aluminum, the grounded conductor shall not be smaller than 12½ percent of the area of the largest service-entrance phase conductor. The grounded conductor of a 3-phase, 3-wire delta service shall have an ampacity not less than that of the ungrounded conductors.

**(2) Parallel Conductors.** Where the service-entrance phase conductors are installed in parallel, the size of the grounded conductor shall be based on the total circular mil area of the parallel conductors as indicated in this section. Where installed in two or more raceways, the size of the grounded conductor in each raceway shall be based on the size of the ungrounded service-entrance conductor in the raceway but not smaller than 1/0 AWG.

FPN: See 310.4 for grounded conductors connected in parallel.

For a multiple raceway or cable service installation, the minimum size for the grounded conductor in each raceway or cable where conductors are in parallel cannot be less than 1/0 AWG. Although the cumulative size of the parallel grounded conductors may be larger than is required by 250.24(C)(1), the minimum 1/0 AWG per raceway or cable correlates with the requirements for parallel conductors contained in 310.4.

**(3) High Impedance.** The grounded conductor on a high-impedance grounded neutral system shall be grounded in accordance with 250.36.

**(D) Grounding Electrode Conductor.** A grounding electrode conductor shall be used to connect the equipment grounding conductors, the service-equipment enclosures, and, where the system is grounded, the grounded service conductor to the grounding electrode(s) required by Part III of this article. This conductor shall be sized in accordance with 250.66.

High-impedance grounded neutral system connections shall be made as covered in 250.36.

FPN: See 250.24(A) for ac system grounding connections.

**(E) Ungrounded System Grounding Connections.** A premises wiring system that is supplied by an ac service that is ungrounded shall have, at each service, a grounding electrode conductor connected to the grounding electrode(s) required by Part III of this article. The grounding electrode conductor shall be connected to a metal enclosure of the service conductors at any accessible point from the load end of the service drop or service lateral to the service disconnecting means.

**250.34 Portable and Vehicle-Mounted Generators.**

**(A) Portable Generators.** The frame of a portable generator shall not be required to be connected to a grounding electrode as defined in 250.52 for a system supplied by the generator under the following conditions:

(1) The generator supplies only equipment mounted on the generator, cord-and-plug-connected equipment through receptacles mounted on the generator, or both, and
(2) The normally non–current-carrying metal parts of equipment and the equipment grounding conductor terminals of the receptacles are connected to the generator frame.

*Portable* describes equipment that is easily carried by personnel from one location to another. *Mobile* describes equipment, such as vehicle-mounted generators, that is capable of being moved on wheels or rollers.

The frame of a portable generator is not required to be connected to earth (ground rod, water pipe, etc.) if the generator has receptacles mounted on the generator panel and the receptacles have equipment grounding terminals bonded to the generator frame.

**(B) Vehicle-Mounted Generators.** The frame of a vehicle shall not be required to be connected to a grounding electrode as defined in 250.52 for a system supplied by a generator located on this vehicle under the following conditions:

(1) The frame of the generator is bonded to the vehicle frame, and
(2) The generator supplies only equipment located on the vehicle or cord-and-plug-connected equipment through receptacles mounted on the vehicle, or both equipment located on the vehicle and cord-and-plug-connected equipment through receptacles mounted on the vehicle or on the generator, and
(3) The normally non–current-carrying metal parts of equipment and the equipment grounding conductor terminals of the receptacles are connected to the generator frame.

Vehicle-mounted generators that provide a neutral conductor and are installed as separately derived systems supplying equipment and receptacles on the vehicle are required to have the neutral conductor bonded to the generator frame and to the vehicle frame. The non–current-carrying parts of the equipment must be bonded to the generator frame.

**(C) Grounded Conductor Bonding.** A system conductor that is required to be grounded by 250.26 shall be connected to the generator frame where the generator is a component of a separately derived system.

FPN: For grounding portable generators supplying fixed wiring systems, see 250.20(D).

Portable and vehicle-mounted generators that are installed as separately derived systems and that provide a neutral conductor (such as 3-phase, 4-wire wye connected; single-phase 240/120 volt; or 3-phase, 4-wire delta connected) are required to have the neutral conductor bonded to the generator frame.

**250.90 General.** Bonding shall be provided where necessary to ensure electrical continuity and the capacity to conduct safely any fault current likely to be imposed.

**250.94 Bonding for Other Systems.** An intersystem bonding termination for connecting intersystem bonding and grounding conductors required for other systems shall be provided external to enclosures at the service equipment and at the disconnecting means for any additional buildings or structures. The intersystem bonding termination shall be accessible for connection and inspection. The intersystem bonding termination shall have the capacity for connection of not less than three intersystem bonding conductors. The intersystem bonding termination device shall not interfere with opening a service or metering equipment enclosure. The intersystem bonding termination shall be one of the following:

(1) A set of terminals securely mounted to the meter enclosure and electrically connected to the meter enclosure. The terminals shall be listed as grounding and bonding equipment.
(2) A bonding bar near the service equipment enclosure, meter enclosure, or raceway for service conductors. The bonding bar shall be connected with a minimum 6 AWG copper conductor to an equipment grounding conductor(s) in the service equipment enclosure, meter enclosure, or exposed nonflexible metallic raceway.
(3) A bonding bar near the grounding electrode conductor. The bonding bar shall be connected to the grounding electrode conductor with a minimum 6 AWG copper conductor.

An external means for intersystem bonding connections is required to be one of the types described in 250.94(1), (2), or (3). Exhibit 250.40 shows acceptable locations for making the connections in an existing building or structure as permitted by the exception. Exhibit 250.41 pictures a device that can be attached to a meter socket enclosure that does not impede the ability to open the enclosure.

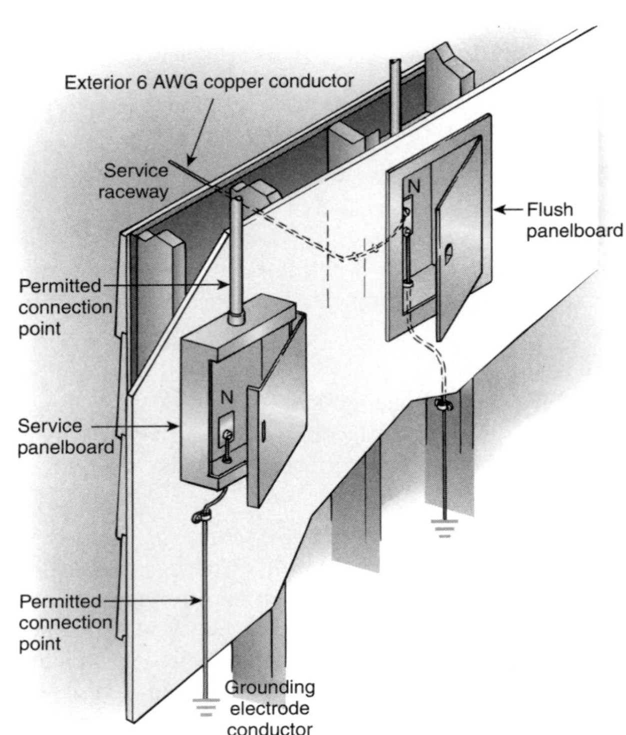


***EXHIBIT 250.40*** *Examples of accessible external means for intersystem bonding, as required by 250.94 for service equipment and building or structure disconnecting means.*

*Exception: In existing buildings or structures where any of the intersystem bonding and grounding conductors required by 770.93, 800.100(B), 810.21(F), 820.100(B), 830.100(B) exist, installation of the intersystem bonding termination is not required. An accessible means external to enclosures for connecting intersystem bonding and grounding electrode conductors shall be permitted at the service equipment and at the disconnecting means for any additional buildings or structures by at least one of the following means:*

*(1) Exposed nonflexible metallic raceways*
*(2) An exposed grounding electrode conductor*
*(3) Approved means for the external connection of a copper or other corrosion-resistant bonding or grounding conductor to the grounded raceway or equipment*

FPN No. 1: A 6 AWG copper conductor with one end bonded to the grounded nonflexible metallic raceway or equipment and with 150 mm (6 in.) or more of the other end made accessible on the outside wall is an example of the approved means covered in 250.94, Exception item (3).

FPN No. 2: See 800.100, 810.21, and 820.100 for bonding and grounding requirements for communications circuits, radio and television equipment, and CATV circuits.

The *Code* requires that separate systems be bonded together to reduce the differences of potential between them due to lightning or accidental contact with power lines. Lightning protection systems, communications, radio and TV, and CATV systems must be bonded together to minimize the potential differences between the systems.

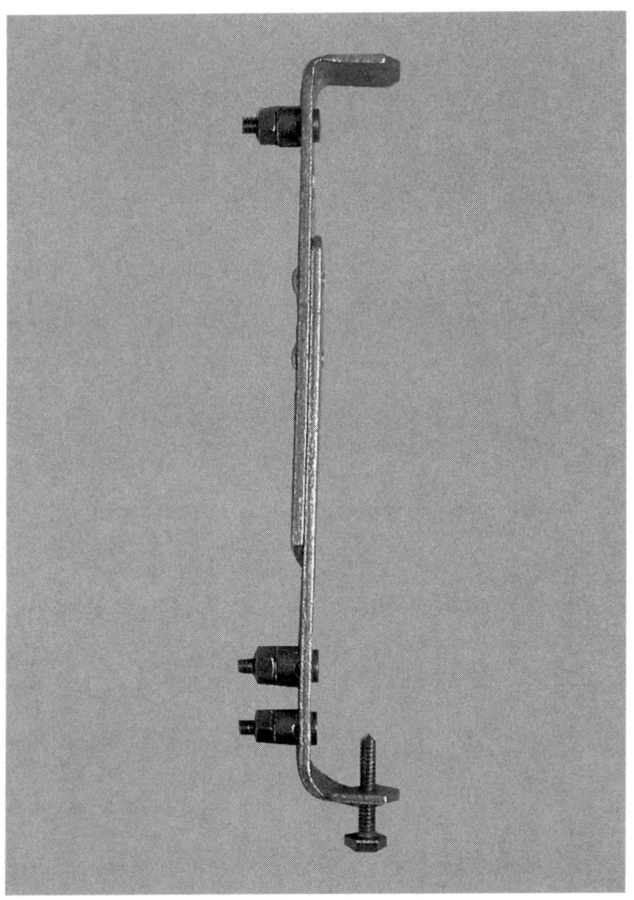

***EXHIBIT 250.41*** *A listed means for making intersystem bonding connections to a grounded meter socket enclosure. (Courtesy of Thomas & Betts Corp.)*

Lack of interconnection can result in a severe shock and fire hazard. The reason for this potential hazard is illustrated in Exhibit 250.42, which shows a CATV cable with its jacket grounded to a separate ground rod and not bonded to the power ground. The cable is connected to the cable decoder and the tuner of a television set. Also connected to the decoder and the television is the 120-volt supply, with one conductor grounded at the service (the power ground). In each case, resistance to ground is present at the grounding electrode. This resistance to ground varies

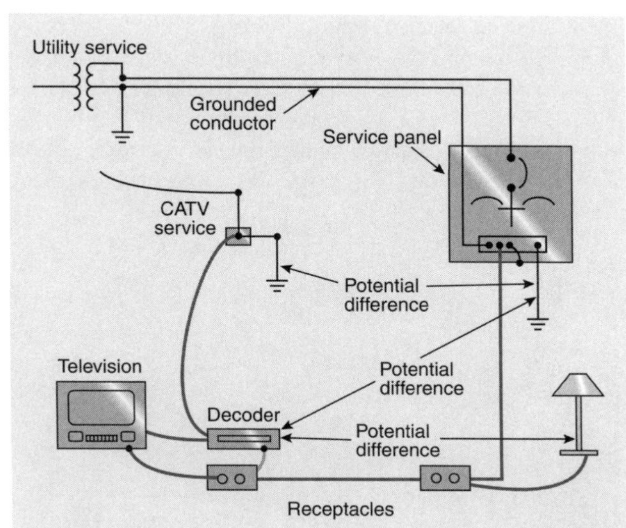


***EXHIBIT 250.42** A CATV installation that does not comply with the Code, illustrating why bonding between different systems is necessary.*

widely, depending on soil conditions and the type of grounding electrode. The resistance at the CATV ground is likely to be higher than the power ground resistance, because the power ground is often an underground metal water piping system or concrete-encased electrode, whereas the CATV ground is commonly a ground rod.

For example, for the CATV installation shown in Exhibit 250.42, assume that a current is induced in the power line by a switching surge or a nearby lightning strike, so that a momentary current of 1000 amperes occurs over the power line to the power line ground. This amount of current is not unusual under such circumstances — the amount could be, and often is, considerably higher. Also assume that the power ground has a resistance of 10 ohms, a very low value in most circumstances (a single ground rod in average soil has a resistance to ground in the neighborhood of 40 ohms).

According to Ohm's law, the current through the equipment connected to the electrical system will be raised momentarily to a potential of 10,000 volts (1000 amperes × 10 ohms). This potential of 10,000 volts would exist between the CATV system and the electrical system and between the grounded conductor within the CATV cable and the grounded surfaces in the walls of the home, such as water pipes (which are connected to the power ground), over which the cable runs. This potential could also appear across a person with one hand on the CATV cable and the other hand on a metal surface connected to the power ground (e.g., a radiator or a refrigerator).

Actual voltage is likely to be many times the 10,000 volts calculated, because extremely low (below normal) values were assumed for both resistance to ground and current. Most insulation systems, however, are not designed to withstand even 10,000 volts. Even if the insulation system does withstand a 10,000-volt surge, it is likely to be damaged, and breakdown of the insulation system will result in sparking.

The same situation would exist if the current surge were on the CATV cable or a telephone line. The only difference would be the voltage involved, which would depend on the individual resistance to ground of the grounding electrodes.

The solution is to bond the two grounding electrode systems together, as shown in Exhibit 250.43, or to connect the CATV cable jacket to the power ground, which is exactly what the *Code* requires. When one system is raised above ground potential, the second system rises to the same potential, and no voltage exists between the two grounding systems.

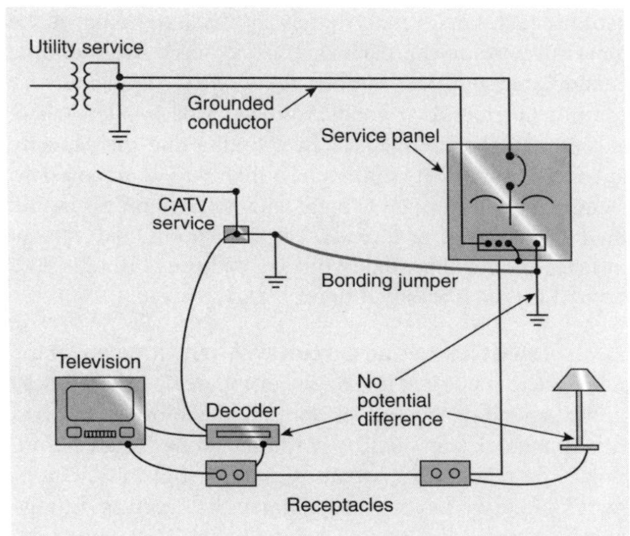


***EXHIBIT 250.43** A cable TV installation that complies with 250.94.*

These bonding rules are provided to address the difficulties that communications and CATV installers encounter in complying with *Code* grounding and bonding requirements. These difficulties arise from the increasing use of plastic for water pipe, fittings, water meters, and service conduit. In the past, bonding between communications, CATV, and power systems was usually achieved by connecting the communications protector grounds or cable shield to an interior metallic water pipe, because the pipe was often used as the power grounding electrode. Thus, the requirement that the power, communications, CATV cable shield, and metallic water piping systems be bonded together was easily satisfied. If the power was grounded to one of the other electrodes permitted by the *Code*, usually by a made electrode such as a ground rod, the bond was connected to the power grounding electrode conductor or to a metallic service raceway, since at least one of these was usually accessible.

With the proliferation of plastic water pipe and the increasing tendency for service equipment (often flush-mounted) to be installed in finished areas, where the grounding electrode conductor is often concealed, as well as the increased use of plastic service-entrance conduit, communications and CATV installers no longer have access to a point for connecting bonding jumpers or grounding conductors. See Exhibits 250.40 and 250.41, and also the commentary following 820.100(D), FPN No. 2.

### 250.96 Bonding Other Enclosures.

**(A) General.** Metal raceways, cable trays, cable armor, cable sheath, enclosures, frames, fittings, and other metal non–current-carrying parts that are to serve as grounding conductors, with or without the use of supplementary equipment grounding conductors, shall be bonded where necessary to ensure electrical continuity and the capacity to conduct safely any fault current likely to be imposed on them. Any nonconductive paint, enamel, or similar coating shall be removed at threads, contact points, and contact surfaces or be connected by means of fittings designed so as to make such removal unnecessary.

**(B) Isolated Grounding Circuits.** Where installed for the reduction of electrical noise (electromagnetic interference) on the grounding circuit, an equipment enclosure supplied by a branch circuit shall be permitted to be isolated from a raceway containing circuits supplying only that equipment by one or more listed nonmetallic raceway fittings located at the point of attachment of the raceway to the equipment enclosure. The metal raceway shall comply with provisions of this article and shall be supplemented by an internal insulated equipment grounding conductor installed in accordance with 250.146(D) to ground the equipment enclosure.

FPN: Use of an isolated equipment grounding conductor does not relieve the requirement for grounding the raceway system.

To reduce electromagnetic interference, 250.96(B) permits electronic equipment to be isolated from the raceway in a manner similar to that for cord-and-plug-connected equipment. Section 250.96(B) specifies that a metal equipment enclosure supplied by a branch circuit is the subject of the requirement and that subsequent wiring, raceways, or other equipment beyond the insulating fitting is not permitted.

Exhibits 250.44 and 250.45 show examples of installations. In Exhibit 250.44, note that the metal raceway is grounded in the usual manner, by attachment to the grounded service enclosure, satisfying the concern men-

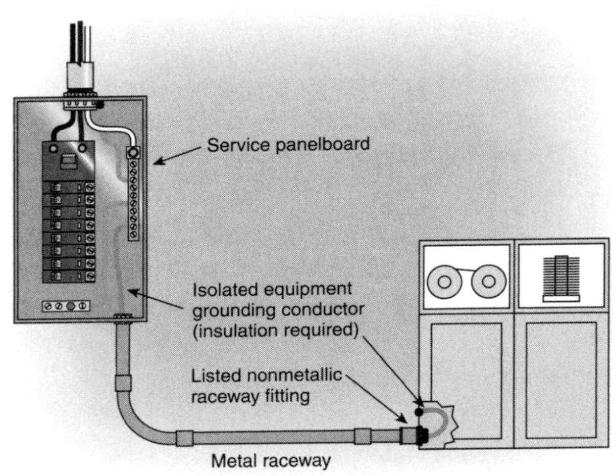


*EXHIBIT 250.44 An installation in which the electronic equipment is grounded through the isolated equipment grounding conductor.*

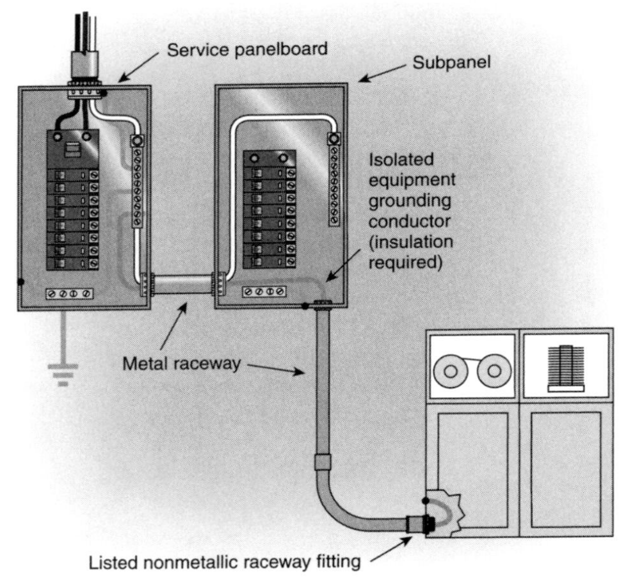


*EXHIBIT 250.45 An installation in which the isolated equipment grounding conductor is allowed to pass through the subpanel without connecting to the grounding bus to terminate at the service grounding bus.*

tioned in the FPN to 250.96(B). In Exhibit 250.45, note that 408.40, Exception, permits, but does not require, the isolated equipment grounding conductor (which is required to be insulated) to pass through the subpanel and run back to the service equipment. The key to this method of grounding electronic equipment is to always ensure that the insulated equipment grounding conductor, regardless of where it terminates in the distribution system, is connected in a manner that creates an effective path for ground-fault current, as required by 250.4(A)(5).

**250.97 Bonding for Over 250 Volts.** For circuits of over 250 volts to ground, the electrical continuity of metal raceways and cables with metal sheaths that contain any conductor other than service conductors shall be ensured by one or more of the methods specified for services in 250.92(B), except for (B)(1).

*Exception: Where oversized, concentric, or eccentric knockouts are not encountered, or where a box or enclosure with concentric or eccentric knockouts is listed to provide a reliable bonding connection, the following methods shall be permitted:*

*(1) Threadless couplings and connectors for cables with metal sheaths*
*(2) Two locknuts, on rigid metal conduit or intermediate metal conduit, one inside and one outside of boxes and cabinets*
*(3) Fittings with shoulders that seat firmly against the box or cabinet, such as electrical metallic tubing connectors, flexible metal conduit connectors, and cable connectors, with one locknut on the inside of boxes and cabinets*
*(4) Listed fittings*

Bonding around prepunched concentric or eccentric knockouts is not required if the enclosure containing the knockouts has been tested and is listed as suitable for bonding. Guide card information from the UL *Guide Information for Electrical Equipment — The White Book* indicates that concentric and eccentric knockouts of all metallic outlet boxes evaluated in accordance with UL 514A, *Metallic Outlet Boxes,* are suitable for bonding in circuits of above or below 250 volts to ground without the use of additional bonding equipment. Metallic outlet boxes are permitted, but not required, to be marked to indicate this condition of use.

The methods in items (1) through (4) in the exception to 250.97 are permitted for circuits over 250 volts to ground only where there are no oversize, concentric, or eccentric knockouts. Note that method (3) permits fittings, such as EMT connectors, cable connectors, and similar fittings with shoulders that seat firmly against the metal of a box or cabinet, to be installed with only one locknut on the inside of the box.

**250.114 Equipment Connected by Cord and Plug** Under any of the conditions described in 250.114(1) through (4), exposed non–current-carrying metal parts of cord-and-plug-connected equipment likely to become energized shall be connected to the equipment grounding conductor.

*Exception: Listed tools, listed appliances, and listed equipment covered in 250.114(2) through (4) shall not be required to be connected to an equipment grounding conductor where protected by a system of double insulation or its equivalent. Double insulated equipment shall be distinctively marked.*

The exception to 250.114 recognizes listed double-insulated appliances, motor-operated handheld tools, stationary and fixed motor-operated tools, and light industrial motor-operated tools as not requiring equipment grounding connections.

(1) In hazardous (classified) locations (see Articles 500 through 517)
(2) Where operated at over 150 volts to ground

*Exception No. 1: Motors, where guarded, shall not be required to be connected to an equipment grounding conductor.*

*Exception No. 2: Metal frames of electrically heated appliances, exempted by special permission, shall not be required to be connected to an equipment grounding conductor, in which case the frames shall be permanently and effectively insulated from ground.*

(3) In residential occupancies:
  a. Refrigerators, freezers, and air conditioners
  b. Clothes-washing, clothes-drying, dish-washing machines; kitchen waste disposers; information technology equipment; sump pumps and electrical aquarium equipment
  c. Hand-held motor-operated tools, stationary and fixed motor-operated tools, and light industrial motor-operated tools
  d. Motor-operated appliances of the following types: hedge clippers, lawn mowers, snow blowers, and wet scrubbers
  e. Portable handlamps
(4) In other than residential occupancies:
  a. Refrigerators, freezers, and air conditioners
  b. Clothes-washing, clothes-drying, dish-washing machines; information technology equipment; sump pumps and electrical aquarium equipment

c. Hand-held motor-operated tools, stationary and fixed motor-operated tools, and light industrial motor-operated tools
d. Motor-operated appliances of the following types: hedge clippers, lawn mowers, snow blowers, and wet scrubbers
e. Portable handlamps
f. Cord-and-plug-connected appliances used in damp or wet locations or by persons standing on the ground or on metal floors or working inside of metal tanks or boilers
g. Tools likely to be used in wet or conductive locations

*Exception: Tools and portable handlamps likely to be used in wet or conductive locations shall not be required to be connected to an equipment grounding conductor where supplied through an isolating transformer with an ungrounded secondary of not over 50 volts.*

Tools must be grounded by an equipment grounding conductor within the cord or cable supplying the tool, except where the tool is supplied by an isolating transformer, as permitted by the exception following 250.114(4). Portable tools and appliances protected by an approved system of double insulation must be listed by a qualified electrical testing laboratory as being suitable for the purpose, and the equipment must be distinctively marked as double insulated.

Cord-connected portable tools or appliances are not intended to be used in damp, wet, or conductive locations unless they are grounded, supplied by an isolation transformer with a secondary of not more than 50 volts, or protected by an approved system of double insulation.

Exhibit 250.48 shows an example of lighting equipment supplied through an isolating transformer operating at 6 or 12 volts that provides safe illumination for work inside boilers, tanks, and similar locations that may be metal or wet.

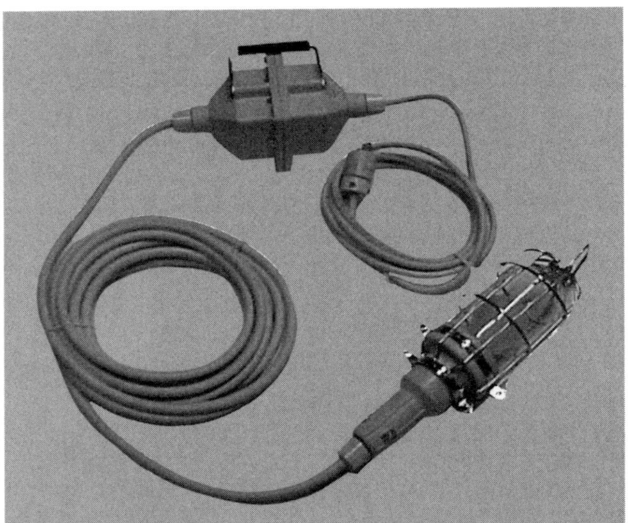

***EXHIBIT 250.48** Lighting equipment supplied through an isolating transformer operating at 6 or 12 volts and therefore not required to be grounded. (Courtesy of Woodhead Industries, a division of Molex Incorporated)*

**250.146 Connecting Receptacle Grounding Terminal to Box.** An equipment bonding jumper shall be used to connect the grounding terminal of a grounding-type receptacle to a grounded box unless grounded as in 250.146(A) through (D). The equipment bonding jumper shall be sized in accordance with Table 250.122 based on the rating of the overcurrent device protecting the circuit conductors.

**(A) Surface-Mounted Box.** Where the box is mounted on the surface, direct metal-to-metal contact between the device yoke and the box or a contact yoke or device that complies with 250.146(B) shall be permitted to ground the receptacle to the box. At least one of the insulating washers shall be removed from receptacles that do not have a contact yoke or device that complies with 250.146(B) to ensure direct metal-to-metal contact. This provision shall not apply to cover-mounted receptacles unless the box and cover combination are listed as providing satisfactory ground continuity between the box and the receptacle. A listed exposed work cover shall be permitted to be the grounding and bonding means when (1) the device is attached to the cover with at least two fasteners that are permanent (such as a rivet) or have a thread locking or screw locking means and (2) when the cover mounting holes are located on a flat non-raised portion of the cover.

The main rule of 250.146 requires an equipment bonding jumper to be installed between the device box and the receptacle grounding terminal. However, 250.146(A) permits the equipment bonding jumper to be omitted where the metal yoke of the device is in direct metal-to-metal contact with the metal device box and at least one of the fiber retention washers for the receptacle mounting screws is removed, as illustrated in Exhibit 250.55.

Cover-mounted wiring devices, such as on 4-in. square covers, are not considered grounded. Section 250.146(A) does not apply to cover-mounted receptacles, such as the one illustrated in Exhibit 250.56. Box-cover and device combinations listed as providing grounding continuity are permitted. The mounting holes for the cover must be located on a *flat, non-raised* portion of the cover to provide the best possible surface-to-surface contact and the recepta-

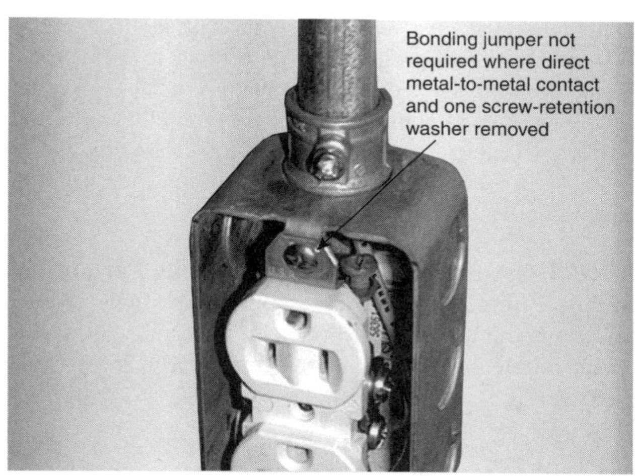


***EXHIBIT 250.55** An example of a box-mounted receptacle attached to a surface box where a bonding jumper is not required provided at least one of the insulating washers is removed.*

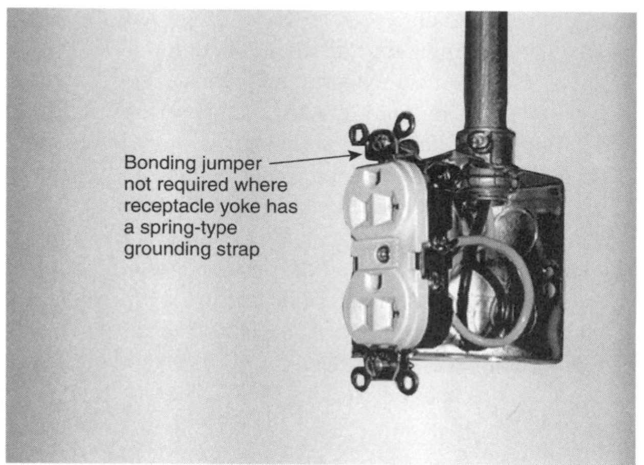


***EXHIBIT 250.57** A receptacle designed with a listed spring-type grounding strap. The strap that holds the mounting screw captive establishes a grounding circuit and eliminates the need to provide a wire-type equipment bonding jumper to the box, in accordance with 250.146(B).*

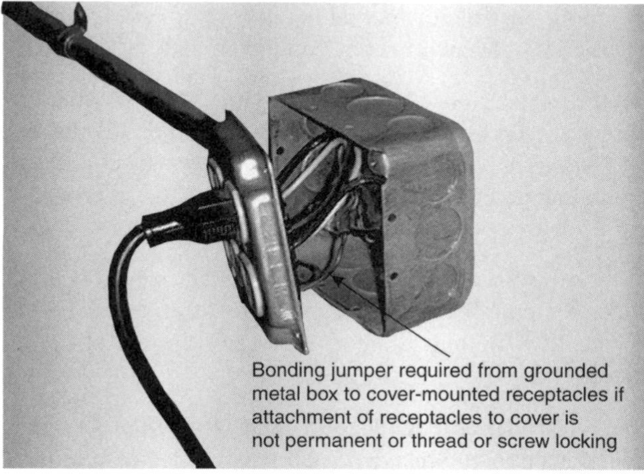


***EXHIBIT 250.56** An example of a cover-mounted receptacle attached to a surface box where a bonding jumper is required.*

cle must be secured to the cover using not less than two rivets or locking means for threaded attachment means.

**(B) Contact Devices or Yokes.** Contact devices or yokes designed and listed as self-grounding shall be permitted in conjunction with the supporting screws to establish the grounding circuit between the device yoke and flush-type boxes.

Section 250.146(B) is illustrated by Exhibit 250.57, which shows a receptacle designed with a spring-type grounding strap for holding the mounting screw and establishing the grounding circuit so that an equipment bonding jumper is not required. Such devices are listed as "self-grounding."

**(C) Floor Boxes.** Floor boxes designed for and listed as providing satisfactory ground continuity between the box and the device shall be permitted.

**(D) Isolated Receptacles.** Where installed for the reduction of electrical noise (electromagnetic interference) on the grounding circuit, a receptacle in which the grounding terminal is purposely insulated from the receptacle mounting means shall be permitted. The receptacle grounding terminal shall be connected to an insulated equipment grounding conductor run with the circuit conductors. This equipment grounding conductor shall be permitted to pass through one or more panelboards without a connection to the panelboard grounding terminal bar as permitted in 408.40, Exception, so as to terminate within the same building or structure directly at an equipment grounding conductor terminal of the applicable derived system or service. Where installed in accordance with the provisions of this section, this equipment grounding conductor shall also be permitted to pass through boxes, wireways, or other enclosures without being connected to such enclosures.

FPN: Use of an isolated equipment grounding conductor does not relieve the requirement for grounding the raceway system and outlet box.

Section 250.146(D) allows an isolated-ground–type receptacle to be installed without a bonding jumper between the metal device box and the receptacle grounding terminal. An insulated equipment grounding conductor, as shown in Exhibit 250.58, is installed with the branch-circuit conductors. This conductor may originate in the service panel, pass through any number of subpanels without being connected to the equipment grounding bus, and terminate at the isolated-ground–type receptacle ground terminal. However, this does not exempt the metal device box from being grounded. The metal device box must be grounded either by an equipment grounding conductor run with the circuit conductors or by a wiring method that serves as an equipment grounding conductor. See 250.118 for types of equipment grounding conductors.

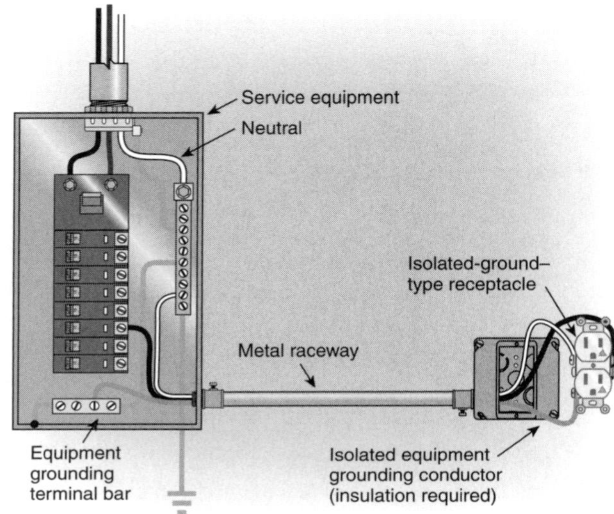


***EXHIBIT 250.58** An isolated-ground–type receptacle with an insulated equipment grounding conductor and with the device box grounded through the metal raceway.*

According to 250.146(D), where isolated-ground–type receptacles are used, the isolated equipment grounding conductor can terminate at an equipment grounding terminal of the applicable service or derived system in the same building as the receptacle. If the isolated equipment grounding conductor terminates at a separate building, a large voltage difference may exist between buildings during lightning transients. Such transients could cause damage to equipment connected to an isolated-ground–type receptacle and present a shock hazard between the isolated equipment frame and other grounded surfaces.

The fine print note to 250.146(D) is a reminder that metallic raceways and boxes are still required to be grounded by one of the usual required methods. This could require a separate grounding conductor, for example, to ground a metal box in a nonmetallic raceway system or to ground a metal box supplied by flexible metal conduit. Where an ordinary grounding-type receptacle is being replaced with an isolated-ground–type receptacle, use of an existing insulated equipment grounding conductor as the isolated receptaclegrounding conductor could effectively defeat or seriously compromise the required box or raceway equipment ground.

**250.188 Grounding of Systems Supplying Portable or Mobile Equipment.** Systems supplying portable or mobile high-voltage equipment, other than substations installed on a temporary basis, shall comply with 250.188(A) through (F).

*Portable* describes equipment that is easily carried from one location to another. *Mobile* describes equipment that is easily moved on wheels, treads, and so on.

**(A) Portable or Mobile Equipment.** Portable or mobile high-voltage equipment shall be supplied from a system having its neutral conductor grounded through an impedance. Where a delta-connected high-voltage system is used to supply portable or mobile equipment, a system neutral point and associated neutral conductor shall be derived.

**(B) Exposed Non–Current-Carrying Metal Parts.** Exposed non–current-carrying metal parts of portable or mobile equipment shall be connected by an equipment grounding conductor to the point at which the system neutral impedance is grounded.

**(C) Ground-Fault Current.** The voltage developed between the portable or mobile equipment frame and ground by the flow of maximum ground-fault current shall not exceed 100 volts.

**(D) Ground-Fault Detection and Relaying.** Ground-fault detection and relaying shall be provided to automatically de-energize any high-voltage system component that has developed a ground fault. The continuity of the equipment grounding conductor shall be continuously monitored so as to de-energize automatically the high-voltage circuit to the portable or mobile equipment upon loss of continuity of the equipment grounding conductor.

**(E) Isolation.** The grounding electrode to which the portable or mobile equipment system neutral impedance is connected shall be isolated from and separated in the ground by at least 6.0 m (20 ft) from any other system or equipment grounding electrode, and there shall be no direct connection between the grounding electrodes, such as buried pipe and fence, and so forth.

**(F) Trailing Cable and Couplers.** High-voltage trailing cable and couplers for interconnection of portable or mobile equipment shall meet the requirements of Part III of Article 400 for cables and 490.55 for couplers.

**250.190 Grounding of Equipment.** All non–current-carrying metal parts of fixed, portable, and mobile equipment and associated fences, housings, enclosures, and supporting structures shall be grounded.

*Exception: Where isolated from ground and located so as to prevent any person who can make contact with ground from contacting such metal parts when the equipment is energized.*

Equipment grounding conductors not an integral part of a cable assembly shall not be smaller than 6 AWG copper or 4 AWG aluminum.

FPN: See 250.110, Exception No. 2, for pole-mounted distribution apparatus.

## Article 300 Wiring Methods

**300.37 Aboveground Wiring Methods.** Aboveground conductors shall be installed in rigid metal conduit, in intermediate metal conduit, in electrical metallic tubing, in rigid nonmetallic conduit, in cable trays, as busways, as cablebus, in other identified raceways, or as exposed runs of metal-clad cable suitable for the use and purpose. In locations accessible to qualified persons only, exposed runs of Type MV cables, bare conductors, and bare busbars shall also be permitted. Busbars shall be permitted to be either copper or aluminum.

In transformer vaults, switch rooms, and similar areas restricted to qualified personnel, any suitable wiring method may be used. Exposed wiring using bare or insulated conductors on insulators is commonly employed, as is rigid metal conduit and rigid nonmetallic conduit (changed to rigid polyvinyl chloride conduit for the 2008 *Code*). Throughout 300.37 as well as 300.39, the term *exposed* was substituted for the term *open*. *Exposed* is preferred because it is a defined term in Article 100.

**300.39 Braid-Covered Insulated Conductors — Exposed Installation.** Exposed runs of braid-covered insulated conductors shall have a flame-retardant braid. If the conductors used do not have this protection, a flame-retardant saturant shall be applied to the braid covering after installation. This treated braid covering shall be stripped back a safe distance at conductor terminals, according to the operating voltage. Where practicable, this distance shall not be less than 25 mm (1 in.) for each kilovolt of the conductor-to-ground voltage of the circuit.

**300.40 Insulation Shielding.** Metallic and semiconducting insulation shielding components of shielded cables shall be removed for a distance dependent on the circuit voltage and insulation. Stress reduction means shall be provided at all terminations of factory-applied shielding. Metallic shielding components such as tapes, wires, or braids, or combinations thereof, shall be connected to a grounding conductor, grounding busbar, or a grounding electrode.

Although no technical changes were made to this section for the 2008 *Code*, the grounding terms used in this requirement were made more specific.

**300.42 Moisture or Mechanical Protection for Metal-Sheathed Cables.** Where cable conductors emerge from a metal sheath and where protection against moisture or physical damage is necessary, the insulation of the conductors shall be protected by a cable sheath terminating device.

## Article 312 Cabinets, Cutout Boxes, and Meter Socket Enclosures

**312.5 Cabinets, Cutout Boxes, and Meter Socket Enclosures.** Conductors entering enclosures within the scope of this article shall be protected from abrasion and shall comply with 312.5(A) through (C).

**(A) Openings to Be Closed.** Openings through which conductors enter shall be adequately closed.

**(B) Metal Cabinets, Cutout Boxes, and Meter Socket Enclosures.** Where metal enclosures within the scope of this article are installed with messenger-supported wiring, open wiring on insulators, or concealed knob-and-tube wiring, conductors shall enter through insulating bushings or, in dry locations, through flexible tubing extending from the last insulating support and firmly secured to the enclosure.

**(C) Cables.** Where cable is used, each cable shall be secured to the cabinet, cutout box, or meter socket enclosure.

The main rule of 312.5(C) prohibits the installation of several cables bunched together and run through a knockout or chase nipple. Individual cable clamps or connectors are required to be used with only one cable per clamp or connector, unless the clamp or connector is identified for more than a single cable.

*Exception: Cables with entirely nonmetallic sheaths shall be permitted to enter the top of a surface-mounted enclosure through one or more nonflexible raceways not less than 450 mm (18 in.) and not more than 3.0 m (10 ft) in length, provided all of the following conditions are met:*

*(a) Each cable is fastened within 300 mm (12 in.), measured along the sheath, of the outer end of the raceway.*

*(b) The raceway extends directly above the enclosure and does not penetrate a structural ceiling.*

*(c) A fitting is provided on each end of the raceway to protect the cable(s) from abrasion and the fittings remain accessible after installation.*

*(d) The raceway is sealed or plugged at the outer end using approved means so as to prevent access to the enclosure through the raceway.*

*(e) The cable sheath is continuous through the raceway and extends into the enclosure beyond the fitting not less than 6 mm (1/4 in.).*

*(f) The raceway is fastened at its outer end and at other points in accordance with the applicable article.*

*(g) Where installed as conduit or tubing, the allowable cable fill does not exceed that permitted for complete conduit or tubing systems by Table 1 of Chapter 9 of this Code and all applicable notes thereto.*

This exception allows multiple nonmetallic cables such as Type NM, NMC, NMS, UF, SE, and USE to enter the top of a surface-mounted enclosure through a nonflexible raceway sleeve or nipple. These sleeves or nipples are permitted to be between 18 in. and 10 ft in length. However, if the nipple length exceeds 24 in., the ampacity adjustment factors of 310.15(B)(2) apply.

FPN: See Table 1 in Chapter 9, including Note 9, for allowable cable fill in circular raceways. See 310.15(B)(2)(a) for required ampacity reductions for multiple cables installed in a common raceway.

**312.11 Spacing.** The spacing within cabinets and cutout boxes shall comply with 312.11(A) through (D).

**(A) General.** Spacing within cabinets and cutout boxes shall be sufficient to provide ample room for the distribution of wires and cables placed in them and for a separation between metal parts of devices and apparatus mounted within them in accordance with (A)(1), (A)(2), and (A)(3).

**(1) Base.** Other than at points of support, there shall be an airspace of at least 1.59 mm (0.0625 in.) between the base of the device and the wall of any metal cabinet or cutout box in which the device is mounted.

**(2) Doors.** There shall be an airspace of at least 25.4 mm (1.00 in.) between any live metal part, including live metal parts of enclosed fuses, and the door.

*Exception: Where the door is lined with an approved insulating material or is of a thickness of metal not less than 2.36 mm (0.093 in.) uncoated, the airspace shall not be less than 12.7 mm (0.500 in.).*

**(3) Live Parts.** There shall be an airspace of at least 12.7 mm (0.500 in.) between the walls, back, gutter partition, if of metal, or door of any cabinet or cutout box and the nearest exposed current-carrying part of devices mounted within the cabinet where the voltage does not exceed 250. This spacing shall be increased to at least 25.4 mm (1.00 in.) for voltages of 251 to 600, nominal.

*Exception: Where the conditions in 312.11(A)(2), Exception, are met, the airspace for nominal voltages from 251 to 600 shall be permitted to be not less than 12.7 mm (0.500 in.).*

**(B) Switch Clearance.** Cabinets and cutout boxes shall be deep enough to allow the closing of the doors when 30-ampere branch-circuit panelboard switches are in any position, when combination cutout switches are in any position, or when other single-throw switches are opened as far as their construction permits.

**(C) Wiring Space.** Cabinets and cutout boxes that contain devices or apparatus connected within the cabinet or box to more than eight conductors, including those of branch circuits, meter loops, feeder circuits, power circuits, and similar circuits, but not including the supply circuit or a continuation thereof, shall have back-wiring spaces or one or more side-wiring spaces, side gutters, or wiring compartments.

**(D) Wiring Space — Enclosure.** Side-wiring spaces, side gutters, or side-wiring compartments of cabinets and cutout boxes shall be made tight enclosures by means of covers, barriers, or partitions extending from the bases of the devices contained in the cabinet, to the door, frame, or sides of the cabinet.

*Exception: Side-wiring spaces, side gutters, and side-wiring compartments of cabinets shall not be required to be made tight enclosures where those side spaces contain only conductors that enter the cabinet directly opposite to the devices where they terminate.*

Partially enclosed back-wiring spaces shall be provided with covers to complete the enclosure. Wiring spaces that are required by 312.11(C) and are exposed when doors are open shall be provided with covers to complete the enclosure. Where adequate space is provided for feed-through conductors and for splices as required in 312.8, additional barriers shall not be required.

## Article 400 Flexible Cords and Cables

### 400.7 Uses Permitted.

**(A) Uses.** Flexible cords and cables shall be used only for the following:

(1) Pendants
(2) Wiring of luminaires
(3) Connection of portable luminaires, portable and mobile signs, or appliances
(4) Elevator cables
(5) Wiring of cranes and hoists
(6) Connection of utilization equipment to facilitate frequent interchange
(7) Prevention of the transmission of noise or vibration
(8) Appliances where the fastening means and mechanical connections are specifically designed to permit ready removal for maintenance and repair, and the appliance is intended or identified for flexible cord connection
(9) Connection of moving parts
(10) Where specifically permitted elsewhere in this *Code*

**(B) Attachment Plugs.** Where used as permitted in 400.7(A)(3), (A)(6), and (A)(8), each flexible cord shall be equipped with an attachment plug and shall be energized from a receptacle outlet.

*Exception: As permitted in 368.56.*

**400.8 Uses Not Permitted.** Unless specifically permitted in 400.7, flexible cords and cables shall not be used for the following:

(1) As a substitute for the fixed wiring of a structure
(2) Where run through holes in walls, structural ceilings, suspended ceilings, dropped ceilings, or floors
(3) Where run through doorways, windows, or similar openings
(4) Where attached to building surfaces

*Exception to (4): Flexible cord and cable shall be permitted to be attached to building surfaces in accordance with the provisions of 368.56(B)*

Section 368.56(B) provides the requirements for the installation of flexible cords installed as branches from busways.

(5) Where concealed by walls, floors, or ceilings or located above suspended or dropped ceilings
(6) Where installed in raceways, except as otherwise permitted in this *Code*

The flexible cords and cables referred to in Article 400 are not limited to use with portable equipment. They may not be used, however, as a substitute for the fixed wiring of a structure or where concealed behind building walls, floors, or ceilings (including structural, suspended, or dropped-type ceilings). See 240.5, 590.4(B), and 590.4(C) for the uses of multiconductor flexible cords for feeder and branch-circuit installations and for overcurrent protection requirements for flexible cord. See 410.30 for cord-connected luminaires.

(7) Where subject to physical damage

**400.9 Splices.** Flexible cord shall be used only in continuous lengths without splice or tap where initially installed in applications permitted by 400.7(A). The repair of hard-service cord and junior hard-service cord (see Trade Name column in Table 400.4) 14 AWG and larger shall be permitted if conductors are spliced in accordance with 110.14(B) and the completed splice retains the insulation, outer sheath properties, and usage characteristics of the cord being spliced.

The requirements of 400.9 are intended to ensure that flexible cords and cables first installed under any of the uses permitted in 400.7(A)(1) through (A)(10) are in their original or near-original condition. Damage to a cord can occur under the sometimes extreme conditions of use to which the cord is subjected. The provisions of this section permit repair of a cord in such a manner that the cord will retain its original operating and use integrity. However, if the repaired cord is reused or reinstalled at a new location, the in-line repair is no longer permitted, and the cord can be used only in lengths that do not contain a splice.

**400.10 Pull at Joints and Terminals.** Flexible cords and cables shall be connected to devices and to fittings so that tension is not transmitted to joints or terminals.

*Exception: Listed portable single-pole devices that are intended to accommodate such tension at their terminals shall be permitted to be used with single-conductor flexible cable.*

FPN: Some methods of preventing pull on a cord from being transmitted to joints or terminals are knotting the cord, winding with tape, and fittings designed for the purpose.

**400.11 In Show Windows and Showcases.** Flexible cords used in show windows and showcases shall be Types S, SE, SEO, SEOO, SJ, SJE, SJEO, SJEOO, SJO, SJOO, SJT, SJTO, SJTOO, SO, SOO, ST, STO, STOO, SEW, SEOW, SEOOW, SJEW, SJEOW, SJEOOW, SJOW, SJOOW, SJTW, SJTOW, SJTOOW, SOW, SOOW, STW, STOW, or STOOW.

*Exception No. 1: For the wiring of chain-supported luminaires.*

*Exception No. 2: As supply cords for portable luminaires and other merchandise being displayed or exhibited.*

Flexible cords listed for hard usage or extra-hard usage should be used in show windows and showcases because such cords may come in contact with combustible materials, such as fabrics or paper products, usually present at these locations and because they are exposed to wear and tear from continual housekeeping and display changes. Flexible cords used in show windows and showcases should be maintained in good condition.

**400.12 Minimum Size.** The individual conductors of a flexible cord or cable shall not be smaller than the sizes in Table 400.4.

*Exception: The size of the insulated ground-check conductor of Type G-GC cables shall be not smaller than 10 AWG.*

**400.13 Overcurrent Protection.** Flexible cords not smaller than 18 AWG, and tinsel cords or cords having equivalent characteristics of smaller size approved for use with specific appliances, shall be considered as protected against overcurrent by the overcurrent devices described in 240.5.

**400.14 Protection from Damage.** Flexible cords and cables shall be protected by bushings or fittings where passing through holes in covers, outlet boxes, or similar enclosures.

A variety of bushings and fittings are available for protecting flexible cords and cables, both insulated and noninsulated. Some bushings or fittings include pull-relief means, as required in 400.10. Many insulating bushings are listed by Underwriters Laboratories Inc. in the following product categories:

1. Conduit fittings (bushings and fittings for use on the ends of conduit in boxes and gutters)
2. Insulating devices and materials
3. Outlet bushings and fittings (for use on the ends of conduit, electrical metallic tubing, or armored cable where a change to open wiring is made)

## Article 404 Switches

### 404.6 Position and Connection of Switches.

**(A) Single-Throw Knife Switches.** Single-throw knife switches shall be placed so that gravity will not tend to close them. Single-throw knife switches, approved for use in the inverted position, shall be provided with an integral mechanical means that ensures that the blades remain in the open position when so set.

**(B) Double-Throw Knife Switches.** Double-throw knife switches shall be permitted to be mounted so that the throw is either vertical or horizontal. Where the throw is vertical, integral mechanical means shall be provided to hold the blades in the open position when so set.

Sections 404.6(A) and (B) clarifiy that an integral "mechanical means" that does not necessarily have to be a "locking device" is required for single-throw and double-throw switches to ensure the switch blades remain disengaged regardless of their orientation when the switch is in the "off" (open) position. New switch designs incorporate mechanical means other than a catch or a latch to ensure that the blades cannot accidentally close from the off position.

**(C) Connection of Switches.** Single-throw knife switches and switches with butt contacts shall be connected such that their blades are de-energized when the switch is in the open position. Bolted pressure contact switches shall have barriers that prevent inadvertent contact with energized blades. Single-throw knife switches, bolted pressure contact switches, molded case switches, switches with butt contacts, and circuit breakers used as switches shall be connected so that the terminals supplying the load are de-energized when the switch is in the open position.

Bolted pressure switches that have energized blades when open, such as bottom feed designs, must be provided with barriers or a means to guard against inadvertent contact with the energized blades. This requirement is intended to provide protection against accidental contact with live parts when personnel are working on energized equipment.

*Exception: The blades and terminals supplying the load of a switch shall be permitted to be energized when the switch is in the open position where the switch is connected to circuits or equipment inherently capable of providing a backfeed source of power. For such installations, a permanent sign shall be installed on the switch enclosure or immediately adjacent to open switches with the following words or equivalent: WARNING — LOAD SIDE TERMINALS MAY BE ENERGIZED BY BACKFEED.*

Batteries, generators, PV systems, and double-ended switchboard ties are typical backfeed sources. These sources may cause the load side of the switch or circuit breaker to be energized when it is in the open position, a condition inherent to the circuitry.

**404.7 Indicating.** General-use and motor-circuit switches, circuit breakers, and molded case switches, where mounted in an enclosure as described in 404.3, shall clearly indicate whether they are in the open (off) or closed (on) position. Where these switch or circuit breaker handles are operated vertically rather than rotationally or horizontally, the up position of the handle shall be the (on) position.

*Exception No. 1: Vertically operated double-throw switches shall be permitted to be in the closed (on) position with the handle in either the up or down position.*

*Exception No. 2: On busway installations, tap switches employing a center-pivoting handle shall be permitted to be open or closed with either end of the handle in the up or down position. The switch position shall be clearly indicating and shall be visible from the floor or from the usual point of operation.*

Exception No. 2 clarifies the off and on positions and the operation to turn the switch off. Some busway switches are designed with a center pivot so that, at any time, one end of the switch handle is in the up position and the other side is down. This allows the switch to be pulled down to turn it off and also pulled down to turn it on, a configuration not in accordance with the requirement in the main rule. The added exception allows this time-proven method of operating busway switches.

**404.8 Accessibility and Grouping.**

**(A) Location.** All switches and circuit breakers used as switches shall be located so that they may be operated from a readily accessible place. They shall be installed such that the center of the grip of the operating handle of the switch or circuit breaker, when in its highest position, is not more than 2.0 m (6 ft 7 in.) above the floor or working platform.

*Exception No. 1: On busway installations, fused switches and circuit breakers shall be permitted to be located at the same level as the busway. Suitable means shall be provided to operate the handle of the device from the floor.*

*Exception No. 2: Switches and circuit breakers installed adjacent to motors, appliances, or other equipment that they supply shall be permitted to be located higher than 2.0 m (6 ft 7 in.) and to be accessible by portable means.*

*Exception No. 3: Hookstick operable isolating switches shall be permitted at greater heights.*

**(B) Voltage Between Adjacent Devices.** A snap switch shall not be grouped or ganged in enclosures with other snap switches, receptacles, or similar devices, unless they are arranged so that the voltage between adjacent devices does not exceed 300 volts, or unless they are installed in enclosures equipped with identified, securely installed barriers between adjacent devices.

Barriers are required between switches that are ganged in a box and used to control 277-volt lighting on 480Y/277-volt systems where two or more phase conductors enter the box. Permanent barriers would be required between devices fed from two different phases of this system because the voltage between the phase conductors would be 480 volts, nominal, and would exceed the 300-volt limit. Barriers are required even if one device space is left empty because the two remaining devices fed from different phase conductors still would be adjacent to each other. This requirement now applies to switches ganged together with any wiring device where the voltage between adjacent conductors exceeds 300 volts.

**(C) Multipole Snap Switches.** A multipole, general-use snap switch shall not be permitted to be fed from more than a single circuit unless it is listed and marked as a two-circuit or three-circuit switch, or unless its voltage rating is not less than the nominal line-to-line voltage of the system supplying the circuits.

The requirement for listed multiple-pole snap switches was added to the 2008 *Code*. It requires that nonlisted switches be listed for switching multiple circuits or have a rating not less than the line-to-line system voltage where more than one circuit is being switched.

**404.9 Provisions for General-Use Snap Switches.**

**(A) Faceplates.** Faceplates provided for snap switches mounted in boxes and other enclosures shall be installed so as to completely cover the opening and, where the switch is flush mounted, seat against the finished surface.

**(B) Grounding.** Snap switches, including dimmer and similar control switches, shall be connected to an equipment grounding conductor and shall provide a means to connect metal faceplates to the equipment grounding conductor, whether or not a metal faceplate is installed. Snap switches shall be considered to be part of an effective ground-fault current path if either of the following conditions is met:

(1) The switch is mounted with metal screws to a metal box or metal cover that is connected to an equipment grounding conductor or to a nonmetallic box with integral means for connecting to an equipment grounding conductor.
(2) An equipment grounding conductor or equipment bonding jumper is connected to an equipment grounding termination of the snap switch.

*Exception to (B): Where no means exists within the snap-switch enclosure for connecting to the equipment grounding conductor or where the wiring method does not include or provide an equipment grounding conductor, a snap switch without a connection to an equipment grounding conductor shall be permitted for replacement purposes only. A snap switch wired under the provisions of this exception and located within reach of earth, grade, conducting floors, or other conducting surfaces shall be provided with a faceplate of nonconducting, noncombustible material or shall be protected by a ground-fault circuit interrupter.*

The provisions of 404.9(B) specify that switching devices, including snap switches, dimmers, and similar control devices, must be grounded. Although the non–current-carrying metal parts of these devices typically are not subject to contact by personnel, there is concern about the use of metal faceplates, which do pose a shock hazard if they become energized. Therefore, the switch must provide a means for connection of an equipment grounding conductor to ground the metal faceplate whether or not one is installed.

The requirements in 404.9(B)(1) and (B)(2) describe the provisions to satisfy the main requirement. Switch plates in existing installations attached to switches in boxes without an equipment grounding conductor must be made of insulating material. See Exhibit 404.1 for an example of the typical method by which a metal faceplate is grounded.

The exception to 404.9(B) requires GFCI protection for the circuits terminating in an ungrounded device box where a metal faceplate that can be touched by persons in contact with a grounded surface is installed. This exception provides additional safety to persons where older electrical installations did not provide an equipment grounding conductor in the wiring method.

**(C) Construction.** Metal faceplates shall be of ferrous metal not less than 0.76 mm (0.030 in.) in thickness or of nonferrous metal not less than 1.02 mm (0.040 in.) in thickness. Faceplates of insulating material shall be noncombustible and not less than 2.54 mm (0.010 in.) in thickness, but they shall be permitted to be less than 2.54 mm (0.010 in.) in thickness if formed or reinforced to provide adequate mechanical strength.

### 404.10 Mounting of Snap Switches.

**(A) Surface-Type.** Snap switches used with open wiring on insulators shall be mounted on insulating material that separates the conductors at least 13 mm (½ in.) from the surface wired over.

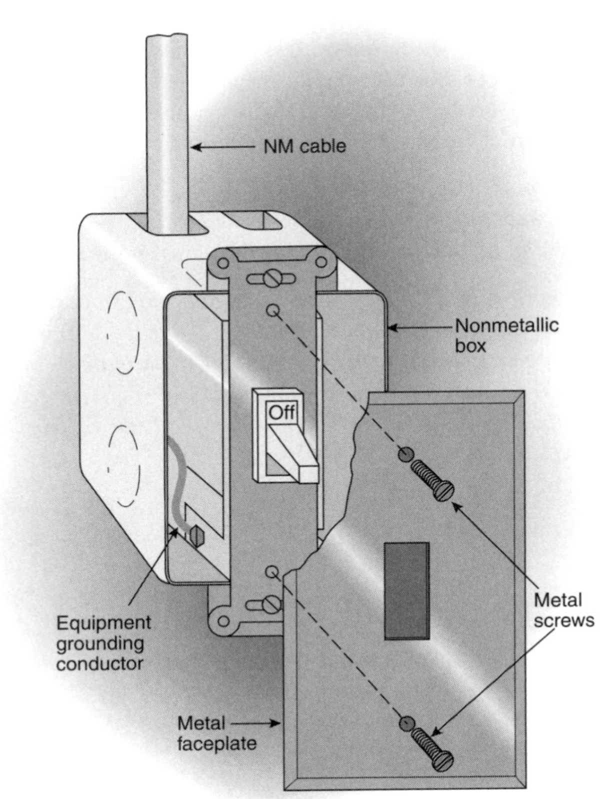


*EXHIBIT 404.1 Grounding of a metal faceplate through attachment to the grounded yoke of a snap switch.*

**(B) Box Mounted.** Flush-type snap switches mounted in boxes that are set back of the finished surface as permitted in 314.20 shall be installed so that the extension plaster ears are seated against the surface. Flush-type snap switches mounted in boxes that are flush with the finished surface or project from it shall be installed so that the mounting yoke or strap of the switch is seated against the box.

Section 404.10(B) makes it clear that regardless of where a box for a flush-type switch is installed (in a wall or in a ceiling), the switch yoke or strap must be seated against the box, or where the box is set back, the yoke or strap must be against the finished surface of the wall, ceiling, or other location in which the box is installed.

Cooperation is necessary among the building trades (carpenters, drywall installers, plasterers, and so on) so that electricians can properly set device boxes flush with the finish surface, thereby ensuring a secure seating of the switch yoke and permitting the maximum projection of switch handles through the installed switch plate.

**404.11 Circuit Breakers as Switches.** A hand-operable circuit breaker equipped with a lever or handle, or a power-operated circuit breaker capable of being opened by hand in the event of a power failure, shall be permitted to serve as a switch if it has the required number of poles.

FPN: See the provisions contained in 240.81 and 240.83.

Circuit breakers that are capable of being hand operated must clearly indicate whether they are in the open (off) or closed (on) position. See 404.7 for details on handle positions. See 240.83(D) for SWD and HID marking for circuit breakers used as switches for 120-volt and 277-volt fluorescent and high-intensity discharge lighting circuits.

**404.14 Rating and Use of Snap Switches.** Snap switches shall be used within their ratings and as indicated in 404.14(A) through (E).

FPN No. 1: For switches on signs and outline lighting, see 600.6.

FPN No. 2: For switches controlling motors, see 430.83, 430.109, and 430.110.

**(A) Alternating-Current General-Use Snap Switch.** A form of general-use snap switch suitable only for use on ac circuits for controlling the following:

(1) Resistive and inductive loads, including electric-discharge lamps, not exceeding the ampere rating of the switch at the voltage involved
(2) Tungsten-filament lamp loads not exceeding the ampere rating of the switch at 120 volts
(3) Motor loads not exceeding 80 percent of the ampere rating of the switch at its rated voltage

**(B) Alternating-Current or Direct-Current General-Use Snap Switch.** A form of general-use snap switch suitable for use on either ac or dc circuits for controlling the following:

(1) Resistive loads not exceeding the ampere rating of the switch at the voltage applied.
(2) Inductive loads not exceeding 50 percent of the ampere rating of the switch at the applied voltage. Switches rated in horsepower are suitable for controlling motor loads within their rating at the voltage applied.
(3) Tungsten-filament lamp loads not exceeding the ampere rating of the switch at the applied voltage if T-rated.

**(C) CO/ALR Snap Switches.** Snap switches rated 20 amperes or less directly connected to aluminum conductors shall be listed and marked CO/ALR.

**(D) Alternating-Current Specific-Use Snap Switches Rated for 347 Volts.** Snap switches rated 347 volts ac shall be listed and shall be used only for controlling the loads permitted by (D)(1) and (D)(2).

**(1) Noninductive Loads.** Noninductive loads other than tungsten-filament lamps not exceeding the ampere and voltage ratings of the switch.

**(2) Inductive Loads.** Inductive loads not exceeding the ampere and voltage ratings of the switch. Where particular load characteristics or limitations are specified as a condition of the listing, those restrictions shall be observed regardless of the ampere rating of the load.

The ampere rating of the switch shall not be less than 15 amperes at a voltage rating of 347 volts ac. Flush-type snap switches rated 347 volts ac shall not be readily interchangeable in box mounting with switches identified in 404.14(A) and (B).

Although not commonly used in the United States, 600Y/347-volt systems are permitted by the *Code*. In accordance with 210.6 and 225.7(D), these systems can be used to supply installations of outdoor lighting. For the purposes of controlling lighting circuits on these systems, 404.14(D)(2) permits a relatively new type of ac specific-use snap switch that is 347-volt rated. Such switches, unless specifically restricted, are permitted to be used on circuits of a lower voltage, such as 277- and 120-volt circuits.

**(E) Dimmer Switches.** General-use dimmer switches shall be used only to control permanently installed incandescent luminaires unless listed for the control of other loads and installed accordingly.

General-use dimmers are not permitted to control receptacles or cord-and-plug-connected table and floor lamps. Section 404.14(E) does not apply to commercial dimmers or theater dimmers that can be used for fluorescent lighting and portable lighting. If a dimmer that has been evaluated only for the control of incandescent luminaires is used, the potential for connecting incompatible equipment such as a cord-and-plug-connected motor-operated appliance or a portable fluorescent lamp is increased by using the dimmer to control a receptacle(s).

## Article 406 Receptacles, Cord Connectors, and Attachment Plugs (Caps)

**406.3 General Installation Requirements.** Receptacle outlets shall be located in branch circuits in accordance with Part III of Article 210. General installation requirements shall be in accordance with 406.3(A) through (F).

**(A) Grounding Type.** Receptacles installed on 15- and 20-ampere branch circuits shall be of the grounding type.

Grounding-type receptacles shall be installed only on circuits of the voltage class and current for which they are rated, except as provided in Table 210.21(B)(2) and Table 210.21(B)(3).

*Exception: Nongrounding-type receptacles installed in accordance with 406.3(D).*

**(B) To Be Grounded.** Receptacles and cord connectors that have equipment grounding conductor contacts shall have those contacts connected to an equipment grounding conductor.

*Exception No. 1: Receptacles mounted on portable and vehicle-mounted generators in accordance with 250.34.*

*Exception No. 2: Replacement receptacles as permitted by 406.3(D).*

**(C) Methods of Grounding.** The equipment grounding conductor contacts of receptacles and cord connectors shall be grounded by connection to the equipment grounding conductor of the circuit supplying the receptacle or cord connector.

FPN: For installation requirements for the reduction of electrical noise, see 250.146(D).

The branch-circuit wiring method shall include or provide an equipment grounding conductor to which the equipment grounding conductor contacts of the receptacle or cord connector are connected.

FPN No. 1: See 250.118 for acceptable grounding means.

FPN No. 2: For extensions of existing branch circuits, see 250.130.

**(D) Replacements.** Replacement of receptacles shall comply with 406.3(D)(1), (D)(2), and (D)(3) as applicable.

**(1) Grounding-Type Receptacles.** Where a grounding means exists in the receptacle enclosure or an equipment grounding conductor is installed in accordance with 250.130(C), grounding-type receptacles shall be used and shall be connected to the equipment grounding conductor in accordance with 406.3(C) or 250.130(C).

**(2) Ground-Fault Circuit Interrupters.** Ground-fault circuit-interrupter protected receptacles shall be provided where replacements are made at receptacle outlets that are required to be so protected elsewhere in this *Code*.

**(3) Non–Grounding-Type Receptacles.** Where attachment to an equipment grounding conductor does not exist in the receptacle enclosure, the installation shall comply with (D)(3)(a), (D)(3)(b), or (D)(3)(c).

(a) A non–grounding-type receptacle(s) shall be permitted to be replaced with another non–grounding-type receptacle(s).

(b) A non–grounding-type receptacle(s) shall be permitted to be replaced with a ground-fault circuit interrupter-type of receptacle(s). These receptacles shall be marked "No Equipment Ground." An equipment grounding conductor shall not be connected from the ground-fault circuit-interrupter-type receptacle to any outlet supplied from the ground-fault circuit-interrupter receptacle.

(c) A non–grounding-type receptacle(s) shall be permitted to be replaced with a grounding-type receptacle(s) where supplied through a ground-fault circuit interrupter. Grounding-type receptacles supplied through the ground-fault circuit interrupter shall be marked "GFCI Protected" and "No Equipment Ground." An equipment grounding conductor shall not be connected between the grounding-type receptacles.

**(E) Cord-and-Plug-Connected Equipment.** The installation of grounding-type receptacles shall not be used as a requirement that all cord-and-plug-connected equipment be of the grounded type.

FPN: See 250.114 for types of cord-and-plug-connected equipment to be grounded.

**(F) Noninterchangeable Types.** Receptacles connected to circuits that have different voltages, frequencies, or types of current (ac or dc) on the same premises shall be of such design that the attachment plugs used on these circuits are not interchangeable.

**406.4 Receptacle Mounting.** Receptacles shall be mounted in boxes or assemblies designed for the purpose, and such boxes or assemblies shall be securely fastened in place unless otherwise permitted elsewhere in this *Code*.

This section notes that boxes in which receptacles are installed are not always securely fastened in place. Receptacles in pendant boxes are permitted, provided the box is supported from the flexible cord in accordance with 314.23(H)(1). A pendant box that is properly suspended is not required to be securely fastened in place.

**(A) Boxes That Are Set Back.** Receptacles mounted in boxes that are set back from the finished surface as permitted in 314.20 shall be installed such that the mounting yoke or strap of the receptacle is held rigidly at the finished surface.

This rule applies to all finished surfaces where the device box is set back from the finished surface of a wall, a ceiling, or other location within a building.

**(B) Boxes That Are Flush.** Receptacles mounted in boxes that are flush with the finished surface or project therefrom shall be installed such that the mounting yoke or strap of the receptacle is held rigidly against the box or box cover.

To comply with 406.4(B), the outlet box used to enclose a receptacle must be rigidly and securely supported according to 314.23(B) or (C). In addition, mounting outlet boxes with the proper setback, according to 314.20, requires the cooperation of other construction trades (drywall installers, plasterers, and carpenters) and the building designers.

The intent of 406.4(A) through (C) is to allow attachment plugs to be inserted or removed without moving the receptacle. Additionally, by restricting movement of the receptacle, effective grounding continuity can be maintained for contact devices or receptacle yokes where the box is installed flush with the wall surface or where it projects therefrom. The proper installation of receptacles helps ensure that attachment plugs can be fully inserted, thus providing a better contact.

**(C) Receptacles Mounted on Covers.** Receptacles mounted to and supported by a cover shall be held rigidly against the cover by more than one screw or shall be a device assembly or box cover listed and identified for securing by a single screw.

Receptacles mounted on raised covers, such as the receptacle illustrated in Exhibit 406.1, are not permitted to be secured by a single screw unless listed and identified for the use.

*EXHIBIT 406.1 A receptacle mounted on a raised cover.*

**(D) Position of Receptacle Faces.** After installation, receptacle faces shall be flush with or project from faceplates of insulating material and shall project a minimum of 0.4 mm (0.015 in.) from metal faceplates.

The reason for requiring receptacles to project from metal faceplates is to prevent faults between the blades of attachment plugs and metal faceplates. Proper mounting of faceplates ensures that attachment plugs can be fully inserted, thus providing a better contact. The *NEC* does not specify the position (blades up or blades down) of a common vertically mounted 15- or 20-ampere duplex receptacle. Although many drawings in this handbook, such as Exhibit 406.1, show the slots for blades up, the receptacle may be installed with the slots for blades down. Receptacles can also be installed horizontally as well as vertically. Refer to 406.8(B) for information on receptacles installed in wet locations.

*Exception: Listed kits or assemblies encompassing receptacles and nonmetallic faceplates that cover the receptacle face, where the plate cannot be installed on any other receptacle, shall be permitted.*

The exception to 406.4(D) allows the use of listed kits that include the receptacle and a nonmetallic faceplate and that have been evaluated by a recognized testing laboratory to ensure that sufficient blade contact is achieved by the attachment plug when inserted in the receptacle. In addition, a nonmetallic faceplate would not fit the standard style receptacle. The second exception permitting nonmetallic faceplates that cover the receptacle face was deleted in the 2008 *Code*.

**(E) Receptacles in Countertops and Similar Work Surfaces in Dwelling Units.** Receptacles shall not be installed in a face-up position in countertops or similar work surfaces.

**(F) Exposed Terminals.** Receptacles shall be enclosed so that live wiring terminals are not exposed to contact.

**406.5 Receptacle Faceplates (Cover Plates).** Receptacle faceplates shall be installed so as to completely cover the opening and seat against the mounting surface.

**(A) Thickness of Metal Faceplates.** Metal faceplates shall be of ferrous metal not less than 0.76 mm (0.030 in.) in thickness or of nonferrous metal not less than 1.02 mm (0.040 in.) in thickness.

**(B) Grounding.** Metal faceplates shall be grounded.

Section 406.5(B) requires that metal receptacle faceplates be grounded. Generally, this requirement is easily met by grounding the metal box. However, isolated ground receptacles installed in nonmetallic boxes are problematic because grounding the receptacle in this case does not ground the faceplate. Section 406.2(D)(2) contains two solutions concerning the receptacle faceplate. First, the general solution is to use only nonmetallic faceplates. Second, the exception to 406.2(D)(2) allows a nonmetallic box manufacturer to add a feature or accessory to accomplish effective grounding of a metal faceplate.

**(C) Faceplates of Insulating Material.** Faceplates of insulating material shall be noncombustible and not less than 2.54 mm (0.10 in.) in thickness but shall be permitted to be less than 2.54 mm (0.10 in.) in thickness if formed or reinforced to provide adequate mechanical strength.

**406.6 Attachment Plugs, Cord Connectors, and Flanged Surface Devices.** All attachment plugs, cord connectors, and flanged surface devices (inlets and outlets) shall be listed and marked with the manufacturer's name or identification and voltage and ampere ratings.

Section 406.6 includes requirements governing the use of flanged surface inlet devices or motor base inlet plugs, often called motor plugs.

An energized cord cap is often improperly used to supply power to a building from a portable generator when a power failure occurs. Section 406.6 prohibits the improper use of a cord cap, where the blades are exposed and energized, to supply power to a cord body or plug into a receptacle to backfeed it. Prongs or blades that are exposed to contact by persons must not be energized unless an energized cord connector is installed in a flanged inlet device. Exhibit 406.2 illustrates a flanged inlet device.

*EXHIBIT 406.2 Flanged inlet device. (Courtesy of Pass & Seymour/Legrand®)*

**(A) Construction of Attachment Plugs and Cord Connectors.** Attachment plugs and cord connectors shall be constructed so that there are no exposed current-carrying parts except the prongs, blades, or pins. The cover for wire terminations shall be a part that is essential for the operation of an attachment plug or connector (dead-front construction).

**(B) Connection of Attachment Plugs.** Attachment plugs shall be installed so that their prongs, blades, or pins are not energized unless inserted into an energized receptacle or cord connectors. No receptacle shall be installed so as to require the insertion of an energized attachment plug as its source of supply.

The design requirements found in 406.6(B) (referred to as dead-front construction) minimize the occurrence of electrical faults between metal plates and attachment plugs with terminal screws exposed on the face of the plug.

The requirements in 406.6(B) were originally found in product information only. However, as an aid to the inspection community, these requirements are now clearly stated in the *NEC*. A live attachment plug cap can be a dangerous situation. Attachment plug caps should never be installed so as to allow the blades to be energized without being plugged into a device.

**(C) Attachment Plug Ejector Mechanisms.** Attachment plug ejector mechanisms shall not adversely affect engagement of the blades of the attachment plug with the contacts of the receptacle.

Section 406.6(C) permits a device that reduces the likelihood of damage to the cord when the cord is pulled to remove the plug. This device is designed for use by persons with mobility or visual impairment.

**(D) Flanged Surface Inlet.** A flanged surface inlet shall be installed such that the prongs, blades, or pins are not energized unless an energized cord connector is inserted into it.

Section 406.6(D) covers flanged inlet devices and prohibit the prongs, blades, or pins of motor base inlet plugs from being energized before a cord body is inserted in it.

**406.7 Noninterchangeability.** Receptacles, cord connectors, and attachment plugs shall be constructed such that receptacle or cord connectors do not accept an attachment plug with a different voltage or current rating from that for which the device is intended. However, a 20-ampere T-slot receptacle or cord connector shall be permitted to accept a 15-ampere attachment plug of the same voltage rating. Non–grounding-type receptacles and connectors shall not accept grounding-type attachment plugs.

For information on receptacle and attachment cap configurations, see NEMA WD 6-2002, *Wiring Devices — Dimensional Requirements*, available for download at www.nema.org.

### 406.8 Receptacles in Damp or Wet Locations.

The requirements of 406.8(A) and (B) as they apply to the covers that are typically used with lower rated receptacles (15 through 60 amperes) are summarized in Exhibit 406.3.

**(A) Damp Locations.** A receptacle installed outdoors in a location protected from the weather or in other damp locations shall have an enclosure for the receptacle that is weatherproof when the receptacle is covered (attachment plug cap not inserted and receptacle covers closed).

The requirement for listed weather-resistant type 15- and 20-ampere receptacles for both damp and wet locations was added to the 2008 *Code*. Studies indicated that normal receptacles were inadequate because covers were either broken off or not closed properly. The major differences between WR and non-WR receptacles are that the WR has additional corrosion protection, UV resistance, and cold impact resistance. A typical WR receptacle is shown in Exhibit 406.4.

An installation suitable for wet locations shall also be considered suitable for damp locations.

A receptacle shall be considered to be in a location protected from the weather where located under roofed open porches, canopies, marquees, and the like, and will

| Damp and Wet Receptacle Locations | Receptacle Cover (Enclosure) Type Requirements | |
|---|---|---|
| | Cover that *is not* weatherproof, with attachment plug cap inserted into receptacle | Cover that *is* weatherproof, with attachment plug cap inserted into receptacle ("in-use" type) |
| 406.8(A): Outdoor damp locations | Minimum type required<br>*Note: "In-use" type covers permitted* | Permitted |
| 406.8(A): Indoor damp locations | Minimum type required<br>*Note: "In-use" type covers permitted* | Permitted |
| 406.8(B)(1)&(2): Outdoor wet locations | Required for receptacle types other than those rated 15 and 20 amperes, 125 and 250 volts, where the tool, appliance, or other utilization equipment plugged into the receptacle *is* attended while in use.<br>*Note: "In-use" type covers permitted* | (a) Required for receptacles rated 15 and 20 amperes, 125 and 250 volts<br>(b) Required for receptacles other than those rated 15 and 20 amperes, 125 and 250 volts, where the tool, appliance, or other utilization equipment plugged into the receptacle *is not* attended while in use. |
| 406.8(B)(2): Indoor wet locations | Required for receptacle types other than those rated 15 and 20 amperes, 125 and 250 volts, where the tool, appliance, or other utilization equipment plugged into the receptacle *is* attended while in use.<br>*Note: "In-use" type covers permitted* | (a) Required for receptacles rated 15 and 20 amperes, 125 and 250 volts<br>(b) Required for receptacles other than those rated 15 and 20 amperes, 125 and 250 volts, where the tool, appliance, or other utilization equipment plugged into the receptacle *is not* attended while in use. |

***EXHIBIT 406.3*** *Requirements for receptacle cover (enclosure) types.*

**EXHIBIT 406.4** *A two-gang weatherproof cover suitable for use in wet locations. (Courtesy of Pass & Seymour/ Legrand®)*

**EXHIBIT 406.5** *A single-gang weatherproof cover suitable for use in wet locations. (Courtesy of Thomas & Betts Corp.)*

not be subjected to a beating rain or water runoff. All 15- and 20-ampere, 125- and 250-volt nonlocking receptacles shall be a listed weather-resistant type.

FPN: The types of receptacles covered by this requirement are identified as 5-15, 5-20, 6-15, and 6-20 in ANSI/NEMA WD 6-2002, National Electrical Manufacturers Association *Standard for Dimensions of Attachment Plugs and Receptacles.*

**(B) Wet Locations.**

**(1) 15- and 20-Ampere Receptacles in a Wet Location.** 15- and 20-ampere, 125- and 250-volt receptacles installed in a wet location shall have an enclosure that is weatherproof whether or not the attachment plug cap is inserted. All 15- and 20-ampere, 125- and 250-volt nonlocking receptacles shall be listed weather-resistant type.

To ensure the weatherproof integrity of the cord-and-plug connection to receptacles located in a wet location,406.8(B)(1) requires receptacle covers that provide a weatherproof enclosure at all times regardless of whether the plug is inserted. The requirement for this type of cover is not contingent on the anticipated use of the receptacle. This requirement applies to all 15- and 20-ampere, 125- and 250-volt receptacles that are installed in wet locations, including those receptacle outlets at dwelling units specified by 210.52(E). Exhibit 406.5 is an example of the type of receptacle enclosure required by 406.8(B)(1).

FPN: The types of receptacles covered by this requirement are identified as 5-15, 5-20, 6-15, and 6-20 in ANSI/NEMA WD 6-2002, National Electrical Manufacturers Association *Standard for Dimensions of Attachment Plugs and Receptacles.*]

*Exception: 15- and 20-ampere, 125- through 250-volt receptacles installed in a wet location and subject to routine high-pressure spray washing shall be permitted to have an enclosure that is weatherproof when the attachment plug is removed.*

**(2) Other Receptacles.** All other receptacles installed in a wet location shall comply with (B)(2)(a) or (B)(2)(b).

Section 406.8(B)(2)(a) applies to receptacles other than those rated 15 and 20 amperes, 125 and 250 volts, that supply cord-and-plug-connected equipment likely to be used outdoors or in a wet location for long periods of time. A portable pumpmotor is an example of such equipment. Receptacles for this application should remain weatherproof while they are in use.

Section 406.8(B)(2)(b) applies to receptacles other than those rated 15 and 20 amperes, 125 and 250 volts, that supply cord-and-plug-connected portable tools or other portable equipment likely to be used outdoors for a specific purpose and then removed.

(a) A receptacle installed in a wet location, where the product intended to be plugged into it is not attended while in use, shall have an enclosure that is weatherproof with the attachment plug cap inserted or removed.

(b) A receptacle installed in a wet location where the product intended to be plugged into it will be attended while in use (e.g., portable tools) shall have an enclosure that is weatherproof when the attachment plug is removed.

**(C) Bathtub and Shower Space.** Receptacles shall not be installed within or directly over a bathtub or shower stall.

Section 406.8(C) prohibits the installation of receptacles inside bathtub and shower spaces or above their footprint, even if the receptacles are installed in a weatherproof enclosure. Prohibiting such installation helps minimize the use of shavers, radios, hair dryers, and so on, in these areas.

The unprotected-line side of GFCI-protected receptacles installed in bathtub and shower spaces could possibly become wet and therefore create a shock hazard by energizing surrounding wet surfaces.

**(D) Protection for Floor Receptacles.** Standpipes of floor receptacles shall allow floor-cleaning equipment to be operated without damage to receptacles.

**(E) Flush Mounting with Faceplate.** The enclosure for a receptacle installed in an outlet box flush-mounted in a finished surface shall be made weatherproof by means of a weatherproof faceplate assembly that provides a watertight connection between the plate and the finished surface.

**406.9 Grounding-Type Receptacles, Adapters, Cord Connectors, and Attachment Plugs.**

**(A) Grounding Poles.** Grounding-type receptacles, cord connectors, and attachment plugs shall be provided with one fixed grounding pole in addition to the circuit poles. The grounding contacting pole of grounding-type plug-in ground-fault circuit interrupters shall be permitted to be of the movable, self-restoring type on circuits operating at not over 150 volts between any two conductors or any conductor and ground.

**(B) Grounding-Pole Identification.** Grounding-type receptacles, adapters, cord connections, and attachment plugs shall have a means for connection of an equipment grounding conductor to the grounding pole.

A terminal for connection to the grounding pole shall be designated by one of the following:

(1) A green-colored hexagonal-headed or -shaped terminal screw or nut, not readily removable.
(2) A green-colored pressure wire connector body (a wire barrel).
(3) A similar green-colored connection device, in the case of adapters. The grounding terminal of a grounding adapter shall be a green-colored rigid ear, lug, or similar device. The equipment grounding connection shall be so designed that it cannot make contact with current-carrying parts of the receptacle, adapter, or attachment plug. The adapter shall be polarized.

Section 406.9(B)(3) requires the grounding terminal of an adapter to be a green-colored ear, lug, or similar device, thereby prohibiting use of an adapter with an attached pigtail grounding wire, which had been used for many years.

(4) If the terminal for the equipment grounding conductor is not visible, the conductor entrance hole shall be marked with the word *green* or *ground*, the letters *G* or *GR*, a grounding symbol, or otherwise identified by a distinctive green color. If the terminal for the equipment grounding conductor is readily removable, the area adjacent to the terminal shall be similarly marked.

FPN: See FPN Figure 406.9(B)(4).

**(C) Grounding Terminal Use.** A grounding terminal shall not be used for purposes other than grounding.

**(D) Grounding-Pole Requirements.** Grounding-type attachment plugs and mating cord connectors and receptacles shall be designed such that the equipment grounding connection is made before the current-carrying connections. Grounding-type devices shall be so designed that grounding poles of attachment plugs cannot be brought into contact with current-carrying parts of receptacles or cord connectors.

The grounding blade of the attachment plug cap of most grounding-type combinations is longer than the circuit conductor blades and is used to ensure a "make-first, break-last" grounding connection. In some non-ANSI-approved pin-and-sleeve-type connections, the grounding contact of the receptacle is closer to the face of the receptacle than it is to other contacts, serving the same purpose.

**(E) Use.** Grounding-type attachment plugs shall be used only with a cord having an equipment grounding conductor.

FPN: See 200.10(B) for identification of grounded conductor terminals.

**406.10 Connecting Receptacle Grounding Terminal to Box.** The connection of the receptacle grounding terminal shall comply with 250.146.

## Article 408 Switchboards and Panelboards

**408.16 Switchboards in Damp or Wet Locations** Switchboards in damp or wet locations shall be installed in accordance with 312.2.

**408.17 Location Relative to Easily Ignitible Material.** Switchboards shall be placed so as to reduce to a minimum the probability of communicating fire to adjacent combustible materials. Where installed over a combustible floor, suitable protection thereto shall be provided.

One way to comply with the requirement of 408.17 is to form and attach a piece of sheet steel or other suitable noncombustible material to the floor under the electrical equipment.

**408.18 Clearances.**

**(A) From Ceiling.** For other than a totally enclosed switchboard, a space not less than 900 mm (3 ft) shall be provided between the top of the switchboard and any combustible ceiling, unless a noncombustible shield is provided between the switchboard and the ceiling.

**(B) Around Switchboards.** Clearances around switchboards shall comply with the provisions of 110.26.

Sufficient access and working space are required to permit safe operation and maintenance of switchboards. Table 110.26(A)(1) indicates minimum working clearances from 0 to 600 volts, and Table 110.34(A) is used for voltages over 600 volts.

**408.20 Location of Switchboards.** Switchboards that have any exposed live parts shall be located in permanently dry locations and then only where under competent supervision and accessible only to qualified persons. Switchboards shall be located such that the probability of damage from equipment or processes is reduced to a minimum.

## Article 410 Luminaires, Lampholders, and Lamps

**410.5 Live Parts.** Luminaires, portable luminaires, lampholders, and lamps shall have no live parts normally exposed to contact. Exposed accessible terminals in lampholders and switches shall not be installed in metal luminaire canopies or in open bases of portable table or floor luminaires.

*Exception: Cleat-type lampholders located at least 2.5 m (8 ft) above the floor shall be permitted to have exposed terminals.*

**410.10 Luminaires in Specific Locations.**

A pamphlet entitled *Luminaires Marking Guide*, available from Underwriters Laboratories Inc., was developed to help the authority having jurisdiction quickly determine whether common types of UL-listed fluorescent, high-intensity discharge, and incandescent fixtures are installed correctly.

**(A) Wet and Damp Locations.** Luminaires installed in wet or damp locations shall be installed such that water cannot enter or accumulate in wiring compartments, lampholders, or other electrical parts. All luminaires installed in wet locations shall be marked, "Suitable for Wet Locations." All luminaires installed in damp locations shall be marked "Suitable for Wet Locations" or "Suitable for Damp Locations."

Where luminaires are exposed to the weather or subject to water saturation, they must be of a type marked "Suitable for Wet Locations." Correct design, construction, and installation of these luminaires will prevent the entrance of rain, snow, ice, and dust. Outdoor parks and parking lots, outdoor recreational areas (tennis, golf, baseball, etc.), car wash areas, and building exteriors are examples of wet locations.

Locations protected from the weather and not subject to water saturation but still exposed to moisture, such as the following, may be considered damp locations:

1. The underside of store or gasoline station canopies or theater marquees
2. Some cold-storage warehouses
3. Some agricultural buildings
4. Some basements
5. Roofed open porches and carports

Luminaires used in these locations must be marked "Suitable for Wet Locations" or "Suitable for Damp Locations." See the definitions of *location, damp; location, dry;* and *location, wet* in Article 100.

**(B) Corrosive Locations.** Luminaires installed in corrosive locations shall be of a type suitable for such locations.

**(C) In Ducts or Hoods.** Luminaires shall be permitted to be installed in commercial cooking hoods where all of the following conditions are met:

(1) The luminaire shall be identified for use within commercial cooking hoods and installed such that the temperature limits of the materials used are not exceeded.
(2) The luminaire shall be constructed so that all exhaust vapors, grease, oil, or cooking vapors are excluded

from the lamp and wiring compartment. Diffusers shall be resistant to thermal shock.

(3) Parts of the luminaire exposed within the hood shall be corrosion resistant or protected against corrosion, and the surface shall be smooth so as not to collect deposits and to facilitate cleaning.

(4) Wiring methods and materials supplying the luminaire(s) shall not be exposed within the cooking hood.

FPN: See 110.11 for conductors and equipment exposed to deteriorating agents.

The requirement in 410.10(C)(4) was initially taken from NFPA 96, *Standard for Ventilation Control and Fire Protection of Commercial Cooking Operations*. NFPA 96 provides the minimum fire safety requirements (preventive and operative) related to the design, installation, operation, inspection, and maintenance of all public and private cooking operations, except in single-family residential dwellings. This coverage includes, but is not limited to, all cooking equipment, exhaust hoods, grease removal devices, exhaust ductwork, exhaust fans, dampers, fire-extinguishing equipment, and all other auxiliary or ancillary components or systems that are involved in the capture, containment, and control of grease-laden cooking effluent.

NFPA 96 is intended to include residential cooking equipment where used for purposes other than residential family use — such as employee kitchens or break areas and church and meeting hall kitchens — regardless of frequency of use.

Grease may cause the deterioration of conductor insulation, resulting in short circuits or ground faults in wiring, hence the requirement prohibiting wiring methods and materials (raceways, cables, lampholders) within ducts or hoods. Conventional enclosed and gasketed-type luminaires located in the path of travel of exhaust products are not permitted because a fire could result from the high temperatures on grease-coated glass bowls or globes enclosing the lamps. Recessed or surface gasketed-type luminaires intended for location within hoods must be identified as suitable for the specific purpose and should be installed with the required clearances maintained. Note that wiring systems, including rigid metal conduit, are not permitted to be run exposed within the cooking hood.

For further information, refer to UL 710, *Standard for Safety for Exhaust Hoods for Commercial Cooking Equipment*.

**(D) Bathtub and Shower Areas.** No parts of cord-connected luminaires, chain-, cable-, or cord-suspended luminaires, lighting track, pendants, or ceiling-suspended (paddle) fans shall be located within a zone measured 900 mm (3 ft) horizontally and 2.5 m (8 ft) vertically from the top of the bathtub rim or shower stall threshold. This zone is all encompassing and includes the space directly over the tub or shower stall. Luminaires located within the actual outside dimension of the bathtub or shower to a height of 2.5 m (8 ft) vertically from the top of the bathtub rim or shower threshold shall be marked for damp locations, or marked for wet locations where subject to shower spray.

The last sentence of 410.10(D) clarifies that securely fastened luminaires installed in or on the ceiling or wall are permitted to be located in the bathtub or shower area. Where they are subject to shower spray, the luminaires must be listed for a wet location. Luminaires installed in the tub or shower zone and not subject to shower spray are required to be listed for use in a damp location. GFCI protection is required only where specified in the installation instructions for the luminaire.

The intent of 410.10(D) is to keep cord-connected, chain-hanging, or pendant luminaires and suspended fans out of the reach of an individual standing on a bathtub rim. The list of prohibited items recognizes that the same risk of electric shock is present for each one.

Exhibit 410.1 illustrates the restricted zone in which the specified luminaires, lighting track, and paddle fans are prohibited. This requirement applies to hydromassage bathtubs, as defined in 680.2, as well as other bathtub types and shower areas. See 680.43 for installation requirements for spas and hot tubs (as defined in 680.2) installed indoors.

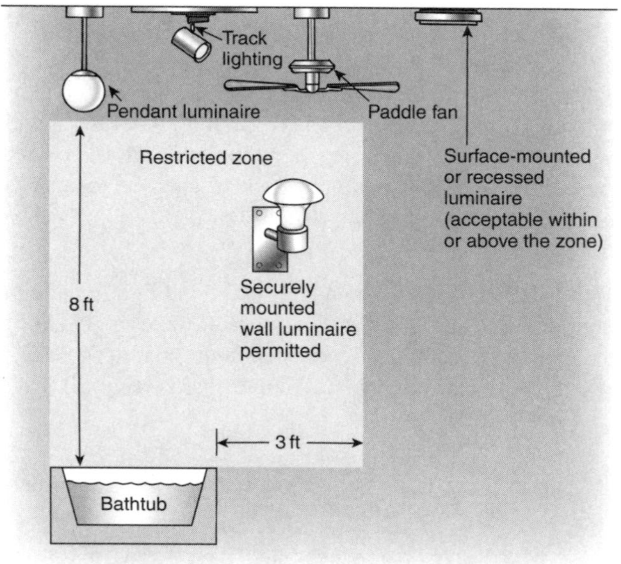


***EXHIBIT 410.1** Luminaires, lighting track, and suspended (paddle) fan located near a bathtub.*

**(E) Luminaires in Indoor Sports, Mixed-Use, and All-Purpose Facilities.** Luminaires subject to physical damage, using a mercury vapor or metal halide lamp, installed in playing and spectator seating areas of indoor sports, mixed-use, or all-purpose facilities shall be of the type that protects the lamp with a glass or plastic lens. Such luminaires shall be permitted to have an additional guard.

Instances of accidental breakage of mercury or metal halide lamp outer jackets have been reported in open luminaires in sports facilities and other similar locations. If the lamp is damaged, glass shards can fall on players or spectators. If the envelope is damaged, the arc tube continues to operate even though the outer jacket may be cracked or missing. This section requires luminaires to have their lamps protected by a glass or plastic lens, and it also permits an additional protective guard over the lens cover.

**410.11 Luminaires Near Combustible Material.** Luminaires shall be constructed, installed, or equipped with shades or guards so that combustible material is not subjected to temperatures in excess of 90°C (194°F).

Nearly every fire requires an initial heat source, an initial fuel source, and an action that brings them together. The requirements of 410.11, 410.12, 410.14, and 410.16 regulate only the placement of the heat source. It is important to remember that successful fire prevention is most likely to come about if the initial heat source and initial fuel source are treated with due care. Tests have shown that hot particles from broken incandescent lamps can ignite combustibles below the lamps.

**410.12 Luminaires over Combustible Material.** Lampholders installed over highly combustible material shall be of the unswitched type. Unless an individual switch is provided for each luminaire, lampholders shall be located at least 2.5 m (8 ft) above the floor or shall be located or guarded so that the lamps cannot be readily removed or damaged.

Section 410.12 refers to pendants and fixed lighting equipment installed above highly combustible material. If a lamp cannot be located out of reach, the requirement can be met by equipping the lamp with a suitable guard. Section 410.12 does not apply to portable lamps.

## Article 422 Appliances

**422.4 Live Parts.** Appliances shall have no live parts normally exposed to contact other than those parts functioning as open-resistance heating elements, such as the heating element of a toaster, which are necessarily exposed.

**422.10 Branch-Circuit Rating.** This section specifies the ratings of branch circuits capable of carrying appliance current without overheating under the conditions specified.

Conductors that form integral parts of appliances are tested as part of the listing or labeling process.

**(A) Individual Circuits.** The rating of an individual branch circuit shall not be less than the marked rating of the appliance or the marked rating of an appliance having combined loads as provided in 422.62.

The rating of an individual branch circuit for motor-operated appliances not having a marked rating shall be in accordance with Part II of Article 430.

The branch-circuit rating for an appliance that is a continuous load, other than a motor-operated appliance, shall not be less than 125 percent of the marked rating, or not less than 100 percent of the marked rating if the branch-circuit device and its assembly are listed for continuous loading at 100 percent of its rating.

Branch circuits and branch-circuit conductors for household ranges and cooking appliances shall be permitted to be in accordance with Table 220.55 and shall be sized in accordance with 210.19(A)(3).

**(B) Circuits Supplying Two or More Loads.** For branch circuits supplying appliance and other loads, the rating shall be determined in accordance with 210.23.

**422.17 Protection of Combustible Material.** Each electrically heated appliance that is intended by size, weight, and service to be located in a fixed position shall be placed so as to provide ample protection between the appliance and adjacent combustible material.

**422.31 Disconnection of Permanently Connected A ppliances.**

**(A) Rated at Not over 300 Volt-Amperes or ⅛ Horsepower.** For permanently connected appliances rated at not over 300 volt-amperes or ⅛ hp, the branch-circuit overcurrent device shall be permitted to serve as the disconnecting means.

**(B) Appliances Rated over 300 Volt-Amperes or ⅛ Horsepower.** For permanently connected appliances rated over 300 volt-amperes or ⅛ hp, the branch-circuit switch or circuit breaker shall be permitted to serve as the disconnecting means where the switch or circuit breaker is within sight from the appliance or is capable of being locked in the open position. The provision for locking or adding a lock to the disconnecting means shall be installed on or at the switch or circuit breaker used as the disconnecting means and shall remain in place with or without the lock installed.

Section 422.31(B) requires a special locking device for service and maintenance personnel. A device that is attached to the circuit breaker handle by a set screw is not an acceptable means to serve as a safe method of locking the device in the off position. The device must have provisions for placement of a lock on it to secure the device in the off position. The lock-out device must be part of the disconnect assembly and must remain in place after the padlock is removed, whether it is a fused disconnect switch, a single circuit breaker, or a circuit breaker in a panelboard. See 422.33(B) for electric ranges.

FPN: For appliances employing unit switches, see 422.34.

**422.32 Disconnecting Means for Motor-Driven Appliance.** If a switch or circuit breaker serves as the disconnecting means for a permanently connected motor-driven appliance of more than ⅛ hp, it shall be located within sight from the motor controller and shall comply with Part IX of Article 430.

*Exception: If a motor-driven appliance of more than ⅛ hp is provided with a unit switch that complies with 422.34(A), (B), (C), or (D), the switch or circuit breaker serving as the other disconnecting means shall be permitted to be out of sight from the motor controller.*

**422.33 Disconnection of Cord-and-Plug-Connected Appliances.**

**(A) Separable Connector or an Attachment Plug and Receptacle.** For cord-and-plug-connected appliances, an accessible separable connector or an accessible plug and receptacle shall be permitted to serve as the disconnecting means. Where the separable connector or plug and receptacle are not accessible, cord-and-plug-connected appliances shall be provided with disconnecting means in accordance with 422.31.

**(B) Connection at the Rear Base of a Range.** For cord-and-plug-connected household electric ranges, an attachment plug and receptacle connection at the rear base of a range, if it is accessible from the front by removal of a drawer, shall be considered as meeting the intent of 422.33(A).

**(C) Rating.** The rating of a receptacle or of a separable connector shall not be less than the rating of any appliance connected thereto.

*Exception: Demand factors authorized elsewhere in this Code shall be permitted to be applied to the rating of a receptacle or of a separable connector.*

## Article 424 Fixed Electric Space-Heating Equipment

**424.28 Nameplate.**

**(A) Marking Required.** Each unit of fixed electric space-heating equipment shall be provided with a nameplate giving the identifying name and the normal rating in volts and watts or in volts and amperes.

Electric space-heating equipment intended for use on alternating current only or direct current only shall be marked to so indicate. The marking of equipment consisting of motors over ⅛ hp and other loads shall specify the rating of the motor in volts, amperes, and frequency, and the heating load in volts and watts or in volts and amperes.

**(B) Location.** This nameplate shall be located so as to be visible or easily accessible after installation.

**424.36 Clearances of Wiring in Ceilings.** Wiring located above heated ceilings shall be spaced not less than 50 mm (2 in.) above the heated ceiling and shall be considered as operating at an ambient temperature of 50°C (122°F). The ampacity of conductors shall be calculated on the basis of the correction factors shown in the 0–2000 volt ampacity tables of Article 310. If this wiring is located above thermal insulation having a minimum thickness of 50 mm (2 in.), the wiring shall not require correction for temperature.

**424.65 Location of Disconnecting Means.** Duct heater controller equipment shall be either accessible with the disconnecting means installed at or within sight from the controller or as permitted by 424.19(A).

## Article 426 Fixed Outdoor Electric Deicing and Snow-Melting Equipment

**426.50 Disconnecting Means.**

**(A) Disconnection.** All fixed outdoor deicing and snow-melting equipment shall be provided with a means for simultaneous disconnection from all ungrounded conductors. Where readily accessible to the user of the equipment, the branch-circuit switch or circuit breaker shall be permitted to serve as the disconnecting means. The disconnecting means shall be of the indicating type and be provided with a positive lockout in the "off" position.

The disconnect must indicate when it is in the on or off position. A means must also be provided for locking the disconnect in the off position. The disconnecting means is allowed to be the branch circuit switch or circuit breaker

*NOTICE: For full text, see the 2008 edition of the NECH*

where it is "readily accessible"; in that case, it must have a means for a positive lockout in the off position. Like 424.19, the requirement that the disconnecting means simultaneously open all the ungrounded conductors was added to the 2008 *Code*. This condition prevents the practice of disconnecting one conductor at a time at terminal blocks or similar devices.

**(B) Cord-and-Plug-Connected Equipment.** The factory-installed attachment plug of cord-and-plug-connected equipment rated 20 amperes or less and 150 volts or less to ground shall be permitted to be the disconnecting means.

**426.51 Controllers.**

**(A) Temperature Controller with "Off" Position.** Temperature controlled switching devices that indicate an "off" position and that interrupt line current shall open all ungrounded conductors when the control device is in the "off" position. These devices shall not be permitted to serve as the disconnecting means unless capable of being locked in the open position.

**(B) Temperature Controller Without "Off" Position.** Temperature controlled switching devices that do not have an "off" position shall not be required to open all ungrounded conductors and shall not be permitted to serve as the disconnecting means.

**(C) Remote Temperature Controller.** Remote controlled temperature-actuated devices shall not be required to meet the requirements of 426.51(A). These devices shall not be permitted to serve as the disconnecting means.

**(D) Combined Switching Devices.** Switching devices consisting of combined temperature-actuated devices and manually controlled switches that serve both as the controller and the disconnecting means shall comply with all of the following conditions:

(1) Open all ungrounded conductors when manually placed in the "off" position
(2) Be so designed that the circuit cannot be energized automatically if the device has been manually placed in the "off" position
(3) Be capable of being locked in the open position

**426.54 Cord-and-Plug-Connected Deicing and Snow-Melting Equipment.** Cord-and-plug-connected deicing and snow-melting equipment shall be listed.

According to the UL *General Information for Electrical Equipment Directory — The White Book*, category KOBQ, UL listed deicing and snow-melting equipment is provided with means for permanent wiring connection, except the equipment rated 20 amperes or less and 150 volts or less to ground may be of cord-and-plug-connected construction. See the definition of *listed* in Article 100.

## Article 427 Fixed Electric Heating Equipment for Pipelines and Vessels

**427.36 Personnel Protection.** Induction coils that operate or may operate at a voltage greater than 30 volts ac shall be enclosed in a nonmetallic or split metallic enclosure, isolated, or made inaccessible by location to protect personnel in the area.

**427.55 Disconnecting Means.**

**(A) Switch or Circuit Breaker.** Means shall be provided to simultaneously disconnect all fixed electric pipeline or vessel heating equipment from all ungrounded conductors. The branch-circuit switch or circuit breaker, where readily accessible to the user of the equipment, shall be permitted to serve as the disconnecting means. The disconnecting means shall be of the indicating type and shall be provided with a positive lockout in the "off" position.

The requirement that the disconnecting means simultaneously open all the ungrounded conductors was added to the 2008 *Code* to prevent the practice of disconnecting one conductor at a time at terminal blocks or similar devices.

**(B) Cord-and-Plug-Connected Equipment.** The factory-installed attachment plug of cord-and-plug-connected equipment rated 20 amperes or less and 150 volts or less to ground shall be permitted to be the disconnecting means.

## Article 430 Motors, Motor Circuits, and Controllers

**430.101 General.** Part IX is intended to require disconnecting means capable of disconnecting motors and controllers from the circuit.

FPN No. 1: See Figure 430.1.

FPN No. 2: See 110.22 for identification of disconnecting means.

**430.102 Location.**

**(A) Controller.** An individual disconnecting means shall be provided for each controller and shall disconnect the controller. The disconnecting means shall be located in sight from the controller location.

The installation shown in Exhibit 430.16 is an example of compliance with the main requirement of 430.102(A).

*EXHIBIT 430.16 The disconnecting means for each controller, which must be within sight of the controller location per 430.102(A). (Courtesy of the International Association of Electrical Inspectors)*

*Exception No. 1: For motor circuits over 600 volts, nominal, a controller disconnecting means capable of being locked in the open position shall be permitted to be out of sight of the controller, provided the controller is marked with a warning label giving the location of the disconnecting means.*

*Exception No. 2: A single disconnecting means shall be permitted for a group of coordinated controllers that drive several parts of a single machine or piece of apparatus. The disconnecting means shall be located in sight from the controllers, and both the disconnecting means and the controllers shall be located in sight from the machine or apparatus.*

*Exception No. 3: The disconnecting means shall not be required to be in sight from valve actuator motor (VAM) assemblies containing the controller where such a location introduces additional or increased hazards to persons or property and conditions (a) and (b) are met.*

*(a) The valve actuator motor assembly is marked with a warning label giving the location of the disconnecting means.*

*(b) The provision for locking or adding a lock to the disconnecting means shall be installed on or at the switch or circuit breaker used as the disconnecting means and shall remain in place with or without the lock installed.*

**(B) Motor.** A disconnecting means shall be provided for a motor in accordance with (B)(1) or (B)(2).

**(1) Separate Motor Disconnect.** A disconnecting means for the motor shall be located in sight from the motor location and the driven machinery location.

**(2) Controller Disconnect.** The controller disconnecting means required in accordance with 430.102(A) shall be permitted to serve as the disconnecting means for the motor if it is in sight from the motor location and the driven machinery location.

*Exception to (1) and (2): The disconnecting means for the motor shall not be required under either condition (a) or condition (b), provided the controller disconnecting means required in accordance with 430.102(A) is individually capable of being locked in the open position. The provision for locking or adding a lock to the controller disconnecting means shall be installed on or at the switch or circuit breaker used as the disconnecting means and shall remain in place with or without the lock installed.*

*(a) Where such a location of the disconnecting means for the motor is impracticable or introduces additional or increased hazards to persons or property*

*(b) In industrial installations, with written safety procedures, where conditions of maintenance and supervision ensure that only qualified persons service the equipment*

FPN No. 1: Some examples of increased or additional hazards include, but are not limited to, motors rated in excess of 100 hp, multimotor equipment, submersible motors, motors associated with adjustable speed drives, and motors located in hazardous (classified) locations.

FPN No. 2: For information on lockout/tagout procedures, see NFPA 70E-2004, *Standard for Electrical Safety in the Workplace*.

The main rules of 430.102(A) and (B) require that the disconnecting means be in sight of the controller, the motor location, and the driven-machinery location. For motors over 600 volts, the controller disconnecting means may be out of sight of the controller, provided the controller has a warning label indicating the location and identification of the disconnecting means, which must be capable of being locked in the open position.

A single disconnecting means may be located adjacent to a group of coordinated controllers, as illustrated in Exhibit 430.17, where the controllers are mounted on a multimotor continuous process machine.

The remote location of disconnects for valve actuator motors (Exception No. 3) was added to the 2008 *Code*.

Per the exception to 430.102(B), the disconnecting means may only be out of sight of the motor, as illustrated in Exhibit 430.18, if the disconnecting means complying with 430.102(A) is individually capable of being locked in the open position and meets the criterion of either (a) or (b) of the exception. If locating the disconnecting means close to the motor location and driven machinery is impracticable due to the type of machinery, the type of facility, lack

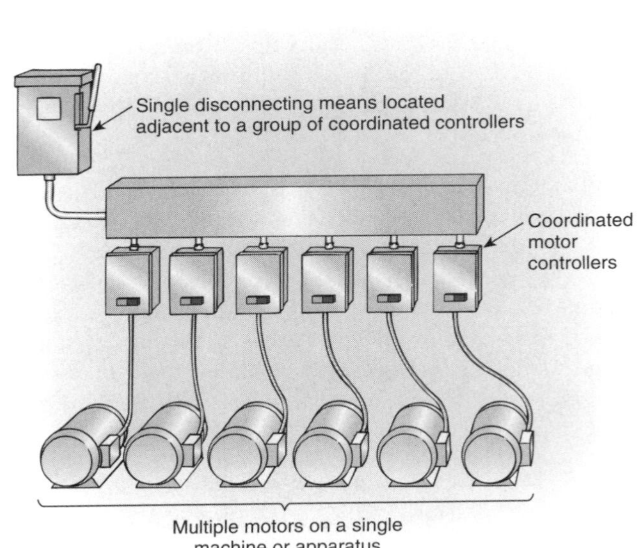


***EXHIBIT 430.17** A single disconnecting means located adjacent to a group of coordinated controllers mounted on a multimotor continuous process machine.*

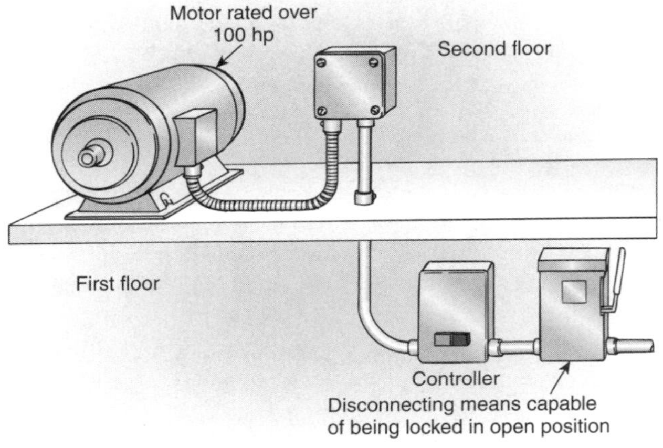


***EXHIBIT 430.18** A controller disconnecting means that is out of sight of the motor — only for cases that meet the requirements of (a) or (b) of 430.102(B), Exception.*

of space for locating large equipment such as disconnecting means rated over 600 volts, or any increased hazard to persons or property, the disconnecting means is permitted to be located remotely.

Industrial facilities that comply with OSHA, CFR 1910.147, *The Control of Hazardous Energy (Lockout/Tagout)*, are permitted to have the disconnecting means located remotely.

Section 430.102 clearly requires that individual disconnect switches or circuit breakers must be capable of being locked in the open position. Disconnect switches or circuit breakers that are located only behind the locked door of a panelboard or within locked rooms do not comply with the requirements of 430.102. The provision for locking or attaching a lock to the disconnecting means must be part of the disconnect and a permanent component of the switch or circuit breaker.

Fine Print Note No. 2 points out an important consideration and reference standard for employee safety in the workplace. NFPA 70E-2004, *Standard for Electrical Safety in the Workplace*, 120.2(A), requires in part that "All electrical circuit conductors and circuit parts shall not be considered to be in an electrically safe condition until all sources of energy are removed, the disconnecting means is under lockout/tagout, [and] the absence of voltage is verified by an approved voltage testing device." Further, it states, "Lockout/tagout requirements shall apply to fixed, permanently installed equipment, to temporarily installed equipment, and to portable equipment." The principles and procedures set forth in NFPA 70E establish strict work rules requiring locking off and tagging out of disconnect switches.

**430.103 Operation.** The disconnecting means shall open all ungrounded supply conductors and shall be designed so that no pole can be operated independently. The disconnecting means shall be permitted in the same enclosure with the controller. The disconnecting means shall be designed so that it cannot be closed automatically.

FPN: See 430.113 for equipment receiving energy from more than one source.

The *Code* requires that a switch, circuit breaker, or other device serve as a disconnecting means for both the controller and the motor, thereby providing safety during maintenance and inspection shutdown periods. The disconnecting means also disconnects the controller; therefore, it cannot be a part of the controller. However, separate disconnects and controllers may be mounted on the same panel or contained in the same enclosure, such as combination fused-switch, magnetic-starter units.

Depending on the size of the motor and other conditions, the type of disconnecting means required may be a motor circuit switch, a circuit breaker, a general-use switch, an isolating switch, an attachment plug and receptacle, or a branch-circuit short-circuit and ground-fault protective device, as specified in 430.109.

If a motor stalls or is under heavy overload and the motor controller fails to properly open the circuit, the disconnecting means, which must be rated to interrupt locked-rotor current, can be used to open the circuit. For motors larger than 100 hp ac or 40 hp dc, the disconnecting means is, in accordance with 430.109(E), permitted to be a general-use or an isolating switch where plainly marked "Do not operate under load."

**430.104 To Be Indicating.** The disconnecting means shall plainly indicate whether it is in the open (off) or closed (on) position.

**430.107 Readily Accessible.** At least one of the disconnecting means shall be readily accessible.

**430.232 Where Required.** Exposed live parts of motors and controllers operating at 50 volts or more between terminals shall be guarded against accidental contact by enclosure or by location as follows:

(1) By installation in a room or enclosure that is accessible only to qualified persons
(2) By installation on a suitable balcony, gallery, or platform, elevated and arranged so as to exclude unqualified persons
(3) By elevation 2.5 m (8 ft) or more above the floor

*Exception: Live parts of motors operating at more than 50 volts between terminals shall not require additional guarding for stationary motors that have commutators, collectors, and brush rigging located inside of motor-end brackets and not conductively connected to supply circuits operating at more than 150 volts to ground.*

**430.233 Guards for Attendants.** Where live parts of motors or controllers operating at over 150 volts to ground are guarded against accidental contact only by location as specified in 430.232, and where adjustment or other attendance may be necessary during the operation of the apparatus, suitable insulating mats or platforms shall be provided so that the attendant cannot readily touch live parts unless standing on the mats or platforms.

FPN: For working space, see 110.26 and 110.34.

## Article 440 Air-Conditioning and Refrigerating Equipment

**440.14 Location.** Disconnecting means shall be located within sight from and readily accessible from the air-conditioning or refrigerating equipment. The disconnecting means shall be permitted to be installed on or within the air-conditioning or refrigerating equipment.

The disconnecting means shall not be located on panels that are designed to allow access to the air-conditioning or refrigeration equipment or to obscure the equipment nameplate(s).

*Exception No. 1: Where the disconnecting means provided in accordance with 430.102(A) is capable of being locked in the open position, and the refrigerating or air-conditioning equipment is essential to an industrial process in a facility with written safety procedures, and where the conditions of maintenance and supervision ensure that only qualified persons service the equipment, a disconnecting means within sight from the equipment shall not be required. The provision for locking or adding a lock to the disconnecting means shall be installed on or at the switch or circuit breaker and shall remain in place with or without the lock installed.*

Exception No. 1 accommodates special conditions associated with process refrigeration equipment. Typically, this equipment is very large, so rated disconnects may not be available. Additionally, this equipment may be in hazardous locations, and locating disconnecting means within sight of the motor may introduce additional hazards. The provision for locking or attaching a lock to the disconnecting means must be part of the disconnect and a permanent component of the switch or circuit breaker. The phrase "and shall remain in place with or without the lock installed" is used to preclude portable or transferable-type lockout devices from being used as the method to provide the ability to lock the switch or circuit breaker in the open (off) position. Examples of this type of locking hardware are shown in Exhibit 440.1.

*Exception No. 2: Where an attachment plug and receptacle serve as the disconnecting means in accordance with 440.13, their location shall be accessible but shall not be required to be readily accessible.*

FPN: See Parts VII and IX of Article 430 for additional requirements.

The references to Parts VII and IX of Article 430 in the fine print note are intended to call attention to the additional disconnect location requirements in 430.102, 430.107, and 430.113. Because 440.3(A) makes the requirements in Article 440 in addition to or amendatory of the provisions of Article 430, the requirement of 440.14 mandates that the equipment disconnecting means be within sight from and readily accessible from the equipment, even if there is also a remote disconnect capable of being locked in the open position under the provision of 430.102(B), Exception.

This special requirement for air-conditioning and refrigeration equipment covered by Article 440 is more stringent than the provisions in Article 430, to provide protection for service personnel working on equipment located in attics, on roofs, or outside in a remote location

*EXHIBIT 440.1 Examples of two different types of locking hardware that are not readily removable or transferable. (Courtesy of Square D/Schneider Electric)*

where it is difficult to gain access to a remote lockable disconnect. See 440.14, Exception No. 1.

## Article 450 Transformers and Transformer Vaults (Including Secondary Ties)

**450.8 Guarding.** Transformers shall be guarded as specified in 450.8(A) through (D).

**(A) Mechanical Protection.** Appropriate provisions shall be made to minimize the possibility of damage to transformers from external causes where the transformers are exposed to physical damage.

One method of providing mechanical protection is to strategically place bollards around the transformer. This practice provides a degree of protection from vehicles.

**(B) Case or Enclosure.** Dry-type transformers shall be provided with a noncombustible moisture-resistant case or enclosure that provides protection against the accidental insertion of foreign objects.

**(C) Exposed Energized Parts.** Switches or other equipment operating at 600 volts, nominal, or less and serving only equipment within a transformer enclosure shall be permitted to be installed in the transformer enclosure if accessible to qualified persons only. All energized parts shall be guarded in accordance with 110.27 and 110.34.

**(D) Voltage Warning.** The operating voltage of exposed live parts of transformer installations shall be indicated by signs or visible markings on the equipment or structures.

**450.9 Ventilation.** The ventilation shall be adequate to dispose of the transformer full-load losses without creating a temperature rise that is in excess of the transformer rating.

FPN No. 1: See ANSI/IEEE C57.12.00-1993, *General Requirements for Liquid-Immersed Distribution, Power, and Regulating Transformers*, and ANSI/IEEE C57.12.01-1989, *General Requirements for Dry-Type Distribution and Power Transformers*.

FPN No. 2: Additional losses may occur in some transformers where nonsinusoidal currents are present, resulting in increased heat in the transformer above its rating. See ANSI/IEEE C57.110-1993, *Recommended Practice for Establishing Transformer Capability When Supplying Nonsinusoidal Load Currents*, where transformers are utilized with nonlinear loads.

Transformers with ventilating openings shall be installed so that the ventilating openings are not blocked by walls or other obstructions. The required clearances shall be clearly marked on the transformer.

Section 450.9 is intended to clarify that transformers are not permitted to be installed directly against walls or other obstructions that block openings for ventilation and that the required clearances should be clearly marked on the transformer (see 450.11).

Fine Print Note No. 2 following 450.9 warns of increased heating of transformers. See the commentary following 450.3, FPN No. 2, and the commentary following 310.15(B)(4) for additional information concerning nonlinear loads.

**450.13 Accessibility.** All transformers and transformer vaults shall be readily accessible to qualified personnel for inspection and maintenance or shall meet the requirements of 450.13(A) or 450.13(B).

Transformers are not accessible if wiring methods or other equipment obstruct the access of a worker or prevent removal of the covers for inspection or maintenance. Practical clearance considerations required for removal and replacement of the transformer are also important.

**(A) Open Installations.** Dry-type transformers 600 volts, nominal, or less, located in the open on walls, columns, or structures, shall not be required to be readily accessible.

**(B) Hollow Space Installations.** Dry-type transformers 600 volts, nominal, or less and not exceeding 50 kVA shall be permitted in hollow spaces of buildings not permanently closed in by structure, provided they meet the ventilation requirements of 450.9 and separation from combustible materials requirements of 450.21(A). Transformers so installed shall not be required to be readily accessible.

Section 450.13(B) permits the installation of dry-type transformers rated 600 volts or less and not exceeding 50 kVA in hollow spaces of hung ceiling areas, provided these spaces are fire resistant, ventilated, and accessible. According to 300.22(C)(2), transformers are permitted to be installed in hollow spaces where the space is used for environmental air, provided the transformer is in a metal enclosure (ventilated or nonventilated) and the transformer is suitable for the ambient air temperature within the hollow space. Of course, the requirement of 450.13(B) applies to transformer installations in "other space used for environmental air" per 300.22(C).

**450.41 Location.** Vaults shall be located where they can be ventilated to the outside air without using flues or ducts wherever such an arrangement is practicable.

**450.42 Walls, Roofs, and Floors.** The walls and roofs of vaults shall be constructed of materials that have adequate structural strength for the conditions with a minimum fire resistance of 3 hours. The floors of vaults in contact with the earth shall be of concrete that is not less than 100 mm (4 in.) thick, but where the vault is constructed with a vacant space or other stories below it, the floor shall have adequate structural strength for the load imposed thereon and a minimum fire resistance of 3 hours. For the purposes of this section, studs and wallboard construction shall not be acceptable.

*Exception: Where transformers are protected with automatic sprinkler, water spray, carbon dioxide, or halon, construction of 1-hour rating shall be permitted.*

FPN No. 1: For additional information, see ANSI/ASTM E119-1995, *Method for Fire Tests of Building Construction and Materials*, and NFPA 251-2006, *Standard Methods of Tests of Fire Resistance of Building Construction and Materials*.

FPN No. 2: A typical 3-hour construction is 150 mm (6 in.) thick reinforced concrete.

Vaults are intended primarily as passive fire protection. The need for vaults is dictated by the combustibility of the dielectric media and the size of the transformer. Transformers insulated with mineral oil have the greatest need for passive protection, to prevent the spread of burning oil to other combustible materials.

Although construction of a 3-hour-rated wall may be possible using studs and wallboard, this construction method is not permitted for transformer vaults. A reduction in fire-resistance rating from 3 hours to 1 hour is permitted for vaults equipped with an automatic fire suppression system.

Askarel is no longer manufactured as a transformer-insulating fluid. Askarel-insulated transformers of less than 35,000 volts do not require vaults, because askarel is considered a noncombustible fluid. Transformers with a listed less-flammable liquid insulation may be installed without a vault, as permitted in 450.23. See the commentary following 450.23(B)(2), which relates to Type I and Type II building construction.

**450.43 Doorways.** Vault doorways shall be protected in accordance with 450.43(A), (B), and (C).

**(A) Type of Door.** Each doorway leading into a vault from the building interior shall be provided with a tight-fitting door that has a minimum fire rating of 3 hours. The authority having jurisdiction shall be permitted to require such a door for an exterior wall opening where conditions warrant.

*Exception: Where transformers are protected with automatic sprinkler, water spray, carbon dioxide, or halon, construction of 1-hour rating shall be permitted.*

FPN: For additional information, see NFPA 80-2007, *Standard for Fire Doors and Other Opening Protectives*.

**(B) Sills.** A door sill or curb that is of sufficient height to confine the oil from the largest transformer within the vault shall be provided, and in no case shall the height be less than 100 mm (4 in.).

**(C) Locks.** Doors shall be equipped with locks, and doors shall be kept locked, access being allowed only to qualified persons. Personnel doors shall swing out and be equipped with panic bars, pressure plates, or other devices that are normally latched but open under simple pressure.

Section 450.43 prohibits the use of conventional rotation-type door knobs on transformer vault doors. It is believed that an injured worker attempting to escape from a transformer vault may not be able to operate a rotating-type door knob but would be able to escape through a door equipped with panic-type hardware.

**450.45 Ventilation Openings.** Where required by 450.9, openings for ventilation shall be provided in accordance with 450.45(A) through (F).

**(A) Location.** Ventilation openings shall be located as far as possible from doors, windows, fire escapes, and combustible material.

**(B) Arrangement.** A vault ventilated by natural circulation of air shall be permitted to have roughly half of the total area of openings required for ventilation in one or more openings near the floor and the remainder in one or more openings in the roof or in the sidewalls near the roof, or all of the area required for ventilation shall be permitted in one or more openings in or near the roof.

**(C) Size.** For a vault ventilated by natural circulation of air to an outdoor area, the combined net area of all ventilating openings, after deducting the area occupied by screens, gratings, or louvers, shall not be less than 1900 $mm^2$ (3 $in.^2$) per kVA of transformer capacity in service, and in no case shall the net area be less than 0.1 $m^2$ (1 $ft^2$) for any capacity under 50 kVA.

**(D) Covering.** Ventilation openings shall be covered with durable gratings, screens, or louvers, according to the treatment required in order to avoid unsafe conditions.

**(E) Dampers.** All ventilation openings to the indoors shall be provided with automatic closing fire dampers that operate in response to a vault fire. Such dampers shall possess a standard fire rating of not less than 1½ hours.

FPN: See ANSI/UL 555-1995, *Standard for Fire Dampers.*

**(F) Ducts.** Ventilating ducts shall be constructed of fire-resistant material.

## Article 460 Capacitors

### 460.2 Enclosing and Guarding.

**(A) Containing More Than 11 L (3 gal) of Flammable Liquid.** Capacitors containing more than 11 L (3 gal) of flammable liquid shall be enclosed in vaults or outdoor fenced enclosures complying with Article 110, Part III. This limit shall apply to any single unit in an installation of capacitors.

**(B) Accidental Contact.** Where capacitors are accessible to unauthorized and unqualified persons, they shall be enclosed, located, or guarded so that persons cannot come into accidental contact or bring conducting materials into accidental contact with exposed energized parts, terminals, or buses associated with them. However, no additional guarding is required for enclosures accessible only to authorized and qualified persons.

Means are required to drain off the stored charge in a capacitor after the supply circuit has been opened. Otherwise, a person servicing the equipment could receive a severe shock, or damage could occur to the equipment.

Exhibit 460.1, diagram (a), shows a method in which capacitors are connected in a motor circuit so that they may be switched with the motor. In this arrangement, the stored charge drains off through the windings when the circuit is opened. Diagram (b) shows another arrangement in which the capacitor is connected to the line side of the motor starter contacts. An automatic discharge device and a separate disconnecting means are required.

As shown in Exhibit 460.2, capacitors are often equipped with built-in resistors to drain off the stored charge, although this type of capacitor is not needed where connected as shown in Exhibit 460.1, diagram (a).

**460.6 Discharge of Stored Energy.** Capacitors shall be provided with a means of discharging stored energy.

**(A) Time of Discharge.** The residual voltage of a capacitor shall be reduced to 50 volts, nominal, or less within 1 minute after the capacitor is disconnected from the source of supply.

**(B) Means of Discharge.** The discharge circuit shall be either permanently connected to the terminals of the capacitor or capacitor bank or provided with automatic means, of connecting it to the terminals of the capacitor bank on removal of voltage from the line. Manual means of switching or connecting the discharge circuit shall not be used.

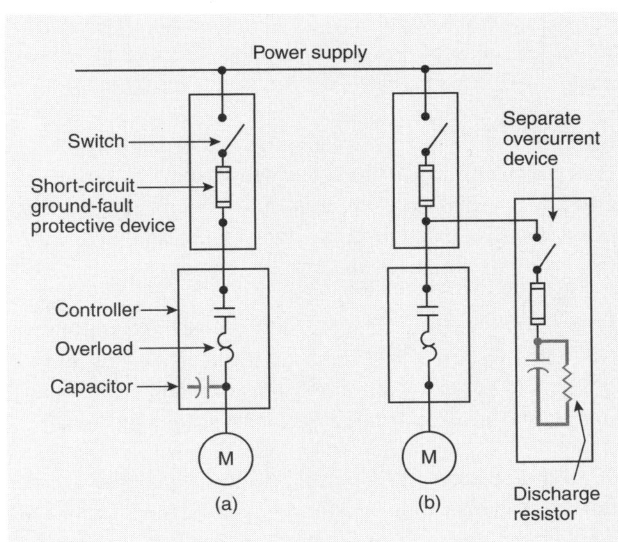


*EXHIBIT 460.1 Methods of connecting capacitors in induction motor circuit for power factor correction.*

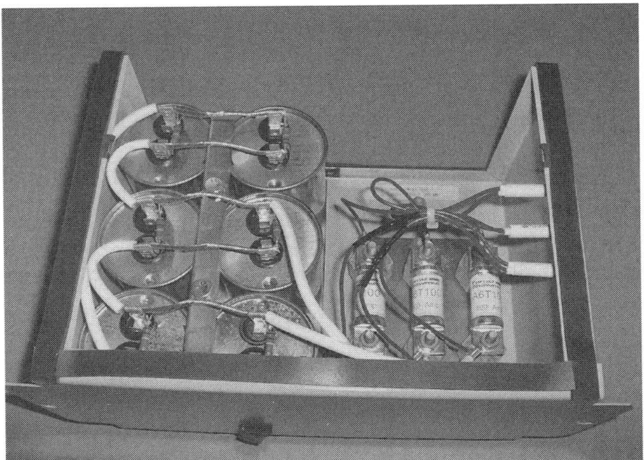

*EXHIBIT 460.2 Power factor correction capacitors with internal discharge resistors (blue) and overcurrent protection. (Courtesy of GE Energy)*

**460.24 Switching.**

**(A) Load Current.** Group-operated switches shall be used for capacitor switching and shall be capable of the following:

(1) Carrying continuously not less than 135 percent of the rated current of the capacitor installation
(2) Interrupting the maximum continuous load current of each capacitor, capacitor bank, or capacitor installation that will be switched as a unit
(3) Withstanding the maximum inrush current, including contributions from adjacent capacitor installations
(4) Carrying currents due to faults on capacitor side of switch

**(B) Isolation.**

**(1) General.** A means shall be installed to isolate from all sources of voltage each capacitor, capacitor bank, or capacitor installation that will be removed from service as a unit. The isolating means shall provide a visible gap in the electrical circuit adequate for the operating voltage.

**(2) Isolating or Disconnecting Switches with No Interrupting Rating.** Isolating or disconnecting switches (with no interrupting rating) shall be interlocked with the load-interrupting device or shall be provided with prominently displayed caution signs in accordance with 490.22 to prevent switching load current.

**(C) Additional Requirements for Series Capacitors.** The proper switching sequence shall be ensured by use of one of the following:

(1) Mechanically sequenced isolating and bypass switches
(2) Interlocks
(3) Switching procedure prominently displayed at the switching location

**460.28 Means for Discharge.**

**(A) Means to Reduce the Residual Voltage.** A means shall be provided to reduce the residual voltage of a capacitor to 50 volts or less within 5 minutes after the capacitor is disconnected from the source of supply.

**(B) Connection to Terminals.** A discharge circuit shall be either permanently connected to the terminals of the capacitor or provided with automatic means of connecting it to the terminals of the capacitor bank after disconnection of the capacitor from the source of supply. The windings of motors, transformers, or other equipment directly connected to capacitors without a switch or overcurrent device interposed shall meet the requirements of 460.28(A).

## Article 490 Equipment, Over 600 Volts, Nominal

**490.21 Circuit-Interrupting Devices.**

**(A) Circuit Breakers.**

**(1) Location.**

(a) Circuit breakers installed indoors shall be mounted either in metal-enclosed units or fire-resistant cell-mounted units, or they shall be permitted to be open-mounted in locations accessible to qualified persons only.

(b) Circuit breakers used to control oil-filled transformers shall either be located outside the transformer vault or be capable of operation from outside the vault.

(c) Oil circuit breakers shall be arranged or located so that adjacent readily combustible structures or materials are safeguarded in an approved manner.

**(2) Operating Characteristics.** Circuit breakers shall have the following equipment or operating characteristics:

(1) An accessible mechanical or other approved means for manual tripping, independent of control power
(2) Be release free (trip free)
(3) If capable of being opened or closed manually while energized, main contacts that operate independently of the speed of the manual operation
(4) A mechanical position indicator at the circuit breaker to show the open or closed position of the main contacts
(5) A means of indicating the open and closed position of the breaker at the point(s) from which they may be operated

**(3) Nameplate.** A circuit breaker shall have a permanent and legible nameplate showing manufacturer's name or trademark, manufacturer's type or identification number, continuous current rating, interrupting rating in megavolt-amperes (MVA) or amperes, and maximum voltage rating. Modification of a circuit breaker affecting its rating(s) shall be accompanied by an appropriate change of nameplate information.

**(4) Rating.** Circuit breakers shall have the following ratings:

(1) The continuous current rating of a circuit breaker shall not be less than the maximum continuous current through the circuit breaker.
(2) The interrupting rating of a circuit breaker shall not be less than the maximum fault current the circuit breaker will be required to interrupt, including contributions from all connected sources of energy.
(3) The closing rating of a circuit breaker shall not be less than the maximum asymmetrical fault current into which the circuit breaker can be closed.
(4) The momentary rating of a circuit breaker shall not be less than the maximum asymmetrical fault current at the point of installation.
(5) The rated maximum voltage of a circuit breaker shall not be less than the maximum circuit voltage.

**(B) Power Fuses and Fuseholders.**

**(1) Use.** Where fuses are used to protect conductors and equipment, a fuse shall be placed in each ungrounded conductor. Two power fuses shall be permitted to be used in parallel to protect the same load if both fuses have identical ratings and both fuses are installed in an identified common mounting with electrical connections that divide the current equally. Power fuses of the vented type shall not be used indoors, underground, or in metal enclosures unless identified for the use.

**(2) Interrupting Rating.** The interrupting rating of power fuses shall not be less than the maximum fault current the fuse is required to interrupt, including contributions from all connected sources of energy.

**(3) Voltage Rating.** The maximum voltage rating of power fuses shall not be less than the maximum circuit voltage. Fuses having a minimum recommended operating voltage shall not be applied below this voltage.

**(4) Identification of Fuse Mountings and Fuse Units.** Fuse mountings and fuse units shall have permanent and legible nameplates showing the manufacturer's type or designation, continuous current rating, interrupting current rating, and maximum voltage rating.

**(5) Fuses.** Fuses that expel flame in opening the circuit shall be designed or arranged so that they function properly without hazard to persons or property.

**(6) Fuseholders.** Fuseholders shall be designed or installed so that they are de-energized while a fuse is being replaced.

*Exception: Fuses and fuseholders designed to permit fuse replacement by qualified persons using equipment designed for the purpose without de-energizing the fuseholder shall be permitted.*

**(7) High-Voltage Fuses.** Metal-enclosed switchgear and substations that utilize high-voltage fuses shall be provided with a gang-operated disconnecting switch. Isolation of the fuses from the circuit shall be provided by either connecting a switch between the source and the fuses or providing roll-out switch and fuse-type construction. The switch shall be of the load-interrupter type, unless mechanically or electrically interlocked with a load-interrupting device arranged to reduce the load to the interrupting capability of the switch.

*Exception: More than one switch shall be permitted as the disconnecting means for one set of fuses where the switches are installed to provide connection to more than one set of supply conductors. The switches shall be mechanically or electrically interlocked to permit access to the fuses only when all switches are open. A conspicuous sign shall be placed at the fuses identifying the presence of more than one source.*

**(C) Distribution Cutouts and Fuse Links — Expulsion Type.**

**(1) Installation.** Cutouts shall be located so that they may be readily and safely operated and re-fused, and so that the exhaust of the fuses does not endanger persons. Distribution cutouts shall not be used indoors, underground, or in metal enclosures.

**(2) Operation.** Where fused cutouts are not suitable to interrupt the circuit manually while carrying full load, an approved means shall be installed to interrupt the entire load. Unless the fused cutouts are interlocked with the switch to prevent opening of the cutouts under load, a conspicuous sign shall be placed at such cutouts identifying that they shall not be operated under load.

**(3) Interrupting Rating.** The interrupting rating of distribution cutouts shall not be less than the maximum fault current the cutout is required to interrupt, including contributions from all connected sources of energy.

**(4) Voltage Rating.** The maximum voltage rating of cutouts shall not be less than the maximum circuit voltage.

**(5) Identification.** Distribution cutouts shall have on their body, door, or fuse tube a permanent and legible nameplate or identification showing the manufacturer's type or designation, continuous current rating, maximum voltage rating, and interrupting rating.

**(6) Fuse Links.** Fuse links shall have a permanent and legible identification showing continuous current rating and type.

**(7) Structure Mounted Outdoors.** The height of cutouts mounted outdoors on structures shall provide safe clearance between lowest energized parts (open or closed position) and standing surfaces, in accordance with 110.34(E).

**(D) Oil-Filled Cutouts.**

**(1) Continuous Current Rating.** The continuous current rating of oil-filled cutouts shall not be less than the maximum continuous current through the cutout.

**(2) Interrupting Rating.** The interrupting rating of oil-filled cutouts shall not be less than the maximum fault current the oil-filled cutout is required to interrupt, including contributions from all connected sources of energy.

**(3) Voltage Rating.** The maximum voltage rating of oil-filled cutouts shall not be less than the maximum circuit voltage.

**(4) Fault Closing Rating.** Oil-filled cutouts shall have a fault closing rating not less than the maximum asymmetrical fault current that can occur at the cutout location, unless suitable interlocks or operating procedures preclude the possibility of closing into a fault.

**(5) Identification.** Oil-filled cutouts shall have a permanent and legible nameplate showing the rated continuous current, rated maximum voltage, and rated interrupting current.

**(6) Fuse Links.** Fuse links shall have a permanent and legible identification showing the rated continuous current.

**(7) Location.** Cutouts shall be located so that they are readily and safely accessible for re-fusing, with the top of the cutout not over 1.5 m (5 ft) above the floor or platform.

**(8) Enclosure.** Suitable barriers or enclosures shall be provided to prevent contact with nonshielded cables or energized parts of oil-filled cutouts.

**(E) Load Interrupters.** Load-interrupter switches shall be permitted if suitable fuses or circuit breakers are used in conjunction with these devices to interrupt fault currents. Where these devices are used in combination, they shall be coordinated electrically so that they will safely withstand the effects of closing, carrying, or interrupting all possible currents up to the assigned maximum short-circuit rating.

Where more than one switch is installed with interconnected load terminals to provide for alternate connection to different supply conductors, each switch shall be provided with a conspicuous sign identifying this hazard.

**(1) Continuous Current Rating.** The continuous current rating of interrupter switches shall equal or exceed the maximum continuous current at the point of installation.

**(2) Voltage Rating.** The maximum voltage rating of interrupter switches shall equal or exceed the maximum circuit voltage.

**(3) Identification.** Interrupter switches shall have a permanent and legible nameplate including the following information: manufacturer's type or designation, continuous current rating, interrupting current rating, fault closing rating, maximum voltage rating.

**(4) Switching of Conductors.** The switching mechanism shall be arranged to be operated from a location where the operator is not exposed to energized parts and shall be arranged to open all ungrounded conductors of the circuit simultaneously with one operation. Switches shall be arranged to be locked in the open position. Metal-enclosed switches shall be operable from outside the enclosure.

*NOTICE: For full text, see the 2008 edition of the NECH*

**(5) Stored Energy for Opening.** The stored-energy operator shall be permitted to be left in the uncharged position after the switch has been closed if a single movement of the operating handle charges the operator and opens the switch.

**(6) Supply Terminals.** The supply terminals of fused interrupter switches shall be installed at the top of the switch enclosure, or, if the terminals are located elsewhere, the equipment shall have barriers installed so as to prevent persons from accidentally contacting energized parts or dropping tools or fuses into energized parts.

See Exhibits 490.1 and 490.2 for an example of a fused interrupter switch and the fuseholder components.

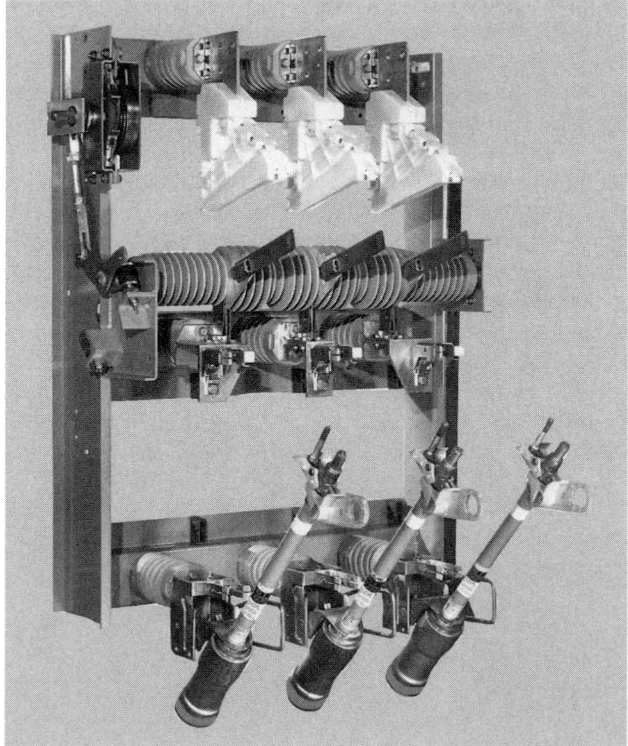

*EXHIBIT 490.1 Group-operated interrupter-switch and powerfuse combination rated at 13.8 kV, 600 amperes continuous and interrupting, 40,000 amperes momentary, 40,000 amperes fault closing. (Courtesy of Schweitzer and Conrad Electric Co.)*

**490.22 Isolating Means.** Means shall be provided to completely isolate an item of equipment. The use of isolating switches shall not be required where there are other ways of de-energizing the equipment for inspection and repairs, such as draw-out-type metal-enclosed switchgear units and removable truck panels.

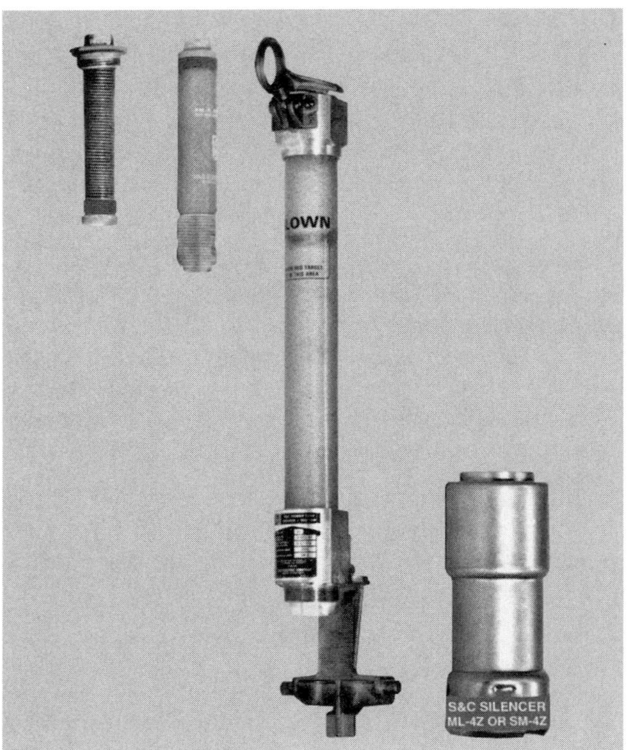


*EXHIBIT 490.2 Components of the indoor solid-material (SM) power fuseholder (boric-acid arc-extinguishing type) with a 14.4 kV, 400E-ampere maximum, 40,000-ampere rms asymmetrical interrupting rating. Shown here are the spring and cable assembly, refill unit, holder, and snuffler. (Courtesy of Schweitzer and Conrad Electric Co.)*

Isolating switches not interlocked with an approved circuit-interrupting device shall be provided with a sign warning against opening them under load.

A fuseholder and fuse, designed for the purpose, shall be permitted as an isolating switch.

**490.24 Minimum Space Separation.** In field-fabricated installations, the minimum air separation between bare live conductors and between such conductors and adjacent grounded surfaces shall not be less than the values given in Table 490.24. These values shall not apply to interior portions or exterior terminals of equipment designed, manufactured, and tested in accordance with accepted national standards.

**490.51 General.**

**(A) Covered.** The provisions of this part shall apply to installations and use of high-voltage power distribution and utilization equipment that is portable, mobile, or both, such as substations and switch houses mounted on skids, trailers, or cars; mobile shovels; draglines; cranes; hoists; drills; dredges; compressors; pumps; conveyors; underground excavators; and the like.

**(B) Other Requirements.** The requirements of this part shall be additional to, or amendatory of, those prescribed in Articles 100 through 725 of this *Code*. Special attention shall be paid to Article 250.

**(C) Protection.** Adequate enclosures, guarding, or both, shall be provided to protect portable and mobile equipment from physical damage.

**(D) Disconnecting Means.** Disconnecting means shall be installed for mobile and portable high-voltage equipment according to the requirements of Part VIII of Article 230 and shall disconnect all ungrounded conductors.

## Article 500 Hazardous (Classified) Locations, Classes I, II, and III, Divisions 1 and 2

### 500.4 General.

**(A) Documentation.** All areas designated as hazardous (classified) locations shall be properly documented. This documentation shall be available to those authorized to design, install, inspect, maintain, or operate electrical equipment at the location.

One type of documentation consists of area classification drawings. This type of documentation provides the necessary information for installers, service personnel, and authorities having jurisdiction to ensure that electrical equipment installed or maintained in classified areas is of the proper type. See the fine print note to 505.4(A). Also see the fourth paragraph of the commentary following 500.1 for a suggested team to develop the hazardous area diagram.

**(B) Reference Standards.** Important information relating to topics covered in Chapter 5 may be found in other publications.

The NFPA and ANSI standards referenced in Articles 500 through 517 are essential for proper application of those articles. The following NFPA codes, standards, and recommended practices not listed in 500.4(B) FPN No. 2 include information on hazardous (classified) locations and the extent of hazardous (classified) locations in specific occupancies or industries:

NFPA 30A, *Code for Motor Fuel Dispensing Facilities and Repair Garages*

NFPA 51, *Standard for the Design and Installation of Oxygen–Fuel Gas Systems for Welding, Cutting, and Allied Processes*

NFPA 51A, *Standard for Acetylene Cylinder Charging Plants*

NFPA 52, *Vehicular Fuel Systems Code*

NFPA 54, *Natural Fuel Gas Code*

NFPA 59A, *Standard for the Production, Storage, and Handling of Liquefied Natural Gas (LNG)*

NFPA 61, *Standard for the Prevention of Fires and Dust Explosions in Agricultural and Food Products Facilities*

NFPA 85, *Boiler and Combustion Systems Hazards Code*

NFPA 88A, *Standard for Parking Structures*

NFPA 99, *Standard for Health Care Facilities*

NFPA 407, *Standard for Aircraft Fuel Servicing*

NFPA 409, *Standard on Aircraft Hangars*

NFPA 495, *Explosive Materials Code*

NFPA 496, *Standard for Purged and Pressurized Enclosures for Electrical Equipment*

NFPA 654, *Standard for the Prevention of Fire and Dust Explosions from the Manufacturing, Processing, and Handling of Combustible Particulate Solids*

NFPA 655, *Standard for Prevention of Sulfur Fires and Explosions*

FPN No. 1: It is important that the authority having jurisdiction be familiar with recorded industrial experience as well as with the standards of the National Fire Protection Association (NFPA), the American Petroleum Institute (API), and the Instrumentation, Systems, and Automation Society (ISA) that may be of use in the classification of various locations, the determination of adequate ventilation, and the protection against static electricity and lightning hazards.

FPN No. 2: For further information on the classification of locations, see NFPA 30-2008, *Flammable and Combustible Liquids Code*; NFPA 32-2007, *Standard for Drycleaning Plants*; NFPA 33-2007, *Standard for Spray Application Using Flammable or Combustible Materials*; NFPA 34-2007, *Standard for Dipping and Coating Processes Using Flammable or Combustible Liquids*; NFPA 35-2005, *Standard for the Manufacture of Organic Coatings*; NFPA 36-2004, *Standard for Solvent Extraction Plants*; NFPA 45-2004, *Standard on Fire Protection for Laboratories Using Chemicals*; NFPA 55-2005, *Standard for the Storage, Use, and Handling of Compressed Gases and Cryogenic Fluids in Portable and Stationary Containers, Cylinders, and Tanks*; NFPA 58-2008, *Liquefied Petroleum Gas Code*; NFPA 59-2004, *Utility LP-Gas Plant Code*; NFPA 497-2004, *Recommended Practice for the Classification of Flammable Liquids, Gases, or Vapors and of Hazardous (Classified) Locations for Electrical Installations in Chemical Process Areas*; NFPA 499-2004, *Recom-*

*mended Practice for the Classification of Combustible Dusts and of Hazardous (Classified) Locations for Electrical Installations in Chemical Process Areas*; NFPA 820-2008, *Standard for Fire Protection in Wastewater Treatment and Collection Facilities*; ANSI/API RP500-1997, *Recommended Practice for Classification of Locations of Electrical Installations at Petroleum Facilities Classified as Class I, Division 1 and Division 2*; ISA-12.10-1988, *Area Classification in Hazardous (Classified) Dust Locations*.

FPN No. 3: For further information on protection against static electricity and lightning hazards in hazardous (classified) locations, see NFPA 77-2007, *Recommended Practice on Static Electricity*; NFPA 780-2008, *Standard for the Installation of Lightning Protection Systems*; and API RP 2003-1998, *Protection Against Ignitions Arising Out of Static Lightning and Stray Currents*.

FPN No. 4: For further information on ventilation, see NFPA 30-2008, *Flammable and Combustible Liquids Code*; and API RP 500-1997, *Recommended Practice for Classification of Locations for Electrical Installations at Petroleum Facilities Classified as Class I, Division 1 and Division 2*.

FPN No. 5: For further information on electrical systems for hazardous (classified) locations on offshore oil- and gas-producing platforms, see ANSI/API RP 14F-1999, *Recommended Practice for Design and Installation of Electrical Systems for Fixed and Floating Offshore Petroleum Facilities for Unclassified and Class I, Division 1 and Division 2 Locations*.

## Article 522 Control Systems for Permanent Amusement Attractions

**522.24 Conductors of Different Circuits in the Same Cable, Cable Tray, Enclosure, or Raceway.** Control circuits shall be permitted to be installed with other circuits as specified in 522.24(A) and (B).

**(A) Two or More Control Circuits.** Control circuits shall be permitted to occupy the same cable, cable tray, enclosure, or raceway without regard to whether the individual circuits are alternating current or direct current, provided all conductors are insulated for the maximum voltage of any conductor in the cable, cable tray, enclosure, or raceway.

**(B) Control Circuits with Power Circuits.** Control circuits shall be permitted to be installed with power conductors as specified in 522.24(B)(1) through (B)(3).

**(1) In a Cable, Enclosure, or Raceway.** Control circuits and power circuits shall be permitted to occupy the same cable, enclosure, or raceway only where the equipment powered is functionally associated.

**(2) In Factory- or Field-Assembled Control Centers.** Control circuits and power circuits shall be permitted to be installed in factory- or field-assembled control centers.

**(3) In a Manhole.** Control circuits and power circuits shall be permitted to be installed as underground conductors in a manhole in accordance with one of the following:

(1) The power or control circuit conductors are in a metal-enclosed cable or Type UF cable.
(2) The conductors are permanently separated from the power conductors by a continuous firmly fixed nonconductor, such as flexible tubing, in addition to the insulation on the wire.
(3) The conductors are permanently and effectively separated from the power conductors and securely fastened to racks, insulators, or other approved supports.
(4) In cable trays, where the control circuit conductors and power conductors not functionally associated with them are separated by a solid fixed barrier of a material compatible with the cable tray, or where the power or control circuit conductors are in a metal-enclosed cable.

**522.25 Ungrounded Control Circuits.** Separately derived ac and 2-wire dc circuits and systems 50 volts or greater shall be permitted to be ungrounded, provided that all the following conditions are met:

(1) Continuity of control power is required for orderly shutdown.
(2) Ground detectors are installed on the control system.

**522.28 Control Circuits in Wet Locations.** Where wet contact is likely to occur, ungrounded 2-wire direct-current control circuits shall be limited to 30 volts maximum for continuous dc or 12.4 volts peak for direct current that is interrupted at a rate of 10 to 200 Hz.

## Article 525 Carnivals, Circuses, Fairs, and Similar Events

**525.21 Rides, Tents, and Concessions.**

**(A) Disconnecting Means.** Each portable structure shall be provided with a disconnect switch located within sight of and within 1.8 m (6 ft) of the operator's station. The disconnecting means shall be readily accessible to the operator, including when the ride is in operation. Where accessible to unqualified persons, the enclosure for the switch or circuit breaker shall be of the lockable type. A shunt trip device that opens the fused disconnect or circuit breaker when a switch located in the ride operator's console is closed shall be a permissible method of opening the circuit.

**(B) Portable Wiring Inside Tents and Concessions.** Electrical wiring for lighting, where installed inside of tents and concessions, shall be securely installed and, where subject to physical damage, shall be provided with mechanical protection. All lamps for general illumination shall be protected from accidental breakage by a suitable luminaire or lampholder with a guard.

**525.22 Portable Distribution or Termination Boxes.** Portable distribution or termination boxes shall comply with 525.22(A) through (D).

Portable distribution or termination equipment must be mounted so that the bottom of the enclosure is at least 6 in. above the ground. This requirement prevents excessive moisture from entering the equipment and allows for proper radius of bend on conductors entering and exiting the equipment from below.

**(A) Construction.** Boxes shall be designed so that no live parts are exposed to accidental contact. Where installed outdoors, the box shall be of weatherproof construction and mounted so that the bottom of the enclosure is not less than 150 mm (6 in.) above the ground.

**(B) Busbars and Terminals.** Busbars shall have an ampere rating not less than the overcurrent device supplying the feeder supplying the box. Where conductors terminate directly on busbars, busbar connectors shall be provided.

**(C) Receptacles and Overcurrent Protection.** Receptacles shall have overcurrent protection installed within the box. The overcurrent protection shall not exceed the ampere rating of the receptacle, except as permitted in Article 430 for motor loads.

**(D) Single-Pole Connectors.** Where single-pole connectors are used, they shall comply with 530.22.

**525.23 Ground-Fault Circuit-Interrupter (GFCI) Protection.**

**(A) Where GFCI Protection Is Required.** The ground-fault circuit interrupter shall be permitted to be an integral part of the attachment plug or located in the power-supply cord, within 300 mm (12 in.) of the attachment plug. Listed cord sets incorporating ground-fault circuit interrupter for personnel shall be permitted.

(1) 125-volt, single-phase, 15- and 20-ampere non-locking-type receptacles used for disassembly and reassembly or readily accessible to the general public
(2) Equipment that is readily accessible to the general public and supplied from a 125-volt, single-phase, 15- or 20-ampere branch circuit

**(B) Where GFCI Protection Is Not Required.** Receptacles that only facilitate quick disconnecting and reconnecting of electrical equipment shall not be required to be provided with GFCI protection.

These receptacles shall be of the locking type.

**(C) Where GFCI Protection Is Not Permitted.** Egress lighting shall not be protected by a GFCI.

Section 525.23 provides three categories: where GFCIs are required, where GFCIs are not required, and where GFCIs are not permitted to be installed. GFCI protection is not allowed on circuits that supply means-of-egress illumination.

## Article 590 Temporary Installations

**590.2 All Wiring Installations.**

**(A) Other Articles.** Except as specifically modified in this article, all other requirements of this *Code* for permanent wiring shall apply to temporary wiring installations.

Temporary installations of electrical equipment must be installed in accordance with all applicable permanent installation requirements except as modified by the rules in this article. For example, the requirements of 300.15 specify that a box or other enclosure must be used where splices are made. This rule is amended by 590.4(G, which, for construction sites, permits splices to be made in multiconductor cords and cables without the use of a box.

**(B) Approval.** Temporary wiring methods shall be acceptable only if approved based on the conditions of use and any special requirements of the temporary installation.

The provisions of 590.2(B) require that all temporary wiring methods be approved based on criteria such as length of time in service, severity of physical abuse, exposure to weather, and other special requirements. Special requirements may range from tunnel construction projects and tent cities constructed after a natural disaster to flammable hazardous material reclamation projects.

**590.3 Time Constraints.**

**(A) During the Period of Construction.** Temporary electric power and lighting installations shall be permitted during the period of construction, remodeling, maintenance, repair, or demolition of buildings, structures, equipment, or similar activities.

**(B) 90 Days.** Temporary electric power and lighting installations shall be permitted for a period not to exceed 90 days for holiday decorative lighting and similar purposes.

Note that the 90-day time limit in 590.3(B) applies only to temporary electrical installations associated with holiday displays. Construction and emergency and test temporary wiring installations are not bound by this time limit.

**(C) Emergencies and Tests.** Temporary electric power and lighting installations shall be permitted during emergencies and for tests, experiments, and developmental work.

**(D) Removal.** Temporary wiring shall be removed immediately upon completion of construction or purpose for which the wiring was installed.

Due to the modifications permitted by Article 590, temporary wiring installations may not meet all of the requirements for a permanent installation. Therefore, all temporary wiring not only must be disconnected but also must be removed from the building, structure, or other location of installation.

**590.4 General.**

**(A) Services.** Services shall be installed in conformance with Parts I through VIII of Article 230, as applicable.

**(B) Feeders.** Overcurrent protection shall be provided in accordance with 240.4, 240.5, 240.100, and 240.101. Feeders shall originate in an approved distribution center. Conductors shall be permitted within cable assemblies or within multiconductor cords or cables of a type identified in Table 400.4 for hard usage or extra-hard usage. For the purpose of this section, Type NM and Type NMC cables shall be permitted to be used in any dwelling, building, or structure without any height limitation or limitation by building construction type and without concealment within walls, floors, or ceilings.

Section 590.4(B) allows Type NM and Type NMC cable to be used in any building or structure regardless of building height and construction type in which the cable is used.

Temporary feeders are permitted to be cable assemblies, multiconductor cords, or single-conductor cords. Cords used as feeders must be identified for hard or extra-hard usage according to Table 400.4. Individual conductors, as described in Table 310.13, are not permitted as open conductors but, rather, must be part of a cable assembly or used in a raceway system. Open or individual conductor feeders are permitted only during emergencies or tests.

All temporary wiring methods must be approved by the authority having jurisdiction. [See 590.2(B).]

*Exception: Single insulated conductors shall be permitted where installed for the purpose(s) specified in 590.3(C), where accessible only to qualified persons.*

**(C) Branch Circuits.** All branch circuits shall originate in an approved power outlet or panelboard. Conductors shall be permitted within cable assemblies or within multiconductor cord or cable of a type identified in Table 400.4 for hard usage or extra-hard usage. Conductors shall be protected from overcurrent as provided in 240.4, 240.5, and 240.100. For the purposes of this section, Type NM and Type NMC cables shall be permitted to be used in any dwelling, building, or structure without any height limitation or limitation by building construction type and without concealment within walls, floors, or ceilings.

Type NM and Type NMC cable are permitted as temporary wiring in any building or structure regardless of the height or construction type of the building in which the cable is used.

The basic requirement for safety in 590.4(C) is that temporary wiring be located and installed so that it will not be physically damaged. In accordance with 590.2(A), temporary wiring must be installed in accordance with the appropriate Chapter 3 article for the wiring method employed (unless modified in Article 590).

Note that hard-usage or extra-hard-usage extension cords are permitted to be laid on the floor.

*Exception: Branch circuits installed for the purposes specified in 590.3(B) or 590.3(C) shall be permitted to be run as single insulated conductors. Where the wiring is installed in accordance with 590.3(B), the voltage to ground shall not exceed 150 volts, the wiring shall not be subject to physical damage, and the conductors shall be supported on insulators at intervals of not more than 3.0 m (10 ft); or, for festoon lighting, the conductors shall be so arranged that excessive strain is not transmitted to the lampholders.*

**(D) Receptacles.** All receptacles shall be of the grounding type. Unless installed in a continuous metal raceway that qualifies as an equipment grounding conductor in accordance with 250.118 or a continuous metal-covered cable that qualifies as an equipment grounding conductor in accordance with 250.118, all branch circuits shall include a separate equipment grounding conductor, and all receptacles shall be electrically connected to the equipment grounding conductor(s). Receptacles on construction sites shall not be installed on branch circuits that supply temporary lighting. Receptacles shall not be connected to the same ungrounded conductor of multiwire circuits that supply temporary lighting.

The intent of the branch-circuit provisions in 590.4(D) is to require separate ungrounded conductors for lighting and receptacle loads so that the activation of a fuse, circuit breaker, or GFCI, due to a fault or equipment overload, does not de-energize the lighting circuit. This section was revised for the 2008 *Code*. Metal cables or raceways must be continuous and qualify as an equipment grounding conductor. If the metal raceway or metal cable is not continuous or does not qualify as an equipment grounding conductor, a separate equipment grounding conductor must be installed.

**(E) Disconnecting Means.** Suitable disconnecting switches or plug connectors shall be installed to permit the disconnection of all ungrounded conductors of each temporary circuit. Multiwire branch circuits shall be provided with a means to disconnect simultaneously all ungrounded conductors at the power outlet or panelboard where the branch circuit originated. Identified handle ties shall be permitted.

**(F) Lamp Protection.** All lamps for general illumination shall be protected from accidental contact or breakage by a suitable luminaire or lampholder with a guard.

Brass shell, paper-lined sockets, or other metal-cased sockets shall not be used unless the shell is grounded.

**(G) Splices.** On construction sites, a box shall not be required for splices or junction connections where the circuit conductors are multiconductor cord or cable assemblies, provided that the equipment grounding continuity is maintained with or without the box. See 110.14(B) and 400.9. A box, conduit body, or terminal fitting having a separately bushed hole for each conductor shall be used wherever a change is made to a conduit or tubing system or a metal-sheathed cable system.

**(H) Protection from Accidental Damage.** Flexible cords and cables shall be protected from accidental damage. Sharp corners and projections shall be avoided. Where passing through doorways or other pinch points, protection shall be provided to avoid damage.

Unlike the requirement in 400.8, flexible cords and cables, because of the nature of their use, are permitted to pass through doorways, in accordance with 590.4(H).

**(I) Termination(s) at Devices.** Flexible cords and cables entering enclosures containing devices requiring termination shall be secured to the box with fittings designed for the purpose.

**(J) Support.** Cable assemblies and flexible cords and cables shall be supported in place at intervals that ensure that they will be protected from physical damage. Support shall be in the form of staples, cable ties, straps, or similar type fittings installed so as not to cause damage. Vegetation shall not be used for support of overhead spans of branch circuits or feeders.

Section 590.4(J), Exception allows holiday lighting to be installed and supported by trees for a period of not more than 90 days, provided the wiring is arranged with proper strain relief devices, tension take-up devices, or other means to prevent damage to the conductors from the tree swaying.

According to 590.4(J), temporary wiring methods do not have to be supported in accordance with the permanent installation requirements (from Chapter 3) for the particular wiring method. It should be noted that the temporary wiring must be removed upon completion of construction and adequate support is needed only to minimize the possibility of damage to the wiring method during its temporary period of use. It is not permitted to use vegetation as a support structure for overhead spans of branch-circuit and feeder conductors.

*Exception: For holiday lighting in accordance with 590.3(B), where the conductors or cables are arranged with proper strain relief devices, tension take-up devices, or other approved means to avoid damage from the movement of the live vegetation, trees shall be permitted to be used for support of overhead spans of branch-circuit conductors or cables.*

**590.5 Listing of Decorative Lighting.** Decorative lighting used for holiday lighting and similar purposes, in accordance with 590.3(B), shall be listed.

**590.6 Ground-Fault Protection for Personnel.** Ground-fault protection for personnel for all temporary wiring installations shall be provided to comply with 590.6(A) and (B). This section shall apply only to temporary wiring installations used to supply temporary power to equipment used by personnel during construction, remodeling, maintenance, repair, or demolition of buildings, structures, equipment, or similar activities. This section shall apply to power derived from an electric utility company or from an on-site-generated power source.

**(A) Receptacle Outlets.** All 125-volt, single-phase, 15-, 20-, and 30-ampere receptacle outlets that are not a part of the permanent wiring of the building or structure and that are in use by personnel shall have ground-fault circuit-interrupter protection for personnel. If a receptacle(s) is installed or exists as part of the permanent wiring of the building or structure and is used for temporary electric power, ground-fault circuit-interrupter protection for personnel shall be provided. For the purposes of this section, cord sets or devices incorporating listed ground-fault circuit-interrupter protection for personnel identified for portable use shall be permitted.

*Exception: In industrial establishments only, where conditions of maintenance and supervision ensure that only qualified personnel are involved, an assured equipment grounding conductor program as specified in 590.6(B)(2) shall be permitted for only those receptacle outlets used to supply equipment that would create a greater hazard if power were interrupted or having a design that is not compatible with GFCI protection.*

**(B) Use of Other Outlets.** Receptacles other than 125-volt, single-phase, 15-, 20-, and 30-ampere receptacles shall have protection in accordance with (B)(1) or the assured equipment grounding conductor program in accordance with (B)(2).

**(1) GFCI Protection.** Ground-fault circuit-interrupter protection for personnel.

**(2) Assured Equipment Grounding Conductor Program.** A written assured equipment grounding conductor program continuously enforced at the site by one or more designated persons to ensure that equipment grounding conductors for all cord sets, receptacles that are not a part of the permanent wiring of the building or structure, and equipment connected by cord and plug are installed and maintained in accordance with the applicable requirements of 250.114, 250.138, 406.3(C), and 590.4(D).

(a) The following tests shall be performed on all cord sets, receptacles that are not part of the permanent wiring of the building or structure, and cord-and-plug-connected equipment required to be connected to an equipment grounding conductor:

  (1) All equipment grounding conductors shall be tested for continuity and shall be electrically continuous.

  (2) Each receptacle and attachment plug shall be tested for correct attachment of the equipment grounding conductor. The equipment grounding conductor shall be connected to its proper terminal.

  (3) All required tests shall be performed as follows:

    a. Before first use on site
    b. When there is evidence of damage
    c. Before equipment is returned to service following any repairs
    d. At intervals not exceeding 3 months

(b) The tests required in item (2)(a) shall be recorded and made available to the authority having jurisdiction.

Due to the more severe environmental conditions often encountered by personnel using temporary wiring while performing activities such as construction, remodeling, maintenance, repair, and demolition, there is generally an elevated exposure to electric shock or electrocution hazards. The requirement of 590.6(A) for GFCI protection of all temporarily installed, 125-volt, single-phase, 15-, 20-, and 30-ampere receptacles is intended to protect personnel using these receptacles from shock hazards that may be encountered during construction and maintenance activities.

The exception to 590.6(A) is limited in scope and application. The exception applies only to those industrial occupancies in which qualified persons will be using 125-volt, single-phase, 15-, 20-, and 30-ampere receptacles. Additionally, either the nature of the equipment being supplied by these receptacles has to be of such importance that the hazard of power interruption outweighs the benefits of GFCI protection or the equipment has been demonstrated to be incompatible with the proper operation of GFCI protective devices. In those instances where the conditions specified by the exception are present, the use of the assured equipment grounding conductor program specified in 590.6(B)(2) is permitted. An electrically operated air supply for personnel working in toxic environments is an example of where the loss of power is the greater hazard. Some electrically operated testing equipment has proved to be incompatible with GFCI protection.

Receptacle configurations, other than the 125-volt, single-phase, 15-, 20-, and 30-ampere types, must be GFCI protected or installed and maintained in accordance with the assured equipment grounding conductor program of 590.6(B)(2).

According to OSHA 29 CFR 1926.404(b)(1)(iii):

> The employer shall establish and implement an assured equipment grounding conductor program on construction sites covering all cord sets, receptacles which are not a part of the building or structure, and equipment connected by cord and plug which are available for use or used by employees. This program shall comply with the following minimum requirements:
>
> (A) A written description of the program, including the specific procedures adopted by the employer, shall be available at the jobsite for inspection and copying by the Assistant Secretary and any affected employee.
>
> (B) The employer shall designate one or more competent persons.

These OSHA requirements are very similar to the present *NEC* requirements for an assured grounding program.

GFCI protection for construction or maintenance personnel using receptacles that are part of the permanent wiring and that are not GFCI protected may be provided by using cord sets or listed portable GFCIs identified for portable use. An example of a GFCI cord set that is identified for portable use is shown in Exhibit 590.1.

Exhibits 590.1 through 590.4 show some examples of ways to implement the temporary wiring requirements of 590.6.

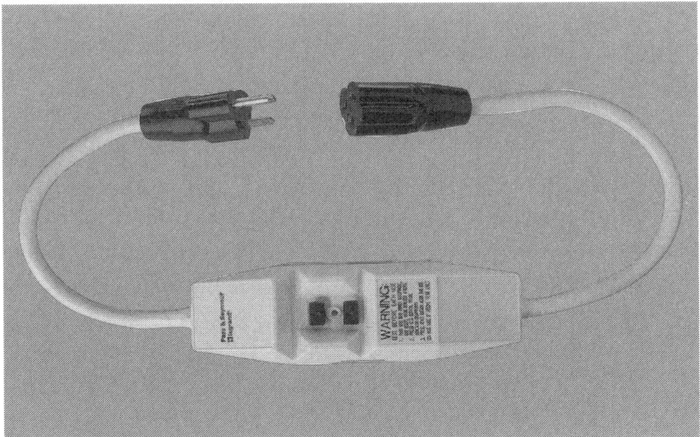

***EXHIBIT 590.1*** *A raintight GFCI with open neutral protection that is designed for use on the line end of a flexible cord. (Courtesy of Pass & Seymour/Legrand®)*

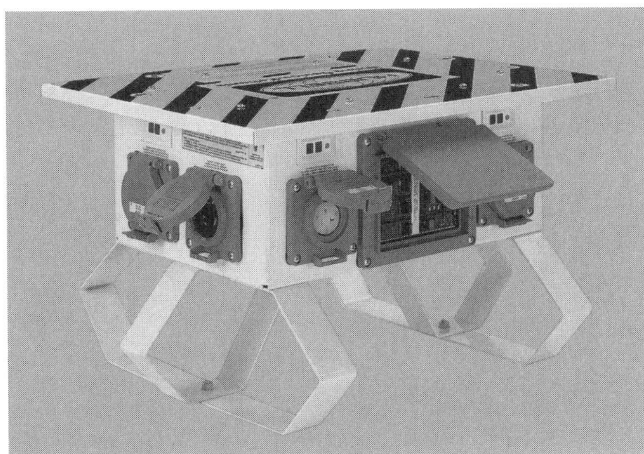

***EXHIBIT 590.2*** *A temporary power outlet unit commonly used on construction sites with a variety of configurations, including GFCI protection. (Courtesy of Hubbell Wiring Device–Kellems)*

**590.7 Guarding.** For wiring over 600 volts, nominal, suitable fencing, barriers, or other effective means shall be provided to limit access only to authorized and qualified personnel.

***EXHIBIT 590.3*** *A watertight plug and connector used to prevent tripping of GFCI protective devices in wet or damp weather. (Courtesy of Hubbell Wiring Device–Kellems)*

***EXHIBIT 590.4*** *A 15-ampere duplex receptacle with integral GFCI that also protects downstream loads. (Courtesy of Pass & Seymour/Legrand®)*

## Article 600 Electric Signs and Outline Lighting

**600.6 Disconnects.** Each sign and outline lighting system, or feeder circuit or branch circuit supplying a sign or outline lighting system, shall be controlled by an externally operable switch or circuit breaker that will open all ungrounded conductors. Signs and outline lighting systems located within fountains shall have the disconnect located in accordance with 680.12.

*Exception No. 1: A disconnecting means shall not be required for an exit directional sign located within a building.*

*Exception No. 2: A disconnecting means shall not be required for cord-connected signs with an attachment plug.*

**(A) Location.**

**(1) Within Sight of the Sign.** The disconnecting means shall be within sight of the sign or outline lighting system that it controls. Where the disconnecting means is out of the line of sight from any section that is able to be energized, the disconnecting means shall be capable of being locked in the open position. The provision for locking or adding a lock to the disconnecting means must remain in place at the switch or circuit breaker whether the lock is installed or not. Portable means for adding a lock to the switch or circuit breaker shall not be permitted.

Section 600.6(A)(1) covers sign installations where the branch circuit or feeder is run directly to the sign. Each branch circuit or feeder supplying a sign must have an externally operable switch or circuit breaker to open the ungrounded conductors. Two options are permitted for locating the sign disconnecting means. The disconnecting means is required either to be located within sight of the sign or to be equipped with the provision to lock it in the open position. Exhibit 600.2 depicts a sign with two supply circuits, which could be feeders or branch circuits. Each circuit is provided with an externally operable switch located within sight of the sign.

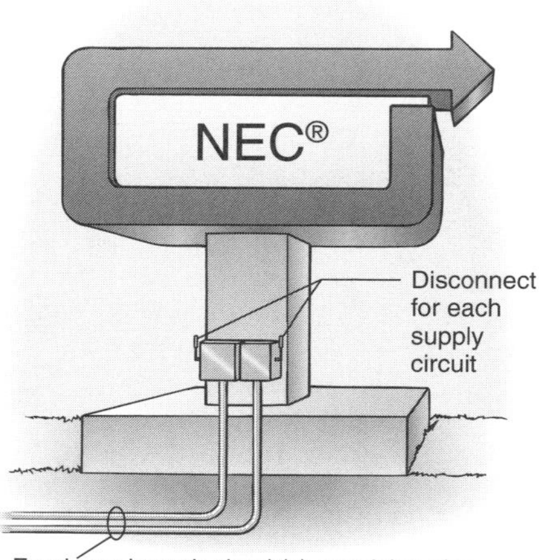


***EXHIBIT 600.2*** *Supply circuit disconnecting means located at or on an electric sign.*

Exhibit 600.3 illustrates three compliant alternatives. The supply circuit disconnecting means shown in Example 1 is externally operable and located at and within sight of the sign. The disconnecting means in Example 2 is externally operable, and its location, though not at or on the sign, is acceptable because it meets the Article 100 definition of *within sight*. Where the disconnecting means is not located within sight of the sign, as shown in Example 3, it is required to be located within sight of the controller and must be capable of being locked in the open position.

**(2) Within Sight of the Controller.** The following shall apply for signs or outline lighting systems operated by electronic or electromechanical controllers located external to the sign or outline lighting system:

(1) The disconnecting means shall be permitted to be located within sight of the controller or in the same enclosure with the controller.
(2) The disconnecting means shall disconnect the sign or outline lighting system and the controller from all ungrounded supply conductors.
(3) The disconnecting means shall be designed such that no pole can be operated independently and shall be capable of being locked in the open position. The provisions for locking or adding a lock to the disconnecting means must remain in place at the switch or circuit breaker whether the lock is installed or not. Portable means for adding a lock to the switch or circuit breaker shall not be permitted.

**(B) Control Switch Rating.** Switches, flashers, and similar devices controlling transformers and electronic power supplies shall be rated for controlling inductive loads or have a current rating not less than twice the current rating of the transformer.

FPN: See 404.14 for rating of snap switches.

## Article 610 Cranes and Hoists

### 610.3 Special Requirements for Particular Locations.

**(A) Hazardous (Classified) Locations.** All equipment that operates in a hazardous (classified) location shall conform to Article 500.

**(1) Class I Locations.** Equipment used in locations that are hazardous because of the presence of flammable gases or vapors shall conform to Article 501.

**(2) Class II Locations.** Equipment used in locations that are hazardous because of combustible dust shall conform to Article 502.

**(3) Class III Locations.** Equipment used in locations that are hazardous because of the presence of easily ignitible fibers or flyings shall conform to Article 503.

See the commentary following 503.155(D).

**(B) Combustible Materials.** Where a crane, hoist, or monorail hoist operates over readily combustible material, the resistors shall be located as permitted in the following:

(1) A well ventilated cabinet composed of noncombustible material constructed so that it does not emit flames or molten metal

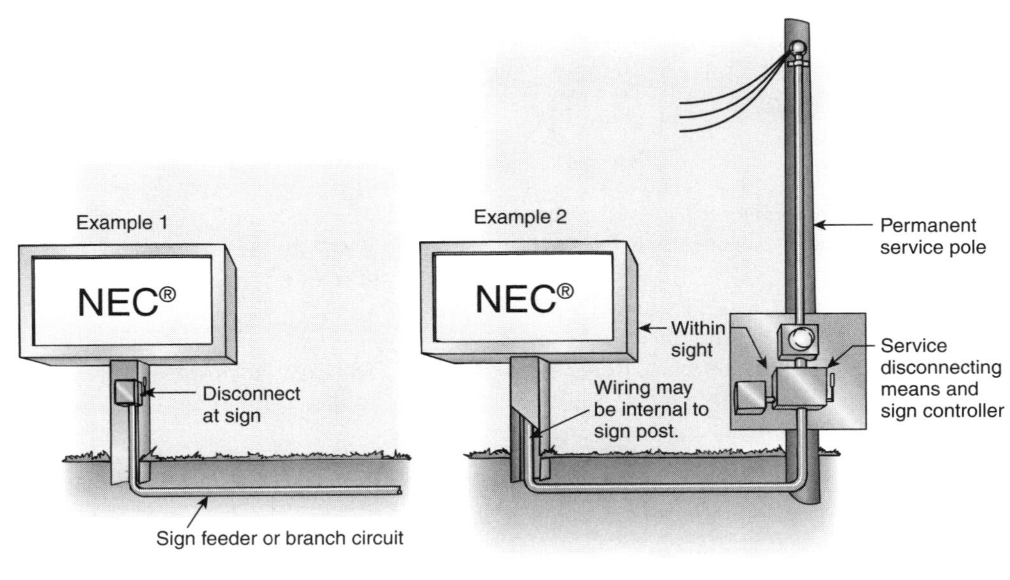


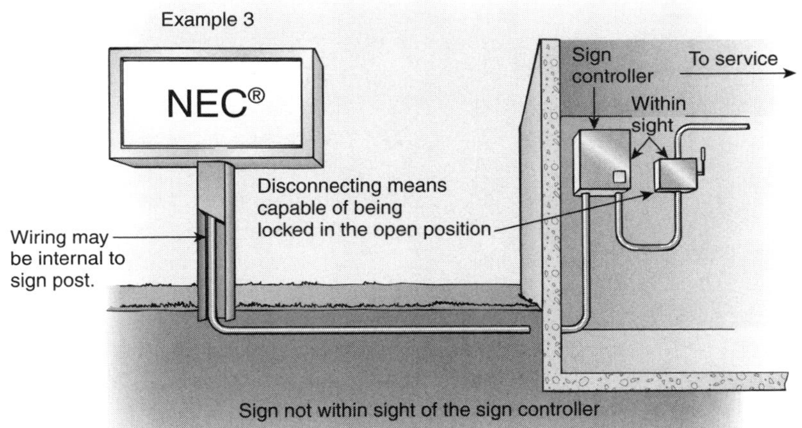


***EXHIBIT 600.3** Three acceptable methods of providing the disconnecting means for an electric sign.*

(2) A cage or cab constructed of noncombustible material that encloses the sides of the cage or cab from the floor to a point at least 150 mm (6 in.) above the top of the resistors

**(C) Electrolytic Cell Lines.** See 668.32.

Special precautions are necessary on electrolytic cell lines to prevent the introduction of exposed grounded parts, as described in 668.32.

**610.31 Runway Conductor Disconnecting Means.** A disconnecting means that has a continuous ampere rating not less than that calculated in 610.14(E) and (F) shall be provided between the runway contact conductors and the power supply. Such disconnecting means shall consist of a motor-circuit switch, circuit breaker, or molded-case switch. This disconnecting means shall be as follows:

(1) Readily accessible and operable from the ground or floor level.
(2) Capable of being locked in the open position. The provision for locking or adding a lock to the disconnecting means shall be installed on or at the switch or circuit breaker used as the disconnecting means and

shall remain in place with or without the lock installed. Portable means for adding a lock to the switch or circuit breaker shall not be permitted as the means required to be installed at and remain with the equipment.

(3) Open all ungrounded conductors simultaneously.
(4) Placed within view of the runway contact conductors.

**610.32 Disconnecting Means for Cranes and Monorail Hoists.** A motor-circuit switch, molded-case switch, or circuit breaker shall be provided in the leads from the runway contact conductors or other power supply on all cranes and monorail hoists. The disconnecting means shall be capable of being locked in the open position. The provision for locking or adding a lock to the disconnecting means shall be installed on or at the switch or circuit breaker used as the disconnecting means and shall remain in place with or without the lock installed. Portable means for adding a lock to the switch or circuit breaker shall not be permitted.

Where a monorail hoist or hand-propelled crane bridge installation meets all of the following, the disconnecting means shall be permitted to be omitted:

(1) The unit is controlled from the ground or floor level.
(2) The unit is within view of the power supply disconnecting means.
(3) No fixed work platform has been provided for servicing the unit.

Where the disconnecting means is not readily accessible from the crane or monorail hoist operating station, means shall be provided at the operating station to open the power circuit to all motors of the crane or monorail hoist.

Many crane installations are not arranged so that the unit is within view of the power supply disconnecting means. Therefore, a disconnecting means (lock-open type) must be provided in the contact conductors, and the means to lock the disconnecting means in the open position must be an accessory that is identified for this use and has to be in place on the disconnecting means at all times. The portable type lock-out mechanisms cannot be used to comply with this requirement. The chance that the portable-type device could easily be removed from the disconnecting means location compromises the safety benefit of having a disconnecting means that is at all times equipped to be locked off or open. However, personnel should be aware that when one crane is being serviced, another unit on the same system could remain energized and be run into the person performing maintenance on the locked-out unit.

**610.33 Rating of Disconnecting Means.** The continuous ampere rating of the switch or circuit breaker required by 610.32 shall not be less than 50 percent of the combined short-time ampere rating of the motors or less than 75 percent of the sum of the short-time ampere rating of the motors required for any single motion.

## Article 620 Elevators, Dumbwaiters, Escalators, Moving Walks, Platform Lifts, and Stairway Chairlifts

**620.4 Live Parts Enclosed.** All live parts of electrical apparatus in the hoistways, at the landings, in or on the cars of elevators and dumbwaiters, in the wellways or the landings of escalators or moving walks, or in the runways and machinery spaces of platform lifts and stairway chairlifts shall be enclosed to protect against accidental contact.

FPN: See 110.27 for guarding of live parts (600 volts, nominal, or less).

**620.5 Working Clearances.** Working space shall be provided about controllers, disconnecting means, and other electrical equipment. The minimum working space shall be not less than that specified in 110.26(A).

Where conditions of maintenance and supervision ensure that only qualified persons examine, adjust, service, and maintain the equipment, the clearance requirements of 110.26(A) shall be waived as permitted in 620.5(A) through (D).

**(A) Flexible Connections to Equipment.** Electrical equipment in (A)(1) through (A)(4) shall be permitted to be provided with flexible leads to all external connections so that it can be repositioned to meet the clear working space requirements of 110.26(A):

(1) Controllers and disconnecting means for dumbwaiters, escalators, moving walks, platform lifts, and stairway chairlifts installed in the same space with the driving machine
(2) Controllers and disconnecting means for elevators installed in the hoistway or on the car
(3) Controllers for door operators
(4) Other electrical equipment installed in the hoistway or on the car

Due to the physical constraints of the locations where this equipment is typically installed and the necessity of performing diagnostic work on it while it is energized,

620.5(A) permits flexible leads on equipment so it can be moved to a location that meets the working clearance requirements of 110.26(A).

**(B) Guards.** Live parts of the electrical equipment are suitably guarded, isolated, or insulated, and the equipment can be examined, adjusted, serviced, or maintained while energized without removal of this protection.

FPN: See definition of *Exposed* in Article 100.

**(C) Examination, Adjusting, and Servicing.** Electrical equipment is not required to be examined, adjusted, serviced, or maintained while energized.

**(D) Low Voltage.** Uninsulated parts are at a voltage not greater than 30 volts rms, 42 volts peak, or 60 volts dc.

**620.51 Disconnecting Means.** A single means for disconnecting all ungrounded main power supply conductors for each unit shall be provided and be designed so that no pole can be operated independently. Where multiple driving machines are connected to a single elevator, escalator, moving walk, or pumping unit, there shall be one disconnecting means to disconnect the motor(s) and control valve operating magnets.

The disconnecting means for the main power supply conductors shall not disconnect the branch circuit required in 620.22, 620.23, and 620.24.

The branch circuits that supply elevator car lighting, receptacles, ventilation, air conditioning, and heating are required to be independent of the control portion of the elevator. In addition, the branch circuits supplying hoistway pit lighting and receptacles and machine room or control room lights and receptacles are not permitted to be disconnected by the main elevator power disconnect. This requirement provides for passenger safety and comfort and for the safety of elevator maintenance personnel during an inadvertent or emergency shutdown of the main power circuit to the elevator.

**(A) Type.** The disconnecting means shall be an enclosed externally operable fused motor circuit switch or circuit breaker capable of being locked in the open position. The provision for locking or adding a lock to the disconnecting means shall be installed on or at the switch or circuit breaker used as the disconnecting means and shall remain in place with or without the lock installed. Portable means for adding a lock to the switch or circuit breaker shall not be permitted as the means required to be installed at and remain with the equipment.

The disconnecting means shall be a listed device.

FPN: For additional information, see ASME A17.1-2004, *Safety Code for Elevators and Escalators.*

*Exception No. 1: Where an individual branch circuit supplies a platform lift, the disconnecting means required by 620.51(C)(4) shall be permitted to comply with 430.109(C). This disconnecting means shall be listed and shall be capable of being locked in the open position. The provision for locking or adding a lock to the disconnecting means shall be installed on or at the switch or circuit breaker used as the disconnecting means and shall remain in place with or without the lock installed. Portable means for adding a lock to the switch or circuit breaker shall not be permitted as the means required to be installed at and remain with the equipment.*

*Exception No. 2: Where an individual branch circuit supplies a stairway chairlift, the stairway chairlift shall be permitted to be cord-and-plug-connected, provided it complies with 422.16(A) and the cord does not exceed 1.8 m (6 ft) in length.*

**(B) Operation.** No provision shall be made to open or close this disconnecting means from any other part of the premises. If sprinklers are installed in hoistways, machine rooms, control rooms, machinery spaces, or control spaces, the disconnecting means shall be permitted to automatically open the power supply to the affected elevator(s) prior to the application of water. No provision shall be made to automatically close this disconnecting means. Power shall only be restored by manual means.

FPN: To reduce hazards associated with water on live elevator electrical equipment.

ASME A17.1-2004, *Safety Code for Elevators and Escalators*, Rule 2.8.2.3, requires that where sprinklers are installed in hoistways, machine rooms, or machinery spaces, a means must be provided to automatically disconnect the main line power supply to the affected elevator(s) upon or prior to the application of water. Water on elevator electrical equipment can result in hazards such as uncontrolled car movement (wet machine brakes), movement of elevator with open doors (water on safety circuits bypassing car and/or hoistway door interlocks), and shock hazards.

Automatic disconnection of the main line power supply is not required by ASME A17.1 where hoistways and machine rooms are not sprinklered. NFPA 13, *Standard for the Installation of Sprinkler Systems*, provides requirements for the installation of sprinklers in machine rooms, hoistways, and pits.

Elevator shutdown is generally accomplished through the use of heat detectors located near sprinkler heads. The heat detectors are designed to actuate and generate an alarm signal prior to water discharge from the sprinkler heads. An output control relay powered by the fire alarm system then provides a monitored output to the main line disconnecting means control circuit, which activates the shunt trip. This practice ensures that all components have secondary power and are monitored for integrity, as required by *NFPA 72®*, *National Fire Alarm Code®*. Stand-alone heat detectors connected directly to the elevator disconnecting means control circuit are not monitored for integrity, have no secondary power supply, and are not permitted by *NFPA 72*.

Elevator shutdown can occur even if the car is not at a landing. However, to avoid trapping occupants in the car(s), it is highly desirable to recall the car(s) to the designated landing prior to disconnecting the main line power. Most fires produce detectable quantities of smoke before there is sufficient heat to activate a sprinkler head. Therefore, ASME A17.1, Rule 102.2(c), requires smoke detectors to be installed in hoistways that are sprinklered for the purposes of recalling the elevator car(s) before the main line power is disconnected. See 6.16.4 of *NFPA 72-2007* for additional requirements relating to the fire alarm system and elevator shutdown.

Exhibit 620.3 illustrates a typical method of supervising control power using a fire alarm system. Loss of control power produces a supervisory signal at the fire alarm control unit that then would be investigated.

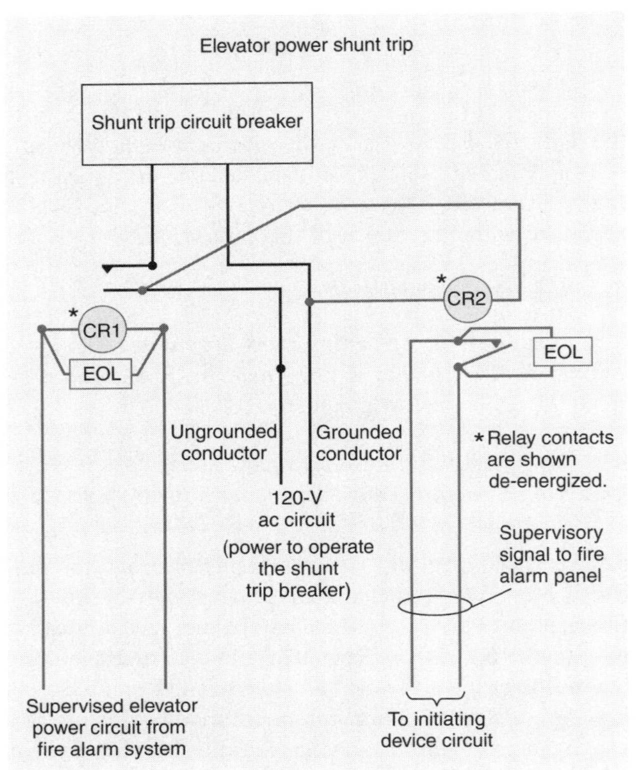


***EXHIBIT 620.3*** *Typical method of control power supervision using a fire alarm control unit.*

**(C) Location.** The disconnecting means shall be located where it is readily accessible to qualified persons.

**(1) On Elevators Without Generator Field Control.** On elevators without generator field control, the disconnecting means shall be located within sight of the motor controller. Where the motor controller is located in the elevator hoistway, the disconnecting means required by 620.51(A) shall be located in a machinery space, machine room, control space or control room outside the hoistway; and an additional, non-fused enclosed externally operable motor circuit switch capable of being locked in the open position to disconnect all ungrounded main power-supply conductors shall be located within sight of the motor controller. The additional switch shall be a listed device and shall comply with 620.91(C).

The provision for locking or adding a lock to the disconnecting means, required by this section, shall be installed on or at the switch or circuit breaker used as the disconnecting means and shall remain in place with or without the lock installed. Portable means for adding a lock to the switch or circuit breaker shall not be permitted.

Driving machines or motion and operation controllers not within sight of the disconnecting means shall be provided with a manually operated switch installed in the control circuit to prevent starting. The manually operated switch(es) shall be installed adjacent to this equipment.

Where the driving machine of an electric elevator or the hydraulic machine of a hydraulic elevator is located in a remote machine room or remote machinery space, a single means for disconnecting all ungrounded main power-supply conductors shall be provided and be capable of being locked in the open position.

Exhibit 620.4 illustrates the requirement on disconnecting means for driving machines or motion and operation controllers not within sight of the main line disconnecting means.

**(2) On Elevators with Generator Field Control.** On elevators with generator field control, the disconnecting means shall be located within sight of the motor controller for the driving motor of the motor-generator set. Driving machines, motor-generator sets, or motion and operation controllers not within sight of the disconnecting means shall be provided with a manually operated switch installed in the control circuit to prevent starting. The manually

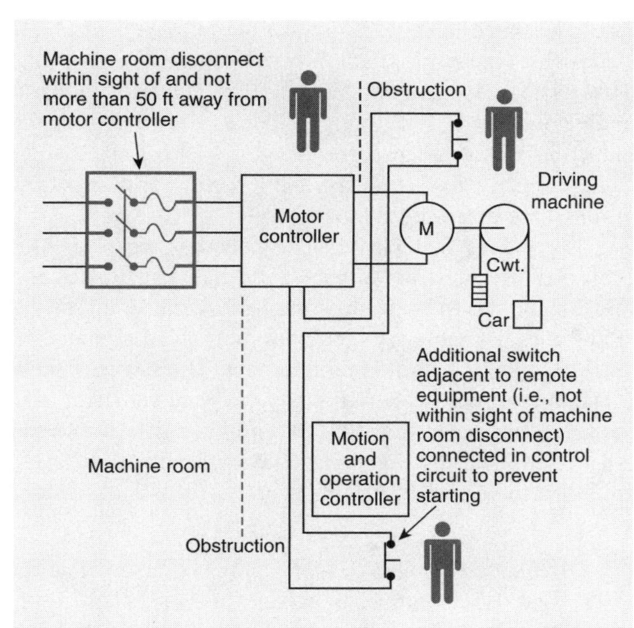


*EXHIBIT 620.4 Disconnecting means for driving machines or motion and operation controllers not within sight of the main line disconnecting means. (Redrawn courtesy of ASME)*

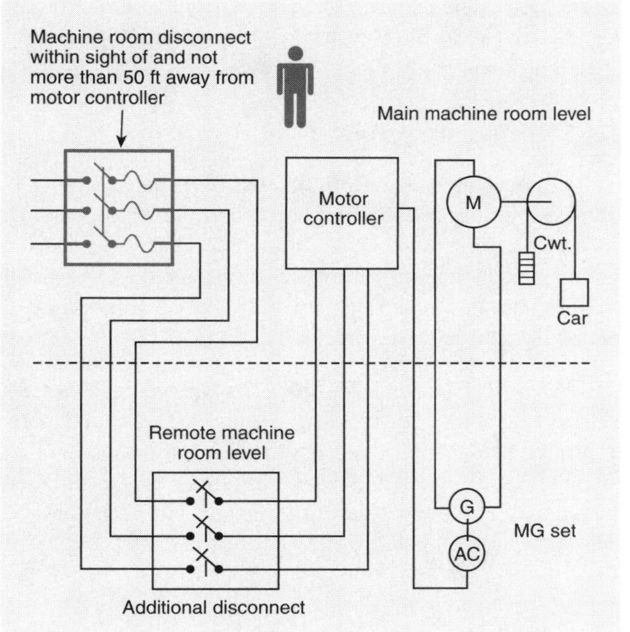


*EXHIBIT 620.5 Disconnecting means for a motor-generator (MG) set in a remote location. (Redrawn courtesy of ASME)*

operated switch(es) shall be installed adjacent to this equipment.

Where the driving machine or the motor-generator set is located in a remote machine room or remote machinery space, a single means for disconnecting all ungrounded main power-supply conductors shall be provided and be capable of being locked in the open position.

See Exhibits 620.5 and 620.6 for examples of disconnecting means for a motor-generator set and for driving machines in remote locations.

**(3) On Escalators and Moving Walks.** On escalators and moving walks, the disconnecting means shall be installed in the space where the controller is located.

**(4) On Platform Lifts and Stairway Chairlifts.** On platform lifts and stairway chairlifts, the disconnecting means shall be located within sight of the motor controller.

**(D) Identification and Signs.** Where there is more than one driving machine in a machine room, the disconnecting means shall be numbered to correspond to the identifying number of the driving machine that they control.

The disconnecting means shall be provided with a sign to identify the location of the supply side overcurrent protective device.

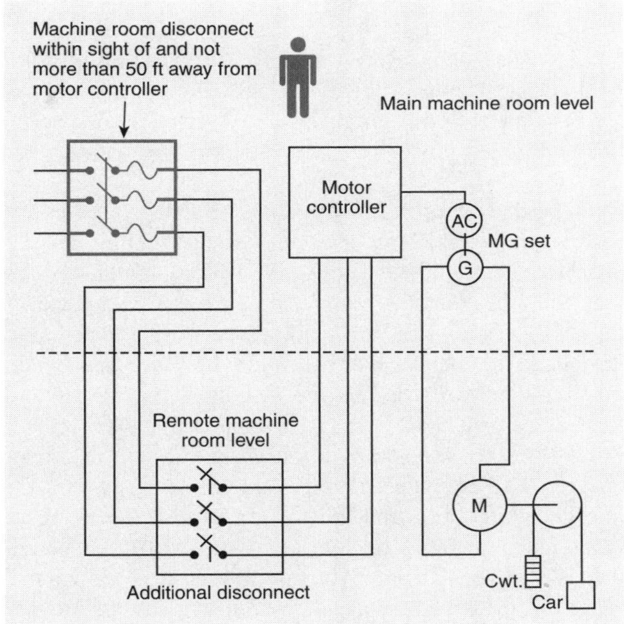


*EXHIBIT 620.6 Disconnecting means for driving machines in a remote location. (Redrawn courtesy of ASME)*

Sign requirements for the location of supply-side overcurrent devices assist the elevator mechanic in troubleshooting during a power loss.

**620.52 Power from More Than One Source.**

**(A) Single-Car and Multicar Installations.** On single-car and multicar installations, equipment receiving electrical power from more than one source shall be provided with a disconnecting means for each source of electrical power. The disconnecting means shall be within sight of the equipment served.

**(B) Warning Sign for Multiple Disconnecting Means.** - Where multiple disconnecting means are used and parts of the controllers remain energized from a source other than the one disconnected, a warning sign shall be mounted on or next to the disconnecting means. The sign shall be clearly legible and shall read as follows:

WARNING
PARTS OF THE CONTROLLER ARE NOT
DE-ENERGIZED BY THIS SWITCH.

**(C) Interconnection Multicar Controllers.** Where interconnections between controllers are necessary for the operation of the system on multicar installations that remain energized from a source other than the one disconnected, a warning sign in accordance with 620.52(B) shall be mounted on or next to the disconnecting means.

**620.53 Car Light, Receptacle(s), and Ventilation Disconnecting Means.** Elevators shall have a single means for disconnecting all ungrounded car light, receptacle(s), and ventilation power-supply conductors for that elevator car.

The disconnecting means shall be an enclosed externally operable fused motor circuit switch or circuit breaker capable of being locked in the open position and shall be located in the machine room or control room for that elevator car. The provision for locking or adding a lock to the disconnecting means shall be installed on or at the switch or circuit breaker used as the disconnecting means and shall remain in place with or without the lock installed. Portable means for adding a lock to the switch or circuit breaker shall not be permitted as the means required to be installed at and remain with the equipment. Where there is no machine room or control room, the disconnecting means shall be located in a machinery space or control space outside the hoistway that is readily accessible to only qualified persons.

This requirement specifies the location of the disconnecting means for lighting, receptacle, and ventilation branch circuits associated with elevators that do not have a machine room. This type of installation includes those designs using drive systems located on the car, on the counterweight, or in the hoistway. Such designs include screw drive or linear induction motor drives. See ASME A17.1-2004, *Safety Code for Elevators and Escalators*, for more information on this type of arrangement.

The means to lock the disconnecting means in the open position must be an accessory that is identified for this use and has to be in place on the disconnecting means at all times. Portable-type lock-out mechanisms cannot be used to comply with this requirement. The chance that a portable-type device could easily be removed from the disconnecting means location compromises the safety benefit of having a disconnecting means that is at all times equipped to be locked off or open.

Disconnecting means shall be numbered to correspond to the identifying number of the elevator car whose light source they control.

The disconnecting means shall be provided with a sign to identify the location of the supply side overcurrent protective device.

**620.54 Heating and Air-Conditioning Disconnecting Means.** Elevators shall have a single means for disconnecting all ungrounded car heating and air-conditioning power-supply conductors for that elevator car.

The disconnecting means shall be an enclosed externally operable fused motor circuit switch or circuit breaker capable of being locked in the open position and shall be located in the machine room or control room for that elevator car. The provision for locking or adding a lock to the disconnecting means shall be installed on or at the switch or circuit breaker used as the disconnecting means and shall remain in place with or without the lock installed. Portable means for adding a lock to the switch or circuit breaker shall not be permitted as the means required to be installed at and remain with the equipment. Where there is no machine room or control room, the disconnecting means shall be located in a machinery space or control space outside the hoistway that is readily accessible to only qualified persons.

Where there is equipment for more than one elevator car in the machine room, the disconnecting means shall be numbered to correspond to the identifying number of the elevator car whose heating and air-conditioning source they control.

The disconnecting means shall be provided with a sign to identify the location of the supply side overcurrent protective device.

**620.55 Utilization Equipment Disconnecting Means.** Each branch circuit for other utilization equipment shall have a single means for disconnecting all ungrounded conductors. The disconnecting means shall be capable of being locked in the open position and shall be located in the machine room or control room/ machine space or control space. The provision for locking or adding a lock to the disconnecting means shall be installed on or at the switch or circuit breaker used as the disconnecting means and shall remain in place with or without the lock installed. Portable means for adding a lock to the switch or circuit breaker shall not be permitted as the means required to be installed at and remain with the equipment.

Where there is more than one branch circuit for other utilization equipment, the disconnecting means shall be numbered to correspond to the identifying number of the equipment served. The disconnecting means shall be provided with a sign to identify the location of the supply side overcurrent protective device.

## Article 625 Electric Vehicle Charging System

**625.23 Disconnecting Means.** For electric vehicle supply equipment rated more than 60 amperes or more than 150 volts to ground, the disconnecting means shall be provided and installed in a readily accessible location. The disconnecting means shall be capable of being locked in the open position. The provision for locking or adding a lock to the disconnecting means shall be installed on or at the switch or circuit breaker used as the disconnecting means and shall remain in place with or without the lock installed. Portable means for adding a lock to the switch or circuit breaker shall not be permitted.

## Article 630 Electric Welders

**630.13 Disconnecting Means.** A disconnecting means shall be provided in the supply circuit for each arc welder that is not equipped with a disconnect mounted as an integral part of the welder.
The disconnecting means shall be a switch or circuit breaker, and its rating shall be not less than that necessary to accommodate overcurrent protection as specified under 630.12.

**630.33 Disconnecting Means.** A switch or circuit breaker shall be provided by which each resistance welder and its control equipment can be disconnected from the supply circuit. The ampere rating of this disconnecting means shall not be less than the supply conductor ampacity determined in accordance with 630.31. The supply circuit switch shall be permitted as the welder disconnecting means where the circuit supplies only one welder.

## Article 645 Information Technology Equipment

**645.10 Disconnecting Means.** An approved means shall be provided to disconnect power to all electronic equipment in the information technology equipment room or in designated zones within the room. There shall also be a similar approved means to disconnect the power to all dedicated HVAC systems serving the room or designated zones and shall cause all required fire/smoke dampers to close. The control for these disconnecting means shall be grouped and identified and shall be readily accessible at the principal exit doors. A single means to control both the electronic equipment and HVAC systems in the room or in a zone shall be permitted. Where a pushbutton is used as a means to disconnect power, pushing the button in shall disconnect the power. Where multiple zones are created, each zone shall have an approved means to confine fire or products of combustion to within the zone.

In 645.10, two separate disconnecting means are required, but a single control, such as one pushbutton, is permitted to electrically operate both disconnecting means. The disconnecting means is required to disconnect the conductors of each circuit from their supply source and close all required fire/smoke dampers. (See the definition of *disconnecting means* in Article 100.) The disconnecting means is permitted to be remote-controlled switching devices, such as relays, with pushbutton stations at the principal exit doors. The actuation of the emergency pushbutton(s) is to be accomplished by pushing the button in, rather than pulling it out. The requirement recognizes that in an emergency situation the intuitive reaction to operating the control is to push, not pull, the button.

The requirements of 645.10 and those of 645.7 for sealing penetrations are intended to minimize the passage of smoke or fire to other parts of the building.

*Exception: Installations qualifying under the provisions of Article 685.*

## Article 660 X-Ray Equipment

**660.5 Disconnecting Means.** A disconnecting means of adequate capacity for at least 50 percent of the input required for the momentary rating, or 100 percent of the input required for the long-time rating, of the X-ray equipment, whichever is greater, shall be provided in the supply circuit. The disconnecting means shall be operable from a location readily accessible from the X-ray control. For equipment connected to a 120-volt, nominal, branch circuit of 30 amperes or less, a grounding-type attachment plug cap and receptacle of proper rating shall be permitted to serve as a disconnecting means.

**660.47 General.**

**(A) High-Voltage Parts.** All high-voltage parts, including X-ray tubes, shall be mounted within grounded enclosures. Air, oil, gas, or other suitable insulating media shall be used to insulate the high voltage from the grounded enclosure. The connection from the high-voltage equipment to X-ray tubes and other high-voltage components shall be made with high-voltage shielded cables.

**(B) Low-Voltage Cables.** Low-voltage cables connecting to oil-filled units that are not completely sealed, such as transformers, condensers, oil coolers, and high-voltage switches, shall have insulation of the oil-resistant type.

Grounded enclosures are required to be provided for all high-voltage X-ray equipment, including X-ray tubes. High-voltage shielded cables are required to be used to connect high-voltage equipment to X-ray tubes, and the shield is required to be grounded, as specified in 660.48.

**660.48 Grounding.** Non–current-carrying metal parts of X-ray and associated equipment (controls, tables, X-ray tube supports, transformer tanks, shielded cables, X-ray tube heads, and so forth) shall be grounded in the manner specified in Article 250. Portable and mobile equipment shall be provided with an approved grounding-type attachment plug cap.

*Exception: Battery-operated equipment.*

## Article 665 Induction and Dielectric Heating Equipment

**665.20 Enclosures.** The converting device (excluding the component interconnections) shall be completely contained within an enclosure(s) of noncombustible material.

**665.21 Control Panels.** All control panels shall be of dead-front construction.

**665.22 Access to Internal Equipment.** Access doors or detachable access panels shall be employed for internal access to heating equipment. Access doors to internal compartments containing equipment employing voltages from 150 volts to 1000 volts ac or dc shall be capable of being locked closed or shall be interlocked to prevent the supply circuit from being energized while the door(s) is open. The provision for locking or adding a lock to the access doors shall be installed on or at the access door and shall remain in place with or without the lock installed.

Access doors to internal compartments containing equipment employing voltages exceeding 1000 volts ac or dc shall be provided with a disconnecting means equipped with mechanical lockouts to prevent access while the heating equipment is energized, or the access doors shall be capable of being locked closed and interlocked to prevent the supply circuit from being energized while the door(s) is open. Detachable panels not normally used for access to such parts shall be fastened in a manner that makes them inconvenient to remove.

**665.23 Warning Labels or Signs.** Warning labels or signs that read "DANGER — HIGH VOLTAGE — KEEP OUT" shall be attached to the equipment and shall be plainly visible where persons might come in contact with energized parts when doors are open or closed or when panels are removed from compartments containing over 150 volts ac or dc.

**665.25 Dielectric Heating Applicator Shielding.** Protective cages or adequate shielding shall be used to guard dielectric heating applicators. Interlock switches shall be used on all hinged access doors, sliding panels, or other easy means of access to the applicator. All interlock switches shall be connected in such a manner as to remove all power from the applicator when any one of the access doors or panels is open.

## Article 668 Electrolytic Cells

**668.10 Cell Line Working Zone.**

**(A) Area Covered.** The space envelope of the cell line working zone shall encompass spaces that meet any of the following conditions:

(1) Is within 2.5 m (96 in.) above energized surfaces of electrolytic cell lines or their energized attachments
(2) Is below energized surfaces of electrolytic cell lines or their energized attachments, provided the headroom in the space beneath is less than 2.5 m (96 in.)
(3) Is within 1.0 m (42 in.) horizontally from energized surfaces of electrolytic cell lines or their energized attachments or from the space envelope described in 668.10(A)(1) or (A)(2)

**(B) Area Not Covered.** The cell line working zone shall not be required to extend through or beyond walls, floors, roofs, partitions, barriers, or the like.

**668.13 Disconnecting Means.**

**(A) More Than One Process Power Supply.** Where more than one direct-current cell line process power supply serves the same cell line, a disconnecting means shall be provided on the cell line circuit side of each power supply to disconnect it from the cell line circuit.

**(B) Removable Links or Conductors.** Removable links or removable conductors shall be permitted to be used as the disconnecting means.

**668.20 Portable Electrical Equipment.**

**(A) Portable Electrical Equipment Not to Be Grounded.** The frames and enclosures of portable electrical equipment used within the cell line working zone shall not be grounded.

*Exception No. 1: Where the cell line voltage does not exceed 200 volts dc, these frames and enclosures shall be permitted to be grounded.*

*Exception No. 2: These frames and enclosures shall be permitted to be grounded where guarded.*

**(B) Isolating Transformers.** Electrically powered, hand-held, cord-connected portable equipment with ungrounded frames or enclosures used within the cell line working zone shall be connected to receptacle circuits that have only ungrounded conductors such as a branch circuit supplied by an isolating transformer with an ungrounded secondary.

**(C) Marking.** Ungrounded portable electrical equipment shall be distinctively marked and shall employ plugs and receptacles of a configuration that prevents connection of this equipment to grounding receptacles and that prevents inadvertent interchange of ungrounded and grounded portable electrical equipments.

**668.21 Power-Supply Circuits and Receptacles for Portable Electrical Equipment.**

**(A) Isolated Circuits.** Circuits supplying power to ungrounded receptacles for hand-held, cord-connected equipment shall be electrically isolated from any distribution system supplying areas other than the cell line working zone and shall be ungrounded. Power for these circuits shall be supplied through isolating transformers. Primaries of such transformers shall operate at not more than 600 volts between conductors and shall be provided with proper overcurrent protection. The secondary voltage of such transformers shall not exceed 300 volts between conductors, and all circuits supplied from such secondaries shall be ungrounded and shall have an approved overcurrent device of proper rating in each conductor.

**(B) Noninterchangeability.** Receptacles and their mating plugs for ungrounded equipment shall not have provision for a grounding conductor and shall be of a configuration that prevents their use for equipment required to be grounded.

**(C) Marking.** Receptacles on circuits supplied by an isolating transformer with an ungrounded secondary shall be a distinctive configuration, shall be distinctively marked, and shall not be used in any other location in the plant.

**668.30 Fixed and Portable Electrical Equipment.**

**(A) Electrical Equipment Not Required to Be Grounded.** Alternating-current systems supplying fixed and portable electrical equipment within the cell line working zone shall not be required to be grounded.

**(B) Exposed Conductive Surfaces Not Required to Be Grounded.** Exposed conductive surfaces, such as electrical equipment housings, cabinets, boxes, motors, raceways, and the like, that are within the cell line working zone shall not be required to be grounded.

**(C) Wiring Methods.** Auxiliary electrical equipment such as motors, transducers, sensors, control devices, and alarms, mounted on an electrolytic cell or other energized surface, shall be connected to premises wiring systems by any of the following means:

(1) Multiconductor hard usage cord.
(2) Wire or cable in suitable raceways or metal or nonmetallic cable trays. If metal conduit, cable tray, armored cable, or similar metallic systems are used, they shall be installed with insulating breaks such that they do not cause a potentially hazardous electrical condition.

**(D) Circuit Overcurrent Protection.** Circuit protection shall not be required for control and instrumentation that are totally within the cell line working zone.

**(E) Bonding.** Bonding of fixed electrical equipment to the energized conductive surfaces of the cell line, its attachments, or auxiliaries shall be permitted. Where fixed electrical equipment is mounted on an energized conductive surface, it shall be bonded to that surface.

## Article 675 Electrically Driven or Controlled Irrigation Machines

**675.8 Disconnecting Means.**

**(A) Main Controller.** A controller that is used to start and stop the complete machine shall meet all of the following requirements:

(1) An equivalent continuous current rating not less than specified in 675.7(A) or 675.22(A)
(2) A horsepower rating not less than the value from Table 430.251(A) and Table 430.251(B), based on the equivalent locked-rotor current specified in 675.7(B) or 675.22(B)

*Exception: A listed molded case switch shall not require a horsepower rating.*

A molded case switch used as a motor controller is not required to have a horsepower rating, but it is required to have a continuous current (ampere) rating not less than that specified by 675.7(A) or 675.22(A).

**(B) Main Disconnecting Means.** The main disconnecting means for the machine shall provide overcurrent protection, shall be at the point of connection of electric power to the machine, or shall be visible and not more than 15 m (50 ft) from the machine, and shall be readily accessible and capable of being locked in the open position. The provision for locking or adding a lock to the disconnecting means shall be installed on or at the switch or circuit breaker used as the disconnecting means and shall remain in place with or without the lock installed. This disconnecting means shall have a horsepower and current rating not less than required for the main controller.

In accordance with 675.8(B), the main disconnecting means is permitted to be up to 50 ft from the machine but must be readily accessible and capable of being locked in the open position. This eliminates one set of overcurrent protective devices and one disconnecting means where the circuit originates at the motor control panel for the irrigation pump and the panel is located within 50 ft of the center pivot machine. It also alleviates some potential problems with machines designed to be towed to a second site.

*Exception No. 1: Circuit breakers without marked horsepower ratings shall be permitted in accordance with 430.109.*

*Exception No. 2: A listed molded case switch without marked horsepower ratings shall be permitted.*

**(C) Disconnecting Means for Individual Motors and Controllers.** A disconnecting means shall be provided to simultaneously disconnect all ungrounded conductors for each motor and controller and shall be located as required by Article 430, Part IX. The disconnecting means shall not be required to be readily accessible.

Article 430, Part IX, provides for safety during maintenance and inspection shutdown periods. See the commentary following 430.103.

## Article 680 Swimming Pools, Fountains, and Similar Installations

**680.7 Cord-and-Plug-Connected Equipment.** Fixed or stationary equipment, other than underwater luminaires, for a permanently installed pool shall be permitted to be connected with a flexible cord and plug to facilitate the removal or disconnection for maintenance or repair.

**(A) Length.** For other than storable pools, the flexible cord shall not exceed 900 mm (3 ft) in length.

**(B) Equipment Grounding.** The flexible cord shall have a copper equipment grounding conductor sized in accordance with 250.122 but not smaller than 12 AWG. The cord shall terminate in a grounding-type attachment plug.

**(C) Construction.** The equipment grounding conductors shall be connected to a fixed metal part of the assembly. The removable part shall be mounted on or bonded to the fixed metal part.

In some climates, it is preferable to disconnect and remove a permanent pool's filter pump during cold-weather months. A 3-ft cord is permitted, to facilitate the removal of fixed or stationary equipment for maintenance and storage. The 3-ft cord limitation does not apply to cord-and-plug-connected filter pumps used with storable-type pools (covered in Part III of Article 680), since these pumps are neither fixed nor stationary. Listed filter pumps for use with storable pools are considered portable and are permitted to be equipped with cords longer than 3 ft.

**680.8 Overhead Conductor Clearances.** Overhead conductors shall meet the clearance requirements in this section. Where a minimum clearance from the water level is given, the measurement shall be taken from the maximum water level of the specified body of water.

**(A) Power.** With respect to service drop conductors and open overhead wiring, swimming pool and similar installations shall comply with the minimum clearances given in Table 680.8 and illustrated in Figure 680.8.

FPN: Open overhead wiring as used in this article typically refers to conductor(s) not in an enclosed raceway.

**(B) Communications Systems.** Communication, radio, and television coaxial cables within the scope of Articles 800 through 820 shall be permitted at a height of not less than 3.0 m (10 ft) above swimming and wading pools, diving structures, and observation stands, towers, or platforms.

**(C) Network-Powered Broadband Communications Systems.** The minimum clearances for overhead network-powered broadband communications systems conductors from pools or fountains shall comply with the provisions in Table 680.8 for conductors operating at 0 to 750 volts to ground.

Service drop conductors, conductors of network-powered broadband communications systems, and aerial feeders and branch circuits are permitted to be located above a swimming pool and associated pool structures where provided with the clearances specified in Table 680.8. Overhead conductors of communications systems are required to comply with 680.8(B). These clearances consider such factors as the use of skimmers with aluminum handles and provide sufficient separation between the conductors and the pool. In some instances, locating a swimming pool below electric fixed conductors is unavoidable; for example, on a building lot with limited area or an existing lot where the electric supply lines are already in place. The clearances for conductors from pools and pool structures harmonize the *NEC* with ANSI C2, *National Electrical Safety Code (NESC)*. The maximum water level of the body of water (pool, spa, hot tub, or other) is used to determine compliance with 680.8. For the definition of *maximum water level*, see 680.2.

**680.12 Maintenance Disconnecting Means.** One or more means to simultaneously disconnect all ungrounded conductors shall be provided for all utilization equipment other than lighting. Each means shall be readily accessible and within sight from its equipment and shall be located at least 1.5 m (5 ft) horizontally from the inside walls of a pool, spa, or hot tub unless separated from the open water by a permanently installed barrier that provides a 1.5 m (5 ft) reach path or greater. This horizontal distance is to be measured from the water's edge along the shortest path required to reach the disconnect.

A readily accessible disconnecting means is required to be located within sight of pool, spa, and hot tub equipment in order to provide service personnel with the ability to safely disconnect power while servicing equipment such as motors, heaters, and control panels. Underwater luminaires are not subject to this requirement. The proximity of the disconnecting means to the pool must be not less than 5 ft unless the disconnecting means is separated from the water by a permanent barrier. See Exhibit 680.2.

**680.41 Emergency Switch for Spas and Hot Tubs.** A clearly labeled emergency shutoff or control switch for the purpose of stopping the motor(s) that provide power to the recirculation system and jet system shall be installed at a point readily accessible to the users and not less than 1.5 m (5 ft) away, adjacent to, and within sight of the spa or hot tub. This requirement shall not apply to single-family dwellings.

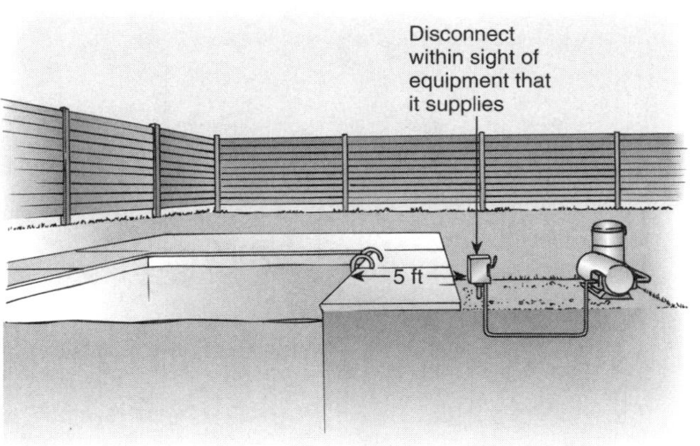


***EXHIBIT 680.2** Required pool equipment disconnect. The disconnect for pool equipment must be located within sight of the pool equipment and at least 5 ft from the pool.*

The provisions of 680.41 require a local disconnecting device for spas and hot tubs that is capable of being used in an emergency. This requirement was added to address entrapment hazards associated with spas and hot tubs. The definitive publication on this issue, *Guideline for Entrapment Hazards: Making Pools and Spas Safer* (Pub. No. 363), is available from the U.S. Consumer Product Safety Commission, Washington, DC 20207, or on-line at www.cpsc.gov.

The emergency shutoff switch must be installed within sight of and at least 5 ft from the spa or hot tub and must be clearly labeled "Emergency Shutoff." See Exhibit 680.17 for an illustration of the switch location. The shutoff switch can be either a line-operated device or a remote-control circuit that causes the pump circuit to open. This requirement does not apply to one-family dwellings.

**680.57 Signs.**

**(A) General.** This section covers electric signs installed within a fountain or within 3.0 m (10 ft) of the fountain edge.

**(B) Ground-Fault Circuit-Interrupter Protection for Personnel.** All circuits supplying the sign shall have ground-fault circuit-interrupter protection for personnel.

**(C) Location.**

**(1) Fixed or Stationary.** A fixed or stationary electric sign installed within a fountain shall be not less than 1.5 m

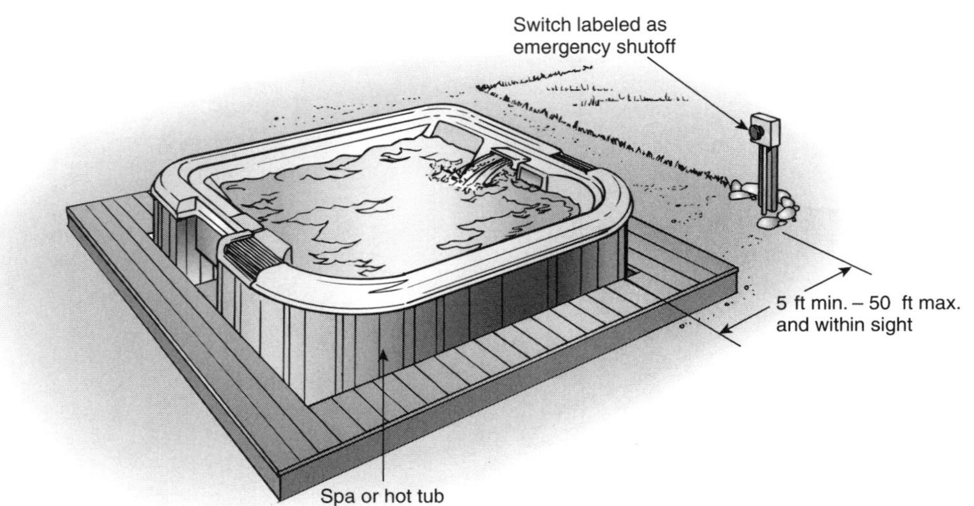


*EXHIBIT 680.17 Location of the emergency shutoff device required by 680.41.*

(5 ft) inside the fountain measured from the outside edges of the fountain.

**(2) Portable.** A portable electric sign shall not be placed within a pool or fountain or within 1.5 m (5 ft) measured horizontally from the inside walls of the fountain.

**(D) Disconnect.** A sign shall have a local disconnecting means in accordance with 600.6 and 680.12.

**(E) Bonding and Grounding.** A sign shall be grounded and bonded in accordance with 600.7.

The use of electric signs in fountains has become increasingly popular. Electric signs in fountains are required to have GFCI protection for personnel. This protection may be provided in the feeder or branch circuit. To prevent contact by persons around the fountain, the sign must be at least 5 ft from the edge of the fountain (see Exhibit 680.18). Disconnecting and bonding requirements in Article 600 apply, and grounding must be provided in accordance with Article 250.

*EXHIBIT 680.18 Electric sign located in a fountain as described in 680.57.*

## Article 682 Natural and Artificially Made Bodies of Water

### 682.14 Disconnecting Means for Floating Structures or Submersible Electrical Equipment.

**(A) Type.** The disconnecting means shall be permitted to consist of a circuit breaker, switch, or both that simultaneously opens all ungrounded circuit conductors, and shall be properly identified as to which structure or equipment it controls.

**(B) Location.** The disconnecting means shall be readily accessible on land and shall be located in the supply circuit ahead of the structure or equipment connection. The disconnecting means shall be within sight of, but not closer than 1.5 m (5 ft) horizontally from, the edge of the shoreline and live parts elevated a minimum of 300 m (12 in.) above the electrical datum plane.

## Article 685 Integrated Electrical Systems

**685.10 Location of Overcurrent Devices in or on Premises.** Location of overcurrent devices that are critical to integrated electrical systems shall be permitted to be accessible, with mounting heights permitted to ensure security from operation by unqualified personnel.

## Article 690 Solar Photovoltaic Systems

**690.4 Installation.**

**(A) Solar Photovoltaic System.** A solar photovoltaic system shall be permitted to supply a building or other structure in addition to any service(s) of another electricity supply system(s).

**(B) Conductors of Different Systems.** Photovoltaic source circuits and photovoltaic output circuits shall not be contained in the same raceway, cable tray, cable, outlet box, junction box, or similar fitting as feeders or branch circuits of other systems, unless the conductors of the different systems are separated by a partition or are connected together.

For example, 690.4(B) does not permit the conductors supplying an exterior luminaire located in close proximity to a roof-mounted PV array to be installed in the same raceway or cable with the conductors of PV source circuits or PV output circuits.

Conductors directly related to a specific PV system, such as those in dc and ac output power circuits, may be contained in the same raceway as PV source and output conductors, providing they meet the requirements of 300.3(C).

**(C) Module Connection Arrangement.** The connections to a module or panel shall be arranged so that removal of a module or panel from a photovoltaic source circuit does not interrupt a grounded conductor to another photovoltaic source circuit. Sets of modules interconnected as systems rated at 50 volts or less, with or without blocking diodes, and having a single overcurrent device shall be considered as a single-source circuit. Supplementary overcurrent devices used for the exclusive protection of the photovoltaic modules are not considered as overcurrent devices for the purpose of this section.

In general, 690.4(C) requires that a jumper be installed between a module terminal or lead and the connection point to the grounded PV source circuit conductor. That way, a module can be removed without interrupting the grounded conductor to other PV source circuits. If interrupted, such conductors, although identified as grounded, would be operating at the system potential with respect to ground, and a shock hazard could result. The reverse-current protection requirement on nearly all PV modules (as indicated by the fuse requirement labeled on the back of each module) generally dictates that each module or string of modules have a series overcurrent device and become a source circuit.

**(D) Equipment.** Inverters, motor generators, photovoltaic modules, photovoltaic panels, ac photovoltaic modules, source-circuit combiners, and charge controllers intended for use in photovoltaic power systems shall be identified and listed for the application.

Equipment listed for marine, mobile, telecommunications, or other applications may not be suitable for installation in permanent PV power systems complying with this *Code*.

**690.17 Switch or Circuit Breaker.** The disconnecting means for ungrounded conductors shall consist of a manually operable switch(es) or circuit breaker(s) complying with all of the following requirements:

(1) Located where readily accessible
(2) Externally operable without exposing the operator to contact with live parts
(3) Plainly indicating whether in the open or closed position
(4) Having an interrupting rating sufficient for the nominal circuit voltage and the current that is available at the line terminals of the equipment

Where all terminals of the disconnecting means may be energized in the open position, a warning sign shall be mounted on or adjacent to the disconnecting means. The sign shall be clearly legible and have the following words or equivalent:

WARNING
ELECTRIC SHOCK HAZARD.
DO NOT TOUCH TERMINALS.
TERMINALS ON BOTH THE LINE
AND LOAD SIDES MAY BE ENERGIZED
IN THE OPEN POSITION.

*Exception: A connector shall be permitted to be used as an ac or a dc disconnecting means, provided that it complies with the requirements of 690.33 and is listed and identified for the use.*

**690.35 Ungrounded Photovoltaic Power Systems.** Photovoltaic power systems shall be permitted to operate with ungrounded photovoltaic source and output circuits where the system complies with 690.35(A) through (G).

**(A) Disconnects.** All photovoltaic source and output circuit conductors shall have disconnects complying with 690, Part III.

**(B) Overcurrent Protection.** All photovoltaic source and output circuit conductors shall have overcurrent protection complying with 690.9.

**(C) Ground-Fault Protection.** All photovoltaic source and output circuits shall be provided with a ground-fault protection device or system that complies with (1) through (3):

(1) Detects a ground fault.
(2) Indicates that a ground fault has occurred
(3) Automatically disconnects all conductors or causes the inverter or charge controller connected to the faulted circuit to automatically cease supplying power to output circuits.

**(D)** The photovoltaic source conductors shall consist of the following:

(1) Nonmetallic jacketed multiconductor cables
(2) Conductors installed in raceways, or
(3) Conductors listed and identified as Photovoltaic (PV) Wire installed as exposed, single conductors.

Three options for PV source output circuits are provided by this section. All cables and conductors installed outdoors and exposed to direct sunlight and wet conditions have to be suitable for these conditions and such suitability is verified by the cables and conductors being listed. Conductors inside raceways installed in wet locations are required to be identified or listed as suitable for this environmental condition. See 310.8(C) for the requirements on conductors installed in wet locations. Open, single conductors are permitted where listed and identified as *Photovoltaic Wire, Photovoltaic Cable, PV Wire,* or *PV Cable.* These conductors are evaluated for use where exposed to direct sunlight and wet conditions. Although not required as a general rule, these conductors can be installed in a raceway at the discretion of the installer.

**(E)** The photovoltaic power system direct-current circuits shall be permitted to be used with ungrounded battery systems complying with 690.71(G).

**(F)** The photovoltaic power source shall be labeled with the following warning at each junction box, combiner box, disconnect, and device where energized, ungrounded circuits may be exposed during service:

WARNING
ELECTRIC SHOCK HAZARD. THE DC
CONDUCTORS OF THIS PHOTOVOLTAIC SYSTEM
ARE UNGROUNDED AND MAY BE ENERGIZED.

PV dc circuits operate in outdoor environments and are expected to be energized for 40 years or more. Aging of the conductors, dust and dirt infiltration, and moisture and water intrusion create leakage paths from the conductors to ground. These high-resistance leakage paths can result in leakage current values less than those detected by the required ground-fault detection device, but they can cause any ungrounded conductor to become a potential shock hazard with respect to ground.

**(G)** The inverters or charge controllers used in systems with ungrounded photovoltaic source and output circuits shall be listed for the purpose.

**690.71 Installation.**

**(A) General.** Storage batteries in a solar photovoltaic system shall be installed in accordance with the provisions of Article 480. The interconnected battery cells shall be considered grounded where the photovoltaic power source is installed in accordance with 690.41.

Batteries in PV power systems are usually grounded when the PV power system is grounded in accordance with Article 690, Part VI.

**(B) Dwellings.**

**(1) Operating Voltage.** Storage batteries for dwellings shall have the cells connected so as to operate at less than 50 volts nominal. Lead-acid storage batteries for dwellings shall have no more than twenty-four 2-volt cells connected in series (48-volts nominal).

*Exception: Where live parts are not accessible during routine battery maintenance, a battery system voltage in accordance with 690.7 shall be permitted.*

**(2) Guarding of Live Parts.** Live parts of battery systems for dwellings shall be guarded to prevent accidental contact by persons or objects, regardless of voltage or battery type.

FPN: Batteries in solar photovoltaic systems are subject to extensive charge–discharge cycles and typically require frequent maintenance, such as checking electrolyte and cleaning connections.

At any voltage, a primary safety concern in battery systems is that a fault (e.g., a metal tool dropped onto a terminal) might cause a fire or an explosion. *Guarded,* as defined in Article 100, describes the best method to reduce this hazard.

**(C) Current Limiting.** A listed, current-limiting, overcurrent device shall be installed in each circuit adjacent to the batteries where the available short-circuit current from a battery or battery bank exceeds the interrupting or withstand ratings of other equipment in that circuit. The installation of current-limiting fuses shall comply with 690.16.

Large banks of storage batteries can deliver significant amounts of short-circuit current. Current-limiting overcurrent devices should be used if necessary.

**(D) Battery Nonconductive Cases and Conductive Racks.** Flooded, vented, lead-acid batteries with more than twenty-four 2-volt cells connected in series (48 volts, nominal) shall not use conductive cases or shall not be installed in conductive cases. Conductive racks used to support the nonconductive cases shall be permitted where no rack material is located within 150 mm (6 in.) of the tops of the nonconductive cases.

This requirement shall not apply to any type of valve-regulated lead-acid (VRLA) battery or any other types of sealed batteries that may require steel cases for proper operation.

Grounded metal trays and cases or containers (as normally required by 250.110) in flooded, lead-acid battery systems operating over 48 volts, nominal, have been shown to be a contributing factor in ground faults. Nonconductive racks, trays, and cases minimize this problem.

**(E) Disconnection of Series Battery Circuits.** Battery circuits subject to field servicing, where more than twenty-four 2-volt cells are connected in series (48 volts, nominal), shall have provisions to disconnect the series-connected strings into segments of 24 cells or less for maintenance by qualified persons. Non–load-break bolted or plug-in disconnects shall be permitted.

**(F) Battery Maintenance Disconnecting Means.** Battery installations, where there are more than twenty-four 2-volt cells connected in series (48 volts, nominal), shall have a disconnecting means, accessible only to qualified persons, that disconnects the grounded circuit conductor(s) in the battery electrical system for maintenance. This disconnecting means shall not disconnect the grounded circuit conductor(s) for the remainder of the photovoltaic electrical system. A non–load-break-rated switch shall be permitted to be used as the disconnecting means.

**(G) Battery Systems of More Than 48 Volts.** On photovoltaic systems where the battery system consists of more than twenty-four 2-volt cells connected in series (more than 48 volts, nominal), the battery system shall be permitted to operate with ungrounded conductors, provided the following conditions are met:

(1) The photovoltaic array source and output circuits shall comply with 690.41.
(2) The dc and ac load circuits shall be solidly grounded.
(3) All main ungrounded battery input/output circuit conductors shall be provided with switched disconnects and overcurrent protection.
(4) A ground-fault detector and indicator shall be installed to monitor for ground faults in the battery bank.

## Article 700 Emergency Systems

### 700.8 Signs.

**(A) Emergency Sources.** A sign shall be placed at the service-entrance equipment, indicating type and location of on-site emergency power sources.

A sign is required at the service that indicates the type and location of an emergency source. See Exhibit 700.2.

***EXHIBIT 700.2** An example of the sign required by 700.8(A).*

*Exception: A sign shall not be required for individual unit equipment as specified in 700.12(F).*

**(B) Grounding.** Where the grounded circuit conductor connected to the emergency source is connected to a grounding electrode conductor at a location remote from the emergency source, there shall be a sign at the grounding location that identifies all emergency and normal sources connected at that location.

Section 700.8(B) requires a sign at the grounding location if the emergency source is a separately derived system and is connected to a grounding electrode conductor at a location that is remote from the emergency source.

**700.16 Emergency Illumination.** Emergency illumination shall include all required means of egress lighting, illuminated exit signs, and all other lights specified as necessary to provide required illumination.

Emergency lighting systems shall be designed and installed so that the failure of any individual lighting element, such as the burning out of a lamp, cannot leave in total darkness any space that requires emergency illumination.

Where high-intensity discharge lighting such as high- and low-pressure sodium, mercury vapor, and metal halide is used as the sole source of normal illumination, the emergency lighting system shall be required to operate until normal illumination has been restored.

*Exception: Alternative means that ensure emergency lighting illumination level is maintained shall be permitted.*

High-intensity discharge (HID) fixtures take some time to start once they are energized. Therefore, if HID fixtures are the sole source of normal illumination in an area, the *Code* requires that the emergency lighting system operate not only until the normal system is returned to service but also until the HID fixtures provide illumination. This may require a timing circuit, photoelectric monitoring system, or the equivalent.

## Article 705 Interconnected Electric Power Production Sources

**705.10 Directory.** A permanent plaque or directory, denoting all electric power sources on or in the premises, shall be installed at each service equipment location and at locations of all electric power production sources capable of being interconnected.

*Exception: Installations with large numbers of power production sources shall be permitted to be designated by groups.*

**705.22 Disconnect Device.** The disconnecting means for ungrounded conductors shall consist of a manually or power operable switch(es) or circuit breaker(s) with the following features:

(1) Located where readily accessible
(2) Externally operable without exposing the operator to contact with live parts and, if power operable, of a type that could be opened by hand in the event of a power-supply failure
(3) Plainly indicating whether in the open (off) or closed (on) position
(4) Having ratings not less than the load to be carried and the fault current to be interrupted. For disconnect equipment energized from both sides, a marking shall be provided to indicate that all contacts of the disconnect equipment might be energized.

FPN No. 1 to (4): In parallel generation systems, some equipment, including knife blade switches and fuses, is likely to be energized from both directions. See 240.40.

FPN No. 2 to (4): Interconnection to an off-premises primary source could require a visibly verifiable disconnecting device.

(5) Simultaneous disconnect of all ungrounded conductors of the circuit
(6) Capable of being locked in the open (off) position

The requirements for disconnects in 705.22 are important. A disconnecting means must serve each generating source. This disconnecting means will be the service-entrance disconnect. Still another disconnecting means may be applied to separate the generating systems.

The basic requirement in 705.22 recognizes the success of applying switches as well as circuit breakers as the disconnecting means for ungrounded conductors. Most safe work practices on the premises use these disconnect devices.

The disconnect at the service entrance is required for disconnecting the premises wiring system from the utility. The utility safe work practices may also use this disconnect device. Utility safe work practices may require a visibly verifiable disconnect device. For this reason, some utility contracts require that a visible break be provided. The second fine print note to 705.22 brings attention to this common utility requirement.

# Index

## Q



## R



## S

# IMPORTANT NOTICES AND DISCLAIMERS CONCERNING NFPA® DOCUMENTS

## NOTICE AND DISCLAIMERS OF LIABILITY CONCERNING THE USE OF NFPA DOCUMENTS

NFPA codes, standards, recommended practices, and guides, including the documents contained herein, are developed through a consensus standards development process approved by the American National Standards Institute. This process brings together volunteers representing varied viewpoints and interests to achieve consensus on fire and other safety issues. While the NFPA administers the process and establishes rules to promote fairness in the development of consensus, it does not independently test, evaluate, or verify the accuracy of any information or the soundness of any judgments contained in its codes and standards.

The NFPA disclaims liability for any personal injury, property or other damages of any nature whatsoever, whether special, indirect, consequential or compensatory, directly or indirectly resulting from the publication, use of, or reliance on these documents. The NFPA also makes no guaranty or warranty as to the accuracy or completeness of any information published herein.

In issuing and making these documents available, the NFPA is not undertaking to render professional or other services for or on behalf of any person or entity. Nor is the NFPA undertaking to perform any duty owed by any person or entity to someone else. Anyone using these documents should rely on his or her own independent judgment or, as appropriate, seek the advice of a competent professional in determining the exercise of reasonable care in any given circumstances.

The NFPA has no power, nor does it undertake, to police or enforce compliance with the contents of these documents. Nor does the NFPA list, certify, test or inspect products, designs, or installations for compliance with these documents. Any certification or other statement of compliance with the requirements of these documents shall not be attributable to the NFPA and is solely the responsibility of the certifier or maker of the statement.

## ADDITIONAL NOTICES AND DISCLAIMERS

### Updating of NFPA Documents

Users of NFPA codes, standards, recommended practices, and guides should be aware that these documents may be superseded at any time by the issuance of new editions or may be amended from time to time through the issuance of Tentative Interim Amendments. An official NFPA document at any point in time consists of the current edition of the document together with any Tentative Interim Amendments and any Errata then in effect. In order to determine whether a given document is the current edition and whether it has been amended through the issuance of Tentative Interim Amendments or corrected through the issuance of Errata, consult appropriate NFPA publications such as the National Fire Codes® Subscription Service, visit the NFPA website at www.nfpa.org, or contact the NFPA at the address listed below.

### Interpretations of NFPA Documents

A statement, written or oral, that is not processed in accordance with Section 6 of the Regulations Governing Committee Projects shall not be considered the official position of NFPA or any of its Committees and shall not be considered to be, nor be relied upon as, a Formal Interpretation.

### Patents

The NFPA does not take any position with respect to the validity of any patent rights asserted in connection with any items which are mentioned in or are the subject of NFPA codes, standards, recommended practices, and guides, and the NFPA disclaims liability for the infringement of any patent resulting from the use of or reliance on these documents. Users of these documents are expressly advised that determination of the validity of any such patent rights, and the risk of infringement of such rights, is entirely their own responsibility.

NFPA adheres to applicable policies of the American National Standards Institute with respect to patents. For further information, contact the NFPA at the address listed below.

### Law and Regulations

Users of these documents should consult applicable federal, state, and local laws and regulations. NFPA does not, by the publication of its codes, standards, recommended practices, and guides, intend to urge action that is not in compliance with applicable laws, and these documents may not be construed as doing so.

### Copyrights

The documents contained in this volume are copyrighted by the NFPA. They are made available for a wide variety of both public and private uses. These include both use, by reference, in laws and regulations, and use in private self-regulation, standardization, and the promotion of safe practices and methods. By making these documents available for use and adoption by public authorities and private users, NFPA does not waive any rights in copyright to these documents.

Use of NFPA documents for regulatory purposes should be accomplished through adoption by reference. The term "adoption by reference" means the citing of title, edition, and publishing information only. Any deletions, additions, and changes desired by the adopting authority should be noted separately in the adopting instrument. In order to assist NFPA in following the uses made of its documents, adopting authorities are requested to notify the NFPA (Attention: Secretary, Standards Council) in writing of such use. For technical assistance and questions concerning adoption of NFPA documents, contact NFPA at the address below.

### For Further Information

All questions or other communications relating to NFPA codes, standards, recommended practices, and guides and all requests for information on NFPA procedures governing its codes and standards development process, including information on the procedures for requesting Formal Interpretations, for proposing Tentative Interim Amendments, and for proposing revisions to NFPA documents during regular revision cycles, should be sent to NFPA headquarters, addressed to the attention of the Secretary, Standards Council, NFPA, 1 Batterymarch Park, Quincy, MA 02169-9101.

For more information about NFPA, visit the NFPA website at www.nfpa.org.

# A Guide to Using the Handbook for Electrical Safety in the Workplace

This 2009 edition of the *Handbook for Electrical Safety in the Workplace* contains the complete text of the 2009 edition of *NFPA 70E®, Standard for Electrical Safety in the Workplace®*, along with explanatory commentary and supplemental material.

ARTICLE 310 — SAFETY-RELATED WORK PRACTICES FOR ELECTROLYTIC CELLS

2. Each article in Chapter 3 stands alone and is not intended to apply to other articles in Chapter 3.

FPN: The *NFPA 70E* Technical Committee might develop additional chapters for other types of special equipment in the future.

**ARTICLE 310**
**Safety-Related Work Practices for Electrolytic Cells**

Article 310 identifies the supplementary or replacement safe work practices workers should use in electrolytic cell line working zones and the special hazards of working with ungrounded dc systems (see Exhibit 310.1).

*EXHIBIT 310.1 An electrolytic cell line. (Photo by David Pace and Michael Petry, courtesy of Olin Corporation)*

Commentary exhibits provide detailed views of Standard concepts. These exhibits are outlined in blue and numbered sequentially throughout each article. The caption is in blue ink, which makes these exhibits easily discernible.

| cell of an electrolytic cell line is a battery and cannot be deenergized
|e electrolyte in the vessel. A cell line is a series of individual cells that
|er electrically. Generally, the process requires a significant amount of di-
|ngrounded.
| electrically safe work condition is not a viable method for avoiding in-
|ded to employees who work in the vicinity of the cells and interconnect-
|lish an understanding of the hazards associated with an unintentional
|of either an individual cell or the interconnecting bus.
| in the interconnecting bus generates a significant magnetic field. Work-
| that the magnetic field might interfere with certain medical devices.
|ble tools and equipment must not be used in the area containing the cells
|us.

|f this chapter shall apply to the electrical safety–related work practices
|electrolytic cell areas.

|Annex L for a typical application of safeguards in the cell line working zone.

|ical Safety in the Workplace 2009

6 ARTICLE 90 — INTRODUCTION

(4) Installations used by the electric utility, such as office buildings, warehouses, garages, machine shops, and recreational buildings, that are not an integral part of a generating plant, substation, or control center

Section 90.2(A) describes facilities that are intended to be within the scope of the standard. Although the scope described in this section is similar to the scope of the *NEC*, the purpose and intent of these documents are considerably different. Whereas the *NEC* applies to equipment and circuits, *NFPA 70E* applies to work practices associated with a workplace that contains electrical hazards both during and after the installation is complete.

Section 90.2(A) clarifies which portions of electric utility facilities are covered by *NFPA 70E*. [See 90.2(B) and the related commentary for information on facilities and specific lighting that are not covered by this standard.] The distinction between electric utility facilities to which this standard does and does not apply is illustrated in Exhibit 90.1.

*EXHIBIT 90.1 Typical electric utility complexes showing examples of facilities covered and not covered by the NEC.*

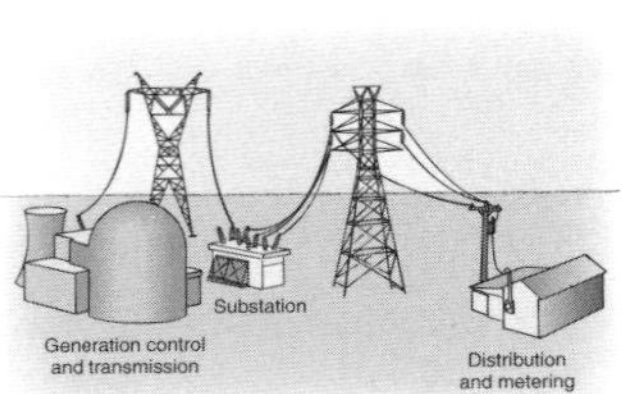


*NEC does not apply.*

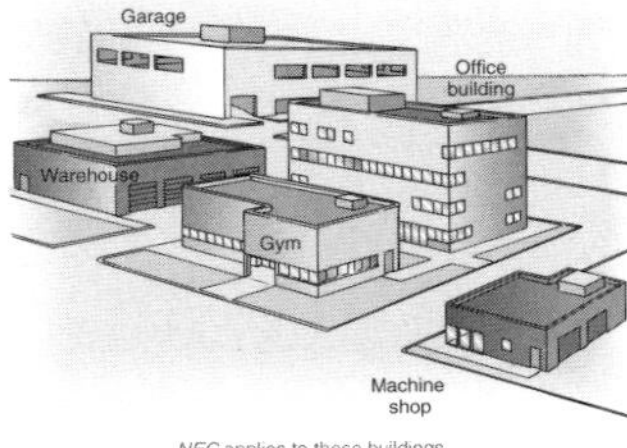


*NEC applies to these buildings.*

Industrial and multibuilding complexes and campus-style wiring often include substations and other installations that employ construction and wiring similar to those of electric utility installations. These installations are on the load side of the service point, and the installation is usually an owner-maintained substation, so their work practices are clearly within the purview of *NFPA 70E*.

*2009 Handbook for Electrical Safety in the Workplace*